Cortical Oscillations in Health and Disease

Frontispiece 1: Eberhard H. Buhl. This book is dedicated to the memory of Professor Eberhard H. Buhl, MD (1959–2003). Eberhard was instrumental in the discovery of persistent gamma oscillations in vitro, and in elucidation of the mechanisms of both gamma and very fast oscillations. The authors of this monograph coauthored, alone or together, 23 research papers with Eberhard. This work would not have been possible without him. He made major contributions to other fields as well.

Frontispiece 2: Professors Mircea Steriade (1924–2006) and Rodolfo R. Llinás. These scientists made fundamental discoveries on neuronal oscillations in vivo, in vitro, and in single cells. As for Eberhard Buhl, our work rests on foundations that they established. This photo was taken at the PhD defense of Dr. Diego Contreras, Université Laval, Québec, 1996 (with thanks to Dr. Farid Hamzei-Sichani).

Cortical Oscillations in Health and Disease

Roger D. Traub, MD
Department of Physical Sciences
IBM T.J. Watson Research Center
Yorktown Heights, NY

Miles A. Whittington, PhD
Professor of Neuroscience
School of Neurology, Neurobiology & Psychiatry
The Medical School
University of Newcastle
Newcastle, UK

UNIVERSITY PRESS

2010

OXFORD
UNIVERSITY PRESS

Oxford University Press, Inc., publishes works that further
Oxford University's objective of excellence
in research, scholarship, and education.

Oxford   New York
Auckland   Cape Town   Dar es Salaam   Hong Kong   Karachi
Kuala Lumpur   Madrid   Melbourne   Mexico City   Nairobi
New Delhi   Shanghai   Taipei   Toronto

With offices in
Argentina   Austria   Brazil   Chile   Czech Republic   France   Greece
Guatemala   Hungary   Italy   Japan   Poland   Portugal   Singapore
South Korea   Switzerland   Thailand   Turkey   Ukraine   Vietnam

Published by Oxford University Press, Inc.
198 Madison Avenue, New York, New York 10016
www.oup.com

Oxford is a registered trademark of Oxford University Press

Library of Congress Cataloging-in-Publication Data

Traub, Roger D.
  Cortical oscillations in health and disease / Roger D. Traub, Miles A. Whittington.
     p. ; cm.
  Includes bibliographical references and index.
  ISBN 978-0-19-534279-6 (alk. paper)
  1. Cerebral cortex—Pathophysiology. 2. Cerebral cortex. 3. Oscillations.
  4. Neurobehavioral disorders.
  I. Whittington, Miles A. II. Title.
  [DNLM: 1. Neocortex—physiology. 2. Periodicity.
  3. Neocortex—physiopathology. WL 307 T777c 2010]
  RC386.2.T73 2010
  616.8—dc22   2009039142

9 8 7 6 5 4 3 2 1
Printed in the United States of America
on acid-free paper

Epigraphs

"Thus each organic body of a living being is a kind of divine machine or natural automaton, which infinitely surpasses all artificial automata. Because a machine which is made by man's art is not a machine in each one of its parts; for example, the teeth of a brass wheel have parts or fragments which to us are no longer artificial and have nothing in themselves to show the special use to which the wheel was intended in the machine. But nature's machines, that is, living bodies, are machines even in the smallest parts, *ad infinitum*. Herein lies the difference between nature and art, that is, between the divine art and ours."

— No. 64, Gottfried Wilhelm Freiherr von Leibniz,
The Monadology (La Monadologie), 1714.

"Der Irrnis und der Leiden Pfade kam ich:
soll ich mich denen jetzt entwunden wähnen,
da dieses Waldes Rauschen
wieder ich vernehme,
dich guten Greisen neu begrüße?
Oder – irr' ich wieder?
Verändert dünkt mich alles."

[I came on paths of error and sorrow:
Shall I consider myself free of them,
As again I hear the murmurs
Of this wood,
Once more greet you, old friend?
Or – am I still wrong?
It all seems changed.]

—Richard Wagner, *Parsifal*, Act 3.

Preface

In the July 9, 1998 issue of *Nature*, there were two letters concerning brain oscillations, back-to-back. One letter came from a group at Oxford led by the late Eberhard Buhl, with lead author André Fisahn, then a Ph.D. student of Eberhard. The lead author of the other letter was Andreas Draguhn, who had been visiting the laboratory of John G.R. Jefferys at the University of Birmingham, with the intention of studying the tetanus toxin model of epilepsy.

Seemingly, the two *Nature* papers had little to do with each other. True, both papers reported results on oscillations obtained from in vitro slices of hippocampus. However, the oscillation frequencies, the cellular mechanisms, and the in vivo correlates were—so it was believed—entirely different.

The Fisahn paper showed that it was possible to produce gamma waves (~40 Hz) in a hippocampal slice by addition of a drug that activated the muscarinic type of acetylcholine receptors. The experimental phenomenon came to be known as "persistent" gamma, because the oscillation lasted for hours. The gamma waves were often superimposed on theta waves (roughly 4–10 Hz), in a pattern that was strikingly reminiscent of the superimposition of gamma on theta in rodent hippocampus in vivo, during exploration, or during certain types of anesthesia. Fisahn et al. found that the gamma oscillation period was determined by phasic synaptic inhibition—which was not in the least surprising. But they in addition found that pyramidal neurons, the principal excitatory neurons of the hippocampus, fired sparsely; that is, each pyramidal cell would fire on only a small percentage of the gamma waves. It was true that this sparseness also occurred in vivo, but was still not expected in the slice, based on the behavior of pyramidal neurons in an earlier in vitro experimental model of gamma oscillations. Sparse firing has been provocative for several groups in trying to understand the network mechanisms of persistent gamma oscillations.

Let us consider next the Draguhn paper. It was relevant to another in vivo signal from rodent hippocampus, now called sharp wave/ripples. Both the sharp wave and the ripples were characterized by the Hungarian-American neuroscientist György Buzsáki and his colleagues. Sharp waves were discovered first, in hippocampal EEG (electroencephalogram) recordings from rats, when the rat was engaged in slow-wave (non-dreaming) sleep, or during immobility, such as associated with behaviors such as drinking or grooming—but not during locomotion, which leads to theta and gamma in the hippocampal EEG. Then, Buzsáki and colleagues reported that "ripples"—low-amplitude oscillations at >100 Hz—were superimposed on the sharp waves.

Andreas Draguhn and colleagues found a way to produce ripples, at about 200 Hz (5 times the frequency of gamma waves) in a slice, without the sharp waves. Further, the Draguhn paper demonstrated that "pure" ripples, at least in vitro, did not require conventional chemical synaptic transmission, with transmitter release and postsynaptic actions. Indeed, the in vitro ripples were seen in media that blocked synaptic transmission. Instead, the in vitro ripples required what is called electrical coupling, or electrotonic coupling, through gap junctions. In other words, neurons (and pyramidal neurons at that) were communicating because electrical current flowed directly from one to another, rather than through classical neurotransmission. Such a notion had been advanced by F. Edward Dudek and Brian MacVicar for excitatory neurons in the hippocampus in the early 1980s, but had become mired in controversy; gap junctions in the brain stem and spinal cord and retina, and in invertebrates, were one thing; in telencephalon quite another. Of course, as is well known, gap junction coupling between inhibitory neurons became a very hot topic indeed in 1999, with the nearly simultaneous publication of papers by Shaul Hestrin and by Barry Connors and their colleagues. But the Draguhn data concerned excitatory neurons, not inhibitory neurons.

Thus Fisahn and coauthors showed gamma waves and synaptic inhibition; Draguhn and co-authors showed ripples and gap junctions. Why imagine a connection? Yet it turns out that there is a deep connection, and—in capsule form—the findings were these: first, gamma oscillations of the sort Fisahn et al. were studying also require gap junctions, as well as synaptic inhibition (and synaptic excitation too, for that matter); and, further, the power spectrum of persistent gamma oscillations contains a small, although robust, high-frequency peak: not at 200 Hz, rather more like 70–100 Hz (but not a harmonic of the gamma). It is as if one could see a ripple oscillation either in pure form, as on top of a sharp wave; or one could see the ripples interrupted by inhibitory synaptic currents, as in a gamma oscillation.

The secret was to suppose that electrical coupling between pyramidal cells took place between axons. That hypothesis led us (working also with Andrea Bibbig) to be able to produce detailed models of both gamma oscillations, and ripples, that were based on straightforward unifying principles. We could account for the high frequencies during gamma, and the sparse firing of pyramidal neurons, for example.

Developing the synthesis between the Fisahn and the Draguhn studies, and at least beginning to understand the implications for clinical problems such as epilepsy—these have taken us 10 years. It is that synthesis that has led to this book. We hope the reader will be excited by the history: synaptic oscillations in the spirit of Ramon y Cajal, unified with nonsynaptic oscillations in the spirit of Camillo Golgi; by the science, including recent ultrastructure; and by the hope for practical applications, as in epilepsy.

Acknowledgments

Foremost, we thank our wives and children.

The following people made critical contributions to the work we discuss here: Andrea Bibbig, Diego Contreras, Mark O. Cunningham, Andreas Draguhn, Rafael Gutiérrez, Fiona E.N. LeBeau, Farid Hamzei-Sichani, Thomas Knöpfel, Nancy Kopell, Mark A. Kramer, Steven J. Middleton, Hannah Monyer, John E. Rash, Anita Roopun, Dietmar Schmitz. Dr. Robert Walkup of IBM provided invaluable assistance with computing.

Our work was supported by NIH/NINDS, the Wellcome Trust, the Medical Research Council (U.K.), the Volkswagen Stiftung, the IBM Corporation, and the Alexander von Humboldt Stiftung. We are very grateful to Dr. Yuan Liu of NIH for her continuing support, and to Dr. Candace Hassall and Professor David Gordon, then of the Wellcome Trust, for their support of RDT when he was working in England (1997–2001).

Contents

Part III Some in vitro oscillations

Part IV Implications for health and disease

Glossary of abbreviations

APV: 2-amino-5-phosphonovaleric acid

DHPG: 3,5-dihydroxyphenylglycine

ECoG: electrocorticogram

EEG: electroencephalogram

EPSC: excitatory postsynaptic conductance

EPSP: excitatory postsynaptic potential

IPSC: inhibitory postsynaptic conductance

IPSP: inhibitory postsynaptic potential

mGluR: metabotropic glutamate receptor

NBQX: 6-nitro-7-sulfamoylbenzo(f)quinolaxine-2,3-dione

VFO: very fast oscillation (>70-80 Hz)

Cortical Oscillations in Health and Disease

PART I

Overview of Normal and Abnormal
Cortical Oscillations

1

Introduction

Recent Developments in Cortical Oscillations. Goals and Summary of this Monograph

Our earlier monograph (Traub, Jefferys, & Whittington, 1999a) was almost entirely concerned with gamma (30–70 Hz) and beta (10–30 Hz) oscillations evoked in hippocampal slices by tetanic stimulation, with a special emphasis on long-range synchrony of the oscillations. The experimental model, which we examined there, was intended to capture aspects of in vivo brain oscillations, produced by sensory stimuli, and proposed to have cognitive significance (Bertrand & Tallon-Baudry, 2000; Engel & Singer, 2001; Frien et al., 1994); the experimental model had implications for memory as well (Traub et al., 1999c; Whittington et al., 1997b). Yet, one could argue that the in vitro experimental model—involving as it did the hippocampus—was in the wrong part of the brain for studying the in vivo and clinical issues. Might we better have studied a sample of sensory cortex in vitro? We now briefly review the in vitro (and computational) model as of 1998, with its strengths, insights, and limitations, and then consider some of advances in the field of cortical oscillations during the last 10 years—advances that have motivated the present study.

The tetanic stimulus, which we applied to the hippocampal slice—consisting of a train of electrical current pulses—besides producing immediate current-induced action potentials, also would lead to release of glutamate and of other neurotransmitter and neuromodulatory substances; alongside these effects, the tetanic stimulus might also lead, secondarily, to significant alterations in the ionic milieu of the neurons, for example, to increases in extracellular [K⁺]. The neurotransmitter/neuromodulatory substances would then act on a variety of receptors, both ionotropic (phasic) and metabotropic (slower, or tonic). Large, slow, excitatory postsynaptic potentials (EPSPs) develop in pyramidal neurons as well as in interneurons, mediated in large part by metabotropic glutamate receptors (Whittington, et al., 1997a; Fig. 1.1A); superimposed on the slow EPSPs are trains of fast inhibitory postsynaptic potentials (IPSPs). Locally synchronized network oscillations develop after a latent period of 50 to several hundred milliseconds, at gamma frequencies, and last for hundreds of milliseconds. Both the latent period and the duration of the electrically evoked gamma oscillations are similar to those observed in visual neocortical gamma oscillations in vivo, evoked by a stimulus such as a moving grating at appropriate orientation (Gray & Singer, 1989; Gray et al., 1992).

Of particular interest was the observation that tetanically evoked gamma oscillations could synchronize, on a millisecond time scale, when two sites were stimulated together, even when the sites were up to 3 or more millimeters apart, and expected axon conduction delays were 6 ms or more. In vitro synchrony was intriguing because gamma oscillation synchrony was also observed in vivo, between neocortical sites several millimeters (or even more) apart (Eckhorn et al., 1988; Engel et al., 1991a, 1991b; Fries et al., 1997; Gray et al., 1989; Roelfsema et al., 1997); and it been suggested, and continues as a working hypothesis, that oscillation synchrony provides a cortical mechanism for solving the "binding problem," whereby cortical regions responding to diverse aspects or features of a given "object in the world" can be functionally— albeit perhaps transiently—fastened together (Engel & Singer, 2001; Engel et al., 1997; Fries et al., 2002; Melloni et al., 2007; Singer & Gray, 1995).

It was also possible to develop a network model of the synchronization mechanism in vitro, based primarily on the rapid excitatory conductance time course in interneurons (Geiger et al., 1995, 1997), and the resultant ability of interneurons to fire tight spike doublets when pyramidal cells fired single-action potentials. The first spike in an interneuron was mostly elicited by synaptic excitation from nearby pyramidal cells, and the second spike in the doublet was triggered by excitatory input arriving from the longer collaterals from more distant pyramidal cells. It turned out that the spacing of action potentials within doublets could enhance synchronization of the two oscillating sites by local inhibitory feedback (Bibbig et al., 2002; Ermentrout and Kopell, 1998; Traub et al., 1996c). Experimentally, and in simulations, interneuron doublets do indeed appear when two stimulated neuronal sites oscillate in synchrony at gamma frequencies, in tetanically activated hippocampal slices (Fig. 1.1); but the doublets do not appear with single-site stimulation, consistent with model predictions.

A basic prediction of the tetanic gamma oscillation model has been verified experimentally in studies of a genetically altered mouse, in which oscillation synchrony is disrupted in the expected fashion (Fuchs et al., 2001). The genetic alteration consisted of a knock-in of GluR-B (also called GluR2) α-amino-3-hydroxyl-5-methyl-4-isoxazole-propionate (AMPA) receptor subunits into fast-spiking interneurons, where they are normally present only to a minimal extent. Excitatory postsynaptic conductances (EPSCs) in fast-spiking interneurons of the genetically altered mice are slower than normal, and the slow kinetics prevents the precise doublet spike timing required for synchronization.

Another basic feature of tetanic gamma oscillations in hippocampal slices—the gating of action potential timing by the underlying train of inhibitory synaptic events—provided insights into a framework of network activity whereby the outputs from populations of interneurons provide the critical temporal structure required to organize outputs of principal cells, almost completely independently of the pyramidal cell firing rates (Fig. 1.2). What is of especial interest in Figure 1.2 is that tetanic gamma oscillations were still associated with gamma-frequency IPSPs under conditions where pyramidal cell (somatic) firing was greatly reduced pharmacologically.

Despite the many similarities between tetanic gamma rhythms in hippocampal slices and their in vivo counterparts, the in vitro experimental model was highly dependent on differences in methods used for tissue preparation and maintenance. Basic alterations in the in vitro environment could lead to major changes in the pharmacology and appearance of the post-tetanic response and a collapse in the inhibition-based gamma rhythm (Bracci et al., 1999; Whittington et al., 2001). This lack of robustness across different laboratories has limited the further usefulness of the tetanic-stimulus approach, although this approach is still used (Song et al., 2005); and the development of a suitable experimental model of transient, sensory-evoked gamma oscillations still remains problematic. However, the field of cortical oscillations has continued to develop along new and unanticipated directions that not only complement and verify many of the fundamental discoveries made with the tetanic model, but also begin to reveal an incredibly rich diversity of mechanisms for rhythm generation in cortical networks.

Indeed, one can say that the field of cortical oscillations has undergone a revolution since 1998, starting with two major developments: first, the experimental discovery of "persistent" gamma oscillations by the late Eberhard Buhl and collaborators, oscillations that depend on both chemical and electrical synapses (Fisahn et al., 1998; Pais et al., 2003); and second, the hypothesis of electrical synapses via gap junctions between axons: an hypothesis supported first by physiological data (Draguhn et al., 1998; Schmitz et al., 2001), and more recently supported by ultrastructure (Hamzei-Sichani et al., 2007). These two developments are intimately related. Axonal coupling allowed us to account for the pharmacological properties of persistent gamma oscillations, as well as to give at least a preliminary explanation of very fast oscillations (>70 Hz) in the hippocampus. The revolution in the field has continued with the experimental discovery, and analysis, of a number of other types of network oscillation in cortex and elsewhere, over a wide range of frequencies (<1 Hz to >200 Hz), many of which depend on electrical coupling, but with

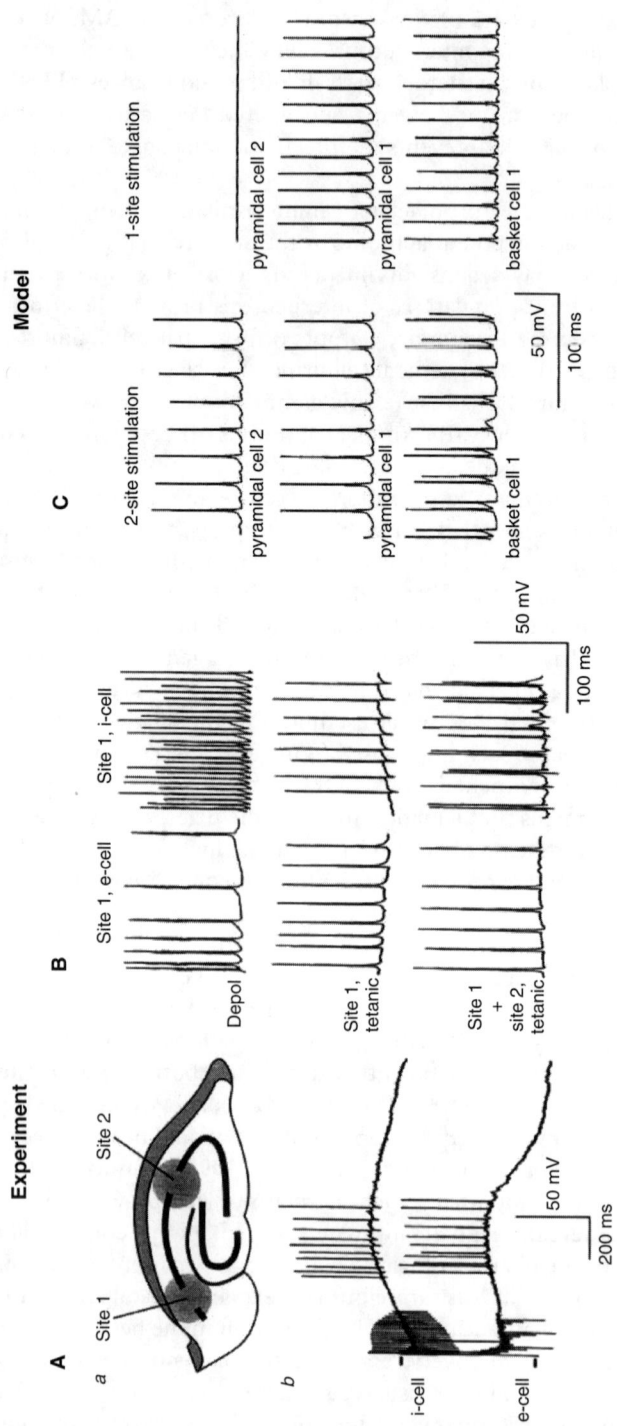

Figure 1.1 Long-range synchrony of gamma oscillations and interneuron doublets. **A:** *a*, Diagram of rat dorsal hippocampal slice. Stimulation sites were both in CA1, either at the subicular end (Site 1), the CA2 end (Site 2), or both; the sites were 1.2 to 4.0 mm apart. *b*, Tetanic stimuli (20 pulses, 100 Hz, in stratum oriens) produce slow depolarizations and action potentials in interneurons (i-cell) and pyramidal cells (e-cell). [The recordings were not simultaneous.] Other experiments in the original paper, and in Traub et al. (1996), show that the stimuli are associated with gamma frequency field potential oscillations, that are locked to the firing of pyramidal cells and interneurons, and that are synchronized between the two sites when both are stimulated together; and the data show that the intracellular depolarizations are primarily mediated by metabotropic glutamate receptors. **B:** In response to depolarizing current pulses, pyramidal cells show accommodating trains of action potentials, while interneurons exhibited fast-spiking (FS) behavior, with little accommodation. Stimulating one site tetanically produced 40–46 Hz gamma oscillations, while stimulating both sites produced a slower (~30 Hz in this case) gamma oscillation, in which the interneuron mostly fired doublets. **C:** Network model recapitulates the experimental findings. The network consisted of a 96×32 array of pyramidal cells, and a 96×4 array of interneurons. Gamma oscillations evoked by one-site stimulation are faster than two-site stimulation, because interneuron doublets tend to slow the network oscillation (62 Hz vs. 41 Hz); FS cells fire singlets with one-site stimulation, but mostly doublets with two-site; the dual oscillations in the two-site case are synchronized (illustrated in the original paper).

(Experimental part reproduced from Whittington et al., 1997, and model part from Traub et al., 1999 with permission.)

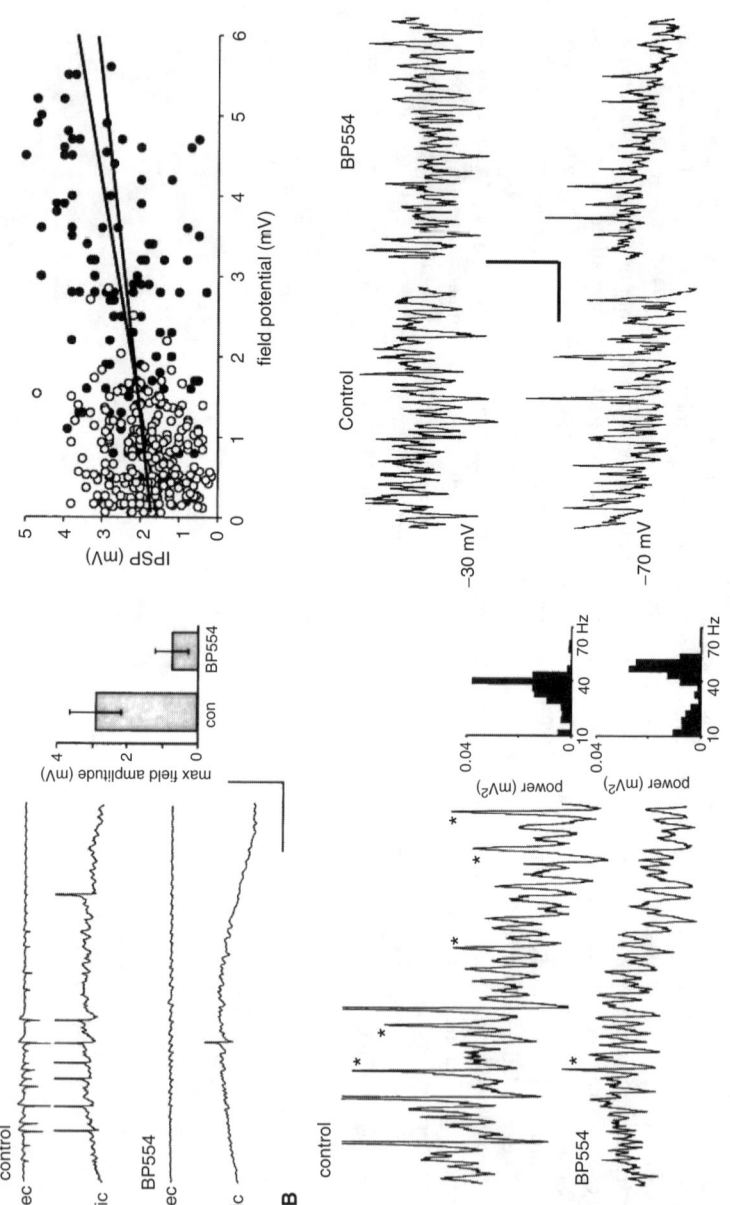

Figure 1.2 IPSP-gated tetanic gamma oscillations under conditions of minimal pyramidal cell firing; the gamma oscillation does not require population spikes. **A**: Tetanic oscillations persist after application of the serotonergic (5HT1$_A$) agonist BP554, which reduces pyramidal cell firing and virtually eliminates population spikes. *ec* is the extracellular field; *ic* is the intracellular recording from a pyramidal cell. **B**: Large EPSPs (asterisks) in pyramidal cells are almost eliminated by BP554, but other oscillating synaptic potentials persist. **C**: There is no significant correlation between the amplitude of field potentials and the amplitude of IPSPs in pyramidal cells. Filled circles are control conditions; open circles, in the presence of BP554. **D**: A fast-spiking interneuron was recorded with QX314 in the pipette to eliminate action potentials. The cell was held at −30 mV, to isolate IPSPs; and at −70 mV to isolate EPSPs. IPSPs were little affected by BP554; EPSPs continue to be present, although smaller. (Reproduced from Whittington et al., 2001 with permission.)

some involving chemical synapses as well. The detailed mechanisms of very fast oscillations are not uniform among all brain regions; and the interplay between effects produced by electrical and chemical synaptic coupling is intricate and beautiful.

The fact that the newly discovered (i.e., since 1998) oscillations can be replicated and analyzed so well, at least in vitro, forces us to consider how the oscillations we study may resemble their in vivo and clinical counterparts, and the medical implications of the data. That is the task of this monograph.

There are at least three reasons to study cortical oscillations. First, they may play a role in normal brain functions, cognition, and especially sleep. Second, they play a role in neuropsychiatric disorders such as epilepsy, schizophrenia, Parkinson's disease, and cerebellar ataxia. Finally, the study of oscillations has catalyzed and enriched cellular neurobiology, for example, with respect to gap junctions, synaptic mechanisms, and intrinsic neuronal membrane physiology. Brain oscillations are also a source of engaging problems to excite the imaginations of mathematicians and mathematical physicists.

The structure of the book is as follows. First, we discuss several neuropsychiatric conditions, for which the study of their associated brain oscillations is relevant to the clinical phenomenology, diagnosis, pathophysiology, or treatment. Next, we present electrophysiological data on the properties of individual cortical (and some other types) of neurons, and how the neurons are organized into multineuronal circuits. This section contains recent data documenting the existence, and importance, of gap junctions between principal (i.e., excitatory) neurons. Hence, in the multineuronal circuits that we describe, neurons are shown to interact with one another both through chemical and electrical synapses, with each type of interaction being critical in its own way. The chapters in this portion of the book review data from the literature as well as our own research; and they describe our approach to modeling single neurons, as well as to modeling complicated circuits of thousands of neurons, where the neurons fall into subtypes having distinct morphological, functional, and interconnection characteristics.

The central portion of the book describes selected types of neuronal population oscillations, where in vitro electrophysiological recordings, combined with detailed network modeling, have contributed to understanding of the cellular mechanisms. For each type of oscillation, we correlate the in vitro data with in vivo data and clinical observations. The particular oscillations that we review are these: *very fast oscillations* (>70–80 Hz), which can be generated in conditions of chemical synaptic blockade, but that occur in other conditions as well, including physiological conditions in vivo, and also as before and during epileptic seizures; *persistent gamma oscillations*, with some initial discussion of how these could be related to abnormal "epileptiform" activity; so-called *beta-2* oscillations, at about 25 Hz, that are generated in large tufted pyramidal neurons of layer 5 of the cerebral cortex; and mixtures of oscillations that occur during epileptogenesis.

What all these diverse types of oscillation have in common is this: electrical coupling via gap junctions.

The last section of the book discusses the implications, for health and disease, of the data outlined above. Concerning health, the most important observation is this: in several sorts of in vitro oscillation, which could well have in vivo correlates, most or all of the somatic action potentials *originate in axons* and then propagate antidromically (as well as orthodromically). The axonal action potentials occur either spontaneously (ectopically), or as a result of action potentials in electrically coupled axons—and not as a result of synaptic input to the soma and dendrites. This is a truly revolutionary concept, because all neural network models (at least those of which we are aware) are based, in one way or other, on the postulate that neuronal firing is a consequence of synaptic inputs to the cells: "synaptic integration," a classic (and very deeply ingrained) concept that pervades all of neuroscience. It is essential to determine the extent to which antidromic firing occurs in the behaving brain, in vivo, and to develop a conceptual neuronal network framework that allows for both synaptic integration and for axonal generation and spread of action potentials.

Concerning disease, we shall have the most to say, by way of novel observations, about epilepsy. Based on the data reviewed later in this chapter, we can address the general question of why so many sorts of oscillations—at widely different frequencies—are associated with the epoch before a seizure, as well as during the seizure itself; and we can offer informed guesses as to how specific types of oscillation might actually initiate a seizure. For schizophrenia, the most relevant observations are these: that sensory stimulation evokes gamma followed by beta oscillations in the intact human brain, as it does in brain slices; but the "habituation" properties of the beta portion are different in schizophrenics as compared with nonschizophrenics. Specifically, repeated sensory stimulation in normal individuals leads to attenuation or disappearance of the beta frequency component of the evoked response (habituation); but this habituation occurs to much lesser degree in schizophrenics. As we have a plausible model of the beta oscillation in slices, we can make informed guesses about what happens in situ.

The implications for Parkinson's disease are related to the observation in humans that symptomatic disease is correlated with enhanced beta oscillations ($\sim$15–30 Hz) in a number of locations within the basal ganglia; and that L-3,4,-dihydroxyphenylalanine (L-DOPA) treatment, which elevates dopamine concentration within the brain, attenuates the beta activity as it relieves symptoms. Basal ganglia beta may be projected from cortex via corticostriate connection; and, as we discuss in the central portion of the book, cortical beta (at least the 20–30 Hz variety) is critically dependent on gap junctions. It also known that dopamine closes gap junctions at a number of sites (e.g., retina), leading us to the hypothesis that at least some clinical aspects of Parkinson's disease are the result of increased amounts of oscillations that depend on gap junctions; and, that dopamine might be clinically effective, at least in part, because of its actions on gap junctions. This hypothesis must be considered critically, because the Piper rhythm (at gamma frequency) is enhanced by L-DOPA treatment, rather than suppressed; the Piper rhythm, however, occurs in the setting of near maximal muscle contractions, while beta rhythms are associated with weaker contractions, or even immobility.

In the course of this monograph, we discuss cerebellar ataxia, although it is premature (in our opinion) to draw firm conclusions about the importance of brain oscillations for ataxia—even as we recognize that there must be some connection. A number of oscillations occur in the cerebellum. The best known one consists of localized regions of ~4–10 Hz rhythms that are generated in the inferior olive and projected to the cerebellum via the climbing fibers. Faster oscillations occur in the cerebellum as well, and have some degree of coherence with oscillations in other parts of the motor system (portions of the thalamus, basal ganglia, and cerebral cortex). As we describe in the central section of this monograph, both gamma and very fast oscillations have also been generated in cerebellar cortex slices, with the very fast oscillations occurring at 100 to 200 Hz, and being produced by electrical coupling. Remarkably, runs of very fast oscillations also occur in the cerebellum in vivo, in a mouse model of Angelman syndrome. The mice are ataxic, as usually are children with this syndrome, and the oscillations in vivo appear to require gap junctions. The human disorder may therefore also, in addition to epilepsy and Parkinson's disease, involve gap junctions as part of the pathway by which signs and symptoms are expressed. Although Angelman syndrome is rare, and the available oscillation data have little to say about the chief deficits in Angelman syndrome, such as the language disorder, still, cerebellar ataxia is a common neurological problem, occurring frequently in prevalent diseases such a multiple sclerosis. We argue that we must open our minds to the notion that disordered oscillations, and disordered gap junctions, could play a role in

Table 1-1 Oscillation Terminology

Oscillation	Frequency Range (Hz)
Slow oscillation	<1 Hz
Delta	2–4 Hz
Theta	4–12 Hz
Alpha	9–13 Hz
Spindle	10–15 Hz
Beta	10–30 Hz
Beta-1	10–20 Hz
Beta-2	20–30 Hz
Gamma	30–70 Hz
VFO (very fast oscillation)	>70 Hz

a wide spectrum of clinical conditions, and that many important, and specifiable, issues thus need to be addressed.

Table 1.1 shows the frequency ranges of the various oscillations to be discussed. Note that there is some overlap in these ranges, and the label attached to the oscillation is determined not just by the frequency, but also by the brain location where the oscillation is recorded, the behavioral state, and by the presumed mechanism. (Thus, e.g., we do not use the term "gamma" for oscillations at frequencies above 70–80 Hz—in contradistinction to many other authors—because of our presumption that oscillations at above 80 Hz are not gated primarily by IPSPs, and our opinion that the term "gamma" should be reserved for oscillations that are so gated).

2

Historical Prelude

The cortex is part of the brain, clearly enough. Thus, a "cortical oscillation" is some sort of brain oscillation. But is it obvious what "a brain oscillation" even is? Let us avoid the issue of how to define an oscillation, leaving that to mathematics and physics, and simply acknowledge that the oscillations we discuss are not, strictly speaking, oscillations in the pure sense: they are not of constant amplitude or of constant frequency. Our "oscillations" are the real thing only in some qualitative sense, and perhaps are best described as alternating fluctuations in cellular excitability. We are here more interested in cells and what they are doing than we are in waves as James Clerk Maxwell or Erwin Schrödinger would have understood them.

So let us call a cortical oscillation something like this: an oscillating level of excitability that we can envision, and preferably can measure with instruments, that is produced by a population of cortical neurons. For purposes of this monograph, the oscillation should not be too slow (say, >~0.1 Hz), nor too fast (say, <1 kHz). As to what exactly the measured oscillating quantity consists of, here we allow ourselves some latitude: it might be an electric potential recorded in a single cell, in brain tissue, on the surface of the brain, or on the scalp; perhaps a magnetic field; perhaps an oscillating voltage or calcium signal recorded with optical means; or some of these in combination: but each of these quantities is related to the excitability of one or more neurons. With some poetic license, we shall call comparable signals, produced by our neuronal network models, "cortical oscillations" as well. In the models, however, the signal in question might in some cases be an electrical potential, but could equally well be something more abstract and difficult to measure experimentally, say the number of axons of a certain cell type exhibiting over-shooting action potentials (as a function of time, of course).

This preface is not the ideal place for philosophizing about time itself, but— following Einstein's work on relativity—it is well to be cautious about assumptions concerning absolute time. On large spatial scales within the universe, the finite speed of light makes us dispense with an absolute time that is valid everywhere; instead, we have to worry about frames of reference. In the brain, neurons, or parts of neurons, are our frames of reference—a point of view required because communication time from one neuronal cell body to another distant one can consume many milliseconds. (It is easy for physiologists to become confused about this. Thus, when one measures potentials at the scalp or on the surface of the brain, communication times between distant sites—say, centimeters apart—are limited only by conduction times of ions through tissue and spinal fluid, and these times are negligible for the frequency ranges of interest here. On the other hand, from the point of view of the actual neurons generating the signals, communication times may be dominated by axonal conduction times and synaptic delays, and these are much longer than the delays imposed by current flow directly through tissue. Thus, with cortical axons sometimes conducting at 0.2 mm/ms or less, 1 cm separation requires 5 ms: a time that is indeed significant for the frequency ranges we discuss.) Given that so much of the interest in oscillations derives from wanting to understand *synchrony* of the oscillations, we have to be very careful about what synchrony means: synchrony in two simultaneously recorded EEG signals means one thing, but synchrony of firing of two widely separated cells means something else.

Why should anyone care about cortical oscillations? No one would ask such a question about the heart, not at least someone who knew that rhythmical contractions of myocardial cells, and their associated electrical trajectories, occur at the same frequency (in general) as the heartbeat itself, and in temporally and spatially coordinated fashion (again, in general) in a manner functionally related to pumping action and to valve operations. Evidently, for the heart, it is important to understand the intrinsic oscillatory mechanisms of individual heart cells, how heart cells communicate with each other (mostly through gap junctions), how the heart beat is initiated and the beat frequency is maintained, and how the design of the heart protects it (for the most part) from disruptive rhythms. All well and good, but the brain does not pump blood.

One might argue, then, that the nervous system contains so-called central pattern generators (CPGs) analogous to the CPGs of invertebrates, and— even in mammals—CPGs control rhythmic motor activities, such as respiration, whisking in rodents, and chewing. The problem is that all of these CPGs are located in the brain stem and not in the cortex (Ballanyi et al., 1999; Cramer & Keller, 2006; Nakamura et al., 1999), even though it must be admitted that the CPGs are influenced by cortical inputs. Our motivation for studying cortical oscillations, then, must derive from elsewhere. The following are a few reasons:

1. History, particularly the history of EEG recordings in humans (Berger, 1929; Jung and Berger, 1979) and the demonstration of the alpha rhythm. [Berger's work followed on the earlier recording of cortical potentials by others, e.g., Richard Caton in the rabbit (Caton, 1875; Haas, 2003).] We shall say more about history later.

2. The appearance of cortical oscillations in defined frequency ranges, and in defined places, as a correlate (and perhaps a cause?) of specific brain states or behaviors: deep sleep without dreaming, awake but attentive mobility, and (of course) the alpha rhythm itself, most prominent over posterior head regions, during relaxed but awake immobility with the eyes closed. We assume—but it is far from proven—that all of these brain oscillations are present to serve some (as yet unknown) function.

3. The association of special sorts of cortical oscillations with altered behaviors. The most striking, diverse, and complicated examples of this phenomenon occur before and during epileptic seizures, where the oscillations in EEG or ECoG (electrocorticogram) range from subtle phenomena hard to distinguish from garden-variety brain waves, to enormous waves never seen in normal physiological states. (One must be wary of the term "normal brain" here, as an epileptic seizure can be induced, with appropriate stimulation, in anyone: an abnormal brain state can occur in an anatomically and functionally normal brain.)

4. The observations made in the late 1980s (Eckhorn et al., 1988; Gray & Singer, 1989) that sensory stimulation evokes cortical oscillations after a delay period, where the oscillation frequency (typically 30–70 Hz) is not related to characteristics of the stimulus. Rather, something about the stimulus may be represented in which particular cells in the cortex are oscillating; and, further, the properties (especially the frequency) of these so-called "induced" oscillations vary over time, on the scale of hundreds of milliseconds or seconds. How are such oscillations generated, and what determines the frequency and synchronization properties? Without answers to those questions, it will be hard to judge, or even formulate, hypotheses concerning function.

5. Finally, and most elusively, most of us are convinced that thought is determined by the activities of large groups (how many? who can say?) of nerve cells. How this comes about is, for now, hard to imagine. Nevertheless, cortical oscillations give us a place to start. First, cortical oscillations can involve many cells, but in ways that are characterized (at least statistically) by observing only a few of the cells at a time, along with the average activity of a larger ensemble (the "field potential"). Second, we know how to replicate a number of cortical oscillations—at least in circuits involving only a few thousand cells—with convincing models: but only when the models take into account known (or, at any rate, investigable) properties of the neurons, and the ways in which the neurons interact with one another (chemical synapses, gap

junctions, the extracellular environment). Because we can replicate oscillations in computer models, it means that the models can make predictions, and scientific progress is catalyzed. Will such endeavors ever lead to understanding thought, rather than just more and more about brain oscillations? The reader must decide.

To put into historical perspective our approach to cortical oscillations, let us consider some of the views expressed by Norbert Wiener in his influential book *Cybernetics*, first published in 1948, but released as a second edition in 1961. The second edition contains, as the last chapter, a section entitled "Brain waves and self-organizing systems." It is constructive to try to recreate the intellectual world in which this chapter was written, and to attempt to summarize advances in neuroscience—experimental discoveries, technical innovations, modeling and conceptual advances—that have changed our views and changed the nature of the questions we ask, even as many of our ideas and much of our style owe a debt to Wiener.

First some words on Norbert Wiener himself. Wiener was a mathematics professor at the Massachusetts Institute of Technology, who died in 1964. His most famous mathematical work was in analysis, especially including harmonic analysis, and what would now be called stochastic processes. He was, however, like his father, a polymath, linguistically gifted (although many found that he neither spoke nor wrote clearly), and one who had studied formal logic and philosophy with the likes of Bertrand Russell. Wiener worked with the purest of the mathematically pure, like G.H. Hardy, but his life and his writings suggest that he drew no clear line between pure and applied mathematics; examples drawn from electrical engineering, information theory, and clinical and experimental physiology pervade *Cybernetics*. He did war work on fire control, the automatic tracking of enemy airplanes, and the automatic aiming of anti-aircraft guns so as to have the best chance of hitting them. Wiener coined the term cybernetics (from the Greek for "steersman") to encompass a mixture of his primary interests—in communication, control, feedback, oscillations, and so forth—with all manner of anticipated applications, including, in particular, understanding the brain.

How is one to understand the neurophysiological world in which the first edition of *Cybernetics* appeared? One type of answer to this question is in terms of what scientists at that time (1948) could *not* do: for example, there was little then in the way of intracellular recording, the techniques of which were being developed only in the 1940s (Brock et al., 1952). There was in the 1940s a wealth of information from clinical observation and neuroanatomy, both macroscopic and microscopic; brain waves had been recorded in humans and in experimental animals (Adrian, 1936, 1942; Adrian & Matthews, 1934; Berger, 1929); single axons had been recorded (Adrian et al., 1931; Gasser & Grundfest, 1939). Charles Sherrington (first edition of his classic book 1906, second edition 1947) had deduced the existence of synaptic inhibition from his studies of spinal reflexes, but the direct demonstration of such inhibition by John Eccles and colleagues, using intracellular recordings in cat spinal motor neurons, was to come (Coombs et al., 1955a). So was the work of Hodgkin and Huxley (1952) on the mechanisms of the action potential in squid axons.

One area of immense progress during the 1940s was in the development of the modern stored-program digital computer, a development no doubt accelerated by military needs—for example, code-breaking during World War II, and the design of nuclear weapons following. Computer concepts heavily influenced how people at the time, including John von Neumann himself, thought about the brain. Consider the classic 1943 paper of McCulloch and Pitts, which argued that neurons could be viewed as logic gates, interconnected by axons carrying bits—not unreasonable, given that an axon is either firing or not firing. McCulloch–Pitts neurons live in the present: they take a simultaneous set of effectively instantaneous "synaptic" inputs, make a decision of 0 or 1, and pass on the result; then they forget everything and become ready to do it again on the next cycle (or time step—the synaptic integration time which they consider is about 0.25 ms). Without possessing intracellular recording data—showing that neuron intrinsic currents, and synaptic currents, often last a long time—one might consider such a logic-device view perfectly sensible. (Knowing that membrane and synaptic currents actually do operate on many different time scales, of course, makes the logic-gate view not sensible at all.) Be that as it may, the McCulloch–Pitts model engaged some of the finest minds of the time (see, e.g., Shannon & McCarthy, 1956). Feedforward neural networks are perhaps a descendant of the McCulloch–Pitts model, but with continuously variable, and modifiable, "synaptic weights," rather than 0's and 1's—and also, of course, without cycles or loops in the architecture (unlike the original McCulloch-Pitts model, which did allow cycles). Feedforward neural networks are still, however, left with the fatal disadvantage of not accounting for multiple time scales in neuronal and synaptic time constants.

Three other strands of thought were also influential on Wiener's ideas on the brain. First was the tendency to think of a neuronal network as a kind of telephone network, in which messages are routed from one place to another. This view led to studies of, for example, optimum encoding of sensory information in trains of action potentials (Rosenblith, 1961), work that continues to the present (de Ruyter van Steveninck et al., 1997). Second, Wiener was familiar with pathological tremors (e.g., in Parkinson's disease) and with cerebellar ataxia and intention tremor. He saw these as a result of disordered feedback mechanisms, as might arise (for example) from excessive axonal conduction delays. (The oscillations to be considered in this book do not, we should mention, take into account sensory feedback; we deal only with oscillations generated autonomously within a relatively localized brain circuit). Finally, Wiener had some ideas about the functional significance of the alpha rhythm, ideas that require a bit of explanation.

Norbert Wiener hypothesized, as others have, that the purpose of the alpha rhythm might be to generate a scan, in a manner analogous to the raster scanning used in cathode ray tubes (with which he was quite familiar); he quoted psychophysical data on reaction times—on the order of a hundred milliseconds, an alpha period—in support of this hypothesis. But what was being scanned? Wiener seemed to imply that the image itself could be scanned; but, in the chapter "Gestalt and Universals," he offered the intriguing notion that the scan was of a much more abstract nature: namely, through a group of transformations of 3-space (stretches, rotations, other deformations) to be applied to an image (see also Pitts & McCulloch, 1947), who show a scheme for actually performing the scan). The idea was this: a circle, for example, considered as a Platonic ideal circle, is recognized visually independently of image size, and independently of other distortions (e.g., seeing the circle from an angle, so that the formed image is what would actually be produced on looking at an ellipse). To make all the necessary comparisons between "possible circle that we are looking at" and "Platonic circle," one can scan through a suitably large number of allowable transformations, and determine if any of these carry the image-as-seen into the idealized image.

Recent data, mainly in vitro, suggest that the visually related alpha rhythm could be generated in the thalamus, and not the cortex, by a subset of thalamo-cortical relay cells that possess high-threshold dendritic calcium channels, and that are electrically coupled with each other via gap junctions (Hughes et al., 2004; Lörincz et al., 2008), whereas other in vitro data suggest the existence of a purely cortical alpha rhythm (M.A. Whittington et al., unpublished data). Whether any sort of scanning can be accomplished by either of these generating mechanisms remains unclear. This example is a perfect case study showing that determining oscillation mechanisms—using, at least in part, in vitro methods—allows us to approach critically ingenious hypotheses on oscillation function; and, further, it is usually easier to weaken an already proposed hypothesis than to come up with an original one.

Concerning the cellular mechanisms of alpha rhythm generation, Wiener had little to say—the data were simply not available to him. He was, nevertheless, able to make a deduction that, in retrospect, seems remarkable: Wiener envisioned a large number of small alpha-oscillators, coupled to one another. However these oscillators worked, and by whatever mechanisms the coupling was exerted, the coupling had to be nonlinear: this is what Wiener proposed, based on the shape of the EEG spectrum around 10 Hz, specifically the relative absence of energy at frequencies just above the alpha peak. There are instances where pure mathematics, properly applied to the brain, can do unexpected things.

Let us now consider—most selectively—how the neurophysiological landscape has changed since *Cybernetics*, concentrating on aspects of special relevance to oscillations:

1. *From whence can we record data, and by what methods?* Data can be recorded now by diverse technologies, from a variety of structures on different scales, and via large numbers of channels. Our instruments can detect extracellular potentials (in brain tissue) from more than 100 separate sites at once,

using electrodes, and even more sites with optical methods. Optical methods also provide access to intracellular calcium concentrations at multiple sites simultaneously. Four or more cells can be recorded together intracellularly in vitro, and almost as many in vivo. Current flow through single membrane channels can be measured, as well as transmembrane potentials in dendritic branches, axons, and even axon terminals. Recorded cells can be filled and reconstructed, sometimes in synaptically connected pairs, and these identified cells then examined ultrastructurally. Magnetic fields can be detected with multiple SQUID detectors from whole brain, and even from brain slices. Transgenic techniques can be used to cause specified cell types to fluoresce, allowing them to be recorded and manipulated in a controlled fashion. Many other technologies exist as well.

So far as preparations go, three particular developments since the 1950s are worth noting: first, in vitro brain slices (Collingridge, 1995; Li & McIlwain, 1957; Yamamoto & McIlwain, 1966) and other acutely isolated preparations, such as isolated spinal cords, or brain stem–cerebellum (Llinás et al., 1981). Such preparations contain thousands, or tens of thousands (or more) neurons, and are often able to exhibit network oscillations, that may involve several distinct cell types. Further, some in vitro preparations can even be correlated with "behavior," in the form of signals propagated along output pathways: an example is so-called fictive swimming in the lamprey spinal cord (Grillner, 1985). At the same time, in vitro preparations allow application of drugs (e.g., blockers of a defined synaptic receptor molecule) to the bath, so that each drug—eventually—reaches a uniform concentration throughout the preparation; such an experiment is, alas, hard or impossible to perform in vivo. One can also, at least sometimes, do intracellular recordings from defined cell types using in vitro preparations. [Of course, as the late Mircea Steriade (2001) and others have pointed out repeatedly, in vitro preparations have disadvantages as well: disconnection from afferents, loss of physiological neuromodulatory inputs, difficulty in correlating cellular activities with behavioral states.]

A second class of preparations developed over the last decades might be described as complex in vivo approaches, in which a behaving animal (or in some cases, a lightly anesthetized animal) receives controlled sensory inputs and perhaps responds all the while, even as brain signals are recorded—signals that could include "local field potentials" (tissue EEG, in effect) and "unit recordings" (signals corresponding to somatic action potentials in individual neurons); or in which an anesthetized and paralyzed animal has recordings taken from many sites simultaneously, both extra- and intracellularly (Steriade, 2001). Data from these types of experiments have been invaluable in defining behavioral correlates of brain oscillations, and in determining relations between oscillations in one part of the brain and another.

Finally, there has been expanding access to recordings from human brain tissue, both in living patients and also in vitro. Of course, scalp EEG was available in Norbert Wiener's time, and pioneering neurosurgeons such as

Wilder Penfield had been recording from, and stimulating, the brains of awake but sedated patients during operative procedures (Penfield & Roberts, 1966; Penfield & Welch, 1949). There has, however, been an enormous increase of activity in this area. Intraoperative recording continues to be used on occasion, in epilepsy patients but also in patients with movement disorders such as Parkinson's disease. In addition, brain electrical activity (with depth electrodes as well as subdural grids and strips) is often recorded for a week or more at a time, while patients are in an in-hospital epilepsy monitoring unit, usually with simultaneous video monitoring. Recordings of this nature provide data with excellent signal-to-noise ratios, a much broader frequency response than attainable with scalp EEG, and the possibility of correlating normal and abnormal brain oscillations with the patient's behavior. Lastly, since the 1970s (Schwartzkroin & Prince, 1976) human brain tissue, removed during operative procedures (often, but not exclusively, performed on patients with a focal epilepsy) has been studied in vitro, with electrophysiology and pharmacology. In this way, it is possible to determine, for example, if oscillatory phenomena investigated in, say, rat experimental models, also occur in the human brain, and with similar properties (Middleton et al., 2008).

2. *How are data stored and analyzed?* As brain oscillations occur over such a broad range of frequencies, and also locations, and as well have such complex relationships with the activities of individual cells, it cannot be overemphasized how important it is to be able record, store, and analyze high-quality data. In particular, one wants to compute spectra and correlations, and to assess the stability and reproducibility of oscillation properties, in time and space. It is all too easy to take for granted the extraordinary advances in the relevant technologies. In Wiener's time, data were stored in analog fashion, on bulky FM tapes, which at least allowed for offline analysis. Chart recorders allowed for the recording of long-duration time series, with a limited frequency response, and extreme difficulty in performing spectral analysis. Short-duration sequences of data could be accurately captured on a cathode-ray tube, but the study of oscillations is hard to accomplish this way. In contrast, today, data can be recorded in digital form from multiple channels, over frequency ranges of tens of kilohertz, for indefinite periods, in quantities of gigabytes or even terabytes. All manners of software application packages and visualization tools are available for offline analysis. [Interestingly, the Cooley–Tukey algorithm for the fast Fourier transform (FFT) was not published until 1965 (Cooley & Tukey, 1965), 17 years after the first edition of *Cybernetics*.] The types of analysis technically possible are limited only by our mathematical abilities and our imagination.

3. *How do we understand the basics of membrane electrophysiology and synaptic function?* Of course, one cannot do justice to this question in a few paragraphs, but we here list some of concepts that have emerged over the last few decades—again, encompassing many ideas that we already tend to take for granted, but should not.

In the time of *Cybernetics*, it was known that action potentials were mediated by ionic fluxes across the membrane, but the quantitative and structural details were unknown. Now, of course, we know that intrinsic membrane currents flow through channel proteins with defined subunit composition, three-dimensional structure, ionic selectivity, and kinetic properties. We know that electrogenesis can occur on the time scale of an action potential, but also on a wide range of slower time scales, due to channels that open on depolarization or on hyperpolarization, that strongly inactivate or contrastingly inactivate minimally. We have phenomenological (but immensely useful) models of channel behavior based on Hodgkin's and Huxley's ideas, but also more refined Markov models of single-channel state transitions. With single-channel patch recordings, antibody staining, localized drug applications, and other methods, we have a rough idea how each channel type is distributed across the membrane of neurons: some at higher concentrations in dendrites, others somatic, and others in the axon. We realize that channels cooperate with one another, not only through modifications of membrane potential, but also because ions flowing through one channel can interact with other channels: for example, calcium ions can act to open up certain types of K^+ channels. Finally, we are beginning to work out how channel properties can be modified by neurotransmitters and neuromodulators acting on slow time scales (seconds or more).

More specifically, we also know that individual neurons can act as oscillators, either producing suprathreshold voltage events at regular intervals, subthreshold oscillations, or a mixture of the two (Gray & McCormick, 1996; Jahnsen & Llinás, 1984a, 1984b; Llinás & Yarom, 1981a, 1981b; Llinás et al., 1991; Nuñez et al., 1992). Intrinsic neuronal oscillations (i.e., oscillations characteristic of individual neurons, capable of being generated when the neuron is isolated from its fellows, provided the membrane potential is set appropriately) can, in at least some instances, make important contributions to neuronal *network* oscillations, even if the network oscillations have properties not entirely determined by the nature of the intrinsic oscillators.

In a corresponding way to intrinsic membrane properties, our understanding of chemical synaptic physiology has extended from original basic concepts—that it is mediated by particular chemical compounds that induce selective types of ionic fluxes—to an understanding of its structural basis (with receptors having defined subunit compositions, ligand and modulator selectivity, voltage sensitivities) and kinetic properties. The biophysics of transmitter release are worked out in great detail. There has been enormous progress in synaptic plasticity, both potentiation and depression, the stimuli that induce it, the modulators that regulate it, the relevant timing of presynaptic and postsynaptic events, the role of calcium ions, and many other features. A clear distinction has been drawn between ligand-gated channels that are also ion channels (ionotropic receptors), and ligand-gated channels that exert their (generally slower) actions through second messengers, by intricate intracellular biochemical pathways. Finally, we recognize that there is a something of a continuum between rapid transmitter actions at well-defined synapses, and much slower effects mediated by modulatory substances released into the extracellular medium that act on extrasynaptic receptors. As we shall explore in this book, increases in the concentration of neuromodulators are a primary means by which long-lasting ("persistent") brain oscillations can be produced. Neurotransmitter actions on the "mere" hundreds-of-millisecond time scale are important in eliciting more transient sorts of oscillations.

In a more general sense, it is obvious that brain oscillations represent collective phenomena, engaging large numbers of cells; and collective phenomena require for their understanding a specification of the signals that each cell sends to other cells, and a specification of how each cell responds to those signals it receives. A truism, to be sure, but the reader must beware of being lulled into the traps that snare so much of the modeling community: for (a) neurons—neuronal somata, that is—are induced to fire, and are inhibited from firing, not only because of phasic (i.e., rapid) synaptic inputs from their fellows, but also because of intrinsic membrane currents, because of slow actions of neuromodulators, and because of retrograde ("antidromic") conduction of spikes along axons; and (b) neurons interact with their fellows not only because of chemical synaptic inputs, but also because of gap junctional coupling.

4. *The pervasiveness of gap junctions in mammals.* Gap junctions are a specialized form of intercellular communication, allowing for the passage of ionic current and small molecules, common throughout the body (epidermis, uterus, liver, ependyma, and many other places); gap junctions are essential to life, allowing for the passage of the cardiac action potential through the cardiac conduction system and from myocyte to myocyte. Gap junctions are formed, typically, by hexamers of proteins inserted into the membrane of each of two nearby cells; the hexamers fit together to form a tunnel. In turn, the coupled hexamers may themselves aggregate into assemblies numbering just a few members up to many thousands, and these assemblies may assume various shapes (plaques, ribbons, and so forth). The constituent junctional proteins derive from various families, including, in mammals, the connexins and the pannexins. A hexamer of connexin molecules is called a "connexon." The term "gap junction" is variously applied to two connexons joined together, to an aggregate of such connexon pairs, or even to the set of all such aggregates connecting a given pair of adjacent cells. The term "electrotonic coupling" is used to refer to the production of voltage changes in one cell by direct current flow from another cell, without the requirement of a chemical mediator or of synaptic release/receptor specializations. The term is not to be confused with "electrotonic current flow" or "electrotonic transmission," which refer to voltage changes within different parts of a given cell, produced by current flow within that particular cell, independent of currents flowing through voltage-gated membrane channels. Electrotonic coupling (also called "electrical coupling") occurs, at least in general, when there is a gap junction, and not otherwise.

Gap junctions and electrotonic coupling between neurons were recognized in invertebrates and in fish and amphibians in the 1950s, 1960s, and 1970s

(Bennett et al., 1963; Furshpan & Potter, 1959; Hagiwara & Morita, 1962; Sotelo & Taxi, 1970; Sotelo et al., 1975).

That gap junctions might be important in the generation of neuronal network oscillations was suggested by several observations: first, the pervasiveness of gap junctions in invertebrate central pattern generating circuits (Marder, 1984); second, the critical role of gap junctions in the medullary pacemaker nucleus of weakly electric fish, a structure that produces high-frequency oscillatory output, that does not contain recurrent chemical synaptic circuitry, and that does contain gap junctions (Dye & Heiligenberg, 1987; Elekes & Szabó, 1985; Moortgat et al., 2000a, 2000b; Sotelo et al., 1975); and, finally, the abundance of gap junctions in the developing nervous system of mammals, a stage when neuronal network oscillations are especially prominent—and at least some of these oscillations do actually depend on gap junctions (Dupont et al., 2006).

Indeed, gap junctions have now been found in the mature mammalian nervous system (including spinal cord, inferior olive, thalamus, hippocampus, cortex), and in a variety of subcellular locations: within chemical synaptic specializations (forming "mixed synapses"), between dendrites, and even between axons. These data are reviewed extensively later in this monograph. For now, suffice it to say that the evidence implicating gap junctions in brain oscillations is extensive; that the concepts involved have emerged since the "classical" period of the 1940s and 1950s; and that neural network theorists almost uniformly neglect to consider gap junctions. The notion that the axons of principal neurons are electrically coupled by gap junctions (Hamzei-Sichani et al., 2007; Schmitz et al., 2001; Traub et al., 1999b), with the obvious possibilities for "cross-talk" between the output lines of neurons, has potentially revolutionary implications for our understanding of brain function—not just for oscillations, but also for many or all types of computations performed by networks of neurons.

5. *The ubiquity of oscillations throughout nervous systems.* We have noted earlier Norbert Wiener's interest in the alpha rhythm. In his time, it was known that high-amplitude rhythmical EEG waves could occur during epileptic seizures, with particular prominence, regularity, and wide distribution in classical absence epilepsy (Gibbs et al., 1935), although Wiener seems not to have pursued this area. There were, as well, quantitative recordings of pathological oscillations (clonus) generated in muscle/spinal cord circuits. Nevertheless, a bewildering array of network oscillations have been discovered and characterized electrophysiologically only since Wiener's time, in virtually all brain regions, at all stages of nervous system development, over an enormous range of frequencies, and occurring in both physiological and pathological conditions. Oscillations occur during sensory processing, during motor output (and also arrest of motion), including the famous hippocampal theta rhythm that is prominent during locomotion but that is not a generator of locomotion per se (Buzsáki, 2002; Green et al., 1960), during memory tasks, and during sleep. Most importantly, motor-related oscillations have components at

the frequency of movement itself, but also other, higher (but nonharmonic), frequencies as well, suggesting some relationship to processing or computation. Advances in this area were motivated, in part, by research into sleep and consciousness (one thinks of the pioneers such as Bremer, Moruzzi, and Magoun, Dement and Kleitman, and later Steriade), into locomotion (Grillner) and coordination (Llinás), and respiration (Ballanyi et al., 1999); but in addition, as noted earlier, by observations on high-frequency oscillations evoked by sensory stimulation. The larger subject of brain oscillations is too vast to cope with in a single monograph—hence our focus on selected regions, particularly cortical regions, in which we ourselves have worked, and our restriction to cases where in vitro data, and simulations, complement the in vivo observations.

6. *Some conceptual models of large network function.* For the theoretical aspects of the present work, perhaps the most important conceptual advances derive from Hodgkin and Huxley on a quantitative accounting of the action potential (1952), and from Wilfrid Rall (1962), who showed that the cable equation, known since the 19th century and a relative of the classic heat equation, could (and should) be applied to the problem of modeling neuronal electrogenesis, as it existed independently of voltage-dependent channels. To the best of our knowledge, Frederick A. Dodge, Jr. and James A. Cooley (of the Cooley–Tukey algorithm), working at the IBM T.J. Watson Research Center in Yorktown Heights, New York, with the support of Hirsh Cohen— were the first (1973) to combine Hodgkin–Huxley ideas with Rall's ideas, in the form of a quantitative "compartmental" model of a spinal motor neuron that contained a cell body, a cylindrical representation of the dendrites, and an axon. This approach to modeling neurons was adopted by one of us (R.D.T.), also working at the Watson Research Center (Traub, 1977), and materially assisted by Dodge; and the approach was then applied to a model network of 100 hippocampal pyramidal neurons, synaptically interconnected, in an attempt to understand synchronized bursting during "epileptiform" events in vitro (Traub & Wong, 1982). Similar approaches were applied, with great sophistication, to analyzing the behavior of invertebrate central pattern generators in cellular terms (Harris-Warrick et al., 1992; Marder & Calabrese, 1996; Selverston, 2005).

Of course, one might argue that none of the above-mentioned approaches is really applicable to function in large neuronal networks. What are some models of function in large networks, besides McCulloch–Pitts and its descendants? We are not capable of cataloging all models that have been proposed, but mention a few that have received considerable attention:

Feedforward neural networks (Rumelhart et al., 1986) have only the most superficial resemblance to actual neuronal networks. As noted by Hecht-Nielsen (1987), a feedforward neural network is a means of approximating an arbitrary continuous real-valued function based (although this may not have been known to the inventors of these methods) on a theorem of Kolmogorov (1957). What is of interest in such networks is that the elements, the "neurons," have inputs and outputs together with a possibly simple dynamics, and that the network has a whole can be modified, for example, with the backpropagation algorithm, in response to experience. Further, feed-forward neural networks have found applications in engineering. Whatever one's views about the usefulness of feedforward networks in understanding the actual brain, it must surely be admitted that such networks are not appropriate for the study of oscillations.

Spin-glass-based models ("attractor models") (Hopfield, 1982; Hopfield & Tank, 1986) have been argued to be a useful characterization of how memory could work, and to have specific connections with features of actual brains: for example, in "Hebbian" (associative) modification of the strength of synapses based on experience and in the representation of "memories" as particular subsets of neurons' having been activated, while the rest of the neurons are suppressed. (Such subsets are now often referred to as "cell assemblies".) The late Daniel Amit, who contributed also to the physics of these models (Amit et al., 1985), was a strong proponent of the biological applicability of attractor models (Amit, 1998), especially to in vivo data obtained from monkey cortex during delayed match-to-sample tasks (Miyashita, 1988; Miyashita & Chang, 1988). Again, whatever one's views on the biological relevance of spin-glass models, such models are not helpful in the study of oscillations, because—at least in basic implementations—neither the "neurons" nor the "synapses" have time constants that are long relative to the basic time step.

Certain, possibly unexpected, subdisciplines of mathematics and mathematical physics have also found applicability to the brain. We mention two: graph theory, especially random graphs (Bollobás, 2000, 2001; Erdös & Rényi, 1960), and cellular automata (Sarkar, 2000). Both have proven useful in analyzing very fast oscillations (Traub et al., 1999b) and may have other consequences as well (see, e.g., Buzsáki, 2006).

7. *Qualitative theory of systems of differential equations and chaos.* Diverse branches of mathematics are represented in *Cybernetics*, but one branch not well-represented consists of the qualitative theory of differential equations, a discipline that seeks to characterize the geometric features of a dynamical system (e.g., the fixed points, the periodic orbits, the strange attractors if these exist, and so forth), and to investigate how these features can change abruptly as a parameter in the dynamical system is varied (leading to so-called bifurcations; Guckenheimer & Holmes, 1983). Qualitative differential equations have a history going back at least to Gauss (1777–1855), with Henri Poincaré (1854–1912) (who was interested in celestial mechanics, among many other things) a real pioneer. There was a burgeoning of interest in this field in the 1960s and 1970s, with the work of Stephen Smale on structural stability and diffeomorphisms, and of Lorenz on chaos, among many other contributors. This theory can, at least at times, give an aesthetically and intellectually thrilling picture of the geometrical properties of a vector flow, the class of all solutions

of a set of differential equations; but that tends to be possible only when the dimension of the system is relatively small. Clearly, the qualitative theory of differential equations would not be reasonably applicable to logical automata of the McCulloch–Pitts type, or to high-dimensional stochastic processes, both models considered by Wiener in *Cybernetics*. We ourselves do not apply this theory either, even though our models are (for the most part) cast in the form of systems of differential equations—because the dimension is too high, in the tens of thousands. Nevertheless, some of our colleagues, especially Nancy Kopell (a former student of Smale), do apply this theory, along with a variety of other mathematical and heuristic ideas, to so-called reduced models, containing small numbers of relatively simplified model neurons, that capture some essential feature of the electrophysiology and of the complex models. As much thought and insight go into determining how to make the reduction, as into the analysis of the reduced system per se. Here we should also note that qualitative methods have given considerable insight into mechanisms of burst generation in single neurons and small ensembles, as well as into coupling between oscillating neurons (Baer et al., 1995; Chow & Kopell, 2000; Kopell & Ermentrout, 2004; Pinsky & Rinzel, 1994).

8. *Advances in computing.* The astonishing advances in computing made since the 1950s are, of course, well known. We may mention, just as examples, the invention of the transistor, of integrated circuits, advances in materials and magnetoresistance, hardware and software for parallel computing, and algorithms. Computing power is essential not only for the conducting of experiments and data analysis, but also for our style of network modeling, which seeks to use as many model neurons (within an order of magnitude) in the simulated network as exist in the preparation (in vitro, naturally)— although not necessarily the same number of synaptic release sites (we tend to approximate a "synaptic connection" between two neurons as consisting of a single release site); and we seek to produce models whose voltage traces look like experimental recordings, which necessitates the use of multiple compartments and multiple ionic conductance types in each neuron. Nevertheless, available computing resources allow such detailed simulations, with thousands of model neurons, to be performed in a matter of hours—too long to explore parameter spaces exhaustively, but short enough that the model is a useful tool for research. Indeed, the bottleneck in the modeling end of the authors' collaboration tends not to be computing per se, but rather conceptual—what to model, how to represent the "relevant" parts of the experimental data, and trying to stay abreast of the relevant experimental literature.

Thus, advances in computing (along with an experimental data base) allow us to go about modeling in a manner not feasible for Wiener and his contemporaries. But would this style have appealed to Wiener et al., even if it were possible for them to implement? Our method concentrates on the physical behavior of neuronal networks, abandoning—at least for now—any resemblance to symbolic logic, to philosophic issues, or to universal questions. Probably our scientific forebears would have found even a (hopefully) temporary relinquishing of goals held so dear to be rather troubling. We can only acknowledge this, with respect to our intellectual ancestors, and forge ahead.

As for our overall approach to oscillations: clearly, even if we are not as smart as Norbert Wiener, and our goals less encompassing, still, we have a plenitude of data, techniques, and ideas, about oscillations, that were not available to him. For all that, there is yet no convincing—but still easily stated—summarizing statement that can be made about brain oscillations, or even about just cortical oscillations. For one thing, there are many different kinds of oscillations, and a frequency taxonomy will not suffice to classify them all: for example, a gamma oscillation, even one produced within a given brain region, can be generated in different ways. One cannot assume that all varieties of oscillations, across frequencies and across brain regions, subserve comparable functions, any more than we can assume that they have the same mechanisms. But most notably, experiments that demonstrate oscillation functions are hard to perform and even harder to interpret (Fuchs et al., 2001, 2007). Our best hope, we believe, is to use transgenic techniques that disrupt an oscillation in vitro and in vivo, taking advantage of our knowledge of cellular mechanisms, and then to examine behavioral effects; but molecular compensations can occur, and the range of cortically mediated behaviors that one can examine in a mouse is limited.

This book, then, represents our best attempt to make sense of a complicated field, drawing on clinical, experimental (in vivo and in vitro) and modeling endeavors. We strive for clarity, but our imperfect understanding imposes unavoidable limits. Imperfect understanding also undoubtedly will cause us, at times, to impose extraneous detail on the reader. While we have separated ourselves from propositional calculus and spin-glasses, and other mathematically or statistical physics-based models of global brain activities, it is not yet clear what theoretical frameworks could, or should, replace them. What is to come requires improvisation.

3

Overview of In Vivo Cortical Oscillations

The experimental study of pathological brain oscillations is inextricably mingled with the study of normal oscillations; it is impossible to disentangle them. One must, however, start somewhere, and it is probably appropriate to begin with normal, physiologically occurring brain rhythms, as they occur in vivo— but with our discussion informed by (one might say prejudiced by) our understanding of the cellular mechanisms: mechanisms as determined by in vitro and simulation analyses, to be elaborated upon in later chapters. Because it is our wish that the discussion be so informed, we are constrained as to which oscillations we can consider; there are more in vitro data relevant to some oscillations than to others. For this reason, we shall concentrate in this chapter on oscillations of the following frequencies: gamma (30–70 Hz), beta (10–30 Hz), and VFO (very fast oscillations, >70 Hz). Biological functions can be reasonably hypothesized for gamma, beta, and VFO. In contrast, the slow oscillation (<1 Hz; Steriade et al., 1993d) seems to arise via biochemical mechanisms rather than—if one may put it thus—neuronal processing mechanisms; one function of the slow oscillation may be to support brief epochs of the other oscillations we want to consider, namely, gamma, beta, and VFO. The most widely known cortical oscillation is probably the alpha rhythm, but in vitro analysis of its cellular mechanisms is yet preliminary; we refer the reader to standard textbooks for the clinical features of the EEG alpha (and related) rhythms (Ebersole & Pedley, 2003; Niedermeyer & Lopes da Silva, 1999), and to recent papers on the cellular mechanisms of thalamically generated alpha-frequency oscillations (Hughes et al., 2004; Lörincz et al., 2008).

Synchronized neuronal network oscillations arise so readily in brain slice systems, in so many different experimental paradigms, that one is tempted to think that oscillations are inevitable, a straightforward consequence of the intrinsic properties of neurons, and the means by which neurons influence one another, through chemical synapses and gap junctions. Such a view can be rejected. Not all slice preparations oscillate. In the living brain, oscillations occur in some places and not other places, in some states of awareness or attention or movement and not others. One needs to consider not only the cellular mechanisms of each particular oscillation—best addressed with in vitro methods—but also the general conditions under which an oscillation can occur in vivo, and how the brain goes about selecting which neurons are to participate—or, alternatively, which neurons are not to participate. We assume that the significance of any brain oscillation is determined by precise (although not necessarily known) physical characteristics: which neurons are involved; the degree to which axons, somata, and dendrites participate suprathreshold or infrathreshold (i.e., whether different portions of the neurons fire or not); the phase relations between individual neuronal activities and the population as a whole; the ways in which the oscillating neurons affect the activities of "downstream" neurons; and the plastic changes induced, or at least facilitated, by the oscillation. Above all, we must avoid being seduced by mystical concepts such as that synchrony or oscillations matter in and of themselves.

The phenomenology of brain oscillations is sufficiently complex that it will help us to have some conceptual framework, which will allow us to formulate specific hypotheses as to what the oscillations are for, and how the different sorts of oscillations are interrelated. As a prerequisite for describing such a framework, however, some general—and counterintuitive—remarks about brain function are required. Specifically, we need to describe two "modes" in which networks of neurons can operate. We call the two modes "standard" and "nonstandard."

The Standard Mode of Neuronal Network Operation

This is the operating mode with which all students of neurophysiology are familiar (Kandel et al., 2000), and incorporates concepts going back to Charles Sherrington (1857–1952; Sherrington, 1947) and Donald Hebb (1904–1985; Hebb, 1949). There are two main ideas (Fig. 3.1A). First, neurons "integrate" their inputs, "adding up" their synaptic inputs [even if the details are actually nonlinear (Llinás, 1975; Rall, 1962)], and generating a sequence of action potentials that are a physical consequence of the synaptic inputs (Llinás, 1975; Rall, 1962) so that action potentials propagate forward along the axon and backward into the dendrites (Stuart & Sakmann, 1994), evoking a transient calcium concentration change (Spruston et al., 1995). Second, this activity-dependent dendritic calcium signal in turn secondarily initiates biochemical

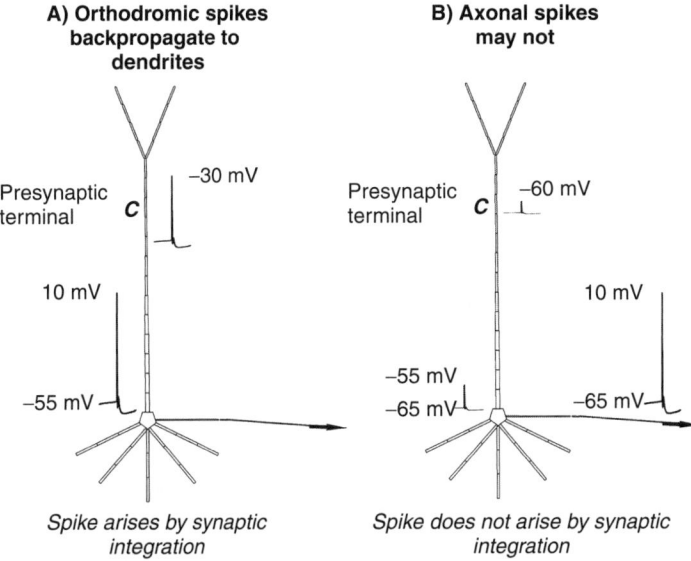

A) Orthodromic spikes backpropagate to dendrites

Presynaptic terminal

−30 mV

10 mV

−55 mV

Spike arises by synaptic integration

B) Axonal spikes may not

Presynaptic terminal

−60 mV

10 mV

−55 mV
−65 mV

−65 mV

Spike does not arise by synaptic integration

Figure 3.1   Two modes of spike initiation: Orthodromic spikes can backpropagate to dendrites, whereas antidromic spikes may block in the proximal axon and not propagate into the dendrites. **A:** "Typically," action potentials are initiated in the axon by somatic depolarization, itself induced by currents flowing from depolarized dendrites. Under these conditions, the somatic spike not only propagates down the axon, but also "backpropagates" into the dendrites (even if with some decrement); in doing so, the backpropagated spike can generate intradendritic signals that can be compared with the synaptic events that helped to trigger the spike: The synapse/neuron system is sensitive to temporal correlations of input and output. **B:** Under some—perhaps "atypical" conditions (but more commonly than realized)— spikes are initiated in the axon through the activities of electrically coupled axons, rather than through synaptically induced somatic depolarization. Such axonal spikes still propagate forward down the axon. Axonal spikes may invade the soma to generate full antidromic spikes that can backpropagate (not shown), but they may also decrement to give only a small "spikelet" in the soma, that decrements even further in the dendrites (Larkum et al., 2008; Schmitz et al., 2001). In neither case is there a meaningful presynaptic signal to which the backpropagated signal can be compared. Temporal correlations between input and output appear not to matter.

processes in synaptic specializations, the net effect of which is to influence plastic changes in synaptic efficacies—and, because the calcium signals are activity dependent—the changes of synaptic efficacies are consequences of the temporal correlations between synaptic inputs and synaptically induced action potential outputs: a process now called "STDP," or spike-time-dependent synaptic plasticity (Amit & Brunel, 1997; Caporale & Dan, 2008; Dan & Poo, 2004; Paulsen & Sejnowski, 2000). Of particular interest and importance is the experimental observation, in accord with Hebb's original intuition, that

potentiation (i.e., increase) in excitatory synaptic strength on pyramidal neurons occurs when the excitatory postsynaptic potentials (EPSPs) in question
precede—i.e., contribute to the causation of—an action potential, although
this time cannot be too long, typically on a time scale of tens of milliseconds
(Bi & Poo, 1998; Markram et al., 1997b).

The Nonstandard Mode of Neuronal Network Operation

In this more recently recognized, and less well known, mode of operation,
action potentials are initiated in axons of pyramidal cells in a manner independent of synaptic inputs, at least excitatory inputs (synaptic inhibition is
able, however, to suppress action potential initiation) (Traub et al., 2003b).
The action potentials may propagate forward to evoke transmitter release,
especially on interneurons, but invade the soma and dendrites only on occasion (Cunningham et al., 2004a; Larkum et al., 2008; Schmitz et al., 2001;
Traub et al., 2003b); mostly (Fig. 3.1B), axonal full spikes are associated only
with "spikelets" in the soma and dendrites. As a result, most axonal outputs
are not expected to correlate with dendritic calcium transients in pyramidal cells;
nor are EPSPs on the pyramidal cells causal for the action potentials in those
cells. Unexpectedly, a number of cortical oscillations, at beta-2 (20–30 Hz),
gamma, and very fast frequencies reflect this nonstandard mode of neuronal
network operation, at least in vitro. This, in turn, raises major questions as to
what the function of cortical oscillations can possibly be.

Neuronal Networks Operating in Standard Mode

Synchronized oscillations have natural consequences for neuronal networks
operating in the standard mode, although these consequences are not always
intuitive. Synchronized network oscillations, by definition, introduce temporal correlations in the firing times of the neuronal somata that participate in
the oscillations. Such temporal correlations in turn have consequences both
for synaptic integration and for spike-time-dependent synaptic plasticity
(STDP), provided that axonal conduction delays are not too large. For synaptic integration, synchrony provides for the possibility of having EPSPs—
generated by distinct presynaptic neurons—adding up in a given postsynaptic
neuron, the addition taking place with precision determined by the tightness
of oscillation synchrony and by the dispersion in conduction delays. This type
of synaptic addition is called coincidence detection, and can be important
in allowing for individually subthreshold EPSPs to contribute to firing a
neuron, especially when EPSPs are rapidly truncated by synaptic inhibition
(Abeles, 1982; Azouz & Gray, 2000; Egger et al., 1999; König et al., 1996; Llinás
et al., 2002; Prescott et al., 2006; Rodgers et al., 2006). For STDP, synchrony
provides for temporally defined relationships between synaptic inputs to,

and outputs from, multiple neurons; this principle can apply in three different ways:

1. For common synaptic inputs to a pool of synchronously firing neurons, there will be correlations in the type of plasticity occurring for these inputs onto *all* of the neurons in the pool.

2. For synaptic interactions between synchronously firing neurons, there will obviously be temporal correlations between the synaptic inputs to, and the action potential output from, each of the neurons in the pool (i.e., SDTP rules should apply) —and so, secondarily, there should be correlations in the plastic changes induced on each of the neurons. [It needs emphasis, however, that the plastic changes may not always be in the direction one expects, if conduction delays within the synchronously oscillating pool are large enough (Fig. 3.2)].

3. Likewise, the synaptic outputs of a synchronously oscillating pool of neurons will have correlated plastic changes on their respective postsynaptic targets, provided the target neurons are themselves synchronized with one another. For the output synaptic contacts to have correlated plastic changes—that is, the synapses all tending to

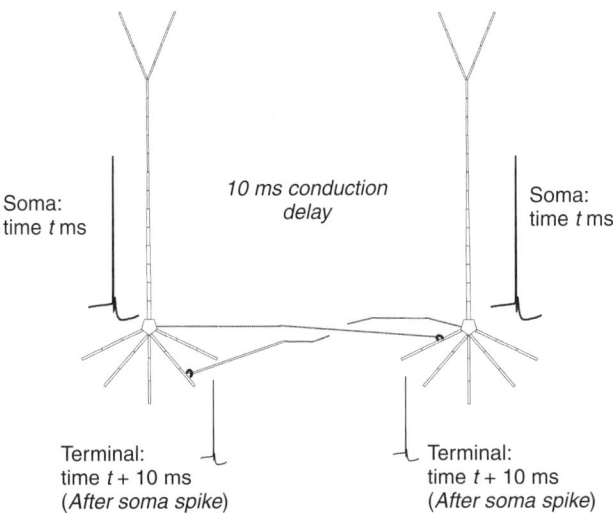

Figure 3.2 Synchrony of somatic firing does not necessarily imply synchrony of axon-terminal and somatic spikes. Schematic of two pyramidal cells that fire synchronously, as determined by somatic measurements. Spike-timing–dependent plasticity (STDP) is expected, however, to depend on the temporal relations between the presynaptic terminal spike, and the dendritically back-propagated somatic spike. In the case illustrated here, where there are 10-ms axon conduction delays, the presynaptic terminal spikes occur 10 ms before the nearby somatic spikes; either no synaptic plasticity, or synaptic depression, might be the result.

change in the same direction, either up or down—it is not required that the presynaptic and postsynaptic pools be synchronized with each other, only among themselves.

The observant reader will have noticed that the aforementioned consequences, for synaptic integration and plasticity, depend—in the standard mode—only on synchrony, and not on oscillation. The concept of synchrony is, logically and physically, distinct from the concept of oscillation. Why handcuff the two concepts together? A truly basic question. It turns out, however, that experimentally—both in vivo and in vitro—synchrony and oscillations do in fact go together; and when one examines the mechanisms that cause many interconnected neurons all to oscillate at about the same frequency (as we shall do later), one sees that these same mechanisms provide for synchrony. The oscillation and the synchronization emerge together.

Relevance of Nonstandard Synchronized Oscillations

If coincidence detection, and correlated synaptic plasticity, operates for synchronized oscillations in the "standard mode,", but not in the "nonstandard mode," why does the brain even have synchronized oscillations in the "nonstandard mode"? This is a paradox. Certainly, coincidence detection—of synaptic inputs—is not relevant when spikes are initiated purely by electrophysiological processes in the axons. Likewise, synaptic plasticity is not likely to occur when synaptic inputs are not causing spike outputs, and where most spikes do not backpropagate into the dendrites. It is not apparent, then, how oscillations generated in "nonstandard" fashion, by axonal and nonsynaptic mechanisms, could be relevant to brain function. While a solution to this paradox is not known definitively, we can offer a simple hypothesis:

Hypothesis: "standard mode" and "nonstandard mode" network behaviors may interact in multiple ways. Consider two networks, *A* and *B*, with *A* nonreciprocally connected to *B*. Standard-mode synaptic activity in downstream network *B* [particularly activity involving gamma-aminobutyric acid (GABA) release] may generate action potentials in the projection axons from network *A*. [This can happen because, as we elaborate upon in later chapters, GABA can excite axons.] These axonal action potentials can antidromically propagate back to network *A* to provide background drive for nonstandard mode patterns of activity. Such "ectopic" action potential generation is seen in CA3-to-CA1 connections in hippocampus (Avoli et al., 1998; Stasheff et al., 1993a,b), in thalamocortical fibers (Pinault & Pumain, 1989), and may be widespread in the peripheral sensory nervous system (Pinault, 1995). Conversely, oscillations in network *A*, generated in "nonstandard" mode, may generate synaptic effects in network *B*. Thus, coincidence detection, and correlated synaptic plasticity, can occur in the downstream network *B* neuronal targets, even if these mechanisms are not taking place in the "primary" oscillating population. Similarly, synaptic input from network *A* may also influence nonstandard network activity in target network *B*. Feedforward inhibition via somatodendritic GABA receptors can terminate axonally generated spike propagation quite potently (see Whittington & Traub, 2003). Standard-mode action potential generation, following dendritic excitation, also produces a large, transient calcium load. Even single somatic action potentials lead to large, long-lasting calcium concentration rises in primary axons and collaterals (Baudoux et al., 2003). Such increases in intracellular calcium could act to block gap junctional conductances required to maintain "nonstandard mode" oscillations.

Properties of Some Selected Cortical Oscillations

Gamma Oscillations Induced by Sensory, Particularly Visual, Input

The approximately simultaneous discovery of visually induced cortical gamma oscillations, by two groups in Germany (Eckhorn et al., 1988; Gray & Singer, 1989), had an immediate—and continuing—excitatory effect on systems neuroscience. The discoveries were not only exciting in themselves, but they seemed to point the way to a vital (and unexpected, at least to most people) link between cognition and cellular electrophysiology. The path seemed open to the materialists' dream: a scientific program for learning about the mind through a defined series of practical neurophysiological experiments. Many were caught up in this dream, including the authors. Although some of the psychological concepts had been discussed earlier, by, for example, von der Marlsburg (von der Malsburg & Buhmann, 1992; von der Malsburg & Schneider, 1986), it is our opinion that much of the credit for the excitement must go to Wolf Singer (b. 1943); Singer saw how the electrophysiological observations took on biological relevance and how this relevance might be best further analyzed. The key idea was to attach meaning to the existence of, or lack of existence of, synchronization of spatially separated neuronal networks, each oscillating at gamma frequency; and to devise an intricate series of in vivo experiments supporting the logical consistency of this postulated meaning. (Of course, the first experimental requirement was to demonstrate that the oscillations existed at all, even at a single cortical site.)

Basic Phenomenology

The classical approach to sensory physiology usually (not always) involves documenting changes in firing rates—in sensory axons or neurons—induced by effective mechanical, visual, or other modality-appropriate stimuli (Hartline, 1938; Rosenblith, 1961). As the information transformations within sensory systems became understood (Hartline & Ratliff, 1958), and as complex—but still biologically relevant—stimuli were used, two avenues of progress developed.

One avenue led to the development of quantitative models of sensory organs, at least simpler ones (Passaglia et al., 1998). The other led to the recognition that cells in sensory regions of the nervous system, distal to the primary receptive organs, responded to special meaningful features of stimuli: for example a moving spot, fly-like, of interest to a frog (Lettvin et al., 1959); or small moving bars, located in particular areas of the visual field, which bars might be viewed as elementary components of larger and more complex objects that a cat would observe (Hubel & Wiesel, 1959, 1962, 1963). And of course, in retrospect, the existence of responses to meaningful stimuli is hardly surprising.

What was shown in the late 1980s and 1990s for the visual system was that populations of cortical neurons responded to small moving bars not only by changes in firing rates, but by generating locally synchronized gamma oscillations: that is to say, one could (a) record a visually evoked oscillating field potential; one could (b) find individual units firing spikes (or brief high-frequency bursts of spikes—"chattering") which oscillated at the same frequency as the field; and (c) the unit oscillations and the field oscillations were tightly phase-locked (Eckhorn et al., 1988; Gray, 1994; Gray & Singer, 1989; Gray & McCormick, 1996; Gray et al., 1990, 1992). Oscillating units occurred mostly in superficial cortical layers, but could also be found in deeper layers (Gray et al., 1990; Livingstone, 1996); and the spatial scale over which oscillations were evoked by small stimuli was a few hundred microns, that is, on the scale of one or a few cortical columns (Eckhorn, 1994; Engel et al., 1991; Maldonado et al., 2000). There are temporal aspects to this local response besides the oscillation itself: the induced gamma oscillation has a latent period of tens of milliseconds to hundreds of milliseconds it lasts only a few hundreds of ms and the latent period and oscillation itself are not phase-locked to temporal aspects of the stimulus (although slower modulation is possible; see later) (Gray et al., 1992). Induced gamma oscillations are not an artifact of anesthesia or of the use of cats as an experimental preparation: they are found in awake cats (Gray & Viana di Prisco, 1997), nonhuman primates (Frien et al., 2000; Kreiter & Singer, 1992), and humans (Lachaux et al., 2000).

Once the brain has the possibility of generating fine temporal structure in its local responses (meaning the response of a pool of nearby neurons to a small visual stimulus), then it has the opportunities of exhibiting temporal structure in its responses on larger spatial scales: larger, that is, both in terms of the size of the stimulus, and in terms of the physical separation of the neurons that respond to the stimulus. One way that larger spatial scale temporal structure could, in principle, be expressed is via the synchronization of two or more pools of neurons, each responding to a "piece" of a large stimulus with its own gamma oscillations. Indeed, this is what happens, in at least some experimental settings (Fig. 3.3). The experiments that show synchronization between two separate cortical sites typically use a long moving bar, so that ends of the bar have distinctly separated visual fields, while still possessing the same orientation. Such a stimulus is spatially distributed, but nevertheless of a simple geometry. When the long moving bar stimulus is perturbed, by

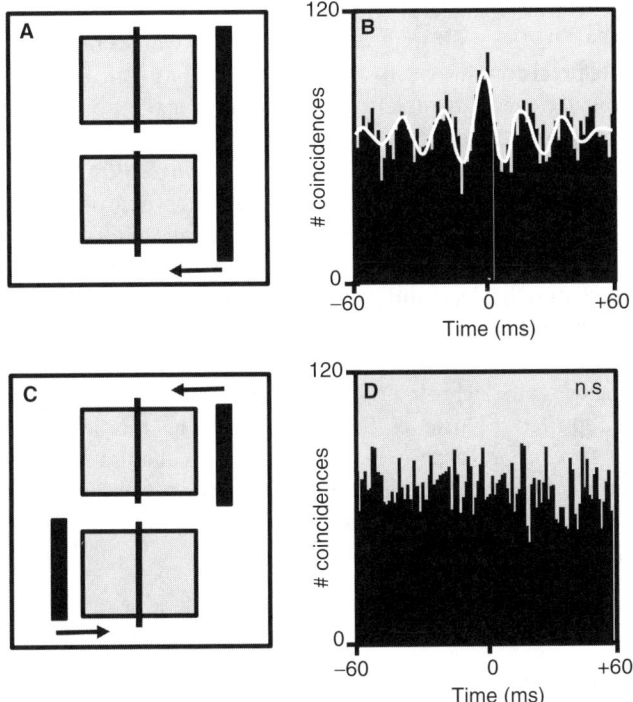

Figure 3.3   Visual induced gamma oscillations depend on "global stimulus properties." Data are from extracellular recordings at two sites in the visual cortex, of an anesthetized cat, responding to either a long moving bar (**A**), or two shorter bars moving in opposite directions (**C**). **A** and **C** also show schematically the receptive fields and orientation preferences (both vertical) of the two sites. With the single long bar, coherent oscillations at ~60 Hz occur at the two sites (**B**), whereas with two oppositely moving bars, there is nonsignificant (n.s.) coherence (**D**). (Reproduced from Engel, Roelfsema et al., 1997 with permission.)

omitting the middle third, or by having the two halves move in opposite directions (Fig. 3.3), then between-site synchronization is lost, although localized sites may still oscillate (Gray et al., 1989).

Binding by Synchrony

Synchronized visually evoked gamma oscillations are found not only with pairs of sites both located in primary visual cortex in one hemisphere; this type of synchronization is also found for sites at homologous locations in opposite hemispheres (Engel et al., 1991a); and in cases where one site is in primary and the other in association visual cortex (Engel et al., 1991b) (although it should be noted that some "association" visual areas also receive direct afferents from the lateral geniculate nucleus (see, e.g., p. 25ff. of Payne & Peters, 2002)). Be that as it may, synchronization, with approximately milli-second-scale precision, has been shown to engage brain areas interconnected by collaterals longer than merely within-column axonal branches. For this and other reasons (see later), enormous importance has been attached to the phenomenon (e.g., Engel et al., 1997 and Engel & Singer, 2001, amongst many other reviews). Specifically, it is postulated that synchronization "binds" together all those brain areas responding to a single discrete object, thus giving that object an existence defined as distinct from all other objects, as well as distinct from the "background"—to use a Gestalt psychological means of expression. Other data (see later) are consistent with such an idea, yet the idea is still, in our opinion, something of an extrapolation from the actual observa-tions (not always a drawback, however); apparently contradictory data have also been reported (Thiele & Stoner, 2003). The binding-by-synchrony hypothesis also needs to contend with the experimental observation that moving stimuli are much more effective at inducing oscillations than are fixed stimuli (Gray et al., 1990)—not in accord with personal experience in distin-guishing the shapes of objects.

Associations of Synchrony with Behavior and Perhaps Awareness

There are two experiments, in particular, that support the existence of such an association, both performed in cats with surgically induced strabismic amblyo-pia (i.e., an extraocular muscle of one eye is sectioned, making it impossible for the cat to direct gaze for the two eyes onto a single object simultaneously—as a result, visual images induced by one eye eventually become "suppressed," as behaviorally confirmed, and as occurs similarly in humans with congenital strabismus). In the first experiment (Roelfsema et al., 1994), it was shown that images projected onto the seeing eye, and onto the nonseeing eye, evoke similar increases in cortical neuronal firing rates; however, only for stimuli projected onto the seeing eye do pairs of neurons synchronize with each other. In the second experiment, performed on awake cats, gratings were projected to one eye or the other, or to both—but when presented to both eyes simulta-neously, the gratings were orthogonal to each other, a so-called "rivalry" con-dition (Fries et al., 1997). Responses during the rivalry condition were compared with responses when only a single eye was stimulated. It was found that visual cortex responding to the seeing eye actually *increased* the degree of synchronized oscillation with rivalry, whereas cortex responding to the non-seeing eye decreased its degree of synchrony. Because, presumably, the responses to the seeing eye alone are "perceived," this experiment establishes a correlation between presumed perception and the extent of synchronization of neuronal oscillations.

The oscillations studied by Fries et al. (1997) and Roelfsema et al. (1994), which are presumed to be related to perception, are gamma oscillations. Other studies, in humans and in nonhuman primates, have also suggested a relation between perception and enhanced synchrony of gamma oscillations, as well as beta oscillations (Doesburg et al., 2005; Melloni et al., 2007; Tallon-Baudry et al., 2005). There is also a study reporting that enhanced interhemispheric coherence at alpha frequencies correlates with object recognition in humans (Mima et al., 2001b).

In Vivo Cellular Mechanisms: The Importance of Synapses and of "Brain Activation"

Extracellular recordings, of local field potentials and of single or multiple units, can take one only so far in understanding the cellular mechanisms of neuronal oscillations, or indeed of any sort of brain activity. Intracellular recordings are also essential (although still not sufficient to tell us all that we wish to know). Because of technical difficulties, however, rather few studies of in vivo visually evoked gamma oscillations have been undertaken. The studies that have been undertaken all demonstrate the critical role played by synaptic interactions in this type of oscillation (although the possible additional role of electrical coupling has not been addressed in vivo). An example of the data that have been obtained is shown in Figure 3.4, taken from Jagadeesh et al. (1992). These data confirm the extracellular observations that action potentials (in cortex oscillating in response to a visual input) are themselves oscillatory; but the data show as well that prominent subthreshold synaptic activity, at gamma frequency, is occurring, and presumably driving the action potentials, suggesting, but not definitively proving, that synaptic interactions actually are producing the oscillation. Even stronger evidence for a critical, and not just epiphenomenal, role of synaptic interactions would be abolition of the gamma oscillations under the influence of appropriate synaptic blockers—a very difficult experiment to do in vivo.

Further elegant intracellular studies have been undertaken to study the role of so-called chattering cells (fast rhythmic bursting cells or FRB cells in another terminology) in generating visually induced gamma. FRB cells exhibit an intrinsic gamma-frequency bursting rhythmicity (Cardin et al., 2005; Gray & McCormick, 1996; Llinás et al., 1991; Nuñez et al., 1992; Steriade et al., 1998) in response to sustained depolarizing currents, with intraburst firing rates at several hundred Hz (Brumberg et al., 2000) (Fig. 3.5). Not only do chattering/FRB cells respond to experimentally induced tonic depolarization with gamma-frequency bursting; at least some FRB pyramidal neurons, in visual cortex, respond to visual inputs in this way, as shown with intracellular recording (Gray & McCormick, 1996). In addition, unit recordings of visually induced gamma oscillations also indicate chattering (fast rhythmic bursting) [see the data of Dr. Charles M. Gray in Fig. 7.4 of Traub et al. (1999a)]. It is natural, then, to ask if FRB neurons are somehow "pacing" the induced gamma oscillation: for example, the visual input might hypothetically induce a slow depolarization in FRB neurons, which synchronize with one another in some undetermined fashion, and then project (so to speak) the gamma

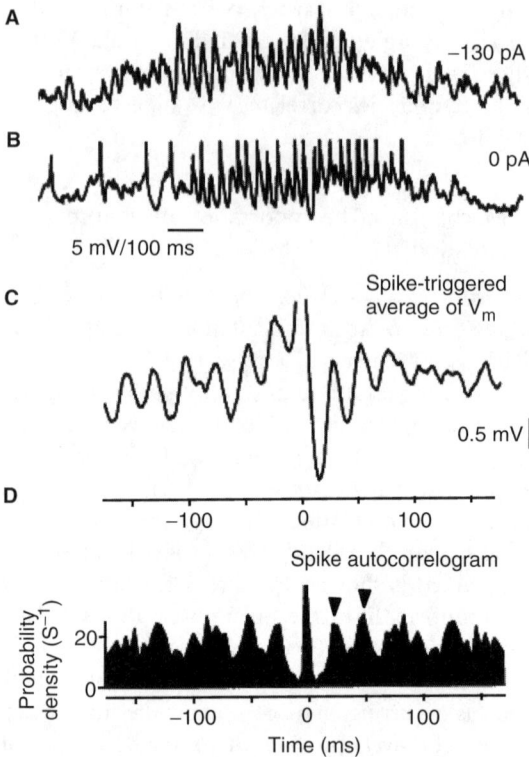

Figure 3.4 Induced gamma oscillations, in cat visual cortex in vivo, are associated with gamma-frequency synaptic potentials. **A, B:** Whole-cell patch clamp recording of a complex cell in visual cortex, as a bar moves across the visual field, at two different holding potentials: this reveals synaptic potentials, together with (truncated) action potentials in **B. C:** Spike-triggered average of 8 traces similar to **B**, with the spikes themselves removed; this also reveals gamma-frequency synaptic oscillations. **D:** Spikes also have a gamma-frequency probability distribution. (Reproduced from Jagadeesh et al., 1992 with permission.)

oscillation synaptically onto other neurons. As Cardin et al. (2005) have shown, however, this hypothesis appears to be false (Fig. 3.6). Instead, the visual cortex neurons that do exhibit gamma-frequency bursting (and these all seem to be so-called "simple cells," a characterization dependent on how they respond to temporally modulated moving gratings) are exhibiting such bursting as a result of oscillatory synaptic inputs, not as a result of tonic depolarization. That is to say, FRB neurons are oscillating as part of a large synaptically interconnected network, and are not acting as pacemakers, at least not in any simple fashion. [It is the case, however, that FRB neurons play an important role in another type of gamma oscillation in vitro (Cunningham et al., 2004b)—more on this in another section.]

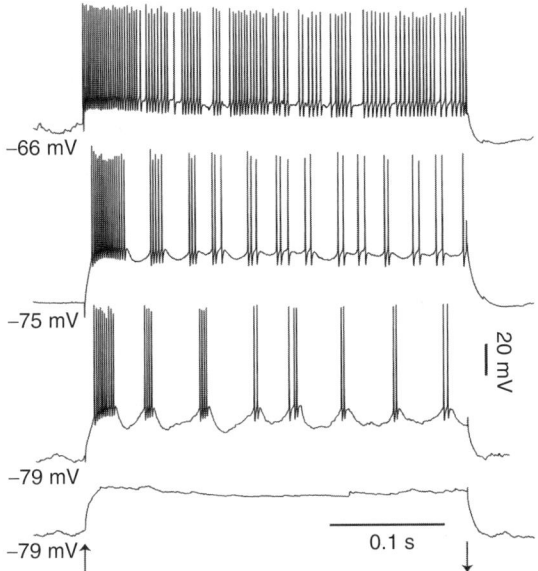

Figure 3.5 Some cortical neurons (chattering cells, or FRB [fast rhythmic bursting] cells) generate gamma oscillations intrinsically, upon steady depolarization. The traces show responses of an FRB neuron in cat primary visual cortex, recorded with a sharp microelectrode, responding to various current pulses (indicated by the arrows) and holding potentials. When depolarized enough (top trace), the cell fires tonically, but at intermediate tonic depolarizations (middle traces), there are beta and gamma frequency bursts; within each burst, the cell fires at several hundred Hertz. (Reproduced from Cardin et al., 2005 with permission.)

Even if tonic depolarized FRB neurons are not driving visually induced gamma in any simple direct fashion, it is still likely that tonic depolarization of cortical neurons, perhaps in combination with a block of intrinsic membrane K⁺ conductances, is nevertheless required for the oscillations to occur. The reason for suspecting that this is the case derives from the potentiating effect on the oscillations, in vivo, of stimulating the mesencephalic reticular formation (Fig. 3.7; see also Herculaneo-Houzel et al., 1999). The mesencephalic reticular formation contains cholinergic neurons and is contiguous with the pedunculopontine tegmentum (PPT; Armstrong et al., 1983). Stimulation of PPT has been shown to induce slow EPSPs in cortical neurons, along with gamma oscillations throughout cortical columns (and reciprocally connected thalamic cells), via a muscarinic receptor-mediated effect (although the muscarinic receptors mediating this effect could lie on cells along a synaptic pathway rather than on the cortical neurons themselves) (Steriade & Amzica, 1996); interestingly, these latter type of gamma oscillations occur with brain stem stimulation *alone*, without the requirement for sensory input—as do the

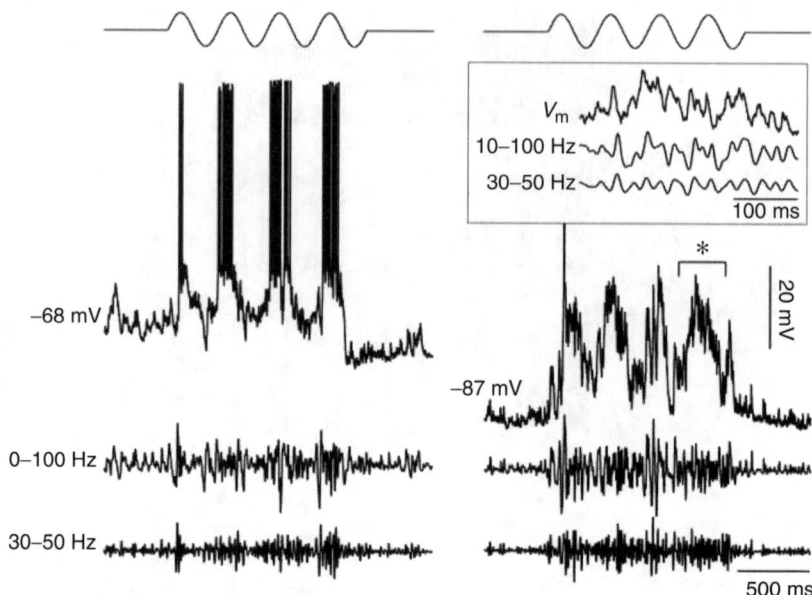

Figure 3.6 Even in FRB cells, visually evoked gamma oscillations are driven by synaptic inputs. An intracellularly recorded simple FRB cell is shown responding to a sinusoidal visual grating (top traces), at two membrane potentials: at rest (left), and significantly hyperpolarized (right). Filtered traces of the intracellular recordings (with spikes removed) are shown below. The grating induces waxing and waning gamma oscillations; these are present about to the same extent when the cell is hyperpolarized enough to prevent firing (right), indicating that the oscillations result from synaptic inputs.

(Reproduced from Cardin et al., 2005 with permission.)

gamma oscillations during sleep (see later). Behavioral attention (however this may be mediated) in awake primates also has a potentiating effect on visually induced gamma oscillations (Fries et al., 2001).

The synaptic mechanisms of in vivo visually evoked gamma are not known in any detail. Such mechanisms have been studied in an in vitro model of induced gamma, and these have been reviewed in our earlier monograph (Traub et al., 1999a; see also Chapter 1 of this book). Unfortunately, that model used hippocampal rather than neocortical slices, and so far has not been adapted to the neocortex, at least to our knowledge. Nevertheless, the in vitro hippocampal model, using tetanic electrical stimulation in lieu of physiological sensory inputs, is the only candidate now available. It is therefore appropriate to outline the basic features of tetanic gamma (Traub et al., 1996c, 1999c; Whittington et al., 1997a, 2001), which we consider as hypotheses concerning the events in vivo:

1. The in vitro tetanic oscillations (Chapter 1, Fig. 1) appear after a latency of tens to hundreds of milliseconds (as in vivo), in association with

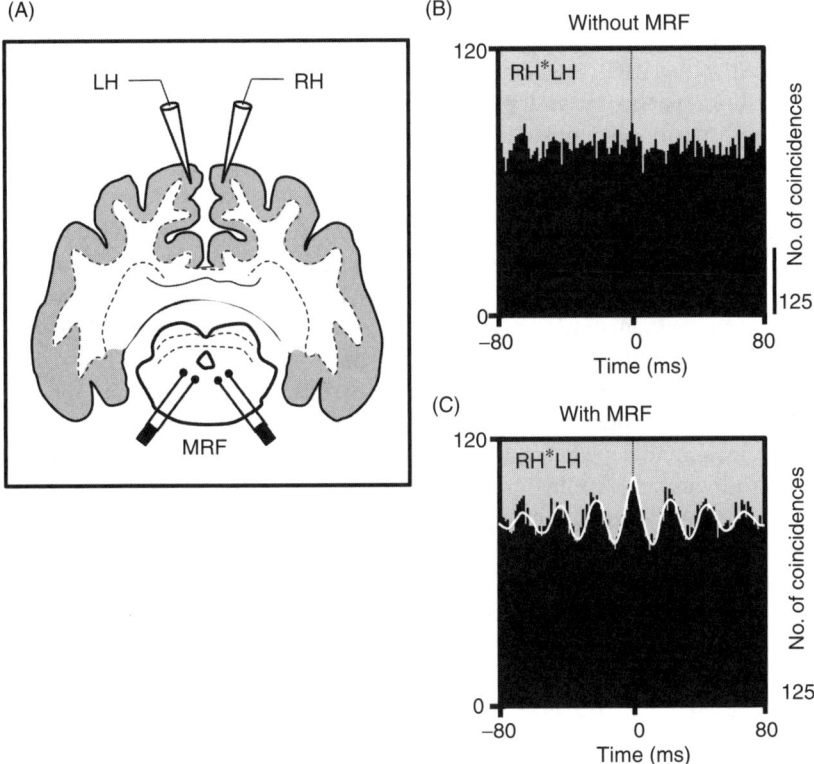

Figure 3.7   For synchronized visually induced oscillations to occur, the cortex must be properly activated. Data from anesthetized cat, with extracellular recording and stimulation. (**A**) Visual responses were recorded in primary visual cortex of the left hemisphere (LH) and right hemisphere (RH); in some cases, stimulation was given to the mesencephalic (midbrain) reticular formation, or MRF. The small boxes in (**B**) and (**C**) show peristimulus spike histograms, in response to visual input, with and without MRF stimulation: The amount of firing is roughly comparable in the two cases. The larger boxes show that gamma-oscillatory synchronization between the two hemispheres is, however, present only with MRF stimulation. Stimulation of the pedunculopontine tegmental nucleus is known to produce slow EPSPs in cortical neurons (Steriade & Amzica, 1996).
(Reproduced from Engel & Singer, 2001 with permission.)

slow intracellular depolarizations in both pyramidal cells (also seen in vivo; see Fig. 3.6) and in interneurons (not, to our knowledge, tested yet in vivo). The in vitro oscillations also have a duration comparable to that seen in vivo, hundreds of milliseconds.

2. The intracellular depolarizations, in vitro, are largely caused by activation of metabotropic glutamate receptors (mGluRs). It is not known whether this is the case in vivo. The mGluR activation seen in vitro is not strictly comparable to the metabotropic actions that would be induced in vivo

by stimulation of brainstem cholinergic nuclei: in the in vitro case, but not necessarily in vivo, the metabotropic activation is induced by the stimulus itself.

3. The synchrony and period of the in vitro gamma oscillations are primarily determined by $GABA_A$ receptor-mediated perisomatic inhibitory postsynaptic potentials (IPSPs). This is suspected to be the case in vivo, but not demonstrated directly.

4. Pyramidal cells and fast-spiking interneurons, in vitro, oscillate in phase.

5. Two in vitro hippocampal gamma-oscillating sites can synchronize tightly, despite axonal conduction delays probably greater than 5 ms. Modeling and mathematical analysis (Bibbig et al., 2001; Ermentrout & Kopell, 1998; Traub et al., 1996c) indicate that synchronization is realized by excitatory synaptic inputs to interneurons, arising from the opposite oscillation site; such inputs provide a temporal corrective signal, so to speak, and depend for their efficacy on the rapid time course of α-amino-3-hydroxyl-5-methyl-4-isoxazole-propionate(AMPA)–receptor-mediated excitatory postsynaptic conductances (EPSCs) in fast-spiking interneurons (Geiger et al., 1995, 1997; Miles, 1990). Experimental and simulated synchronized tetanic gamma oscillations, when evoked at two separate sites, are associated with spike doublets in fast-spiking interneurons (Chapter 1, Fig. 1), and the doublets appear to be essential in actually bringing the two-site synchronization about. When the time course of fast-spiking interneuronal EPSCs is slowed down by a transgenic manipulation, then—as predicted—two-site synchronization in the tetanic model is disrupted (Fuchs et al., 2001). These critical predictions on long-range oscillatory synchronization mechanisms remain to be examined in vivo, in a sensory-evoked oscillation paradigm.

Frequency Alone Is Not the Defining Characteristic of Induced Gamma

Sensory-induced oscillations in the turtle visual system (Prechtl, 1994) and in insect olfactory systems (Laurent, 1996; Laurent & Davidowitz, 1994) occur at beta frequencies, not gamma: there is nothing mystical about gamma, in the sense of some universal absolute. The sensible question posed by Singer and his colleagues—why gamma, in particular?—must be answered in terms of mechanism. We can offer two hypotheses as to why mammals use gamma frequencies for sensory-induced oscillations, and the presumed perceptual binding functions that these oscillations may subserve; each hypothesis is based on its own underlying assumption. First, suppose that what counts for sensory induced oscillations is the ability of two oscillating sites to synchronize with each other, without involving oscillations in intermediate tissue, and to synchronize despite considerable conduction delays between the two sites. Then, oscillations based on $GABA_A$ receptor-mediated IPSPs, generated by fast-spiking interneurons themselves excited by "narrow" EPSCs can do the job; at least this is demonstrated to be true in principle, using in vitro studies. Whether such a mechanism is unique cannot be claimed; but, on the other hand, once evolution has come up with a mechanism that works, alternative mechanisms may not be required. Birds, having wings, do not require helicopter rotors as well. Second, gamma oscillations are believed to interface with synaptic plasticity, based on theoretical grounds, and there is experimental evidence (in vitro) that such an interface actually exists (Whittington et al., 1997b). In that case, then, the period of the gamma oscillation, of order tens of milliseconds, may need to interface with the time constants of processes that determine synaptic potentiation and depression; one of these time constants is that for intradendritic calcium concentration changes—and these have the same order of magnitude as does the gamma oscillation itself (Bibbig et al., 2001; Sabatini et al., 2002).

Other (Nonvisual) Means of Inducing Gamma

The reader must not assume that gamma oscillations are specific to the visual system; at the same time, in reviewing the literature, the reader must be careful to distinguish between gamma oscillations phase-locked to a stimulus ("evoked" oscillations, often examined in the auditory system), and "induced" gamma oscillations not phase-locked to a stimulus. The "induced" visual oscillations discussed in the preceding text are *not* phase-locked to the stimulus. Indeed, there are also auditory induced gamma oscillations that are not phase-locked to the input stimulus (Palva et al., 2002), and some of these auditory induced gamma oscillations have been proposed to be related to learning (Jeschke et al., 2008). Electrical stimulation of nonprimary auditory thalamus has also been shown to induce cortical gamma oscillations (Barth & MacDonald, 1996; Sukov & Barth, 2001). A rather fast (80 Hz) induced gamma oscillation has been reported, following painful median nerve (i.e., somatosensory) stimulation (Chen & Herrmann, 2001). Cortical gamma oscillations have also been reported in a patient with somatic hallucinations (Baldeweg et al., 1998), but the mechanism of these oscillations may be different than for somatosensory induced oscillations; they may, for example, be more similar to the gamma oscillations that normally occur during slow wave sleep (see later). And of course it has long been known that olfactory stimulation evokes gamma oscillations in multiple parts of the olfactory system (Adrian, 1942, 1950; Eeckman & Freeman, 1990; Neville & Haberly, 2003). Induced gamma oscillations are of general biological interest; the deeper principles involved cannot be completely specific to vision.

Critique of the Binding Hypothesis and Gamma Oscillations

Probably the deepest mechanistic and functional analysis of sensory induced oscillations has been undertaken by Gilles Laurent and his colleagues (Bazhenov et al., 2001; Friedrich & Laurent, 2001; Laurent, 1996, 2002;

Laurent et al., 1996; MacLeod & Laurent, 1996; MacLeod et al., 1998; Stopfer et al., 1997; Stopfer & Laurent, 1999; Wehr & Laurent, 1996), in the olfactory system of insects. One of the major thrusts of this work has concerned how distinct odors are classified, in the distributed firing patterns of neurons in one olfactory structure or another: as processing proceeds away from the sensory periphery, the representation becomes sparser, in that fewer cells respond to a given odor, and they do so with fewer action potentials. In addition, recognition of the distinction between odors is degraded when population oscillations are disrupted in the insect olfactory system. Evidently, oscillations are important in the analysis of sensory inputs, and in encoding the results of the analysis in such a way as to be useful to other parts of the nervous system. Still, it is not clear what analog "binding"—in the Gestalt sense—has in the olfactory system. There are, to be sure, many odors to be recognized, and they may be intermixed with each other, but is there an analog of a large "object" constituted of smaller "pieces," so that oscillations generated by each piece need to be synchronized with each other? If extrapolations are possible from the insect olfactory system to the mammalian visual system, then we must consider the possibility that induced oscillations allow for, or contribute to, more effective processing of visual inputs, and for rendering sparser the cellular representation of seen objects; but we must find other tools and models to extend our understanding of the perceptual aspects of synchronized oscillations.

Perhaps an experimental approach would involve a hypothesis such as this: the occurrence of tight synchrony in sensory-induced oscillations allows us to talk about, or write about, what we have seen, now and in the future. To test the hypothesis, we need a method to titrate oscillation synchrony without abolishing consciousness—if such a manipulation is possible.

Fast Oscillations Associated with the Waking State: Expectancy and Short-Term Memory

Visually evoked gamma oscillations occur in anesthetized animals, as well as awake ones. Thus, although synchronized gamma oscillations may be necessary for conscious awareness, the reverse cannot be true. There are, however, other types of fast oscillations, particularly at beta-2 (20–30 Hz) frequencies, that occur during, and in preparation for, motor tasks; and they occur also in other types of cognitive tasks where short-term (lasting seconds) "working" memory is involved. That is to say, the oscillations seem, behaviorally, to be associated with the waking state—even though oscillations having similar morphology, and possibly similar cellular mechanisms, can be found during sleep and even in brain slices.

The behavioral setting of beta-2 oscillations is different than for sensory-induced oscillations: an expectant state before an action, as opposed to possibly passive sensory stimulation; and correspondingly, the cortical locations of beta-2 oscillations are different as well: instead of primary sensory cortices and adjacent association cortex, where sensory-induced gamma tends to be found, beta-2 oscillations are found in primary motor, premotor, and supplementary motor cortex, as well as in lateral inferior parietal cortex—but, and one must be careful here, beta-2 is also found in a sensory region, the somatosensory cortex. One must note, however, that somatosensory cortex is both (1) strongly interconnected with motor cortex and also (2) one of the cortical regions of origin (i.e., besides "motor" regions) of the corticospinal tract (Miller, 1987; Rapisarda et al., 1985; Rathelot & Strick, 2007; Toyoshima & Sakai, 1982). To put it another way, somatosensory cortex is, in some sense, also motor cortex.

Delayed match-to-sample and related tasks are one example of a cognitive paradigm giving rise to beta oscillations. In such a task, a person or an animal is presented with a brief stimulus, let us say a visual pattern, which is chosen from a finite repertoire of possible stimuli. The individual is then supposed to remember what the stimulus was, for some period of time (typically one or a few seconds), so that when a second "test" stimulus is presented, that stimulus can be judged as the same or different than the original stimulus: the judgment is then expressed in the form of a motor action, again chosen from a finite repertoire. Alternatively, the subject may simply be expected to produce a specific action, determined by the original stimulus, but only after the delay period. Either of these paradigms tests whether the original stimulus is remembered, and both require immobility during the delay period.

A key set of observations consisted in the demonstration by Miyashita and colleagues (Miyahsita, 1988; Miyashita & Chang, 1988) that, during the delay period, there was a sustained increase in firing rates within particular localized brain regions (inferior temporal cortex for the Miyashita studies), and there was evidence that there might be stimulus specificity in *which* neurons increased their firing rates, and which did not: thus, stimulus 1 might cause sustained increased firing in neurons A, B, and C; while stimulus 2 would do this for neurons D, E, and F. Working memory was presumably identifiable, then, with a so-called population code: the identity of an object was determined by selecting a subset of neurons out of a group, tonically stimulating this subset, and omitting the stimulation (or actively suppressing) all the other neurons in the group. These experimental observations were subsequently extended from inferior temporal cortex to prefrontal cortex; the involvement of prefrontal cortex in working memory is discussed further in Chapter 5 (Parkinson's disease).

Significantly, the earlier studies of increased firing, during the delay period, concentrated on the activities of single units or relatively small sets of nearby units (so-called multiunit recordings), rather than on local field potential recordings; the latter signals reflect local averaged synaptic currents, and are generally the most sensitive means of demonstrating that a local population is engaged in a collective oscillation. (Of course, in vivo, unlike the case in selected experimental preparations, synaptic conductances are generally operating, and will not have been suppressed by, say, receptor-blockers or by low-calcium media. Thus, synaptic currents—reflecting as they do the summated activities of many cells—are a reliable means of *detecting* the presence of a collective neuronal oscillation, even if the underlying mechanisms that generate the oscillation do not critically depend on the presence of such currents.) In addition, the earlier investigators were not focused on the issue of whether a neuronal population was oscillating or not, but rather instead on increases and decreases in activation of one cell or another. Nevertheless, as Figure 3.8 demonstrates (in data from an awake macaque monkey, performing a task involving the selection of direction in which to make a saccade after a delay or "memory" period), the delay period is associated not just with increased neuronal activity, but with fast population oscillations as well (note especially Fig. 3.8B, left, and Fig. 3.8Cb). It is of special interest that the increased oscillations quickly become reduced at the end of the delay period, once action is actually taken. Similar observations have been reported by other investigators [in monkeys (Lebedev & Wise, 2000; MacKay & Mendonça, 1995); and in the work of Catherine Tallon-Baudry and her colleagues, in humans (Bertrand & Tallon-Baudry, 2000; Tallon-Baudry & Bertand, 1999; Tallon-Baudry et al., 1997, 1998a,b, 1999a,c, 2001)]. Even in rats, somatosensory cortex (specifically, barrel cortex) fast oscillations, at 25 to 45 Hz, precede an episode of exploratory whisking (Hamada et al., 1999).

Although our discussion has been focused on oscillations occurring during a delayed match-to-sample task, it is important to point out that approximately 25 Hz oscillations have been reported to occur in monkey somatosensory cortex, during intentional reaching movements—the retrieval of raisins, that is, motor activity involving active movement, and not expectant immobility (Murthy & Fetz, 1992, 1996a, 1996b); the oscillations were, however, quite brief, with a mean of about 4 cycles per oscillatory epoch. These oscillations appeared to arise in deeper cortical layers (local field potential maxima at 1 to 2 mm cortical depth; and the oscillations could synchronize between left and right motor cortices, presumably by virtue of the callosal connections between these areas (Jenny, 1979).

During the delay period, in a delayed match-to-sample task, the subject is immobile. Are the oscillations recorded during the delay period involved in holding a memory, or rather (or in addition) in maintaining immobility? When human subjects move a manipulandrum, epochs of immobility (when the manipulandrum is held in a fixed position) are associated with beta oscillations in somatosensory and motor cortex (Fig. 3.9). Figure 3.9 is taken from the work of Stuart Baker and his colleagues, who have in addition shown that motor/somatosensory beta oscillations—in this type of paradigm—are coherent with muscle electrical activity (EMG, electromyogram), and so presumably are coherent with oscillations in the spinal cord (oscillations that can generally not be directly recorded in humans) (Baker et al., 1997, 1999; Kilner et al., 2003; see also Donoghue et al., 1998). Other investigators, for example, Mima et al. (2000, 2001a), have demonstrated, in humans, coherence between motor cortex and muscle over the frequency range 14 to 50 Hz, extending into gamma frequencies—coherence presumably mediated, at least in one direction, by the corticospinal tract. The fact that corticospinal tract (pyramidal tract) stimulation can reset motor-cortical beta oscillations is consistent with this notion (Jackson et al., 2002). One must not think exclusively, however, of a unidirectional transmission of motor-cortical beta oscillations to the spinal cord: somatosensory cortex also participates in motor-behavioral beta oscillations in vivo, in synchrony with motor cortex (Witham et al., 2007), and there is evidence that proprioceptive afferents also display oscillatory activity (Baker et al., 2006). With in vitro models, both motor cortex (Yamawaki et al., 2008) and somatosensory cortex (Roopun et al., 2006) have been shown capable of generating beta oscillations, in each case within the deep layers, and involving cells expected to give rise to corticospinal tract fibers (see later chapters). Both in vivo (Baker et al., 2003a) and in vitro, the firing rates of individual somata are lower than the frequency of the field beta: unfortunately, the signal processing within the spinal cord, during motor tasks in humans and nonhuman primates, is little understood. [It is also not known whether motor cortical beta oscillations are specifically transmitted to alpha motor neurons, to gamma (fusimotor) motorneurons, or both—independent pathways to the two pools of motorneurons may exist (Burke et al., 1978; Clough et al., 1971; Koeze, 1973; Koeze et al., 1968; Rothwell et al., 1990).]

There are two issues in the preceding data, on beta oscillations associated with motor activity, which require special consideration. First, not only is motor/somatosensory cortex beta associated with immobility (at least in certain behavioral paradigms), but such beta is increased in Parkinson's disease, a disorder associated with pathological immobility. [Also, interestingly, beta-frequency motor neuron discharge is characteristic of sustained muscle contractions in humans (Grimby et al., 1981).] Could there be a connection? This question is addressed in more detail in Chapter 5. Second, somatosensory association cortex beta-2 oscillations are generated (in vitro) without a requirement for phasic synaptic transmission: in the "nonstandard" mode, to use terminology introduced earlier in this chapter. Assuming that a similar nonstandard mode is operating in vivo (and this remains to be proved), what about the spinal cord, which is interacting with motor/somatosensory cortex: is the spinal cord oscillating in a standard (synaptic) or nonstandard (nonsynaptic) mode? As noted earlier, our hypothesis is that the spinal cord is operating in the standard mode, with synaptic transmission intact, and with the possibilities for synaptic plasticity to take place according to conventional rules; this hypothesis, however, remains to be verified experimentally.

Some Theoretical Notions to Which the Study of Working Memory has Given Rise

A basic question concerning working memory is this: how is it that a defined population of neurons fires at elevated rates? Does the cellular make-up of this population encode just what the memory is? The late Daniel J. Amit

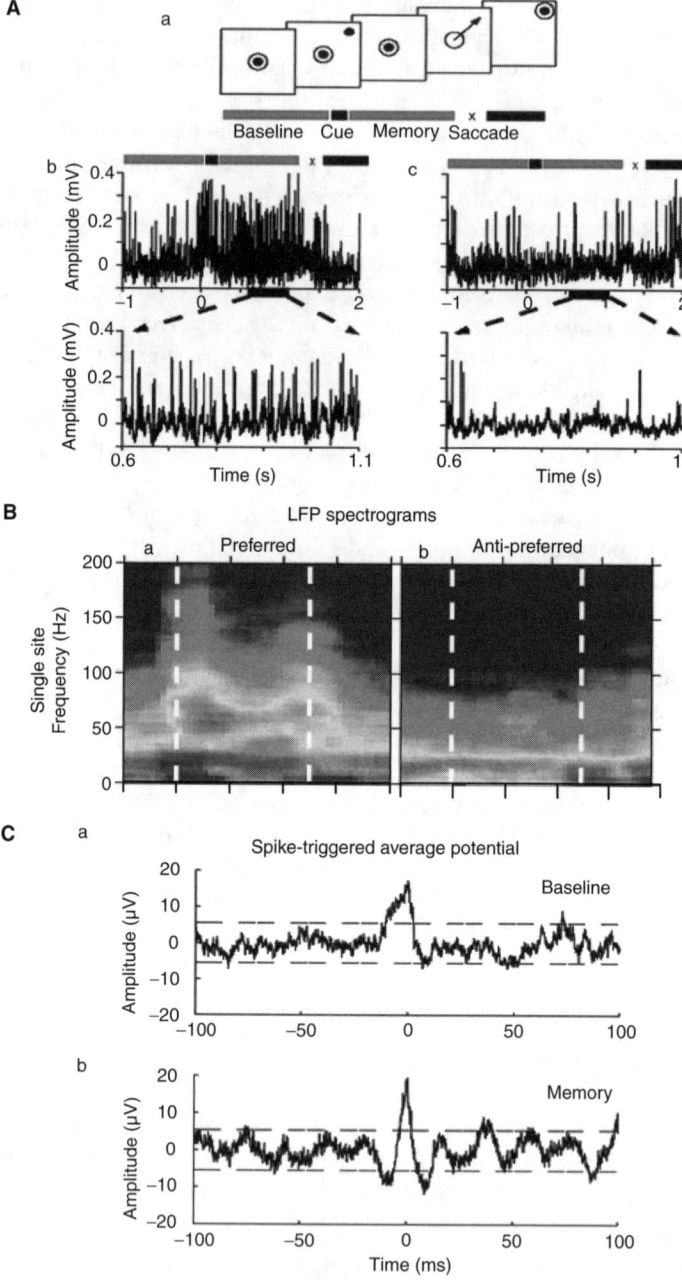

A

a

Baseline Cue Memory Saccade

b

Amplitude (mV)

c

Time (s)

B

LFP spectrograms

a Preferred

b Anti-preferred

Single site Frequency (Hz)

C

a

Spike-triggered average potential

Amplitude (µV)

Baseline

b

Amplitude (µV)

Memory

Time (ms)

Figure 3.8   Beta and gamma (and also very fast) oscillations occur in parietal cortex during the expectancy phase of a memory task. Tetrode-recorded extracellular data from macaque monkey parietal cortex (area LIP). **A:** Schematic of the task. Baseline activity is recorded; then a cue is briefly shown in one of eight directions (Northeast, in this example), a memory phase of 1 second occurs, and then the monkey is to make a saccade in the cued direction. **B:** Field potential activity (polarity reversed), when the saccade is to occur in the "preferred" direction, i.e., in the direction that elicits maximal activity for the particular electrode site. Oscillatory activity is evident in the raw data, in the expanded trace below; **C:** Likewise, but for a saccade in the opposite (anti-preferred) direction. **B:** Time–frequency plot of oscillatory activity (~25 to ~100 Hz) during the memory phase (between the two vertical white dashed lines), when the saccade is to occur in the preferred direction (left), but not in the opposite direction (right). **C:** Oscillatory activity at the single-cell level, as determined by spike-triggered averaging, occurs during the memory phase (below), but not during baseline conditions (above). Please see color insert. (Reproduced from Pesaran et al., 2002 with permission).

(1938–2007) and colleagues, inspired originally by the Miyashita papers on monkey inferior temporal cortex, approached this problem by viewing the relevant neocortical neuronal network in terms of a Hopfield model, with phasically acting recurrent excitatory and inhibitory synaptic connections, and the activated (high firing rate) population appearing as an attractor in the system dynamics (Amit & Brunel, 1997; Griniasty et al., 1993; Miyashita, 1988; Miyashita & Chang, 1988). Another type of model (Camperi & Wang, 1998; Compte et al., 2000; Wang, 1999b, 2001; Wang et al., 2004) depends on N-methyl-D-aspartate (NMDA) receptor–mediated depolarizations in the relevant neurons, possibly augmented by slow intrinsic membrane currents that allow neurons to possess bistable membrane potentials, and with recurrent synaptic inhibition present as well. Such an approach, in common with the Hopfield model, also critically depends on recurrent synaptic connections between the activated neurons, but the connections act predominantly on a slow (hundreds of milliseconds) time course. Selection of the activated subpopulation occurs by transient depolarization [perhaps mediated by a metabotropic glutamate receptor activated inward current (Sidiropoulou et al., 2009)], taking place in particular neurons, rather than emerging solely from the properties of the recurrent synaptic connections. Neither of these types of models accounts for fast (20 Hz and above) network oscillations.

Our own in vitro and modeling work, discussed immediately in the text that follows and also in later chapters, suggests that activation of kainate receptors "turns on" a subpopulation of oscillating neurons, with little participation of recurrent synaptic excitation between the principal neurons—either of brief (AMPA receptor–mediated) or slower (NMDA receptor–mediated) time courses. Unfortunately, the in vitro work does not solve the problem of selection—which exact neurons are to participate in the activated state?—as network activation is brought about by bath application of a drug. One of the reasons why in vivo

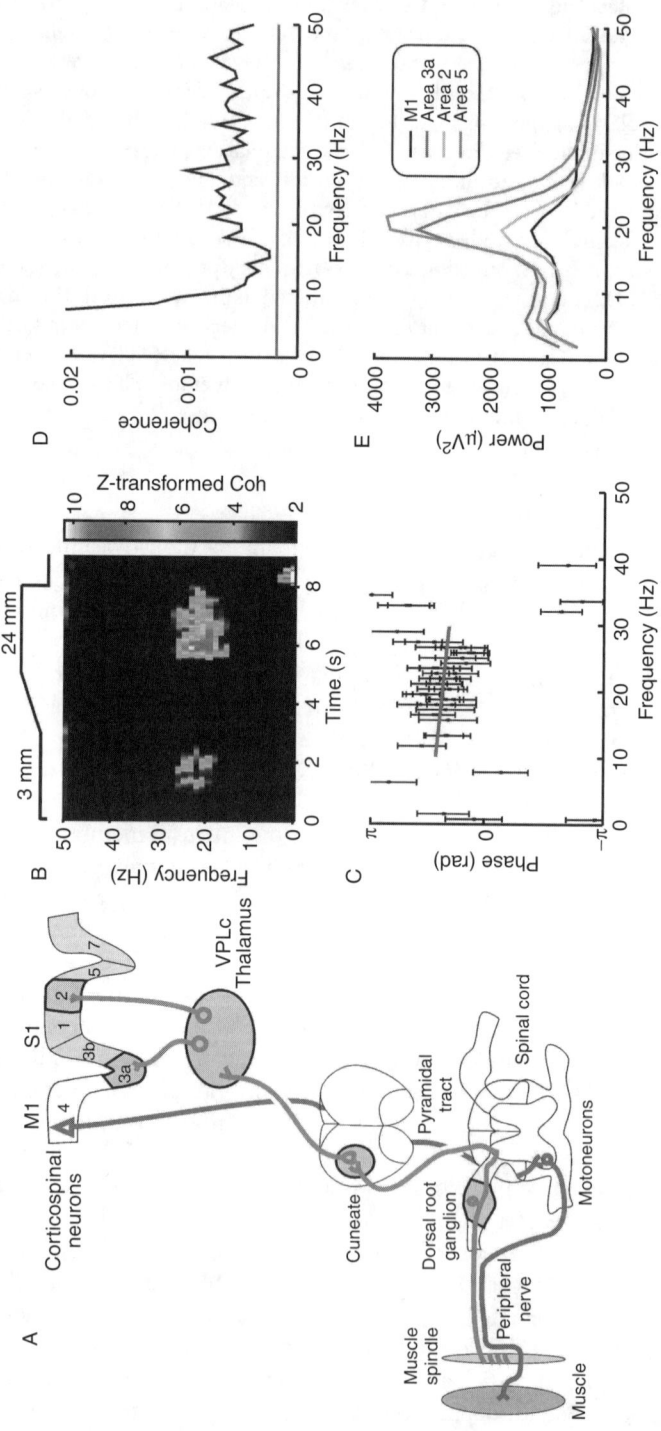

Figure 3.9   Beta oscillations, such as occur in motor and somatosensory cortex during the memory phase of a cognitive task, are coherent between motor cortex and muscle (despite >10 ms conduction delays). (**A**) Principal descending motor (red) and ascending sensorimotor (blue) pathways. (**B**) Coherence between motor cortex and muscle during a precision grip task (human data). Coherence at beta frequencies is present when the grip position is steady, and not otherwise. (**C**) The phase of this corticomuscular coherence is independent of frequency (the red regression line has a slope not significantly different from zero). (**D**) beta and gamma coherence exists between forearm EMG and the discharge of putative muscle spindle afferents (awake, behaving monkey data). The horizontal red line shows the level of significance. (**E**) Beta power in different cortical areas of the monkey: power is higher in primary somatosensory and posterior parietal areas than in primary motor cortex (M1). Please see color insert.

(Reproduced from Baker, 2008 with permission.)

experiments on delayed match-to-sample, and related paradigms, are so critical concerns just this issue: perhaps the in vivo experiments may shed light on how the brain "decides" to provoke an activated state, with or without network oscillations, in a particular subpopulation of neurons.

Fast Oscillations Associated with the Depolarizing Phase of the Slow Oscillation of Sleep

Working memory (in prefrontal, parietal, and inferior temporal cortex) is apparently associated with the activation of selected brain regions; and within these regions, there appears to be an additional selection of some neurons—but not others—that fire at high rates, this subset of neurons presumably encoding what is to be remembered. Similar principles—consisting of localized cortical activation, and selection within the activated region—perhaps apply to cortical control of movement, and possibly even to all cortical operations characteristic of the waking state. Further, as we have seen (Fig. 3.8), localized activation is associated with fast oscillations.

An interesting contrast to *localized* cortical activation, discussed earlier, consists of the *global* cortical activated epochs (each lasting hundreds of milliseconds, with hundreds-of-milliseconds to seconds separation between epochs), that occur during the so-called slow (<1 Hz) oscillation of sleep—a striking phenomenon first discovered by Mircea Steriade and collaborators in the early 1990s (Amzica & Steriade, 1995a, b; Contreras & Steriade, 1995; Steriade et al., 1993a–d). The slow oscillation occurs during slow wave sleep; the term "slow wave sleep" derives, however, not from the slow oscillation, but rather from the delta waves (~2–5 Hz) that occur in this sleep state, the delta waves having been recognized long before the slow oscillation was first discovered (Loomis et al., 1935). Frequencies at 1 Hz and below are ordinarily not recognized in scalp EEG (although they can be, with special methods [Achermann & Borbély, 1997, 1998]): such frequencies are ordinarily removed with high-pass filters, set so as to exclude artifacts caused by slow impedance changes in the EEG electrode-scalp contacts; in contrast, the slow oscillation of sleep was first recognized by Steriade and colleagues with intracortical field potential electrodes, and with intracellular recordings.

Field potential recordings indicate that the whole cortex, and thalamus, participates in the slow oscillation, approximately synchronously, with between-region phase delays in the tens to greater than 100 ms (Amzica & Steriade, 1995a); in humans, there are individual waves of activity that propagate across the cortex at 1.2 to 7.0 m/s (Massimini et al., 2004); and slow waves triggered by transcranial magnetic stimulation (TMS) also propagate across the cortex (Massimini et al., 2007). [Interestingly, however, *transient* TMS responses evoked during slow-wave sleep propagate far less than the corresponding responses evoked in the waking state, and the sleep responses also lose the fast oscillations (20–35 Hz) characteristic of TMS-evoked responses during wakefulness (Massimini et al., 2005).] The global synchronization of the slow oscillation is maintained, at least in part, by subcortical white matter connections (Amzica & Steriade, 1995b).

Figure 3.10 (left) shows the slow oscillation as recorded with an intracellular electrode from a cortical neuron, probably a pyramidal cell; a filtered version of the extracellular field is shown below. (Chapter 4, Fig. 4.7 illustrates the slow oscillation *field*, in an unfiltered signal, as measured with an intracortical extracellular electrode— also in an anesthetized cat.) The data in Figure 3.10 (left) were recorded from a cat anesthetized with ketamine-xylazine: the original discovery of the slow oscillation and most subsequent in vivo studies used this preparation (and, of course, intracellular recordings cannot be obtained from humans in situ). Eighty-eight percent of principal cortical neurons were found, in one study (Steriade et al., 1993a), to participate in the slow oscillation, in a manner similar to the cell shown in Figure 3.10 (left): that is, with an alternating series of large sustained *depolarizations* [containing superimposed action potentials, spikelets (as will be shown later), and synaptic potentials], and large relative *hyperpolarizations*, each lasting hundreds of milliseconds. Interneurons participate in the slow oscillation as well, as shown by the series of IPSPs occurring during the slow depolarization ("upstate" or "activated state") (Steriade et al., 1993a), as well as by direct intracellular recording of interneurons (e.g., Fig. 4 of Steriade et al., 2001). Figure 3.10 demonstrates, or at least hints at, the following critical features of the slow oscillation:

1. *Absence of firing in the hyperpolarized state ("downstate").* Although both principal neurons and interneurons fire during the upstate, both classes of neuron are silent during the downstate. Hence, the downstate is apparently not maintained by active synaptic inhibition. Steriade and colleagues used the term "disfacilitation" to refer to the state of affairs whereby the upstate seems to collapse by loss of recurrent synaptic excitation; consistent with this notion, the input resistance of cells is highest during the downstates, when both EPSPs and IPSPs are in abeyance (Contreras et al., 1996b). David McCormick and colleagues, and others as well, using arguments based on their in vitro model of a cortical slow oscillation, have explained the upstate as reflecting a "balance" between synaptic excitation and inhibition (Haider et al., 2006; Hasenstaub et al., 2005; Mao et al., 2001; Sanchez-Vives & McCormick, 2000). As we shall see later, other in vitro data indicate that the balance notion can be only partly true, as an unexpected intrinsic membrane K^+ current provides a critical contribution toward terminating the upstate.

2. *Fast oscillations occur during the upstate.* Figure 3.10 (left) makes clear that fast oscillations (~20 Hz and above) occur during the upstates, but not during the downstates; Steriade and colleagues elaborated extensively on this finding: one reason was that the occurrence of fast oscillations during slow wave sleep—a brain state (usually) without conscious correlates and producing little residual memory consciously accessible upon awakening—proved that fast oscillations (including, in particular, gamma oscillations) could not be *specific* for the conscious state, or for cognitive awareness. Yet, this issue may not be so simple (see later). In the rat, there appears to be a differential increase in

beta-2 EEG oscillations (roughly 20–30 Hz) in slow-wave sleep, as compared with waking and REM sleep; and also a differential decrease in gamma (Maloney et al., 1997). [In humans, 40 Hz (gamma) oscillations have been reported to characterize REM sleep (i.e., dreaming sleep), and to be reduced during slow-wave sleep (Llinás & Ribary, 1993).]

3. *Fast oscillations are coherent between thalamus and cortex.* During slow wave sleep (or at least the ketamine-xylazine-induced state that approximates it), the slow oscillation occurs in thalamus as well as cortex, and in phase. But in addition, fast oscillations (>~20 Hz) also occur in the thalamus during the upstates, and there is coherence between thalamus and cortex when "reciprocally connected" portions of each are compared (i.e., when one compares pools of thalamocortical relay cells that innervate particular cortical columns, and layers 6 and 5 of these columns contrariwise innervate the relay neurons and nearby nucleus reticularis cells) (Steriade & Amzica, 1996; Steriade, Contreras, et al., 1996). The experimental determination of reciprocal connectivity is achieved by stimulating the cortical region (respectively, thalamic region), and recording evoked synaptic potentials in the thalamic region (respectively, cortical region).

4. *Fast oscillations during slow wave sleep have spatially limited coherence.* Fast oscillations (15–75 Hz) were examined with multiple extracellular electrodes separated by 1 mm, in naturally sleeping and awake cats (Destexhe et al., 1999a). These oscillations were not coherent between distant electrodes, and only rarely and briefly (a few hundred milliseconds) between neighboring electrodes (see also Steriade, Contreras, et al., 1996; Steriade et al., 1995—in the latter study, coherence of fast oscillations was estimated to extend only over about one cortical column, a few hundred microns). Destexhe et al. (1999a) did observe, in one instance, correlations of fast oscillations between electrodes 7 mm apart, but this was during REM sleep.

5. *Sleep spindles occur on the initial phase of the upstate.* The slow oscillation of sleep serves as a kind of reference frame, around which other sleep-associated oscillations are temporally organized (Steriade, 2001, 2003, 2005). This type of temporal organization has even been shown for beta oscillations in humans (Mölle et al., 2002). One of the best-known of these sleep-associated oscillations consists of sleep spindles, which occur on the leading phase of the intracellular depolarization portion of the slow oscillation, and concerning which there is a vast literature about the EEG correlates and in vivo cellular mechanisms: much of this literature is reviewed in the Steriade monographs (2001, 2003). Suffice it to say here that sleep spindles (at roughly 9–15 Hz, depending on species) are generated in the thalamus, rather than the cortex: but whether in nucleus reticularis alone (Steriade et al., 1987), or through synaptic interactions between nucleus reticulars and principal thalamic nuclei, remains somewhat controversial—opinions on this issue tending to be influenced by opinions on what constitutes the most appropriate in vitro model of spindles. Thalamic spindles are synaptically projected to cortex and produce spindle-frequency oscillations there that are readily detectable in the

EEG. Further, reciprocal thalamocortical and corticothalamic synaptic interactions have a major influence on the intrathalamic coherence of spindles (Contreras et al., 1996a).

For in vitro studies of spindle-like network oscillations, the reader is referred to the studies of McCormick and colleagues (Bal et al., 1995a,b; von Krosigk et al., 1993). Modeling issues have been considered, either in nucleus reticularis alone or in reciprocally connected reticularis and principal thalamus, by Bazhenov et al. (2000), Destexhe et al. (1993, 1994, 1996a), Golomb et al. (1994), Traub et al. (2005a), Wang and Rinzel (1993), and Wang et al. (1995), among others.

It is well known that the intrinsic properties of thalamic neurons, particularly low-theshold Ca^{2+} spikes and the h-current, make major contributions to shaping the spindle oscillation (Bal & McCormick, 1993; Contreras et al., 1993; Crunelli et al., 1989; Deschênes et al., 1984; Jahnsen & Llinás, 1984a,b; McCormick & Pape, 1990a); it is perhaps less well appreciated that intrinsic properties of thalamic neurons may contribute to the slow oscillation itself, and not merely spindles, even if the primary "mover" is cortex (Blethyn et al., 2006; Curró Dossi et al., 1992; Hughes et al., 2002b; Leresche et al., 1991; Soltesz et al., 1991; Williams et al., 1997a; Zhu et al., 2006). The latter paper demonstrated that metabotropic glutamate receptor (mGluR) activation could elicit an intrinsic slow oscillation in at least some thalamocortical relay neurons; it should be noted that physiological inputs from the cortex can activate mGluRs in reticularis neurons (Blethyn et al., 2006) and in thalamocortical relay cells (McCormick & von Krosigk, 1992). Consistent with these observations, it has been shown in vivo that removal of overlying cortex prevents the emergence of a thalamic slow oscillation (Timofeev & Steriade, 1996).

6. *Current sinks are different for the slow oscillation and the superimposed fast oscillations.* When field potentials are recorded at numerous cortical depths, it becomes apparent that the slow oscillation, and the superimposed fast oscillations, have quite different properties. The slow oscillation has striking phase reversal in the middle cortical layers, at 0.25- to 0.5-mm depths; whereas fast oscillations do not (in vivo) have any clear phase reversal, and instead exhibit multiple sinks and sources with depth (Steriade et al., 1995). As we shall see later, many regions of cortex can have independent generators of fast oscillations in superficial (layers 2 and 3) and deep (layers 5 and 6) sites. Possibly this is not the case for the slow oscillation, which might rather depend on cells in both superficial and deep layers.

7. *The activated state is unlikely, in itself, to encode memories in the form of an attractor, at least in any obvious way.* As we have discussed in the preceding text, application of Hopfield-type attractor models to working memory is based, in part, on the notion that the identity of the remembered object (or task) is encoded in the identity of those neurons—within a defined brain region—selected to fire at high rates. For this type of memory to work efficiently—meaning, in particular, that many possible memories can be encoded—then any *one* memory is encoded sparsely, that is, with a set of

high-firing neurons that is small relative to the number of neurons in the region (Amit et al., 1985; Amit & Brunel, 1997)—so-called sparse coding. On the other hand, as we have seen in the preceding text (Steriade, Nuñez, & Amzica, 1993), almost 90% of principal cortical neurons are firing during the slow oscillation—hardly sparse! Application of physiological observations on the slow oscillation of sleep, to cognitive issues such as working memory, is then fraught with hazard. We can here identify some of the problems that we consider most pertinent, and which may be addressable with in vitro models of the slow oscillation, and their superimposed fast oscillations: most important, we believe, is the understanding of what determines whether a given pyramidal cell fires at a particular time, during an activated state such as an upstate. "Standard" models of neuronal networks are based on the assumption that timing is determined by membrane depolarization, and the pattern of synaptic inputs—and yet action potentials can be generated in axons, influenced in large part by action potentials in electrically coupled axons. Second, there is the critical issue of what factors determine if a principal cell is to be selected to fire at all during an activated state—is tonic depolarization enough? A given pattern of synaptic inputs? Or a given location in a gap junctional network?

8. *Thalamic neurons participate in the slow oscillation, but cortex can generate a slow oscillation on its own, at least if a sufficient mass of tissue is present.* The cortical slow oscillation persists after destruction of the underlying ipsilateral thalamus (Steriade, Nuñez, & Amzica, 1993); it must be possible for the slow oscillation to be generated within the cortex itself. Nevertheless, there seems to be a critical mass of cortex that is required, at least if the proper rhythmicity of the oscillation is to be maintained (Timofeev et al., 2000): a cortical area of about 2 cm^2 may be required. How these in vivo data can be reconciled with the existence of slow oscillations in neocortical and entorhinal cortical slices in vitro is not altogether clear. It is interesting that in one in vitro model of the slow oscillation, using thalamocortical slices, there was evidence that thalamic input helped to trigger cortical upstates, even though cortex was able to generate at least some upstates without the thalamic part of the preparation (Rigas & Castro-Alamancos, 2007): this may be the situation in vivo as well.

How Is the Slow Oscillation Generated?

In vivo, the slow oscillation of sleep correlates with the occurrence of certain EEG patterns (delta waves); of behavioral sleep (immobility, lack of awareness); and by the absence of signs pointing to REM sleep (i.e., during slow wave sleep, one does not see rapid eye movements, diffuse skeletal muscular paralysis, or continuous EEG fast rhythms); and the slow oscillation can be aborted by stimulation of the cholinergic pedunculopontine tegmental nucleus (Steriade, Amzica, & Nuñez, 1993). These observations, while obviously important, nevertheless do not provide sufficient information to address a number of questions, however; for example, why there are superimposed fast oscillations and whether there is meaning to the patterns of neuronal activity during the slow oscillation. [For example, there are a number of studies purporting to show "replay" of waking firing patterns, recapitulated (so to speak) during sleep, both in hippocampus and in cortex (Euston et al., 2007; Ji & Wilson, 2007). How such replay comes about and what its significance is, however, are not apparent.]

In vitro data (Cunningham et al., 2006b) provide some important clues as to cellular mechanisms of the slow oscillations, data that need to be confirmed with in vivo experiments. We shall discuss some of the data in detail in a later chapter, but mention here the following: our in vitro data provide counterintuitive explanations both for the upstate and for the transition to the downstate. Specifically, our data indicate that the upstate arises through glutamate actions on *kainate* receptors, rather than AMPA or NMDA receptors; and that the transition to the downstate is initiated by a metabolically regulated intrinsic neuronal K+ current, mediated by ATP-gated K+ channels. The mediation of the upstate by kainate provides an experimental underpinning for one of our basic experimental oscillation protocols, as we discuss next.

In Vitro Fast Oscillations Can Occur that are Continuous, Without a Slow Oscillation

In Figure 3.10B, we provide a first look at in vitro data on neocortical fast oscillations. The figure shows recordings of gamma oscillations in a slice of rat secondary auditory cortex, as well as simulation data from a network model (Cunningham et al., 2004b). [The model contained 1,152 pyramidal cells, both regular spiking (RS) and chattering/fast rhythmic bursting (FRB) as well as 192 fast-spiking (FS) interneurons and 96 low-threshold-spiking (LTS) interneurons. Synaptic receptors were of AMPA and GABA_A types, and there was electrical coupling between pyramidal cell axons, and between interneuron dendrites.] Although we expand greatly in later chapters on the cellular mechanisms of cortical gamma oscillations in vitro, the point to be made now is this: that the network model, the in vitro experimental data, and the in vivo data (left part of the figure) resemble each other in numerous ways, particularly in the oscillation frequency and in the subthreshold synaptic potentials— but with one notable difference: the in vivo fast oscillation is broken up at the frequency of the slow oscillation (about 1 Hz), whereas the in vitro oscillation runs continually. We ascribe these respective patterns according to the following hypothesis: the in vitro experiment depends on bath-applied kainate, while in vivo, the upstates are sustained by kainate receptors—as if the slice situation corresponded to a *persistent upstate*; however, in vivo, the upstates are interrupted by intrinsic ATP-gated K+ channels, whereas in vitro (for unknown reasons), the metabolic state of the neurons is stable enough that such ATP-gated K+ channels do not open—hence the absence of the periodic interruptions of the activated state. It is a striking mystery how and why, in vivo, the brain enters into a state where the cortical neurons appear to be incapable of sustaining their activity for more than a few hundred milliseconds.

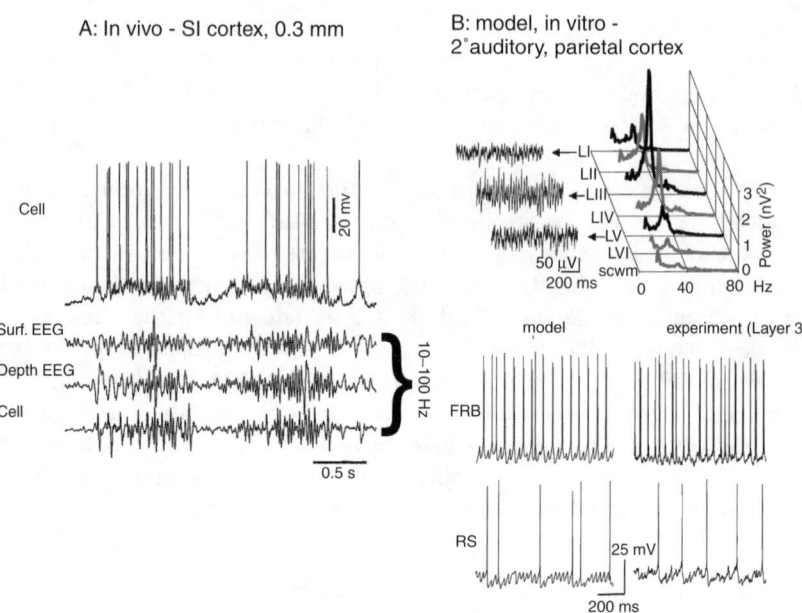

Figure 3.10 Beta and gamma oscillations occur during the intracellular depolarizations of the cortical slow oscillation of sleep; and these oscillations can be mimicked by drug application in vitro. **A:** Intracellular and filtered field potential recordings of the cortical slow oscillation, with fast oscillations superimposed on the "upstates" (data from primary somatosensory cortex, in the ketamine/xylazine-anesthetized cat). The intracellular (~0.3 mm) and field activities (both surface and depth) are at least approximately in register throughout the depth for the slow oscillation; and the fast oscillations (>~20 Hz) are also in register. The intracellular recording shows what appear to be oscillating synaptic potentials (compare Fig. 3.4). **B:** Gamma (30–70 Hz) oscillations in auditory/parietal cortex in vitro (bathed in 400 nM kainate), along with results of a detailed network simulation. The upper part shows that peak gamma power is in layers 2/3. In addition (not well seen here), the fast oscillations are continuous and are not superimposed on and modulated by a slow oscillation. The lower part demonstrates firing behavior in FRB (fast rhythmic bursting) cells, which fire on roughly half the gamma waves; and RS (regular spiking) pyramidal cells, which fire more intermittently. Both types of cell exhibit synaptic potentials at gamma frequency. The in vitro cell potentials resemble those of the in vivo cell in **A** during the "upstates."

[Composite reproduced from Traub, Cunningham, & Whittington, 2008 with permission; Data in A from Steriade, Amzica, & Contreras (1996), reproduced with permission. Data in B from Cunningham et al. (2004), reproduced with permission.]

[The astute reader will have noticed as well that, in the vitro data shown in Fig. 3.10, gamma power is far greater in superficial layers than in deep layers. The distribution of frequency and power between layers depends, as we shall show, on both the region of cortex, and on the pharmacological methods used to activate the cortex.]

To us, the data in Figure 3.10 provide justification for extrapolation of in vitro oscillation data to at least some in vivo contexts, provided the extrapolation is done critically and carefully.

Very Fast Oscillations Superimposed on Sensory Evoked Potentials

Brief sensory stimulation, in any modality, evokes a series of neural (hence electrically recordable) responses in cortical structures, at early (<~150 ms) and at longer (<~500 ms) latencies, typically consisting of waves that last on the order of tens of milliseconds; these responses are produced as the neural "traffic" proceeds along axons, causes cell firing and then synaptic currents, and in turn influences successive pools of neurons. Evoked responses of this type have been of interest not only to sensory physiologists, but also to physiological psychologists, who study the so-called P300 (a positive cortical potential at 300-ms latency), or contingent negative variation (Sutton et al., 1965); and to clinical neurologists who wish to evaluate sensory pathways in patients with suspected (for example) tumors or demyelinating disease (Ebersole & Pedley, 2003; Halliday, 1967). Neural population events having a similar appearance to evoked responses also occur spontaneously: examples are vertex waves in the EEG during sleep (Ebersole & Pedley, 2003), and physiological sharp waves, described first in the hippocampus by György Buzsáki (1986): one may think of such population events as being (so to speak) internally generated "evoked" responses, or "evoked" potentials. Synchronized epileptiform bursts (see Chapter 4) can also be regarded as a type of internally generated "evoked" response, differing from a physiological sharp wave in quantitative parameters, such as the number of cells participating in the response, and the number of action potentials per participating neuron (Buzsáki, 1986).

Neuronal population responses lasting tens of ms often have very fast oscillations superimposed upon them. As far as we are aware, such an association was first made for epileptiform events in the hippocampus in vitro (Schwartzkroin & Prince, 1977; Wong & Traub, 1983), where the superimposed oscillations were typically several hundred Hertz; however, the association is not confined to pathological, or to in vitro, population events. As Buzsáki and colleagues showed (Buzsáki, et al., 1992; Klausberger et al., 2003; Ylinen et al., 1995a), in vivo hippocampal physiological sharp waves also contain superimposed "ripples" at about 200 Hz. It is critical to understand the relation between the slower spontaneous or evoked responses—which are attributed to synchronized synaptic currents—and the very fast oscillations that are superimposed: critical both for cellular mechanisms, and for proposing reasonable hypotheses as to the functional implications of the firing patterns of the constituent neurons.

Figure 3.11, taken from the work of Daniel S. Barth, shows that very fast oscillations can occur superimposed on genuine evoked responses, produced by stimulating a body part: in the case of this figure, via a rapid induced twitch of the whiskers (trimmed and tied together) on one side of an anesthetized

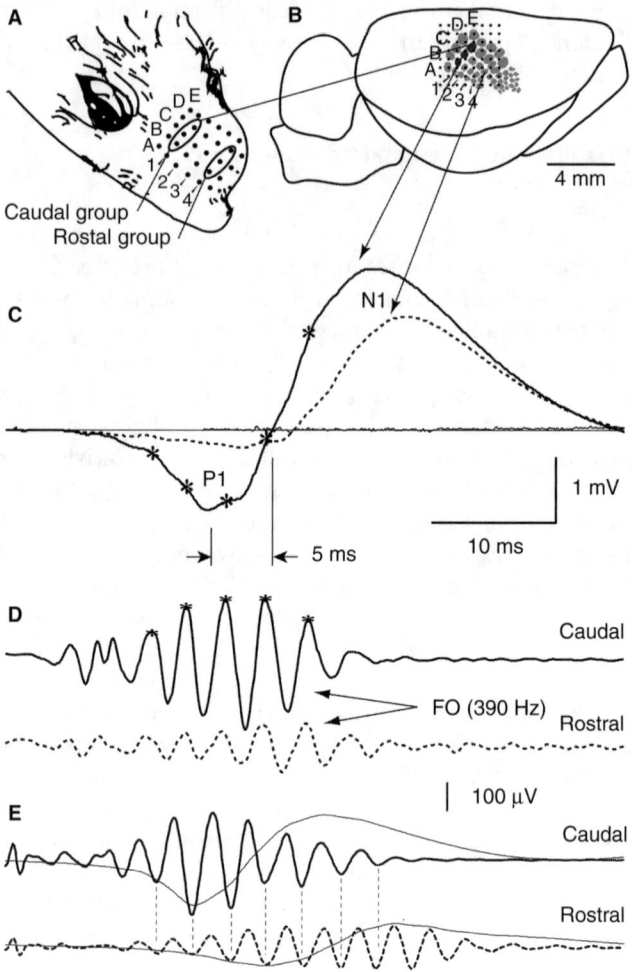

Figure 3.11   Very fast oscillations (~390 Hz in this case) occur superimposed on a somatosensory evoked response in rat barrel cortex. **A**: Arrangement of the vibrissae (large whiskers) on the rat snout. **B**: Corresponding arrangement of "barrels" in barrel cortex, a part of primary somatosensory cortex, with overlaid extracellular recording array. **C**: Classical somatosensory evoked potential (negative upwards), produced by brief displacement of a group of trimmed, tied-together, whiskers. One sees the early positive (P1) and negative (N1) waves, a two sites (solid line and dashed line). **D**: Corresponding ~390 Hz VFO (called FO in the figure), after digital filtering, 200–1000 Hz. **E**: Example of phase-aligned VFO at two sites; this VFO is particularly long-lasting.

(Reproduced from Barth et al., 2003 with permission.)

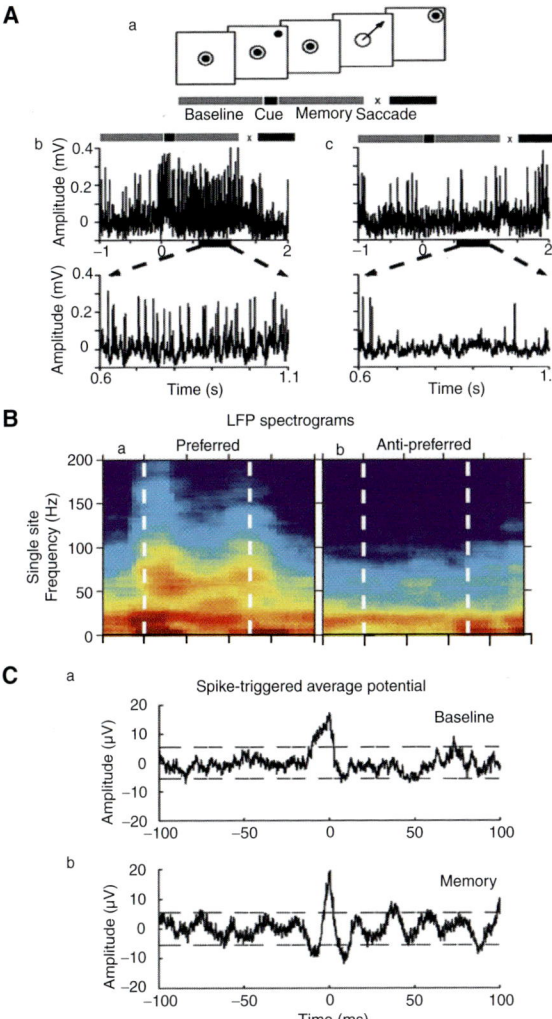

Figure 3.8 Beta and gamma (and also very fast) oscillations occur in parietal cortex during the expectancy phase of a memory task. Tetrode-recorded extracellular data from macaque monkey parietal cortex (area LIP). **A**: Schematic of the task. Baseline activity is recorded; then a cue is briefly shown in one of eight directions (Northeast, in this example), a memory phase of 1 second occurs, and then the monkey is to make a saccade in the cued direction. **B**: Field potential activity (polarity reversed), when the saccade is to occur in the "preferred" direction, i.e., in the direction that elicits maximal activity for the particular electrode site. Oscillatory activity is evident in the raw data, in the expanded trace below; **C**: Likewise, but for a saccade in the opposite (anti-preferred) direction. **B**: Time–frequency plot of oscillatory activity (~25 to ~100 Hz) during the memory phase (between the two vertical white dashed lines), when the saccade is to occur in the preferred direction (left), but not in the opposite direction (right). **C**: Oscillatory activity at the single-cell level, as determined by spike-triggered averaging, occurs during the memory phase (below), but not during baseline conditions (above).

(Reproduced from Pesaran et al., 2002 with permission.)

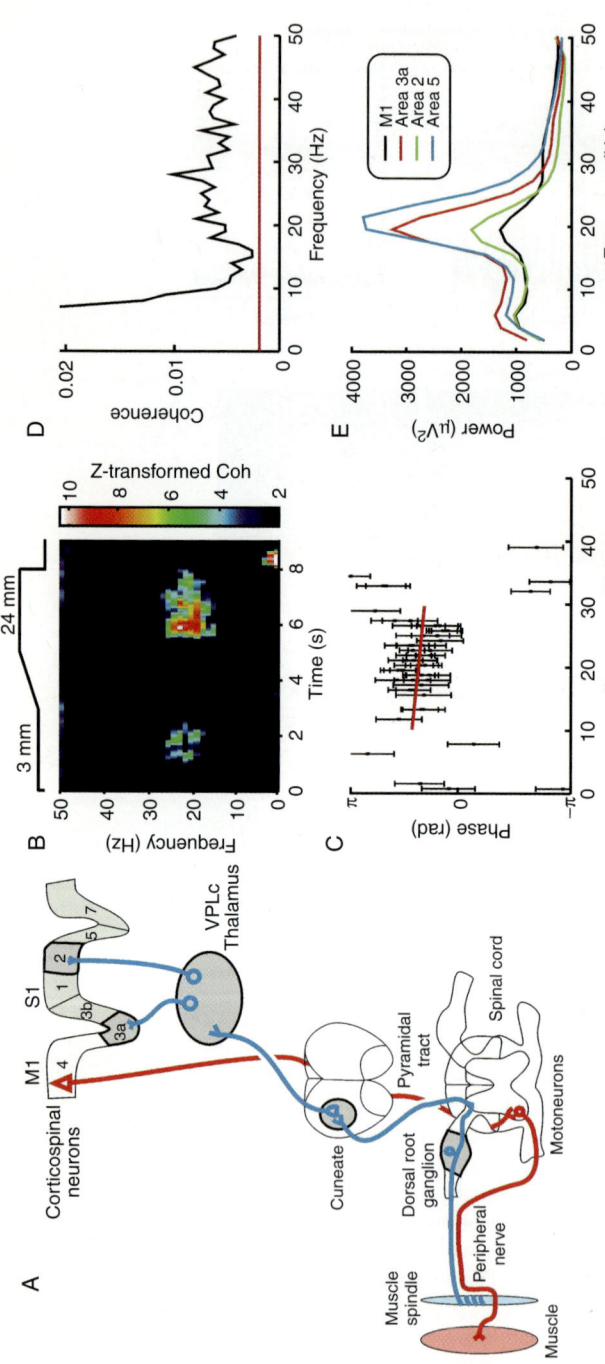

Figure 3.9 Beta oscillations, such as occur in motor and somatosensory cortex during the memory phase of a cognitive task, are coherent between motor cortex and muscle (despite >10 ms conduction delays). (**A**) Principal descending motor (red) and ascending sensorimotor (blue) pathways. (**B**) Coherence between motor cortex and muscle during a precision grip task (human data). Coherence at beta frequencies is present when the grip position is steady, and not otherwise. (**C**) The phase of this corticomuscular coherence is independent of frequency (the red regression line has a slope not significantly different from zero). (**D**) Beta and gamma coherence exists between forearm EMG and the discharge of putative muscle spindle afferents (awake, behaving monkey data). The horizontal red line shows the level of significance. (**E**) Beta power in different cortical areas of the monkey: power is higher in primary somatosensory and posterior parietal areas than in primary motor cortex (M1).

(Reproduced from Baker, 2008 with permission.)

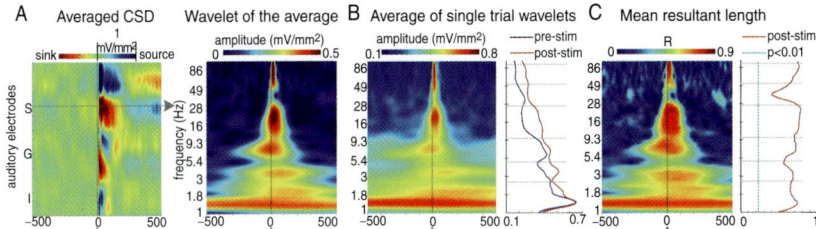

Figure 3.12 Auditory evoked activity contains fast oscillations. Extracellular data from awake macaque monkey, recorded with a multisite probe with >15 contacts throughout all cortical layers; stimulation consisted of pure tones and broad-band noise. **A:** CSD (current source density) of the evoked response; S = supragranular (mainly layers 2 and 3), G = granular (layer 4), I = infragranular. **B:** Frequency components of the evoked signal at one supragranular site (indicated by the gray arrow). The frequency scale is logarithmic. There is power at theta, alpha, beta, gamma, and VFO frequencies. **C:** Demonstration that the phase of the oscillations is not random (vertical blue line n graph at the right), but rather is reproducible trial-to-trial. (Reproduced from Lakatos et al., 2007 with permission.)

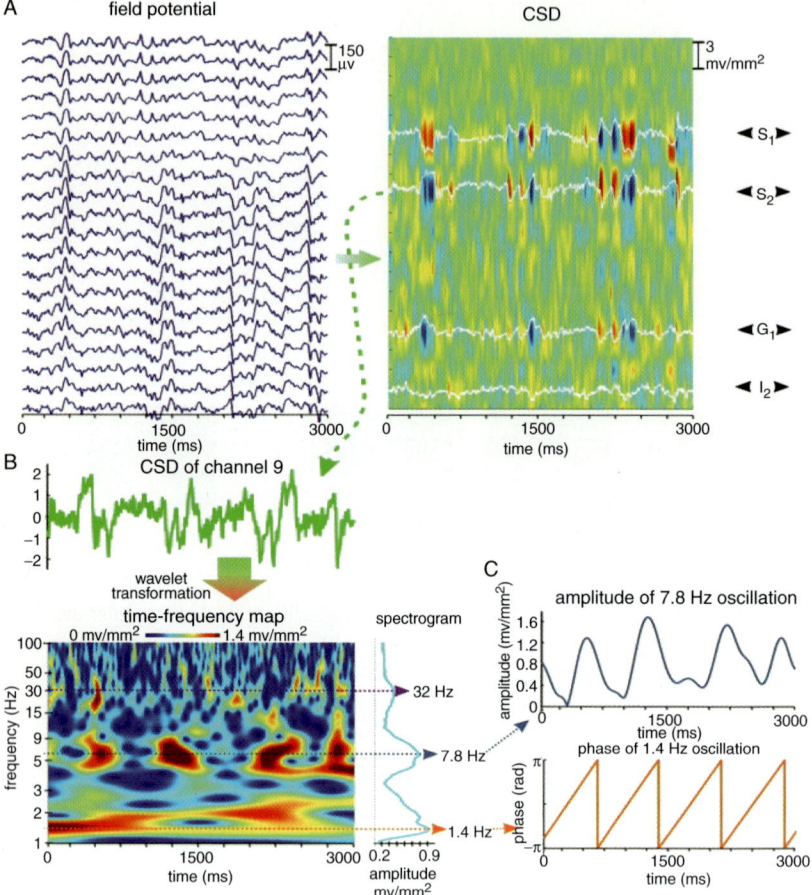

Figure 3.13 Example of one oscillation (delta) modulating the amplitude of another (theta). Spontaneous extracellular field potential data, primary auditory cortex of awake monkey. **A**: Simultaneous field potentials at 20 different cortical depths, recorded with multisite probe (left). Right: current source density (CSD) of this data. S = supragranular, G = granular, I = infragranular, as in Fig. 3.12. **B**: CSD data from on supragranular site, shown as raw signal (green) and in time–frequency plot (below). There is power in delta, theta and gamma ranges. **C**: The phase of delta (1.4 Hz) modulates the amplitude of theta (7.8 Hz). Not shown is the additional modulation of gamma amplitude by theta phase.

(Reproduced from Lakatos et al., 2005 with permission.)

rat's snout. [A note on the nomenclature: the waves are called *Px* ('positive" *x*), where *x* is an integer designating first, second, third, etc.—in this case first; and *Nx*, for corresponding negative waves; however, *x* can also stand for a time in milliseconds, as in P300, meaning the latency in ms when the wave occurs. The "positive" and "negative" refer in turn to the polarity of the signal as measured at the surface of the cortex; and to further confuse matters, signals in the evoked potential literature—following an old EEG tradition—are often plotted inverted (as in Fig. 3.11C), that is with negativity upwards.] Careful study of this so-called somatosensory evoked potential (Fig. 3.11C) shows the superimposed very fast oscillation, which is much easier to see in the filtered signals below. With controlled stimulation of individual whiskers, and pairs of whiskers, it becomes apparent that the very fast oscillations are generated within the cortex, with the thalamic inputs serving as a trigger (Staba et al., 2003); and it becomes apparent that the very fast oscillations are spatially organized within the cortex, in a manner determined by intracortical connections (Barth, 2003; Staba et al., 2005), although whether by synaptic or gap junctional connections, or both, remains to be determined. Notably, layer 4 multiunit neuronal responses can follow rapid mechanical whisker vibrations, 1:1, at frequencies up to 320 Hz (Ewert et al., 2008): thus, very fast oscillations in this rather specialized somatosensory cortex may actually be subserving a direct encoding function.

Ylinen et al. (1995a) had shown that the extracellular field associated with hippocampal sharp wave ripples corresponded faithfully to rhythmical IPSPs in pyramidal cells, motivating the hypothesis that it was networks of interneurons that might actually generate the ripples. There are, however, other ways to account for the observation of Ylinen et al., if one assumes that the pyramidal cell axon plexus is the primary generator of the ripple (Traub & Bibbig, 2000)—we shall return to this critical issue in a later chapter, when we review in vitro data on very fast oscillations. In any case, however, very fast oscillations, superimposed on somatosensory evoked potentials in rat barrel cortex, do not appear to be generated by interneurons: flooding the tissue with GABA does not affect the fast oscillations, although it does abolish the N1 wave (Staba et al., 2004a); and conversely, subconvulsive concentrations of the GABA$_A$ antagonist bicuculline actually enhance the very fast oscillation (in terms of producing more waves), without changing either the amplitude or the frequency (Jones & Barth, 2002).

The time course of synaptic excitation between pyramidal cells is relatively fast [deactivation $\tau = 3$ ms in CA3 pyramids at room temperature (Geiger et al., 1995), with physiological temperatures speeding this up, but with dendritic electrotonic filtering slowing down the time course of actual EPSPs (Miles & Wong, 1986)]; but, on the other hand, obtaining stable network oscillations solely through recurrent synaptic excitation is probably not possible (van Vreeswijk et al., 1994). So if neither synaptic inhibition nor synaptic excitation produce very fast oscillations, superimposed on sensory evoked repsonses, that presumably leaves—by exclusion—gap junctions. Unfortunately, there is as yet little positive evidence, in vivo, to strengthen this idea (although there is a great deal of in vitro evidence in support of the idea). Genetic knockout of the main neuronal gap junction protein, connexin-36, has little effect on hippocampal sharp wave ripples (Buhl et al., 2003) or on barrel cortex very fast oscillations (Daniel S. Barth, personal communication). It is possible, but not proven, that in the connexin-36 knockout mouse, another gap junction protein is upregulated in pyramidal cells. Interestingly, both humans (Curio 2000; Curio et al., 1994) and piglets show approximately 600 Hz oscillatory components in somatosensory evoked responses, and for piglets the signal has been suggested to arise from thalamocortical axons and terminals in layer 4 (Ikeda et al., 2002, 2005), as well as in cortical somata and dendrites (Okada et al., 2005). It is possible that gap junctions exist within and between these terminals and axons (Hamzei-Sichani et al., 2007.). Very fast oscillations have also been recorded from human anterior temporal cortex, during neurosurgical procedures performed with the patient awake, in response to auditory stimuli—especially if the stimulus was unexpected (Edwards et al., 2005); the cellular mechanisms of the human auditory-stimulated oscillations have not been investigated; however, the existence of such responsiveness in human cortex is of vast importance, particularly for the understanding of the initiation of seizures (see Chapters 4, 13).

Temporal Interactions Between Cortical Oscillations at Different Frequencies

We have seen that oscillations can be superimposed on transient neuronal population events (Fig. 3.11), and also on particular phases of another, slower, oscillation (Fig. 3.10). Perhaps the best-known example of one oscillation superimposed on, and amplitude-modulated, by another is the case of gamma oscillations superimposed on the hippocampal theta rhythm—something that occurs in vivo (Bragin et al., 1995; Soltesz & Deschênes, 1993; Ylinen et al., 1995a), and with in vitro models as well (Fisahn et al., 1998; see Chapter 11; Gillies et al., 2002). Understanding how these sorts of interactions take place, between oscillations of different frequencies, may be important for unraveling the functional importance of each frequency, and how the respective distinct functions might be inter-related (Palva et al., 2005). As always, our default hypothesis is that ideas about function are most readily developed when the cellular mechanisms have been spelled out. This is especially important when it comes to considering the cortical beta-1 (~15 Hz) oscillation, which—at least in vitro—appears to be produced by fitting together (rather than phase-resetting or amplitude-modulating) two simpler oscillations (Kramer et al., 2008; Roopun et al., 2008b).

For now, however, suffice it to illustrate, in the auditory cortex of awake behaving monkeys, examples of phase-resetting of multiple oscillation frequencies by a transient somatosensory stimulus (i.e., a brief stimulus in one modality resets oscillations in a part of the brain mostly devoted to a different modality) (Fig. 3.12); and of multiple interactions whereby the amplitude of one oscillation is influenced by the phase of another slower oscillation, during spontaneous activity (Fig. 3.13). This type of data, taken from the work of Peter Lakatos, Charles Schroeder, and colleagues, provides examples of phenomena that may, in principle, be analyzed further with in vitro experimental models. Other examples of cortical oscillatory superimpositions have been reported as well (Canolty et al., 2006).

Cerebellar Oscillations

Although this book's title begins "Cortical Oscillations . . .," meaning "Cerebral Cortical Oscillations . . .," there is evidence—at least for the motor system—of interactions of oscillations between widely dispersed brain regions, in a manner that could be functionally meaningful, and that almost certainly has disease implications. Further, the cellular mechanisms of cerebellar oscillations provide interesting contrasts with neocortical mechanisms of oscillations at comparable frequencies, because the synaptic architecture of the cerebral and cerebellar cortices is so different. For these reasons, we shall introduce the subject of cerebellar oscillations here. These fall into several types, including the following:

1. Oscillations at theta and alpha frequencies, generated among the electrically coupled pool of inferior olivary neurons, and transmitted to the deep cerebellar nuclei and cerebellar cortex via climbing fibers (Blenkinsop & Lang, 2006; Leznik & Llinás, 2005; Martin & Handforth, 2006). Oscillations of this

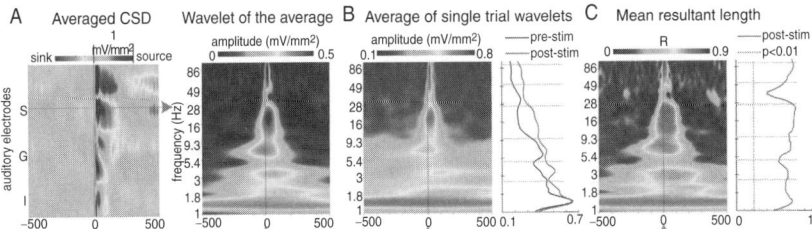

Figure 3.12   Auditory evoked activity contains fast oscillations. Extracellular data from awake macaque monkey, recorded with a multisite probe with >15 contacts throughout all cortical layers; stimulation consisted of pure tones and broad-band noise. **A:** CSD (current source density) of the evoked response; S = supragranular (mainly layers 2 and 3), G = granular (layer 4), I = infragranular. **B:** Frequency components of the evoked signal at one supragranular site (indicated by the gray arrow). The frequency scale is logarithmic. There is power at theta, alpha, beta, gamma, and VFO frequencies. **C:** Demonstration that the phase of the oscillations is not random (vertical blue line n graph at the right), but rather is reproducible trial-to-trial. Please see color insert.

(Reproduced from Lakatos et al., 2007 with permission.)

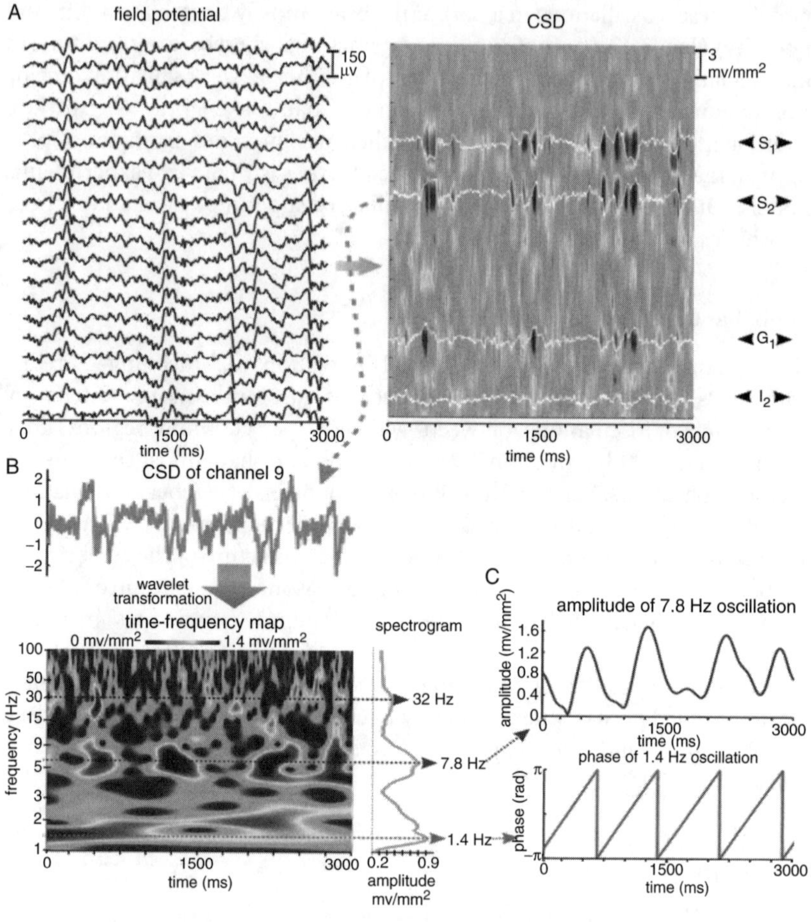

Figure 3.13 Example of one oscillation (delta) modulating the amplitude of another (theta). Spontaneous extracellular field potential data, primary auditory cortex of awake monkey. **A:** Simultaneous field potentials at 20 different cortical depths, recorded with multisite probe (left). Right: current source density (CSD) of this data. S = supragranular, G = granular, I = infragranular, as in Fig. 3.12. **B:** CSD data from on supragranular site, shown as raw signal (green) and in time–frequency plot (below). There is power in delta, theta and gamma ranges. **C:** The phase of delta (1.4 Hz) modulates the amplitude of theta (7.8 Hz). Not shown is the additional modulation of gamma amplitude by theta phase. Please see color insert.

(Reproduced from Lakatos et al., 2005 with permission.)

sort are relevant in motor control and in physiological tremor, but fall outside the scope of this book.

2. Very fast oscillations, which occur in the cerebellar cortex of certain genetically modified, and ataxic, mice (Cheron et al., 2004). These are discussed in Chapter 7.

3. Beta and gamma (as well as lower frequency) oscillations that are generated within the cerebellar cortex. Oscillation coherence has been demonstrated between deep cerebellar nuclei and motor thalamus (ventrolateral and ventral intermediate nuclei), as well as motor cortex, and possibly also the basal ganglia. Such oscillations, involving extensive components of somatosensory and motor systems, have been studied in humans and experimental animals with field potential recordings and with magnetic field measurements. The oscillations are modulated by voluntary movement (usually tending to be suppressed by such activity, with the exception of whisking in rats) and by expectancy (as is the case for cortical beta-2). Coherence between cerebellum and thalamocortical/basal ganglia structures has been reported mostly for alpha and beta frequencies, roughly 8 to 30 Hz (Courtemanche & Lamarre, 2005; Courtemanche et al., 2003; O'Connor et al., 2002; Marsden et al., 2000; Pellerin & Lamarre, 1997; Pollok et al., 2005), but also extending into the gamma range (Soteropoulos & Baker, 2006). Coherence of beta oscillations has also been found between deep cerebellar nuclei and tonically active muscle EMG (Aumann & Fetz, 2004; see Chapter 7).

4

Epilepsy

Epilepsy is defined, clinically, as a condition in which a patient has *repeated, spontaneous or nonprovoked, seizures.* Each seizure has a beginning (possibly after an aura or warning, actually the initial part of the seizure), perhaps some development, and an end. After the seizure—the "postictal" period—there may be some transient abnormalities, although they can last for days or even longer: drowsiness, headache, confusion, irritability, for example. A minority of patients also experience "Todd's phenomenon," a transient focal neurological deficit (e.g., a hemiparesis, or weakness on one side of the body); such a transient deficit points to localized brain pathology—a "focal lesion"—that is probably important for initiating the seizures. The medical conditions that lead to epilepsy are exceedingly numerous. There may or may not be a structural abnormality in the brain, either widespread or localized. There may or may not be a genetic disorder, a metabolic disorder (such as renal failure), or "epileptic syndrome" (defined later).

Let us examine critically the aforementioned definition of epilepsy, specifically the italicized words, noting how ambiguous the definition really is and difficult to apply in individual cases (even if it be the most satisfactory clinical definition available).

• *Repeated.* This word is included in the clinical definition because of the observation that many people have a single seizure, for completely unclear reasons, and most clinicians will not treat a patient who has a single seizure with anticonvulsant medications, unless there is a compelling reason to do so (e.g., some finding in the history, physical examination, EEG, or brain imaging studies that indicate the patient could have additional seizures); the inclination is not to want to label someone who has had a single seizure as having epilepsy. But the "compelling reasons" do sometimes apply. Thus, a teenager might have a *single* generalized seizure, with the history revealing the occurrence of myoclonic jerks on waking up in the morning, and the EEG showing a generalized polyspike-wave pattern. Such a teenager probably has juvenile myoclonic epilepsy (Janz syndrome; Delgado-Escueta & Enrile-Bascal, 1984), and would most likely be treated with an anticonvulsant medication. Thus "repeated" is not an absolute requirement for making the diagnosis of epilepsy.

• *Spontaneous or nonprovoked.* Any living person whose forebrain is at least approximately in working order can be induced to have a convulsion with electric stimulation of the brain. This is even done therapeutically with electroconvulsive therapy (ECT) for life-threatening depression (Greenberg & Kellner, 2005). A patient who has had repeated seizures that were therapeutically induced—that is, that clearly were not spontaneous—is not considered to have epilepsy. In a similar way, some people who take cocaine, or who repeatedly binge on alcohol, may have seizures—seizures that presumably would not have occurred otherwise (although in an individual patient, such a determination may be difficult, given that cocaine and alcohol can both raise the likelihood of a seizure in someone who also has seizures without exposure to these substances) (Koppel et al., 1996; Ng et al., 1988). The problem for the clinician is to determine if one or more seizures are spontaneous. This can be straightforward, if there is a history of ECT or ingestion of known seizure-inducing agents, but in other cases is difficult or impossible to determine. Indeed, one runs into the question: even in a patient for whom every clinician agrees the diagnosis is epilepsy, is it not the case that *every* seizure is induced by something or other, even if we do not know exactly what, and "epilepsy" consists of a propensity for seizures to be induced by these possibly unknown events? This question, too, is controversial. Fernando Lopes da Silva and colleagues, for example, have argued that seizures might, in certain brains, start because of random, statistical variations in background activity (Suffczynski et al., 2005). We shall have more to say on this question later.

• *Seizures.* What it is that constitutes a seizure is actually a complicated issue. Certainly it is the case that some clinical events are, without doubt, seizures, especially when the patient is observed, and examined by, an expert. With a generalized tonic–clonic seizure, a convulsion, it is hard to simulate the initial cry; the salivation, cyanosis, and tongue-biting; the extreme stiffening combined with loss of postural control, then rhythmic jerking movements; the injuries; the sphincter incontinence; the subsequent Babinski signs and confusion. Other clinical seizure types are also highly characteristic. Nevertheless, even experienced epileptologists can have trouble distinguishing a seizure of frontal lobe origin from a so-called pseudoseizure. Seizure-related behaviors in newborns are notoriously difficult to distinguish from the normal movements of an infant. Finally, electrical events can be generated in the brain, called "electrographic" seizures, without any clinical concomitant at all, particularly when they occur during sleep. If the patient happens not to have recording electrodes, in the right place at the right time, how is anyone to know if a seizure has occurred? And how abnormal must the electrical activity be, and how sustained, and involving what volume of tissue, to call it an electrographic seizure? These are questions that can be addressed only through studies of the cellular mechanisms of seizures, and of the boundary—however hard it is to define—of brain activity that lies between the physiological and the pathological.

In summary, then, we have a rough definition of what constitutes epilepsy, from the clinical point of view, one that is sometimes straightforward to apply and sometimes not; and the "not" arises not because of verbal quibbling, but of fundamental difficulties in understanding just what the brain might be doing during a presumed seizure, and in gaining access to the appropriate signals. Let us now consider, in a highly condensed manner, some of the history of how our point of view on epilepsy came to be what it is.

1. In ancient times, and in many different places (Egypt, Babylonia, Greece, India, and other countries), physicians described a disease consisting of attacks that could have motor, sensory, or psychic features. The fact that a commonality could be recognized between these various manifestations is remarkable. The second-century Greek physician Galen attributed some forms of epilepsy to the brain, but others to the body parts that seemed to be affected, a foot for example. (The notion that brain disorders could give rise to involuntary movements, and to hallucinated sensations, either simple or complex, came only in modern times.) Galen also introduced the term "aura" (a "breeze") based on the accounts of two of his patients with what would now be called partial complex seizures. An aura—experienced by some but by no means all patients—was long considered to be a warning of impending seizure, but is now considered the initial manifestation of the seizure itself.

2. Beginning during the Renaissance, but flourishing with vigor from the 19th century, were the notions of cerebral localization—that particular parts of the brain had specific functions (somatic sensation, vision, speech, movement, and so forth). In the case of some functions—primary sensations, for example—it became possible to define "cortical maps," relating small regions of cortex (perhaps on a submillimeter spatial scale) to skin sensation, vision, or pitch, in a particular spot on the body or visual field, or a particular sound frequency. Such notions suggested the concept that the particular manifestations of a seizure—for example, twitching of a finger, or a buzzing sensation on a toe—were caused by abnormal activity in a corresponding, but well-defined, part of the cerebral cortex. This idea was, however, more difficult to apply to alterations of consciousness and emotional states that can occur during a seizure, the issue being how to localize the altered activity. The idea can also be difficult to apply when there are complex patterns of movement, or automatisms. Seizures that produce alterations of consciousness and automatisms

most commonly arise from a temporal lobe, less often a frontal lobe, rarely other locations. Of course, by the time consciousness is altered, widespread brain regions may be participating, directly or indirectly, in the altered brain activity.

3. Robert Bentley Todd (1809–1860) observed that after a generalized convulsion, some patients will have a persisting (but eventually reversible) hemiparesis. This came to be called "Todd's paralysis," but is now more often called "Todd's phenomenon," because the persisting focal neurological deficit may be something other than paralysis, for example, an aphasia (language difficulty with normal acoustic and peripheral speech apparatus), or visual field abnormality. In all such instances, Todd's phenomenon suggests the existence of a localized brain lesion that causes, or is connected with, both the seizures and also the focal neurological deficit. Sommer published evidence in 1880 implicating focal pathology in the hippocampus as a cause of certain types of epilepsy, what would now be called simple and partial complex seizures (Sommer, 1880, quoted by Goldensohn, 2001).

4. In the late 19th century, the outstanding English neurologist John Hughlings Jackson (1835–1911) made meticulous observations of clinical seizure phenomenology. By relating seizure manifestations with focal neurological deficits (hemiparesis, aphasia)—whose cerebral localization was understood, at least approximately—he was able to deduce that seizures originated in cerebral cortex, and predicted that seizures corresponded to "abnormal discharges." However, electrophysiology was in such a primitive state at the time Jackson wrote that is hard to be sure exactly what he meant by "discharges." Jackson also described what he called "dreamy states," corresponding, at least in part, to what are now called partial complex seizures. Dreamy states can begin with epigastric sensations, déja vu or jamais vu, and be associated with masticatory movements or swallowing. Partial complex seizures can also begin, as Jackson recognized, with a hallucinated smell (usually unpleasant); he called these "uncinate fits," because such seizures could originate in or near the uncus, a part of the medial temporal neocortex overlying the amygdala. Finally, Jackson described what is now called "Jacksonian march," in which seizure-associated movements start in one spot, perhaps the hand or foot, and then spread contiguously—the spread corresponding to an enlarging area of epileptic activity in the motor or sensory cortex; Jacksonian march occurs surprisingly slowly, over tens of seconds, much too slow for conventional synaptic transmission, and clearly involving a complex form of collective phenomenon in the brain.

5. What would now be called antiepileptic drug therapy (or anticonvulsant drug therapy) began in the 19th century with Sir Charles Locock (and others; Friedlander, 1986a). Bromides were sedating and not especially effective, as well as having other toxicities (Trump & Hochberg, 1976), and declined in use after phenobarbital was synthesized in 1911 and shown to be an effective anticonvulsant by Hauptmann in 1912 (López-Muñoz et al., 2004; Kumbier & Haack, 2002)—many years before the discovery of the gamma-aminobutyric acid

A ($GABA_A$) receptor, with which phenobarbital interacts. In the 1930s, Tracy Putnam and H. Houston Merritt examined the effects of a series of phenyl compounds on the threshold for eliciting seizures in cats with electrical stimulation (Putnam & Merritt, 1937); they then showed that one of these compounds, phenytoin (diphenylhydantoin; trade name, Dilantin) was an effective anticonvulsant in humans (Merritt & Putnam, 1938; Friedlander, 1986b). Phenytoin was much less sedating than phenobarbital, and was later shown to have a different mechanism of action, interacting with Na^+ channels, particularly persistent Na^+ channels (Catterall, 1999; Lampl et al., 1998). A large number of anticonvulsant medications have been introduced since the landmark contributions of Merritt and Putnam (Aldenkamp et al., 2006; Marson et al., 2007; White et al., 2007).

6. A dietary treatment for epilepsy, used more in children than in adults, was introduced in 1921 by R.M. Wilder: the objective of ketogenic diet, high in fat and low in carbohydrates, was to capture some of the metabolic effects of fasting without producing starvation. The ketogenic diet is still in use in children with intractable forms of epilepsy, such as the Lennox–Gastaut syndrome, with continuing uncertainty, however, as to its mode of action (Huffman & Kossoff, 2006).

7. In the 1930s, within a few years of Berger's fundamental 1929 paper, studies began to appear correlating specific abnormal EEG patterns with the occurrence of defined seizure types, and with the propensity of the brain to initiate seizures; an "interictal spike" is an example of an EEG pattern that indicates seizure propensity ("interictal" means "between-seizure") (Swartz & Goldensohn, 1998). F.A. Gibbs and E.L. Gibbs were pioneers in this area, with Herbert Jasper, Henri Gastaut, Pierre Gloor, and many others making seminal contributions. A methodological issue is important enough to be mentioned here: the correlation of 3-per-second spike-wave and absence seizures ("petit mal") was noted in the 1930s (Gibbs et al., 1935); interictal spikes were also recognized in the 1930s, in experimental contexts (Fisher & Löwenback, 1934) and clinical ones (Gibbs et al., 1936; Jasper, 1936). Patients tend not to move very much during absence seizures or interictal spikes, but the situation is completely different with a major convulsion, drop attacks, or other seizure types associated with movement: for the movement induces artifact into the EEG and makes accurate recording impossible. General anesthesia does not solve this problem, as it (for the most part) suppresses seizure activity. The correlation of certain seizure types with EEG patterns requires either pharmacologic paralysis, in which case part of the clinical phenomenology to be correlated is suppressed; or the use of invasive recording techniques.

EEG is vital both for patient care and for our understanding of epilepsy. For patient care, EEG can provide evidence that seizures are actually occurring or are at least likely (as opposed to, e.g., so-called pseudoseizures, or cardiogenic drop attacks), whether there is a localized site of origin, and also as to the type of seizure, and in particular whether the seizure disorder is part of a so-called "epileptic syndrome" (defined later). This information is important for treatment (an antiepileptic drug that is effective against one type of epilepsy may be ineffective in another, or even make it worse), in guiding further evaluation of a patient, in prognosis, and in suggesting whether genetic studies—or at least further family history—are appropriate.

Let us consider how EEG is vital for understanding epilepsy. EEG recorded at the scalp confirms that seizures are paroxysmal events electrically as well as clinically; and scalp EEG allows the electroencephalographer to correlate seizure-associated signals, emanating from the brain, with states of awareness and with sleep stages (as the latter can also be defined with EEG). In addition, and of critical importance, EEG recording has been extended from measuring scalp signals to measuring signals closer to, or within the brain: with subdural electrodes (placed on the surface of the brain), or with depth electrodes within brain substance. Even microelectrode arrays are now used (Schevon et al., 2008). Further, in animal models, and especially with in vitro techniques, one can record intracellularly from neurons as they participate in seizures or seizure-like events.

The essential point to grasp is this: with extensions in our ability to record signals from the brain, from brain slices, and individual cells (and not just electrical signals, but also magnetic and optical ones), epilepsy has come to assume a dual nature: a clinical nature, defined by experiences and history offered by the patient and by observers of the patient's behavior; and a network-cellular nature defined by observation of signals from nervous tissue. This duality has come about in stages: first, of course, the clinical base; then the observation that certain clinical phenomenology is associated with certain sorts of EEG signals, and extending that observation to animal models of epilepsy. This led to the concept of an "electrographic seizure," an epoch of abnormal EEG activity that could be correlated with a clinical seizure, that is, something apparent to an observer, or that produced a reliable patient report of an unusual subjective experience. In turn, it was recognized that EEG signals that looked like electrographic seizures—as defined in this way—also occurred during sleep, when there was little if any observable consequence, and of course no patient report. Likewise, electrographic seizures might even occur in the waking state without obvious behavioral concomitant. It seemed logical to view such an electrographic seizure as a bona fide seizure—because (a) such events suggest underlying brain or metabolic pathology and (b) a patient showing electrographic seizures might well require treatment, even if more classical behavioral seizures have never been observed. Finally, it became clear that signals looking like electrographic seizures could even occur in brain slices, or in vivo in such small volumes of cortex that a behavioral concomitant is hard to imagine. Indeed, signals looking very much like electrographic seizures can even occur in computer models.

Although the dual nature of epilepsy, clinical versus cellular/network, is unavoidable, and absolutely necessary for scientific progress, it can also lead to conceptual errors, and one always must be cautious. Because this book deals with oscillations, we shall mostly be speaking about epilepsy from a network/cellular point of view—but not always.

The dual nature of epilepsy is difficult to carry over to other neuropsychiatric disorders, as the latter virtually never have the clear bioelectric (or biomagnetic) correlates that epilepsy does; nor have we reached the point where we can speak of, for example, an in vitro model of schizophrenia or Parkinson's disease.

8. Some epilepsy patients are treated, at least in part, by neurosurgeons. Although the use of trephination for epilepsy treatment dates back to ancient times (Gross, 1992), the pioneer in the modern era of epilepsy surgery was Wilder Penfield (1891–1976); Penfield was also the pioneer in invasive (i.e., intracranial, instead of from the scalp) EEG recording (Almeida et al., 2005). The most commonly performed surgical procedures (Duncan, 2007; Lee SK et al., 2005; Zimmerman & Sirven, 2003) consist of (a) the removal (with surrounding tissue) of a structural lesion—perhaps a tumor or a focal cortical dysplasia—identifiable on an imaging procedure such as magnetic resonance imaging, and shown as well by recording methods to be the initiating site of the patient's seizures ("lesionectomy"); (b) the removal of cortical tissue that fulfills electrical criteria for initiating seizures, but without necessarily being visible with imaging procedures ("cortical resection"); and (c) anterior temporal lobectomy or other procedures designed to remove relatively substantial parts of the medial temporal lobe, including hippocampus, where the underlying lesion is likely to be mesial temporal sclerosis—an important pathology underlying partial complex seizures (Benifla et al., 2006; Falconer & Taylor, 1968). "Disconnection" procedures, including section of the corpus callosum, are now performed rarely, as is hemispherectomy. Vagus nerve stimulation may have marginal efficacy, and the therapeutic possibilities of local cortical stimulation are under investigation.

The neurosurgical treatment of a patient with epilepsy involves a great deal more than the neurosurgery itself. If a removal of a portion of the patient's brain is contemplated, two basic questions must be addressed: with that portion of the brain removed, is the patient likely to have many fewer (or none at all) seizures? This usually requires documenting that most or all of the patient's seizures actually originate from within the tissue whose removal is contemplated. Second, will the removal of the tissue likely cause a devastating neurological insult, such as a permanent and severe disorder of language? If so, surgery cannot be performed. To address these questions, certain studies are likely to be done in a modern Epilepsy Center, beyond the usual history and physical examination, family history, scalp EEG, brain imaging studies, medication trials, neuropsychological evaluation, and so forth: these studies involve documenting the origin of the patient's seizures in one brain region or another, and may involve the placement of intracranial recording electrodes (subdural or depth or both), and also the simultaneous monitoring of the patient with intracranial EEG *and* with video recording of the patient's behavior, supplemented of course with the observations of the Epilepsy Center staff. Intracranial recording of this nature, besides being vital for the individual patients, has had a major effect on thinking about epilepsy. First has been widening of the views of just what sort of electrical activity should qualify as "electrographic seizure"; indeed, augmenting of knowledge is producing confusion as well as clarity, as it is not always clear what is "seizure" and what is "normal," especially when it is impossible to use patient behavior or experience as a guide. Second, intracranial recording has demonstrated high-frequency signals—not generally accessible to scalp recording technologies—that are associated with seizures, that precede seizures, and that can occur interictally. These signals—fast and very fast oscillations—offer clues to epilepsy pathophysiology and pathogenesis, and we shall have much to say about them.

When brain tissue is removed from a patient with epilepsy, part of the tissue is automatically sent for neuropathological analysis, typically after fixation, staining, and inspection with the light microscope, to determine, if possible, a structural cause (or at least predisposition) for the patient's seizures, and to determine if a pathological process is present that requires further attention (e.g., a neoplasm). In addition, tissue may also be available for research protocols, including ultrastructure (which requires special fixation) and electrophysiology (which requires keeping the tissue alive, in a transport chamber with oxygenated, glucose-rich, artificial cerebrospinal fluid bathing the sample until it can be sliced and transferred to a special recording chamber). Research studies of course require prior approval by the hospital Institutional Review Board, and may also require an informed consent procedure.

9. It is beyond the scope of this monograph to offer a detailed description of the many clinical types of seizures that have been described, or of the attempts that have been to develop a rational classification; we refer the reader to textbooks (Engel & Pedley, 1998) and review articles (Sirven, 2002). We can only emphasize that logical classification is essential for patient care, for guiding genetic studies, and as a background for the investigation of cellular mechanisms. Two features of all epilepsy classification schemes deserve mention here, however.

First is the distinction between seizures of (at least approximately) generalized onset and those of focal (or sometimes multifocal) onset. (Note, however, that focal seizures may, after some interval, become generalized, affecting the brain bilaterally; this is called "secondary generalization." Such seizures are considered distinct from "primary generalized seizures.") For primary generalized seizures, the onset is, of course, not uniform across the whole cortex on a millisecond time scale; but there is, nevertheless, no reproducible site of onset, and the seizure affects the brain bilaterally almost immediately, and usually with alterations of consciousness and/or of postural tone, and possibly associated with jerking movements (single movements or perhaps rhythmical). Focal seizures, as the name implies, may appear solely as a localized electrographic discharge; or there may be, as well, evidence of brain dysfunction that corresponds to a focal abnormality, for example, visual or olfactory hallucinations, twitching movements or paresthesias in one part of the body, turning the head and eyes to one side, or intense anxiety (so-called "ictal fear"). Generalized seizures are more likely than focal ones to have a genetic basis, but this is by no means absolute (see later); likewise, focal seizures are more likely to have an identifiable structural lesion as their cause. It was long thought that generalized seizures must arise in the brain stem or diencephalon, because of the bilateral expression at onset—for example, there was the "centrencephalic" epilepsy theory (Metrakos & Metrakos, 1961) or the "cortico-reticular" theory of Gloor (Gloor et al., 1973). More recent in vivo data from experimental seizures in cats, however, indicate that at least some generalized seizures originate in the neocortex (Steriade & Contreras, 1995), with suppression of firing in thalamic neurons (see also Crunelli & Leresche, 2002; Pinault et al., 1998). [Nevertheless, data from genetic animal models of certain generalized epilepsies also indicate clear abnormalities in thalamic neurons (X-B Liu, et al., 2007; Shin et al., 2006; Tsakiridou et al., 1995).]

Is there something fundamentally different about the cellular mechanisms of focal, as compared to generalized, epilepsies? This question matters to us, as the data on which we shall be drawing derives from experimental models of both types of epilepsy. One cannot answer the question definitively yet, but we need a hypothesis to proceed; and our hypothesis is that the cellular mechanisms of all of the epilepsies have significant (if not complete) overlap.

Second, over the last few decades, the concept of "epileptic syndrome" has come to be formulated, and its importance recognized particularly in child neurology; we say "last few decades," although it has been pointed out that one epileptic syndrome, now called benign Rolandic epilepsy, was described, at least in part, in the late 16th century (van Huffelen, 1989), childhood absence epilepsy was described in 1705 (Trinka, 2005), and West syndrome in 1841 (West, 1840–1841): of course, without EEG! The term "epileptic syndrome" now refers to a constellation of clinical and EEG features, family history, and natural disease history; correct syndrome diagnosis is, needless to say, essential for treatment recommendations, and advice on prognosis and the advisability of familial or genetic studies. Epileptic syndromes differ widely in their associated seizure types, EEG patterns, severity, and associated morbidities (such as intellectual deficits). A *partial* list of epileptic syndromes would include, besides childhood absence epilepsy, the following:

• *West syndrome, or infantile spasms* (Hrachovy & Frost, 2006; Korff & Nordli, 2006; Millichap et al., 1962). Sometimes on a genetic or structural basis, infants and small children have attacks consisting of head flexion and arm abduction. There is a high incidence of associated mental retardation. The EEG may exhibit a remarkable pattern called hypsarrhythmia, consisting of high-amplitude waves and asymmetric spike-waves; the attacks are associated with a so-called electrodecremental response, with disappearance of the high-amplitude waves and "flattening" of the EEG. In at least some, and perhaps all, instances of the electrodecremental response, the EEG actually displays very fast activity at >50 Hz, detectable with proper techniques (Asano et al., 2005). [It is our contention that very fast activity at these, and higher frequencies, are involved in most or all of the epilepsies.]

• *Lennox-Gastaut syndrome* (Stephani, 2006; Yaqub, 1993) Some children who have had West syndrome go on to develop the Lennox-Gastaut syndrome (LGS) as they get older; LGS can also occur in the setting of other neurological disorders, or in children who were previously normal. The incidence of retardation and behavioral problems in LGS patients is high. Patients have a multiplicity of seizure types, and associated EEG abnormalities, in variable patterns. Seizure types can include: (a) "atypical absence"—called "atypical" because the attacks do not have a clear onset or termination (unlike the case in typical childhood absence epilepsy); (b) tonic spasms, perhaps related to infantile spasms, which may be associated with loss of consciousness or with falling; (c) myoclonic jerks; (d) other seizure types, including generalized tonic–clonic convulsions. The EEG background is abnormal, with excessive slowing, and may also include interictal and ictal electrographic seizure patterns: interictal spikes, spike waves, polyspike waves; the latter two forms can continue rhythmically at approximately 2 Hz, sometimes for long periods, especially during sleep (spike-wave during classical childhood absences tends to be faster, 3 Hz and above). In addition, children with LGS may show another ictal EEG pattern called, variously, tonic seizure patterns or fast runs, consisting of waves at 10 Hz and above, and lasting for some seconds. These patients tend to respond little or not at all to anticonvulsant medications and indeed the illness can be quite devastating. As we discuss further in the text that follows, Mircea Steriade discovered an animal model that captures many of the electrographic, if not clinical, aspects of Lennox–Gastaut syndrome; and this animal model offers a number of clues to pathogenesis.

• *Rolandic epilepsy with centrotemporal EEG spikes* (Beaussart & Faou, 1976; Boor et al., 2007; Camfield & Camfield, 2002; Loiseau et al., 1983; Scheffer et al., 1995). This disorder of childhood, one of the more common childhood epilepsies, is associated with focal motor or sensorimotor seizures involving the face, mouth, and tongue most commonly, perhaps with drooling, that take place during sleep. Secondary-generalized tonic–clonic seizures occur as well, and generalized convulsions can also occur on awakening. The EEG demonstrates characteristic interictal spikes that are generated by synchronous discharges in the face and hand areas of sensorimotor cortex. Remarkably, this form of epilepsy remits at adolescence. It is presumed to be familial with low penetrance, without a specific genetic abnormality having been identified with certainty.

• *Panayiotopoulos syndrome* (Covanis, 2006; Demirbilek & Dervent, 2004; Koutroumanidis, 2007). This is another childhood epilepsy syndrome that remits with maturity, but that has unusual clinical manifestations: autonomic signs and symptoms, such as nausea and vomiting, pallor, and pupillary dilatation; cardiorespiratory arrest has been described, but is rare. Consciousness may be altered, and the attack may evolve into a syncope-like state (i.e., unresponsive and flaccid) or into a generalized convulsion. Not surprisingly, parents and even highly trained physicians may not make the connection between nausea and vomiting on the one hand, and seizure on the other hand; nevertheless, simultaneous EEG/ video monitoring has shown that the autonomic attacks do correspond to cortical epileptiform discharges. Possibly the epileptic activity spreads to the insula or hypothalamus. Attacks can last longer than 30 minutes. The EEG demonstrates interictal spikes at variable locations, more over the posterior head than anterior, with occipital spikes common.

• *Juvenile myoclonic epilepsy (JME), Janz syndrome* (Delgado-Escueta & Enrile-Bascal, 1984; Janz, 1985; Jayalakshmi et al., 2006; Pedersen & Petersen, 1998; Usui et al., 2005). This is a common, presumably familial, generalized seizure disorder with typical onset in the teens (although a few patients also have focal features). One representative case history would consist of an intellectually and neurologically normal youth who begins to have irregular jerking movements, perhaps in a series, of the neck, shoulders and proximal arms ("myoclonic jerks"); and then, perhaps a few years later, begins to have generalized convulsions, possibly preceded by myoclonus. Both the myoclonus and the convulsions occur most commonly, but not exclusively, on awakening; and they are both likely to be exacerbated by sleep deprivation. Some patients have absence attacks as well. The EEG typically shows, interictally (between seizures) generalized interictal spikes, or spike-wave, or polyspike-wave, at frequencies up to about 6 Hz. Ictal patterns are more complex. While most patients have both the myoclonus and the convulsions readily controlled with a single antiepileptic medication, this disorder tends not to remit, unlike Rolandic epilepsy and Panayiotopoulos syndrome; so that it is common for JME patients to be prescribed antiepileptic treatment indefinitely. JME is of particular interest not only because it is so common, but also because of the number of genetic studies devoted to it, and because its typical EEG patterns can be replicated in brain slices and in network models.

• *Childhood absence epilepsy ("petit mal")* (Camfield & Camfield, 2002; Panayiotopoulos, 2001; Wirrell, 2003). This disorder has many variants, but in its most well known form it has onset in childhood, and consists of well-defined (i.e., abrupt onset and termination) episodes of loss of awareness, but without usually loss of postural tone: there may be staring and disconnection from the environment ("absence"), but not falling. The eyes may blink several times per second, and there may be twitching of facial muscles, eyelids, or the hands. In untreated patients, attacks can usually be precipitated by hyperventilation. The EEG concomitant is classic 3-Hz generalized spike-wave, although the EEG ictal rhythm can be slower or faster, and may contain polyspike-wave complexes. Spike-wave paroxysms also occur during slow-wave sleep, in which case, of course, one can not speak of "absence," but rather of electrographic seizures. Patients may suffer from generalized tonic–clonic convulsions as well as absence.

The absence attacks often remit with age, but the generalized convulsions can continue. Children with this epilepsy syndrome usually do not have neurological deficits or abnormal brain imaging studies; and yet difficulties with social adjustment have been described, even when the seizures are controlled (as they usually are). There are instances where absence attacks last surprisingly long, even hours or days ("absence status": note that other seizure patterns can also, in some patients, last this long and constitute "status"). A number of animal models of absence exist (see later), and some investigators believe that absence seizures should be regarded as "thalamocortical" phenomena rather than strictly cortical; this in contr-distinction to the consensus that other epilepsies, and certainly the focal epilepsies, reflect strictly cortical pathology and pathophysiology.

We devote considerable attention to the issue of how the epilepsies are approached at the cellular level, but draw attention to one salient fact here. Animal models exist for a number of different types of seizure, and animal tissue can be studied in vitro, although, for reasons ranging from economic to ethical, in vitro studies are rarely performed with primate tissue. In addition, as we have mentioned previously, human tissue, removed at epilepsy surgery, is also sometimes available for research study. What we wish to emphasize, however, is that epilepsy surgery is performed only on certain patients, usu-ally those with intractable seizures of focal or multifocal onset; and epilepsy surgery is virtually never performed on patients with the epilepsy syndromes discussed previously. Insight into the mechanisms of the epilepsy syndromes therefore derives by making reasonable inferences from animal epilepsy models that have similar electrographic features; or, in a few cases, by identi-fying a genetic abnormality in humans with a given epilepsy syndrome, and then recreating a mouse with a comparable genetic abnormality and studying that mouse. The latter approach has been used with a rare syndrome called "generalized epilepsy with febrile seizures plus" (Gardiner, 2005).

10. Since ancient times, epilepsy has been considered to be at least in part familial. In the 19th century and early 20th century especially, such notions became linked with the eugenics movement, along with uncritically accepted views on the relation between epilepsy and insanity (Friedlander, 2001; Waller, 2002). How views of such a connection arose in the first place are not clear, given the existence of people of extraordinary accomplishments who hap-pened to have epilepsy (Julius Caesar, Fyodor Dostoevski); possibly, such benighted ideas arose from faulty generalizations of observations on selected individuals with genuinely inherited disorders, with either metabolic causes (phenylketonuria) or abnormalities in brain growth and development (tuber-ous sclerosis, neurofibromatosis), wherein coexpression of mental retardation and epilepsy is possible. It may be difficult to appreciate how much has been learned over the last decades, even if, at the same time, we do not begin to know as much as we would like to and need to. On the one hand, there is a much more refined classification of seizure phenomenology (so-called semiology), with corresponding clinical data (EEG, brain imaging, biochemical tests), and

this refined classification greatly facilitates identification of families with common epilepsy characteristics, and prevents (we hope) lumping together families that actually have different underlying disorders; there are extensive epidemiological studies, including twin studies; and there has been an explosion of capabilities with the molecular biology revolution. Modern views of epilepsy genetics, while confusing enough, need to be summarized at least briefly (Berkovic et al., 1998, 2004, 2006; Gardiner, 2005; Greenberg & Pal, 2007; Noebels, 2003a, 2003b; Steinlein, 2004); no treatment of cellular and network mechanisms can ignore the human data—and, conversely, molecular biology cannot, by itself, answer the question as to how a given genetic abnormality actually leads to seizures. Analysis of the behavior of seizing neuronal networks needs to offer at least plausible hypotheses as to how the identified epilepsy genes cause disease. Some of the salient facts are these:

- Twin studies (e.g., comparing seizure incidence in monozygotic twins vs. dizygotic twins) indicate that genetic factors contribute both to generalized and to focal epilepsies, but more to the generalized epilepsies.
- Only some of the familial epilepsies obey (relatively) simple Mendelian genetics. There are examples of defined genes for disorders of Mendelian type, in which epilepsy is part of the expression of a broader illness [tuberous sclerosis (European Chromosome 16 Tuberous Sclerosis Consortium, 1993; Sampson & Harris, 1994), Unverricht-Lundborg syndrome (Pennachio et al., 1996)]. In addition, there is a large number of inherited metabolic diseases in which seizures occur, but are not the most disabling feature of the respective disease (e.g., Tay–Sachs disease).
- There exist inherited disorders in which brain development and cortical structure are specifically affected, with which epilepsy can be associated. Besides tuberous sclerosis, examples include lissencephaly (smooth brain) and subcortical band heterotopia (Güngör et al., 2007; Lian & Sheen, 2006).
- Data from family-genetic studies can be difficult to interpret for a number of reasons, including these: (a) There may be variable expression of the syndrome within a family (different family members may have different types of seizures, or may not have seizures at all but rather a characteristic abnormal EEG)—even though, one presumes, the affected family members all share a particular genotype. (b) Different families, whose disease expression looks similar across and between families, may be found to have mutations in entirely different genes. (c) Conversely, mutations in the same given gene may give rise to distinct patterns of disease expression in different families. Evidently, disease expression is influenced by genetic background—all the other genes in aggregate—and genetic background variability may be operative in different human populations. (d) There are common problems of reproducibility of some of the reported genetic associations, such as single nucleotide polymorphisms, with particular epilepsy syndromes (e.g., juvenile myoclonus epilepsy). Even without issues of reproducibility, the mechanistic significance of reported findings can be unclear (Hempelmann et al., 2006; Mas et al., 2004; Suzuki et al., 2004).

• Considerable progress in identifying specific epilepsy genes has been made with some of the many epilepsy syndromes. In aggregate, the total number of patients with these disorders, where an abnormal gene can be reliably identified, is quite small relative to the total epilepsy population. Nevertheless, the data are of extreme importance, because the operative pathophysiological mechanisms—which remain to be better understood—may be of broad relevance.

• Several of the reported mutations in epilepsy syndromes involve membrane channels, either voltage- or ligand-gated ("channelopathies"): transient Na^+; K^+ (including KCNQ2 and KCNQ3, giving rise to the M-current, of which we shall have much to say later), and KCNA1 (homologous to the shaker K^+ channel gene *Kv1.1*—in one family, a mutation in this gene was associated with focal seizures, episodic ataxia, and myokymia (Zuberi et al., 1999); $GABA_A$ receptor subunits, neuronal nicotinic acetylcholine receptors (Steinlein et al., 1995); and—if we include animal data—Ca^{2+} channels. (The presence of nicotinic acetylcholine receptors in the list is of great interest because, as we discuss later, such receptors are present in axons.) Corresponding genetic abnormalities can, in principle, be produced in mice, and the consequences examined in vivo and in vitro—recognizing, of course, that Nature can come up with compensations for genetic abnormalities. There are surprisingly few data on gap junction abnormalities in putatively heritable epilepsies, but, as we shall discuss, gap junction mediated network oscillations depend critically on the intrinsic excitable properties of the nearby membranes, and not just on the gap junctions themselves in isolation. Considerations of this sort require network models to be made transparent.

Thus, despite a wealth of literature and data, epilepsy genetics in humans (and also animals) is very much work in progress, not only in terms of the genetics per se, but also in terms of relating the genetics to pathophysiology.

11. Brown-Séquard believed that he had developed an experimental model of epilepsy, in guinea pigs, in the mid-19th century (Brown-Séquard, 1856; Koehler, 1994), although his results (he had observed focal and secondarily generalized seizures after spinal cord hemisection) have not been replicated in the modern laboratory. (The Koehler 1994 paper is well worth perusal, as it so vividly illuminates how drastically our thinking about epilepsy has evolved in 150 years; for example, in the mid-19th century, most investigators thought that epilepsy arose in the medulla and spinal cord. The distinction between a body part that moved or where sensation was felt, and the cerebral "representations" thereof—these had not progressed much beyond Galen. How absurd will our current ideas seem in another 150 years?) Today, the study of experimental epilepsy models is a lively subdiscipline of Neuroscience, partly because these models have so changed how we think about epilepsy (i.e., we now often think of it in cellular terms), but also because epilepsy studies have provided so much basic information about the brain, relevant to its normal function. We shall outline some of the approaches that have been

taken, concentrating on techniques. The last part of the chapter will then summarize clinical, experimental, and simulation data that form the groundwork for our subsequent discussion of oscillations.

First, we must address the question: what does it even mean to speak of an "epilepsy model"? There are dual aspects to this question, reflecting the dual aspects of epilepsy itself, discussed earlier. Thus, for an in vivo model, assuming that nonanesthetized animals are being studied, we can speak of behaviors in an animal that look similar to those a patient might exhibit during a seizure. If these behaviors occur repeatedly, and correlate with electrical discharges in the animal brain that also have similar patterns to seizure-associated electrical discharges in an epileptic human, then we may certainly speak of an epileptic animal. The first part of this paradigm is what Brown-Séquard followed; although he did not have access to EEG recordings, he could observe behavior—and no one today would question that his description of what his guinea pigs were actually doing constituted seizures. (What might be questioned now is the relation between his experimental manipulation and the later development of seizures in the animals.) Next, dually, in the case of anesthetized animals, sleeping animals, or brain slices, as we do not have behavior, we must instead rely on electrical recordings—with electrodes near or in the brain, and/or within individual neurons—and again we seek correspondences with analogous signals in humans, in cases where a clinician would speak of an electrographic seizure, an interictal spike or polyspike, or other signal known to precede or superimpose on an epileptiform discharge. Our assumptions concerning the correspondence between animal and human epileptiform events can be checked by performing a given experimental manipulation in a human brain slice (removed at neurosurgery, as described previously), and also in an animal brain slice, and comparing the results. What we cannot yet do, however, is record intracellularly from human neurons in situ; in this instance, it is an act of faith that if we *could* so record, the neurons would exhibit similar patterns of activity to animal neurons that are engaged in an analogous type of seizure. This act of faith, if one may call it that, is part of our more general hypothesis that the difference between animal brains and human brains is not to be found in the behavior of single neurons, or even of circuits with thousands of neurons; but rather is to be found in properties of much larger ensembles of cells: properties which we do not yet understand, but which are probably not relevant to the cellular mechanisms of epileptogenesis.

Let us begin with in vivo models of "true" epilepsy, that is, where there are recurrent spontaneous seizures. Epilepsy is well known in dogs, especially in certain breeds such as the Labrador retriever (Jaggy et al., 1998). Spontaneous and photosensitive seizures also occur in baboons (Naquet et al., 1995; Szabó et al., 2005), and for a time *Papio papio* was used for study of antiepileptic drug mechanisms (White et al., 1992). Nevertheless, for a variety of reasons, rodents (rats, mice, Mongolian gerbils) are the preferred animals for studies of spontaneous seizures; these reasons are economic, ethical, and the feasibility of performing genetic manipulations in mice. The great preponderance of spontaneous epilepsies in rodents are generalized and mostly resemble absence epilepsies (Buchhalter, 1993; Noebels, 2003b).

We consider mice first (Barclay et al., 2001; Osten & Stern-Bach, 2006; Pietrobon, 2002; Steinlein & Noebels, 2000). A number of genetic lines exist, reflecting spontaneous mutations, and designated by evocative nicknames (*ducky, stargazer, tottering*, etc.). Many of the mouse spontaneous genetic epilepsies exhibit absence-like seizures, have an associated ataxia, and result from mutations in the CACNA1A gene, encoding for a subunit of P/Q-type calcium channels [originally discovered in, and having high-density in, Purkinje cells (Hillman et al., 1991), but present elsewhere in the brain as well, including at presynaptic terminals, with effects on synaptic transmission (Li L. et al., 2007)]. Epileptic mouse lines have also been created by gene knockouts, for example of *Kv1.1* (Rho et al., 1999; Smart et al., 1998); double knockouts have also been created (Glasscock et al., 2007), to show one "epilepsy-protective" mutation tending to compensate for an "epilepsy-facilitating" mutation.

The best known spontaneous seizure models in rats also exhibit absence-like seizures. The rat lines include the WAG/Rij strain (Coenen & Van Luijtelaar, 2003) and the GAERS strain ("genetic absence epilepsy in rats from Strasbourg") (Marescaux et al., 1992). In the GAERS rats, T-type calcium currents are increased in GABAergic nucleus reticularis thalami neurons (Tsakiridou et al., 1995); while $GABA_A$-receptor-mediated inhibitory postsynaptic potentials (IPSPs) in these same neurons are larger and faster than in control rats (Bessaïh et al., 2006). These specific abnormalities may be relevant to the generation of the absence seizures (Tóth et al., 2007)—but see also Polack et al. (2007), who provide evidence that seizures in GAERS animals are initiated in neocortex, consistent with observations of Steriade in cats. Other rat strains with absence-like seizures also exist (Inoue et al., 1990). Another genetic epilepsy in rats consists of audiogenic seizures, for example, in Wistar rats (Garcia-Cairasco, 2002; Kiesmann et al., 1988); this epilepsy is of interest in that it begins with the animal running wildly before a tonic–clonic convulsion, and because of the apparently critical role of the inferior colliculus—a subcortical structure, not considered in most classical theories of epilepsy.

Besides in vivo models of spontaneous seizures, occurring without provocations by the experimenter, there are also in vivo models in which some manipulation is performed to induce one or more seizures acutely, or to facilitate the development of pathology such that spontaneous seizures occur after a latent period. In some instances, an experimental protocol produces both effects. (In addition, brain tissue from any of the in vivo seizure models can be studied in vitro, providing data complementing the in vivo observations.) In vivo epilepsy models are used both to study seizure mechanisms, and to provide preparations that can be used for screening antiepileptic medications (following Merritt and Putnam) or other proposed antiepileptic therapies. Let us consider some of the protocols that have been used (Dichter, 2006; Fisher, 1989):

• We have mentioned the acute electrical induction of seizures, as employed in electroconvulsive therapy (ECT) in human patients with depression, and as used by Putnam and Merritt in cats in the 1930s. This method was first employed by Pavel Y. Kaufmann in 1912, in dogs (Kaufmann, 1912, cited by Goldensohn, 2001). Two related variations of the acute electrical induction of seizures exist, whose purpose is to create a state in which seizure threshold is chronically reduced, or in which seizures even occur spontaneously: one type of protocol is to apply repeated, initially subconvulsive, electrical stimulation. This procedure, called kindling, may eventually lead to a state in which the same stimulus, initially subthreshold for seizure induction, eventually becomes capable of seizure initiation; and a state of spontaneous seizures may finally supervene. Alternatively, an extremely prolonged seizure itself—called status epilepticus—may be employed as a variation of experimentally injected electric current. Kindling was originally introduced by Graham Goddard in the late 1960s, using stimulation of the amygdala, as part of a series of experiments to study learning (Goddard, 1969; McIntyre & Goddard, 1973; McIntyre et al., 2002); amygdala stimulation continues to be used, and other kindling sites have also been developed (e.g., the perforant path in the hippocampus, and less frequently neocortical sites (McIntyre, 1979)). Kindling protocols are most often used in rats, but cats have been employed as well (Wada et al., 1974). A second type of protocol employs status epilepticus or at least a flurry of seizures, which can be induced with (a) the muscarinic agonist, pilocarpine (Clifford et al., 1987; Cole & Nicoll, 1984); (b) by injection of sufficiently high concentrations of kainate (Leite et al., 2002), which causes neuronal death by binding to kainate receptors (a collection of subtypes of glutamate receptors); or (c) by injection of tetanus toxin, which acutely diminishes neurotransmitter (particularly GABA) release (Bagetta & Nisticò, 1992; Empson et al., 1993; Williamson et al., 1992) and causes an epileptic encephalopathy that may remit after weeks, or may continue (Louis et al., 1990; Mellanby et al., 1977). [Intriguingly, gap junctions appear to be important in tetanus toxin epileptic encephalopathy (Nilsen et al., 2006), a topic to which we shall return later.] Many or all of these related methods, besides causing loss of neurons, induce reorganization of circuitry—for example, mossy fiber sprouting, with formation of aberrant excitatory synapses onto dentate granule cells, occurs after lesions in the hilus of the hippocampus (Jiao & Nadler, 2007; Okazaki et al., 1995; Sloviter, 1999; Sutula & Dudek, 2007; Tauck & Nadler, 1985). Kindling and status epilepticus induce other forms of reorganization in the dentate granule cell/mossy fiber system as well, including production of new granule cells (Parent et al., 1998; Scott et al., 1998), although there is evidence that neither mossy fiber sprouting, nor neurogenesis, are essential for the development of kindling per se.

• Instead, they may be relevant to the subsequent epileptic state (Osawa et al., 2001). Kindling also leads to release, or at least enhanced release, of GABA from excitatory mossy fiber terminals onto CA3 pyramidal cell dendrites (Gómez-Lira et al., 2002; Treviño & Gutiérrez, 2005; Walker et al., 2001); the functional consequences of this effect for epilepsy remains to be determined.

• Another method for study of epileptogenesis consists of producing a structural lesion of the brain, typically cortex, and then making recordings over time, either in vivo near to the site of the lesion, or from brain slices. For example, noxious substances have been injected, such as an iron salt, in part replicating the effects of intracerebral blood, which is known to be epileptogenic (Willmore et al., 1978), or metallic cobalt powder (Dow et al., 1962). [The cobalt model is of special interest to us, because intracellular recordings often reveal small action potentials (Pumain, 1981)—variously called partial spikes, spikelets, and fast prepotentials—that were originally proposed to originate in dendrites, but many of which are now known to come from axons. The overwhelming significance of axons in epilepsy, and oscillations in general, will be a major theme of this work.]

• Another technique involves sectioning subcortical white matter, perhaps in combination with transcortical cuts: the "undercut cortex," or "partially deafferented cortex" (Hoffman et al., 1994; Jin et al., 2006; Topolnik et al., 2003).

• The injection of metals into the cortex, and cutting through white matter, leads to the development of epileptiform discharges and seizures only after a latent period of days or weeks.

Our brief survey of in vivo epilepsy models concludes with consideration of two types of protocol that lead to seizure discharges within minutes. Both sorts of protocol have been of major importance in the development of views on epileptogenesis, and one them—penicillin—involves observations in humans as well as animals.

Penicillin and Epileptogenesis

The antibiotic penicillin was discovered accidentally by Alexander Fleming in 1928; he noticed that bacterial colonies died if they were too close to contaminating colonies of *Penicillium* molds, and concluded that the molds were secreting a bactericidal compound. A group at Oxford—Howard Florey, Ernest Chain and Norman Heatley—found a method to produce enough penicillin that they could demonstrate (first in mice, then in humans) efficacy against bacterial infections in mammalian hosts (Neushul, 1993). Development of means to produce the antibiotic in mass quantities was motivated by World War II—a fascinating story of culture methods, mutagenesis, and selection of productive strains of the mold (Neushul, 1993; Richards, 1964), of enormous implications for the practice of medicine, but of specific relevance here because, by the end of the war, enough penicillin was available that it could be used in high concentrations, both clinically and experimentally. It was then that it was discovered that clinical use of penicillin could—in highly selective conditions— cause epileptiform EEG abnormalities, clinical and electrographic seizures, and even status epilepticus that was fatal on occasion.

We now know that penicillin is a weak antagonist of $GABA_A$ receptors (Curtis, 1973; Hochner et al., 1976; Lindquist et al., 2004; Macdonald & Barker, 1978), but that knowledge came decades after the epileptogenic effects of the drug were discovered. $GABA_A$ receptor blockade is responsible for most or all of penicillin's convulsant effects, and the drug was of critical importance in developing an experimental epilepsy model in which the detailed cellular mechanisms could be worked out (see later). This model is now called "disinhibition"; investigators at present, however, rarely use penicillin, but instead more specific and effective blockers of $GABA_A$ receptors, such as picrotoxin and bicuculline, both of which are derived from plants rather than microorganisms (Curtis, 1973).

Possible convulsant effects of an antibiotic become apparent in a clinical or experimental context only when sufficiently high concentrations are reached in the brain. How might this come about? First, an experimentalist might apply the drug directly to the brain, or a neurosurgeon might irrigate a wound in the presence of a dural defect (to reduce the risk of a postoperative wound infection), which allows the drug to reach the cortical surface; second, the antibiotic might be administered systemically in very high concentrations, or in lesser concentrations but in the presence of a failure of a patient to metabolize or excrete the drug, so that unexpectedly large amounts of drug build up in the body (in the case of penicillin, the important defect to consider is renal insufficiency, as penicillin is mostly excreted by the kidneys (Beyer, 1947)). Finally, the antibiotic might be administered systemically, but a defect in the blood–brain barrier allows higher concentrations to accrue in the brain than expected. One or all of these factors is at work in the following examples:

Walker et al. (1945) injected penicillin focally into the cortex of macaque monkeys, and observed myoclonic jerks of the contralateral limbs, turning of the head and eyes (with injection into the frontal eye fields), and twitching. Spread of movements to the ipsilateral side, and generalized convulsions then took place. Todd's paralysis, with weakness of contralateral limbs after the convulsion, could then occur. If the monkey were paralyzed pharmacologically, penicillin injection led to local EEG interictal spikes, and electrographic seizures; the latter consisted of rhythmic spikes, spike-wave complexes, and polyspike-wave complexes.

In 1963, Humphries described a focal motor seizure followed by a secondarily generalized convulsion, after application of 5 million units of crystalline penicillin to the brain of a 35-year-old patient at surgery; the patient was undergoing removal of a skull reticulosarcoma.

New and Wells (1965) described two patients with renal insufficiency who had been administered large amounts of penicillin, and developed a syndrome of increased deep tendon reflexes, myoclonic jerks and generalized convulsions. One patient, when not convulsing, also experienced hallucinations.

A group in Canada (Gloor, 1969; Seamans et al., 1968) described a series of patients undergoing cardiopulmonary bypass (postulated to alter the blood–brain barrier, because of microemboli) for heart valve replacements. These patients received large amounts (tens of millions of units) of penicillin as prophylaxis against bacterial endocarditis. The patients developed an encephalopathy with myoclonus, generalized seizures, and even status epilepticus that could be fatal. Providing extremely important observations, some of the patients had intraoperative EEG while they were pharmacologically paralyzed, and the presence of electrographic status epilepticus documented. Detailed EEG findings in these latter cases included interictal spikes and sharp waves, and so-called fast runs (EEG spikes at 10–25 Hz), in a pattern extremely similar to that observed by Steriade and colleagues in cats (see text that follows and later chapters).

Further pioneering experimental studies of penicillin epileptogenesis in rats and cats began in the 1960s, motivated by the monkey data of Walker et al., and by the clinical observations. We say "pioneering" because the experimental studies began to use intracellular recording, in conjunction with a means of inducing seizures acutely; the resulting data eventually led to a significant deepening of our understanding of the basic cellular events in at least the simpler types of seizure phenomena: Sawa et al. (1963) had introduced intracellular recording with electrically induced seizures, and Matsumoto and Ajmone Marsan (1964a,b,c) recorded cortical interictal spikes, induced by topical application of penicillin, using extracellular microelectrodes. Eli Goldensohn and Dominick Purpura, along with David Prince, however, provided the fundamental observation of the "depolarization shift"—a large intracellular depolarization lasting tens of ms—in neurons within the "spike focus" produced by local application of penicillin; and of the "inhibitory surround," that is, slower hyperpolarizations in neurons near to, but not within, the focus (Goldensohn & Purpura, 1963; Prince, 1968; Prince & Wilder, 1967). Marc Dichter and the late W. Alden Spencer then extended these observations to penicillin foci in the in vivo cat hippocampus (Dichter & Spencer, 1969a,b). Explaining the depolarization shift in simple, testable terms became a challenge (Traub & Wong, 1982).

Both Prince and colleagues (Fisher & Prince 1977a,b; Prince & Farrell, 1969), and Gloor and colleagues (Avoli & Gloor, 1981, 1982; Avoli et al., 1983; Giaretta et al., 1987; Quesney et al., 1977) also investigated the effects of penicillin administered systemically to cats. The drug consistently produced electrographic seizures, although not the generalized seizures corresponding to convulsive attacks that had been seen in humans, but rather spike-wave discharges, as might be associated with absence, were the cats unanesthetized. Gloor and his colleagues emphasized that phasic IPSPs were not completely blocked in this syndrome—which does not, of course, contradict a reduction by penicillin of "unitary" IPSCs (the synaptic currents in a pyramidal cell, induced by a single action potential in a single interneuron). Interneurons often fire intensely during epileptiform discharges (Domann et al., 1991), so that "compound" inhibitory potentials may be observable even when unitary inhibitory currents and potentials are small. Interpretation of the findings in generalized penicillin epilepsy is, at least for us, rather confusing. Further progress came from Steriade and colleagues, using a different experimental model (see later), but who were able to perform simultaneous intracellular recordings from multiple sites: cortex and cortex, or cortex and thalamus.

Spontaneous electrographic seizure activity in cats anesthetized with ketamine/xylazine. Mircea Steriade and colleagues made a major advance in understanding in vivo epileptogenesis, in a landmark series of papers published in 1998 (Steriade & Contreras, 1998; Steriade et al., 1998; Neckelmann et al., 1998; Timofeev et al., 1998). Before we can discuss that series, however, it is necessary first to introduce another major discovery from that laboratory, the slow oscillation of sleep (Steriade et al., 1993a,b; Timofeev et al., 2000) (a subject to which we shall return in Part III of this book). The slow oscillation was discovered first in anesthetized cats, but was quickly shown in unanesthetized cats with midbrain transsections, and later in naturally sleeping cats and humans (Steriade & Amzica, 1998). The slow oscillation is observed in EEG and ECoG (electrocorticographic) or EThG (electrothalamographic) recordings during anesthesia and slow-wave [non-dreaming, or non-REM (rapid eye movement)] sleep, at frequencies less than 1 Hz. The slow oscillation can arise in deafferented cortical slabs or in the presence of thalamic destruction: thus, this oscillation is intrinsic to the cortex [although other data indicate that diencephalic networks can also generate a slow oscillation, at least under certain conditions (Zhu et al., 2006)]. What is remarkable about it is that about 90% of cortical neurons, both principal excitatory neurons and inhibitory interneurons, develop, in approximate phase with one another, large depolarizations ("upstates," "activated states") and large hyperpolarizations ("downstates"), each on a time scale of hundreds of milliseconds to seconds. The difference in membrane potential between upstate and downstate can be greater than 20 mV. As the relative depolarization during an upstate is on the same scale as a paroxysmal depolarization shift (PDS) during an interictal spike, one may indeed ask: how can it be that these paroxysmal-looking upstates are not a form of epilepsy—and how can they be a normal phenomenon? It has been proposed (Destexhe et al., 2007) that the upstates are really fragments of the waking activated brain state; if that is so, then it is the downstates that constitute paroxysmal hyperpolarizations, rather than the upstates constituting paroxysmal depolarizations: epilepsy-in-reverse, or antiepilepsy, one might suppose.

Be that as it may, the Steriade group's 1998 series of papers (and subsequent publications, discussed later in this book) showed that in the anesthetized cat, treated with bicuculline or not, electrographic seizure discharges developed out of upstates of the slow oscillation—perhaps not completely surprising, given that it is during the upstates that cortical neurons are depolarized. [The most frequently used anesthesia was combined ketamine and xylazine. Ketamine is a blocker of N-methyl-D-aspartate (NMDA) receptors, but also μ opioid receptors (Villars et al., 2004), while xylazine, a blocker of $α_2$-noradrenergic receptors (Docherty & Starke, 1981) has long been used by veterinarians for sedation. With ketamine/xylazine, many of the cats would develop—for somewhat unclear reasons—spontaneous seizures, without superaddition of other drugs.]

The 1998 Steriade series of papers demonstrated a range of electrographic seizure types, as seen in the Lennox–Gastaut syndrome: spike-wave, polyspike-wave, and fast runs (EEG spikes at 10–25 Hz). They convincingly showed that all of these types of discharge arose in cortex, and did not require thalamic input; such observations do not rule out, however, the notion that interconnected thalamus and cortex will generate somewhat modified patterns as compared to the disconnected state. Rather, the data suggest that one should focus one's attention on cortical abnormalities, to best understand how seizures start. The same group later demonstrated that spontaneous seizures in ketamine/xylazine anesthetized felines began not only during upstates, but specifically in the presence of a very fast rhythm, at approximately 100 Hz (see later).

To summarize: for thousands of years, physicians were preoccupied with the observations of patients and cats, and were trying to determine, based on intuition and without understanding brain mechanisms, what should be called epilepsy and what should not (Eadie & Bladin, 2001). Advances in correlated neurological observation and neuropathology, in basic neurophysiology, and in EEG began to change that and open up new possibilities for progress. In vivo models of epilepsy have taken us well beyond observations of the phenomenology of epilepsy, clinical and electrical, to consideration of structural rearrangements in brain circuits that may contribute to at least some forms of the human disease. Progress in understanding detailed cellular mechanisms of epilepsy has come, however, mostly from in vitro studies (at least in our opinion); we discuss next, at least in a preliminary way, some of the in vitro studies. The reader is also referred to our earlier monographs (Traub & Miles, 1991; Traub et al., 1999a) for a more extensive discussion.

Figure 4.1 shows the fine structure of a brief electrographic seizure in vitro, and in a network model. The experimental data were recorded from a ventral hippocampal slice from a rat (CA3 region), and the epileptiform event was evoked by a shock: thus, we do not have the opportunity to determine if the event is preceded by very fast oscillations (>80 Hz), as so often occurs with spontaneous electrographic seizures (see later). The slice was bathed in bicuculline, to block $GABA_A$ receptors—"disinhibition"—a condition that probably does not occur in vivo, by itself, but which may be approximated in special conditions that we shall consider in later chapters. We have divided the electrographic seizure event into three stages: 1°, 2°, and 3°. The 1° burst

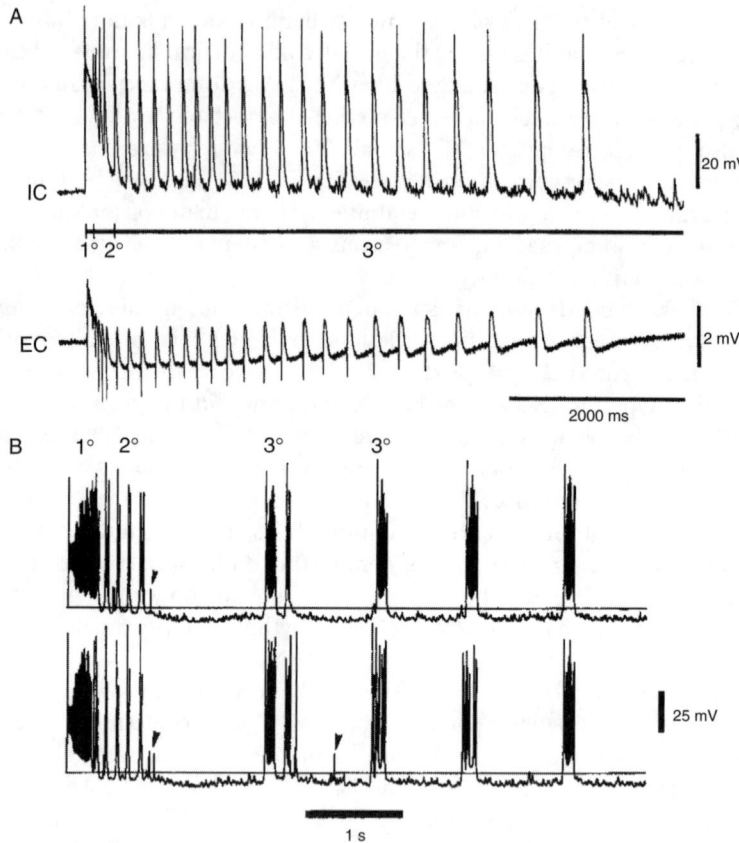

Figure 4.1   Brief electrographic seizure in hippocampal slice and simulation.
A: Data from a rat ventral hippocampal slice, bathed in 10 μM bicuculline to block
$GABA_A$ receptors, with a shock to the perforant path. *IC* is intracellular recording
from a presumed CA3 pyramidal cell, and *EC* is extracellular recording: a local EEG.
The large amplitude extracellular potentials, and the close phase-locking between the
cell and the field, indicate that the population is discharging in synchronous fashion,
characteristic of an epileptiform event. Note the large depolarizations in the neuron,
>20 mV, defining paroxysmal depolarization shifts (PDSs). This brief electrographic
seizure is divided into three phases: a primary (1°) burst, a brief series of secondary
(2°) bursts, and a slower series of tertiary (3°) bursts. The intracellular tertiary bursts
are separated by intervals of synaptic noise. **B**: Corresponding events in a model of
192 CA3 pyramidal cells, interconnected by excitatory chemical synapses, containing
AMPA and desensitizing NMDA receptors, with noise provided by spontaneous
axonal action potentials ("ectopic spikes"). The arrowhead points to a spikelet,
reflecting decremental antidromic conduction of an ectopic spike.
(Reproduced from Traub, Borck et al., 1996 with permission.)

is wide; if it occurred by itself in vivo, it would be designated an interictal spike or sharp wave. This burst shows the characteristic paroxysmal depolarizing shift. Next, there is a brief series of rapid secondary bursts (Miles et al., 1984), which can be missed without careful inspection. Finally, there is a series of 3° bursts, the seizure proper, the portion that could have behavioral consequences if this were going on in the brain of an awake individual. From the cellular data, it is easy for the reader to appreciate why there might be behavioral consequences: the portion of the brain engaged in the illustrated activities is not processing its afferent signals, but is instead completely occupied with an autonomous caricature of brain operations. It is a caricature for two reasons: the firing pattern exhibited by any single neuron is never observed normally (or so it is believed); and the correlations between neuronal firings are extreme—"hypersynchrony".

Our understanding of how the electrographic seizure comes about is summarized in the simulation of Figure 4.1B, which incorporates ideas developed over about 15 years (1981–1996). The simulation—of the behavior of 192 model CA3 pyramidal cells, interconnected by recurrent excitatory chemical synapses—shows two different cells rather than one cell and the field, but the idea is the same: there is synchrony between cells, and the structure of the event is similar. Let us consider the structural features of this model, a prerequisite to understanding the collective behavior:

• First, the individual model cells: these have multiple pieces (compartments) corresponding to sections of dendrites, soma, and axon (Traub et al., 1994b). Membrane channels are distributed nonuniformly, with faster Na^+ channels concentrated in the axon and soma, and slower Ca^{2+} channels in the dendrites (Wong & Prince, 1981). The interplay between these channels, which have different thresholds, kinetics, and inactivation properties, has these features: fast spikes develop in the proximal axon, and can occur through dendritic/somatic depolarization ("orthodromic"), or through voltage fluctuations in the more distal axon ("antidromic" ectopic spikes); in contrast, slower Ca^{2+}-mediated action potentials can arise in the apical dendrites, provided the membrane is not shunted by intrinsic currents or synaptic inhibition (Miles et al., 1996; Traub et al., 1994b). Conditions that favor a dendritic Ca^{2+}-mediated action potential are a strong depolarizing current pulse, or a large sustained dendritic depolarization; in the latter case, a series of rhythmic Ca^{2+}-mediated action potentials can occur (in a pattern observed in many other types of large neurons, including certain neocortical pyramidal cells, and cerebellar Purkinje cells). Axosomatic and dendritic regions interact in what has been called a ping-pong effect (Pinsky & Rinzel, 1994): somatic spikes "backpropagate" into the dendrites and favor dendritic calcium spikes; while a dendritic calcium spike, because it is relatively prolonged, can depolarize the soma and so lead to extra axosomatc spikes—and thus produce a burst of action potentials, a so-called "intrinsic burst". (The persistent, slowly inactivating Na$^+$ current can also contribute to extra axosomatic spikes and bursting; we will return to this in later chapters.) In real pyramidal neurons, the interplay between soma and dendrites is regulated by a number of dendritic channels, such as the "A" type of transient K$^+$ current (Hoffman et al., 1997), and the anomalous rectifier, or h-current (Poolos et al., 2002); these more detailed factors were not included in the model of Figure 4.1B. Bursts are terminated by outward K$^+$ currents, both voltage- and Ca^{2+}-activated, and possibly by inactivation of Ca^{2+} currents. One of the Ca^{2+}-activated outward currents, the AHP (afterhyperpolarization) current lasts hundreds of milliseconds to seconds.

• CA3 pyramidal cells have recurrent collaterals with one another, activating α-amino-3-hydroxyl-5-methyl-4-isoxazole-propionate (AMPA)/ kainate and NMDA types of glutamate receptors. The connections are powerful enough that a burst of action potentials in one pyramidal cell can evoke a burst of action potentials in a monosynaptically connected different pyramidal cell, with a latency of tens of milliseconds; further, each CA3 pyramidal cell contacts more than one other pyramidal cell (Miles & Wong, 1983, 1986, 1987a). Thus, the conditions are right for a chain reaction of bursting to occur once a single pyramidal cell fires a burst, provided synaptic inhibition does not interfere (Miles & Wong, 1987a; Traub et al., 1987).

• The NMDA component of synaptic excitation has three relevant properties: it is enhanced by depolarization, it lasts hundreds of ms, and it desensitizes (Clark et al., 1990; Forsythe & Westbrook, 1988; Jahr & Stevens, 1990a, 1990b; Perouansky & Yaari, 1993).

• The membrane potential between 3º bursts is dominated by membrane noise (Fig. 4.1A; see also Borck & Jefferys, 1999), suggesting spontaneous firing in axons or presynaptic terminals. We simulated this with a high rate of spontaneous ectopic spikes (8 Hz per axon). [As we shall see, if allowance is made for electrical coupling between axons, the spontaneous ectopic rate, per axon, can be made far lower and still lead to comparably high rates of axonal network activity.]

• One factor proposed to limit the rate of burst firing in epilepsy is depletion of glutamate from presynaptic terminal stores (Bains et al., 1999 Staley et al., 1998). We did not allow for that possibility here (but see later chapters); instead, the 3° burst rate is determined by intrinsic conductances and the excitatory synaptic noise.

We can now put the pieces together, and use the simulation to offer a (partial) reconstruction of how the electrographic seizure in Figure 4.1A came about (Traub et al., 1996a): the stimulus causes some of the pyramidal cells to fire, and this firing spreads along the recurrent collaterals (the details of how this happens are shown in Fig. 4.2); this produces the primary burst. During the height of the PDS, all of the pyramidal cells are firing, the presynaptic terminals are releasing glutamate that binds to AMPA/kainate and NMDA

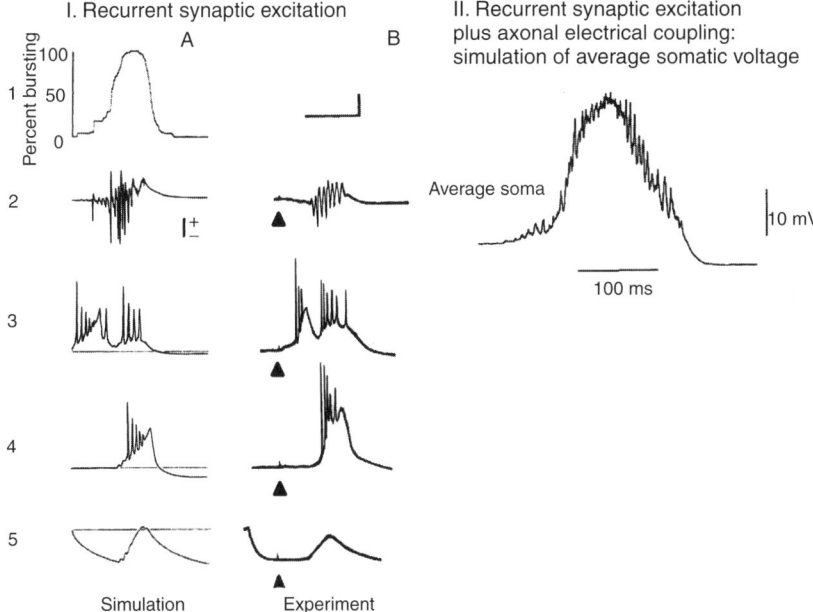

Figure 4.2 Synchronizing mechanisms underlying the interictal spike and superimposed very fast oscillation. Recurrent chemical synaptic excitation can explain the ability of firing in a small number of neurons to recruit a large population, after a characteristic latent interval. It does not; however, explain the high-frequency signal in the epileptiform field potential, a signal that *can* be accounted for by electrical coupling. I: simulation of a network of 100 intrinsically bursting neurons, with recurrent synaptic connections having two properties: (**A**) Each neuron contacted an average of five others; (**B**) when a single presynaptic neuron fired a burst, it would induce bursting in a connected neuron with latency about 10 ms (a property confirmed in later experiments of Miles & Wong, 1986, 1987). There was no synaptic inhibition, corresponding to the experimental conditions. When four cells were stimulated, population activity increased rapidly over a period of tens of ms (line 1), leading to an epileptiform average signal (line 2). Cells receiving the initial stimulus (e.g., line 3) burst immediately, and then a second time, because of synaptic excitation received from the major population activity. Most cells fired a burst after a latent interval (e.g., line 4), corresponding to the growth of activity in the population. Hyperpolarizing such a neuron (line 5) uncovered the giant EPSP induced by the synchronized bursting of most of the population. Similar phenomena were observed in the disinhibited (with penicillin) hippocampal CA2 region in vitro (right column) after a local shock *except* for the obvious high-frequency oscillation during the epileptiform field potential. [Calibrations: 50 ms (simulation), 60 ms (experiment); 4 mV (A2, B2); 25 mV (A3, A4, A5); 20 mV (B3, B4, B5).] II. A later simulation (of a larger network), using both recurrent synaptic excitation, and electrical coupling between axons, shows a high-frequency oscillation superimposed on the average activity. The high-frequency oscillation was not present without the electrical coupling (see original paper).

[Reproduced from Traub & Wong, 1982 (I), and Traub et al., 1999 (II) with permission.]

receptors, and dendrites are depolarized. Hence, large, slow NMDA currents flow into the dendrites, causing trains of dendritic bursts, a consequence of the intrinsic properties of the pyramidal cells, and these dendritic bursts lead to axosomatic bursts of action potentials: the secondary bursts (Traub et al., 1993); the secondary bursts are at least partially synchronized throughout the population because of phasic coupling between the neurons provided by AMPA receptors. As NMDA receptors desensitize and NMDA currents decrease, and as AHP currents increase, the dendrites start to repolarize, dendritic bursts cease, and the 2° bursts terminate. Now begins the phase of synaptic noise, facilitated by extracellular $[K^+]$ increase, and $[Ca^{2+}]$ decrease, caused by the 1° and 2° bursts. The 3° bursts are each recapitulations of the 1° burst, with the synaptic noise and antidromic spikes acting as triggers, and the interburst interval regulated by AHP currents. This scheme predicts that pharmacological blockade of NMDA receptors should lead to narrowing of the 1° and 3° bursts, and disappearance of the 2° bursts; and this is observed experimentally (Traub et al., 1996a).

Figure 4.2-I makes clear how activity spreads through a population of intrinsically bursting neurons, interconnected by sufficiently many recurrent excitatory connections (only AMPA/kainate receptors were used in the simulation, column A, not NMDA receptors); details are in the figure legend and the original publication (Traub & Wong, 1982). This model explains not only the firing patterns, but also the large excitatory synaptic currents that develop during a PDS (Johnston & Brown, 1981). Note, however, a matter that appears, superficially, to be a minor detail—but which will turn out to be important enough: the experimental "epileptiform field potential" (Fig. 4.2-IB, line 2) contains a high-frequency oscillation superimposed on it, that has been compared to a comb. The comb is not apparent in the corresponding simulation (Fig. 4.2IA, line 2). The reason is that the simulation did not include electrical coupling through gap junctions. In subsequent simulations, e.g., Figure 4.2-II, in which electrical coupling was included—between axons—along with recurrent excitatory chemical synapses, one finds that the epileptiform event contained two components: a slower one, tens to hundreds of milliseconds, and also the superimposed high-frequency oscillation. Why is this important? First, it implies that if axonal coupling exists, then action potentials can spread from neuron to neuron by two complimentary pathways: the recurrent excitatory synapses, as described above, and also gap junctions. Second, the occurrence of two types of activity-spread in turn implies that synchronization and epileptiform discharges might develop with weaker than expected chemical excitatory synapses, or even with some degree of synaptic inhibition present, because of there being two distinct and alternative pathways for pyramidal neurons to excite one another. Finally, if electrical coupling between axons were to exist, it might announce its presence in the form of high-frequency oscillations preceding the epileptiform discharge, as well as superimposed on it. The next figures provide examples of such high-frequency oscillations

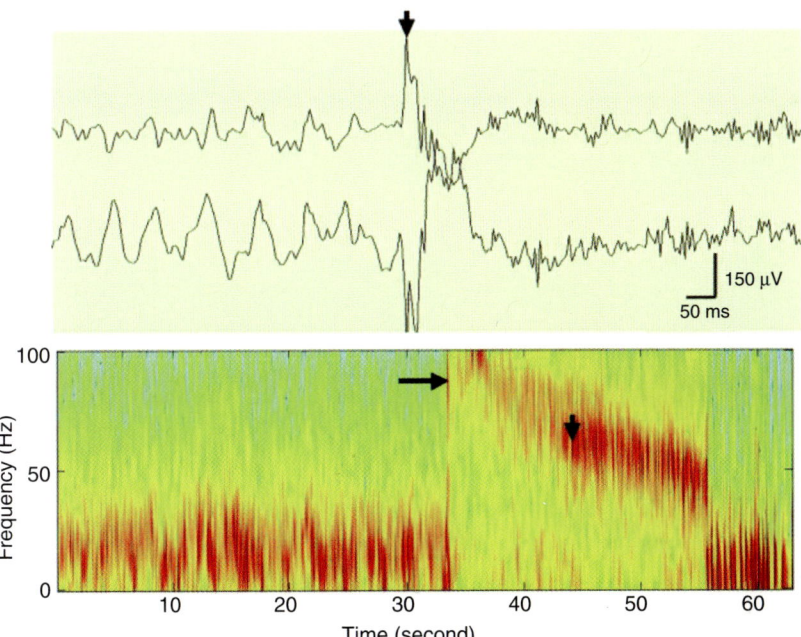

Figure 4.4 Prolonged VFO prior to an electrographic seizure. This patient had seizures of frontal lobe onset. The upper electrocorticographic traces show VFO in association with an EEG spike: Compare Figs. 4.2, 4.3, and 4.6. The color-coded time-frequency plot below shows fast activity, beginning at ~90 Hz (horizontal arrow), that lasts about 15 seconds up to the first evident motor activity (vertical arrow).

(Reproduced from Worrell et al., 2004 with permission.)

["very fast oscillations" (VFO)]. How VFO comes about is a topic that must be deferred until Chapter 10. Figures 4.3 to 4.8 illustrate actual EEG and ECoG signals from human patients.

Allen et al. (1992) and Fisher et al. (1992) described brain oscillations at greater than 80 Hz before seizures of frontal lobe and temporal lobe onset, respectively. Even higher frequency oscillations, up to 500 Hz, have been observed before seizures in human epilepsy of temporal lobe onset (Bragin et al., 1999a,b); beta (10–30 Hz) and even faster oscillations, up to about 70 Hz, have also been observed before seizures in children, including those with periodic spasms and infantile spasms (Akiyama et al., 2005; Gobbi et al., 1987; Kobayashi et al., 2004; Nealis & Duffy, 1978). A group at Great Ormond Street Children's Hospital in London reported on a series of 11 children with focal cortical dysplasias and intractable seizures, investigated with ECoG recordings using subdural grid electrodes, and reported by Traub et al. (2001c). VFO before a frontal lobe seizure in one of the patients is shown in Figure 4.3A and B. Several details are of note: the first striking event is not the VFO itself, but rather a burst of activity lasting about 2 seconds, in some of the electrodes, for example, G08; next, VFO occurs that is of lower amplitude, and is more spatially restricted than the electrographic seizure that follows. Figure 4.3C shows VFO superimposed on an interictal spike recorded from this patient (compare with Fig. 4.2): the superimposition of VFO on a synchronized burst, then, is a general phenomenon, and not one restricted to experimental epileptic events in the in vitro hippocampus.

In recent years, a number of studies have emerged reporting VFO before, and in association with, seizures in humans (Akiyama et al., 2006; Jacobs et al., 2008; Khosravani et al., 2008; Staba et al., 2002; Worrell et al., 2004). It is possible that all seizures begin with VFO, but this is hard to prove. The reason is that VFO generation, at least in experimental animals after kainate-induced lesions, can be exquisitely localized (Bragin et al., 2002a), even to volumes of the order of 1 mm^3. Interictal bursts and sharp waves, in vivo and in vitro, can also be observed in similar small, or even tinier, volumes of brain tissue (Köhling et al., 1998; Schevon et al., 2008). For comparison, subdural grid electrodes are typically about 1 cm apart. Thus, a failure to observe VFO before seizure may reflect use of too gross an electrode in trying to detect too fine and localized a signal. In addition, however, runs of VFO can occur in human data without the immediate onset of a seizure (Worrell et al., 2004); the association between VFO and seizure is not all-or-none.

The example in Figure 4.3 shows VFO continuing for a few seconds before a seizure. There are cases, however, such as illustrated in Figure 4.4, in which VFO proceeds for over ten seconds, although the frequency progressively declines. Such observations may offer clues to pathogenesis: VFO can be a "stable state" so to speak, persisting over a longer time scale than do synaptic currents, at least the usual ones.

Our next example is an experimental one, from the ketamine/xylazine anesthetized cat discussed previously. Recall that seizures in this model begin

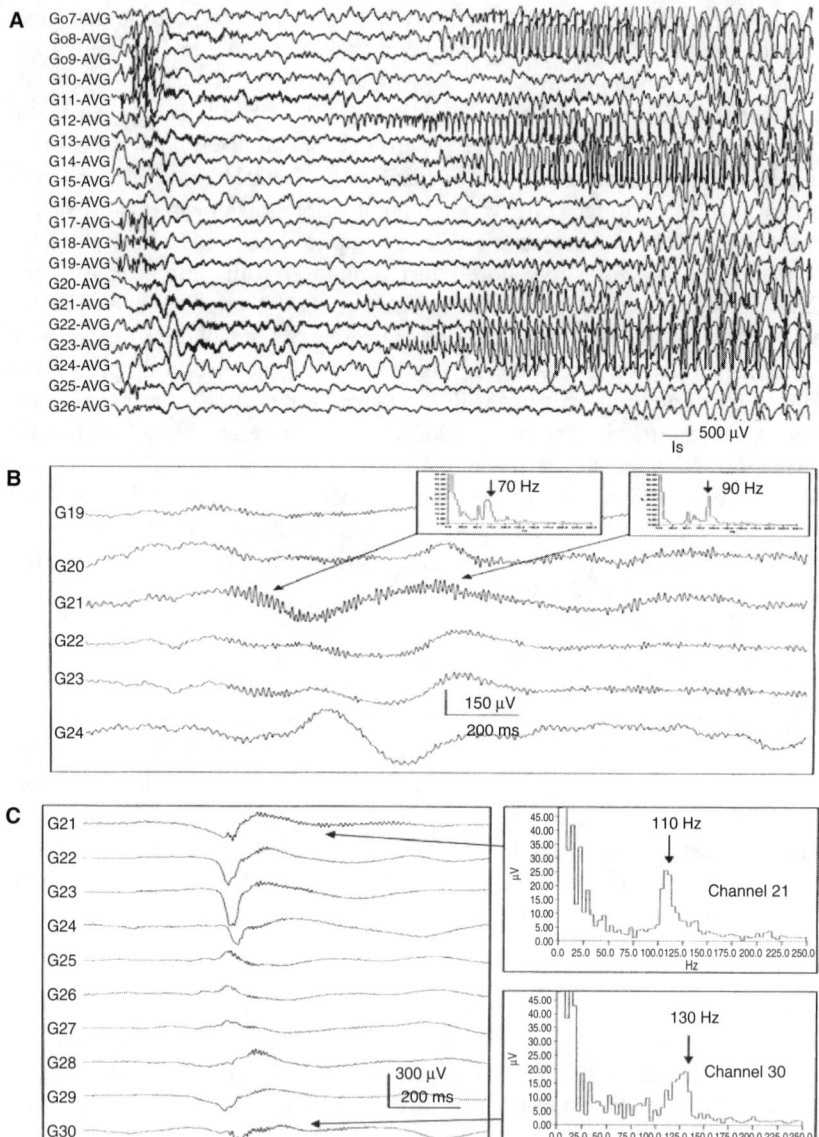

Figure 4.3 Brief, relatively localized, VFO leading into a more extensive electrographic seizure, in a child with a subcortical dysplasia. A focal seizure in a child is preceded by localized very fast oscillatory activity. The patient, age 13 months, had a right frontal subcortical dysplasia, subsequently removed surgically. EEG activity was recorded with a subdural grid of electrodes. **A:** Electrographic seizure preceded by low-amplitude very fast activity, restricted to a few recording sites (including G11, G13, G21–G23). **B:** Using a different recording technique, to give better signal:noise ratio, shows the very fast activity (preceding a seizure) to have frequency components of 70– 90 Hz. **C:** An interictal burst, recorded from the same patient, contains superimposed very fast oscillations at 110–130 Hz: compare with the in vitro and in vivo interictal spikes in Figs. 4.2, 4.4, and 4.6. Signals recorded and analyzed by T. Baldeweg, H. Cross, and S. Boyd.

(Reproduced from Traub et al., 2001 with permission.)

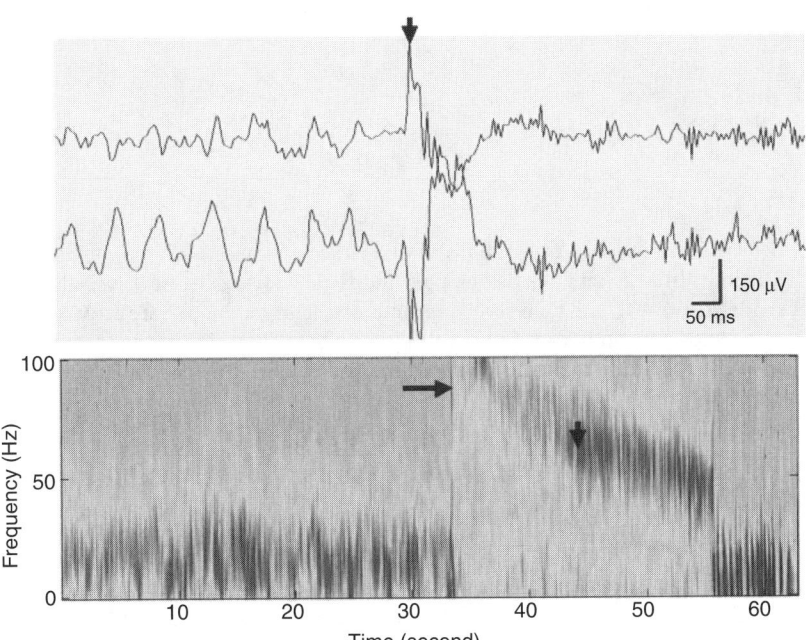

Figure 4.4 Prolonged VFO prior to an electrographic seizure. This patient had seizures of frontal lobe onset. The upper electrocorticographic traces show VFO in association with an EEG spike: Compare Figs. 4.2, 4.3, and 4.6. The color-coded time-frequency plot below shows fast activity, beginning at ~90 Hz (horizontal arrow), that lasts about 15 seconds up to the first evident motor activity (vertical arrow). Please see color insert.

(Reproduced from Worrell et al., 2004 with permission.)

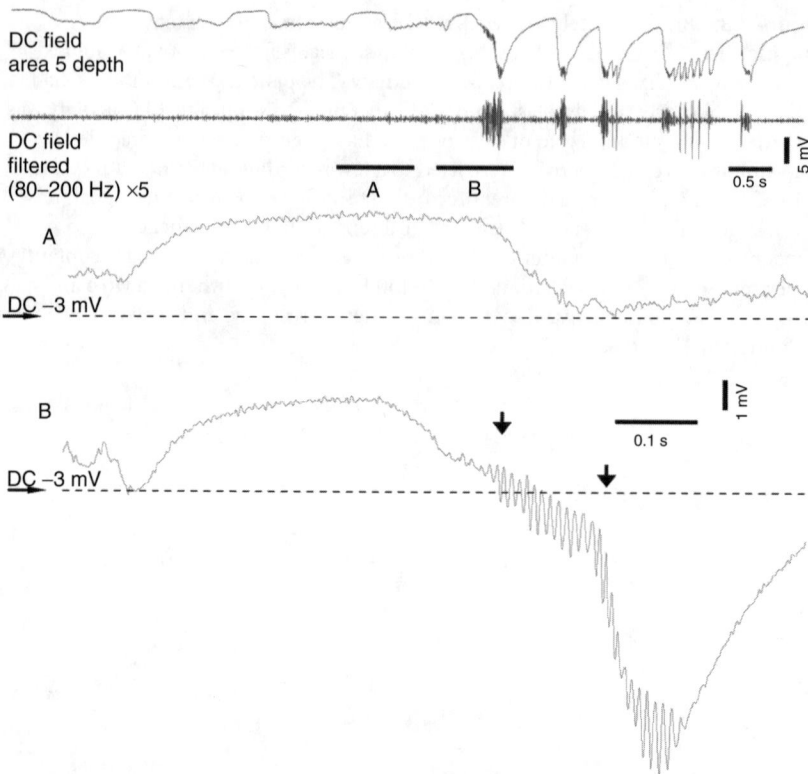

Figure 4.5 Transient VFO (~100 ms) leading into a so-called EEG spike. Upper traces show intracortical EEG, beginning with several waves of the slow oscillation (Steriade et al. 1993), and then several seconds of electrographic seizure activity. The expanded portion **A** shows a wave of the slow oscillation. The expanded portion **B** shows the beginning of the slow wave transforming into VFO that continues into the first EEG "spike."

(Reproduced from Grenier et al., 2003 with permission.)

from up-states. Grenier et al. (2001) documented that VFO could occur, at low amplitude, superimposed on upstates, as could gamma and beta oscillations (Steriade et al., 1995, 1996). As Figure 4.5 demonstrates, immediately before a seizure, and leading into it, VFO occurs during the upstate at increased amplitude. The VFO/seizure transition is more abrupt than in the human, but we presume—until data contradict us—that the underlying mechanisms are similar. Yet another example of a VFO/electrographic seizure combination, this one in vitro, can be found in Traub et al. (2001c), using a tetanic stimulation protocol in a hippocampal slice that has been alkalinized with trimethylamine (TMA), to open gap junctions (Spray et al., 1981). Some of those data are illustrated in Chapter 13.

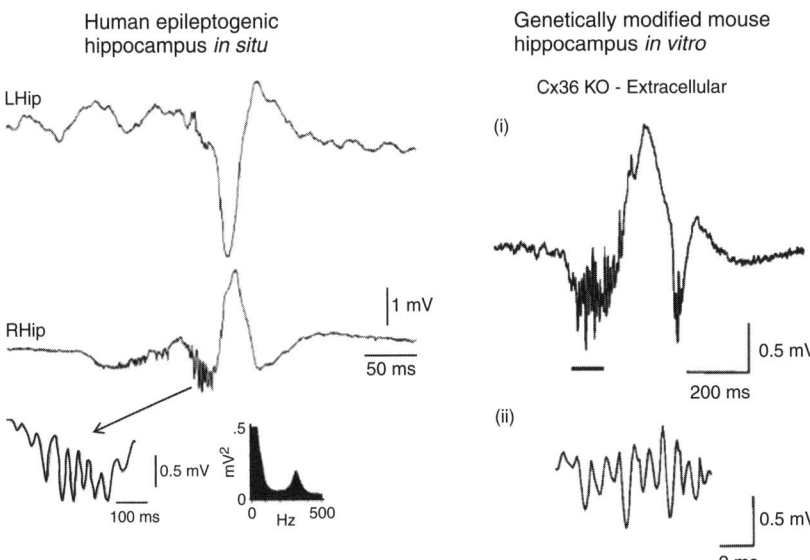

Figure 4.6 Examples from a patient, and from an experimental epilepsy model in vitro, of the association between VFO and so-called interictal spikes. The in vivo data (left) were recorded with depth electrode microwires in the left and right hippocampus; seizures began in the right hippocampus. A sharp wave is shown that was simultaneous in both hippocampi, but which was preceded by VFO (peak ~320 Hz) only on the right. The inset (lower left) shows low-pass-filtered (<600 Hz) VFO. The in vitro data (right) were recorded in a hippocampal slice, bathed in kainate, from a connexin-36 knockout mouse developed by Hannah Monyer; this tissue also generates gamma oscillations. Similar gamma/epileptiform phenomenology is described in Chapter 13, using a different experimental model.

(Patient recording reproduced from Staba et al., 2004b with permission; in vitro data reproduced from Pais et al., 2003 with permission.)

The epileptiform activity that may follow a short run of VFO can be more limited than the electrographic seizure shown in Figure 4.5: a brief run of VFO may simply be followed by an interictal spike. Figure 4.6 (left) shows an example in a human patient, and (right) from a hippocampal slice bathed in nanomolar kainate (to excite the cells without killing them). The slice was taken from a mouse in which connexin-36 was knocked out. Connexin-36 is one of the major mammalian neuronal gap junction proteins; as we discuss in a later chapter, connexin-36 is built into gap junctions that couple interneurons, but also contributes to the coupling of principal neurons, between axons and at mixed synapses. That VFO occurs in a connexin-36 knockout suggests that other gap junction proteins besides connexin-36 can also act to couple principal neurons, at least as a compensation.

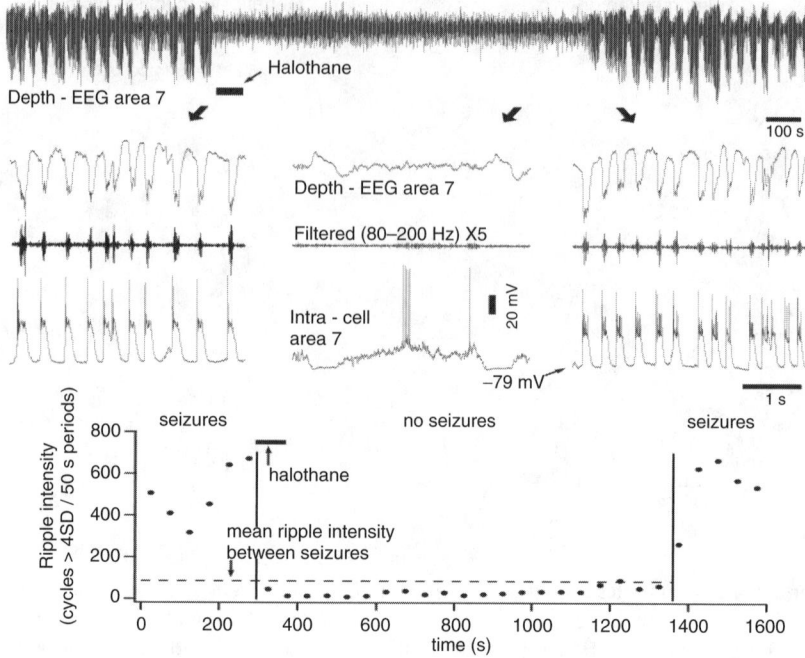

Figure 4.7 Further evidence that VFO is associated with seizures in vivo, and that VFO generation requires electrical coupling via gap junctions. The uppermost traces shows, on a very slow time scale, EEG seizure activity, followed by the reversible suppression of this seizure activity by halothane (a gap-junction-blocking anesthetic). The middle traces show, on a shorter time scale, EEG and intracellular records (in a regular-spiking, RS, neuron) from pre-halothane and post-halothane portions of the data (note the repeating epileptiform bursts), along with corresponding records from the halothane interval (without epileptiform bursts). The graph below quantitates the suppression of VFO ("ripples") during the time when seizures are suppressed. (Reproduced from Grenier et al., 2003 with permission.)

We now want to further strengthen the concept that gap junctions contribute to VFO, as well as to seizures. Figure 4.7 illustrates electrographic seizure activity in a ketamine/xylazine anesthetized cat. When halothane was added to the inhaled gases in this paralyzed ventilated animal, both seizure activity and very fast oscillations were suppressed [halothane is a gap junction blocker (Peracchia, 1991)]. That halothane was not simply suppressing all neuronal activity is proven by the control intracellular recording, which shows normal-amplitude action potentials—but no PDS—in the presence of halothane.

The data shown here, then, lead us to one of our major hypotheses: that electrical coupling, via gap junctions between principal neurons, is responsible for VFO that occurs prior to and during seizures, and may even be involved in producing seizures themselves. And by an extension of this idea, such

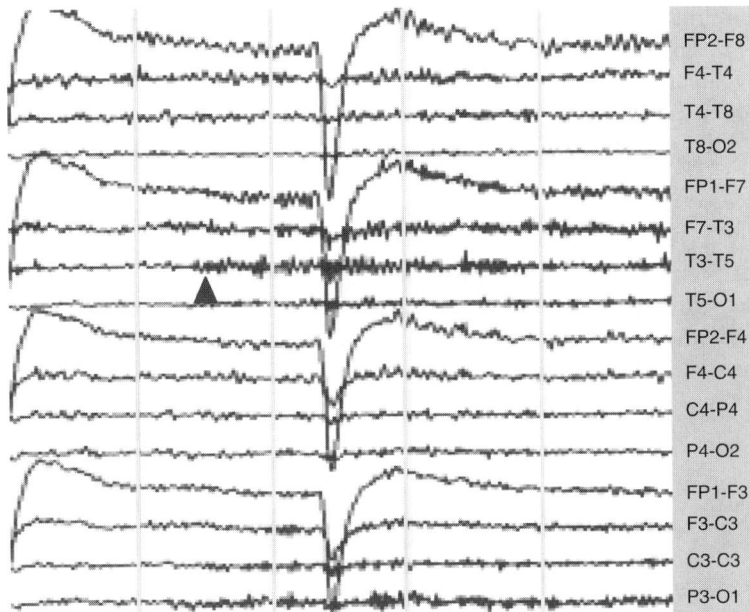

FP2-F8
F4-T4
T4-T8
T8-O2
FP1-F7
F7-T3
T3-T5
T5-O1
FP2-F4
F4-C4
C4-P4
P4-O2
FP1-F3
F3-C3
C3-C3
P3-O1

Figure 4.8   Example of beta-frequency EEG oscillation occurring during status epilepticus. The EEG shows unilateral activity at 16–20 Hz. It is possible (but can not be confirmed) that the EEG activity corresponds to fast runs in populations of cortical neurons. The runs of beta correlated clinically with unresponsiveness, blinking, and gaze deviation. The very large slower potentials are blink artifacts. (Reproduced from Bonati et al., 2006 with permission.)

coupling may contribute to other types of central nervous system disease pathogenesis. Much of the remainder of this book will be devoted to examination of this hypothesis and its consequences.

Finally, we wish to remind the reader that other sorts of fast oscillations, besides VFO, can be harbingers of, or participate in, epileptic seizures. We have cited some instances above, but Figure 4.8 provides a graphic example of a beta oscillation. Unfortunately, the cellular details in this human case are not accessible, only the EEG. Be that as it may, we will need to examine a whole range of oscillations besides VFO if we are to understand epilepsy. Several of the types of oscillation that we consider later are at beta or gamma frequencies—and also depend on gap junctions.

To conclude, and offer perspective, we must note that still approximately one-third of epilepsy patients have seizures that are considered medically intractable (Moran et al., 2004; Sillanpää et al., 1998; Tellez-Zenteno et al., 2004), and only some of these patients are surgical candidates (Lüders, 2008). However much progress there has been, and however proud we are of all that has been achieved and that is racing onwards in the clinical and research communities at present, we alas have a long rough road to travel. Some of the outstanding questions are these: do our experimental models capture all of the cellular phenomenology that exists in human epilepsies (e.g., infantile spasms); how is it that particular structural lesions or molecular alterations in the brain predispose to seizures in the way they actually do, as opposed to the ways that we understand from our models; and, in any given patient, why does a seizure occur at time X and not at time Y? And most importantly, what can we do to prevent epilepsy from occurring at all, once we have identified a predisposing abnormality? And, if complete prevention is not feasible, are there pharmacological approaches that might be developed that have fewer side effects than methods now available, either on brain functions, or on other organ systems (liver, skin, bones, etc.)?

5

Parkinson's Disease

The defining description of Parkinson's disease was published by James Parkinson in 1817 (Parkinson, 1817); the disease was named after its original describer ("malade de Parkinson") by the pioneering French neurologist Jean-Martin Charcot (1825–1893), also known for the original clinicopathological descriptions of multiple sclerosis and amyotrophic lateral sclerosis (ALS), among many other accomplishments. The disease was also called "paralysis agitans," although that term is little used today. James Parkinson's description consisted only of clinical observation, without neuropathology. The characteristic neuropathology was not worked out until the 1930s, and molecular, genetic, and etiological aspects of the disease still remain under active investigation. Perhaps the most critical discovery about Parkinson's disease concerned elucidation of the role of dopamine deficiency in causing many, but not all (Ahlskog, 2007), of the signs and symptoms; this discovery led to a range of related pharmacological treatments that are, on the one hand, close to miraculous; and, on the other hand, ultimately frustrating, because of disease progression, failure to maintain desired clinical actions, and the emergence of troubling side effects. Breakthroughs in understanding at least some of the neuronal network physiology came (1) with the development of experimental models of the disease, in part based on the recognition that a clinically similar syndrome could be induced in humans via ingestion of a toxin—as happened both during a quickly aborted clinical trial, and accidentally; and (2) with electrophysiological recordings from human patients, taken through electrodes as part of therapeutic procedures—either ablative or through an implanted device intended for long-term "deep brain stimulation" (DBS).

Once the neuropathology of Parkinson's disease was understood, and with recognition of other distinct diseases affecting the basal ganglia, and with developments in pharmacological treatments for hypertension and psychiatric disorders, it became clear that at least some of the signs and symptoms of Parkinson's disease could occur without the Parkinson-characteristic neuropathology: for example, the diseases striatonigral degeneration and supranuclear palsy share clinical features with Parkinson's disease; and a patient taking the antipsychotic medication haloperidol may also develop clinical features of Parkinson's disease once the dose is large enough. These latter instances are now designated "Parkinsonism," a clinical constellation of signs and symptoms, with "Parkinson's disease" referring to a characteristic clinical syndrome with a likewise characteristic underlying neuropathology; however, the pathology cannot be unequivocally confirmed before the death of the patient, so that to apply the diagnosis to a living patient requires the clinical exclusion of other conditions known to cause Parkinsonism.

The most characteristic clinical features of Parkinson's disease (Thomas & Beal, 2007) include a rest tremor (that may initially be voluntarily suppressible, and that tends to diminish with intentional movement), muscle rigidity, slowing of movement ("bradykinesia"—in its extreme form, "akinesia"), and a characteristically impaired gait. Later, postural instability, freezing, and falls can constitute major problems. Numerous other features have also been described. With the immobility eventually comes confinement to chair and then bed, as well as difficulty swallowing and clearing secretions, with the medical complications associated therewith. Some patients develop dementia; a related disease, "Lewy body dementia," combines Parkinsonism with dementing and psychotic features, with some unusual behavioral syndromes, such as the REM sleep behavior disorder (Boeve et al., 2001).

The pathology of Parkinson's disease is reviewed in the classical texts, such as Greenfield's *Neuropathology* (Blackwood et al., 1967). Perhaps the cardinal feature, as clearly described by Hassler (1938), is loss of pigmented (melanin-containing) neurons in the pars compacta of the substantia nigra, in the midbrain; it was not recognized until much later that these neurons are dopaminergic, and that there is a fundamental "neurochemical" aspect to the disease that distinguishes it from most other neurological diseases (see later). Within the surviving pars compacta nigral neurons are distinctive hyaline (glassy) inclusions called Lewy bodies, giving rise to the designation "Lewy body disease" sometimes applied to Parkinson's disease. These inclusions were described by Frederic Heinrich Lewy (1885–1950) in Germany, in 1913; Lewy did not, however, recognize the connection between his morphological discovery and Parkinson's disease; that recognition came with the work of Trétiakoff (1919) and Foix and Nicolesco (1925). Chemically, Lewy bodies are now known to consist mainly of modified α-synuclein (a protein so-named because it is found in preSYNaptic terminals and NUCLEar membranes; Jellinger, 2003); but Lewy bodies also contain the protein ubiquitin and other constituents.

With regard to the pathology of Parkinson's disease, it is essential to keep in mind some complicating factors:

1. Neuronal loss is not confined to pars compacta of substantia nigra, nor is it confined to dopaminergic neurons; and indeed, the substantia nigra is not even the initial site of degenerative changes (Blackwood et al., 1967; Braak et al., 2006; Wolters & Braak, 2006). There is cell loss amongst the pigmented noradrenergic neurons of the locus coeruleus and other parts of the brain stem, in anterior olfactory structures, autonomic neurons, and eventually in the globus pallidus and cerebral cortex.

2. Lewy bodies are not restricted to substantia nigra in Parkinson's disease, and further are found in diseases other than Parkinson's disease (Jellinger, 2003; Pearce et al., 1995). Lewy bodies occur, for example, in the olfactory bulb in Parkinson's disease; and they are found in "dementia with Lewy bodies," multiple system atrophy, and Hallervorden–Spatz disease.

The history of how dopamine came to be recognized as a transmitter in the brain, and the importance of dopamine deficiency in Parkinson's disease—as well as the possibility of improving symptoms by intravenous or oral administration of the dopamine precursor L-3,4-dihydroxyphenylalanine (L-dopa)—this history is engagingly recounted in a recent paper by a major participant, the Austrian pharmacologist Oleh Hornykiewicz (born 1926; Hornykiewicz, 2006). The following timeline is extracted from that paper, which contains the appropriate references:

1957: Dopamine was discovered in the brain.

1958: Arvid Carlsson (born 1923, Nobel laureate 2000) et al. noted that reserpine depletes brain noradrenaline, serotonin and dopamine; and administration of L-dopa (which is metabolized to dopamine in the brain) replenishes dopamine. Reserpine, derived from a plant (genus *Rauwolfia*), was used at that time to treat hypertension (the drug is still used, but far less), and Parkinsonism could occur as a side effect. In retrospect, but not in 1958, this association can be interpreted as linking Parkinsonism with brain dopamine deficiency.

1959: Dopamine was recognized to occur at particularly high concentrations in the basal ganglia of the normal brain, especially the caudate nucleus and putamen (collectively designated the "corpus striatum").

1960: Dopamine was found to be decreased in the corpus striatum of patients with Parkinson's disease, but not in the striatum of patients with Huntington's disease (another basal ganglion disease). As the striatum is greatly atrophied in patients with Huntington's disease, this suggested that it was striatal dopamine normally originating from a *striate-external source* that had become diminished in Parkinson's disease—rather than dopamine originating in neurons intrinsic to the corpus striatum itself.

1961: Intravenous L-dopa was administered to 20 patients with Parkinson's disease (most of whom were taking an MAO (monoamine oxidase) inhibitor, expected to increase brain dopamine availability); the L-dopa produced dramatic, but short-lived, clinical improvement.

1962: There was a trial of oral L-dopa for Parkinson's disease.

1967: There was a trial of high-dose L-dopa. L-Dopa is now generally given in combination with one or more other medications, such as carbidopa, that either prevent peripheral metabolism of the drug—so that more L-dopa enters the brain—or that limit dopamine breakdown within the brain. Both actions enhance dopamine availability where it is needed. Without carbidopa or other comparable medication, high doses of L-dopa are generally required to produce beneficial effects that are not simply transient.

L-Dopa, and dopamine receptor agonists, are now the mainstays (but not sole constituents) of Parkinson's disease pharmacotherapy. Nevertheless, as any clinical review makes clear, Parkinson's disease is not solved by pharmacotherapy, despite the transformation of clinical care engendered by such therapy. It is rather like diabetes mellitus and insulin: a dramatic improvement in therapeutic possibilities, but complications of diabetes, or of insulin itself, eventually supervene in many patients. In the case of Parkinson's disease, progression of the underlying pathology eventually leads to the so-called on–off phenomenon (sudden switches between symptomatic improvement and akinesia), to drug-induced dyskinesias (abnormal movements), to postural instability and falling, and often to dopamine-independent signs and symptoms (autonomic insufficiency, psychiatric difficulties).

Thus, as new pharmacological approaches are investigated, several other approaches to the disease are being pursued. First are attempts to prevent degeneration of nigral and other neurons to begin with (Schapira & Olanow, 2004). For this, genetic studies of that minority of patients with familial Parkinson's disease are key, as they may lead to clues as to the molecular pathogenesis—a subject not within the purview of this book. Second are invasive approaches, involving a neurosurgical procedure; these grew out of earlier observations on tremor-reducing effects of lesions in the thalamus and globus pallidus. Finally, and related to the invasive approaches (for technical reasons) are attempts to understand the pathophysiology: how exactly is it that dopamine depletion (and whatever other relevant effects there are of disease-associated degenerative changes) causes the signs and symptoms of Parkinsonism? The latter approach turns out to involve neuronal oscillations, as we shall see. To expand upon these topics—surgical approaches, and brain oscillations relevant to Parkinson's disease—we need first to consider several topics by way of introduction: some of the organizational principles of the basal ganglia, some of the known effects that dopamine exerts on neurons, synapses, and gap junctions, and a few of the experimental models of Parkinson's disease.

Organizational Principles of the Basal Ganglia

It is not so easy to give a concise, but accurate, summary statement of just what the cerebral cortex is doing; but at least one can say that one of the major outputs of the cerebral cortex—the corticospinal tract—passes to the spinal cord and has direct effects (as well as indirect effects) on motorneuronal firing, and hence muscle contraction. Of course, most of the detailed neuronal trans-actions within the cord itself are obscure, and the cord is influenced by other descending tracts from the brain stem (rubrospinal, vestibulospinal, reticulos-pinal): but, still, one can intuit a more-or-less immediate relationship between certain cortical "computations" and the resulting, directly observable, move-ments. Alas, for the basal ganglia (and the cerebellum), such an immediate relationship does not exist. The basal ganglia and cerebellum do not have output tracts that pass to the spinal cord, but only influence movement quite indirectly, and in a manner all too mysterious (Bolam et al., 2000). Certainly it is true that acute lesions—such as infarctions caused by strokes—in these structures can affect movement: chorea or hemiballism in the case of striatum (i.e., the caudate nucleus plus the putamen) and globus pallidus, or subtha-lamic nucleus, respectively (Durán-Ferreras et al., 2006; Hashimoto et al., 2001; Ristic et al., 2002); and infarctions can cause ataxia and intention tremor for the cerebellum (and its input and output pathways; see Chapter 7).

Both the basal ganglia and the cerebellum influence the cortex indirectly, through their respective connections to the ventral thalamus (which in turn excites cortical areas, including primary and supplementary motor cortex). Basal ganglia output is "funneled" through the pars reticulata of the substantia nigra, globus pallidus (Akkal et al., 2007), and the entopeduncular nucleus (Bolam et al., 2000), while most cerebellar output is funneled through the glutamatergic cells of the deep cerebellar nuclei. The nigral (pars reticulata), pallidal, and entopeduncular "output" neurons of the basal ganglia are them-selves mostly GABAergic, although a few appear to be cholinergic (Deniau et al., 2007; Kha et al., 2000; Penney & Young, 1981; Radnikow & Misgeld, 1998); this situation is quite different than for the mostly glutamatergic (excitatory) deep nuclei of the cerebellum: in some sense, basal ganglia and cerebellar effects on the ventral thalamus are complementary, one producing EPSPs, the other IPSPs.

The great nuclear masses of the basal ganglia—the caudate nucleus, puta-men, and globus pallidus—have, as principal neurons, spiny GABAergic cells, and recurrent excitatory synaptic circuits are not to be found there. A corre-sponding statement is true for the cerebellar cortex (although cerebellar granule cells—lying along one of the afferent pathways into the cerebellar cortex—are excitatory glutamatergic neurons, they do not appear to participate in recur-rent excitatory synaptic loops). The synaptic architecture of most of the basal ganglia, and of the cerebellar cortex, could not be more different than the cerebral cortex (and hippocampus). On the other hand, spiny neurons in the corpus striatum of the rat are dye coupled and electrically coupled (as determined with pair recordings; Onn & Grace, 1994; Venance et al., 2004), suggesting that they are gap junctionally coupled, and convincing ultrastructural evidence exists for gap junctions between striatal parvalbumin-positive GABAergic interneurons (Fukuda, 2009); similarly, cerebellar Purkinje neurons are dye coupled (Middleton et al., 2008). It is possible that electrical coupling via gap junctions substitutes, in the basal ganglia, for excitatory synaptic coupling, but this remains to be proved. One would be deeply interested in knowing whether subthalamus neurons are electrically coupled, for reasons soon to become apparent, but, so far as we are aware, the issue is not settled. The basal ganglia do contain reciprocal glutamatergic-GABAergic loops, for example, involving the glutamatergic cells of the subthalamus and reciprocally, synaptically connected, GABAergic cells of the globus pallidus.

By far the predominant synaptic input to the basal ganglia, especially to the corpus striatum and subthalamic nucleus (Kolomiets et al., 2001), is glutamatergic and derives from a very widespread projection from the cerebral cortex. Cortical neurons projecting to the striatum are of two broad types (Reiner et al., 2003): those also projecting to the pyramidal tract, with a collateral into the striatum, and having a large cell body in lower layer 5, and those not projecting to the pyramidal tract but rather with intracortical collaterals (often to the opposite hemisphere) and often connecting to bilateral striatum—these latter cortical neurons had medium-sized cell bodies in layer 3 and upper layer 5. [These anatomical details matter: if, as seems likely, basal ganglia oscillations are actually generated in the cortex (Yamawaki et al., 2008; and see later), then one needs to know which cortical layers might be involved, and how the oscillations are transmitted to the basal ganglia. As we shall see later, specific frequencies of cortical oscillations are often generated in particular layers of the cortex, and by distinctive mechanisms.] In the rat, somatosensory cortical cells that project to the subthalamus have their cell bodies in layer Vb (Canteras et al., 1988), and one presumes this to be true in humans as well.

Some Effects of Dopamine in the Brain

So far as is known, dopamine does not act as a phasic ionotropic neurotransmitter, so that it does not produce synaptic currents that last milliseconds or tens or hundreds of milliseconds—unlike, for example, glutamate and GABA (although both the latter transmitters can *also* exert metabotropic effects, occurring through biochemical steps). Instead, dopamine, after binding to its receptors, acts indirectly, through second-messenger systems and biochemical signaling cascades. Consequently, the presence or absence of dopamine neurons in the brain will not exert effects on oscillating neuronal circuits by direct involvement—or lack of involvement—of their firing patterns in the phasic, temporal structure of the oscillation. Instead—or at least this is what we presume to be the case—what matters will be the presence or absence of tonic dopamine effects on the system parameters of the oscillating circuits. What we must consider, therefore, is the nature of the system parameters in cortical neurons that dopamine can influence (our hypothesis is that most of the abnormal oscillations occurring in Parkinson's disease actually arise in the cortex, even though they may at times be recorded in basal ganglia).

If cortical circuits oscillate abnormally in Parkinson's disease, we must also address the question of how this might occur, as the substantia nigra itself is not the source of release of dopamine into the cortex, instead sending its output to basal ganglia; rather, dopaminergic input to the cerebral cortex derives from cells in the ventral tegmental area (VTA) of the midbrain (Berger et al., 1974, 1976; Oades & Halliday, 1987; Ohara et al., 2003; Thierry et al., 1973). We must allow for the possibility of at least some degeneration of midbrain tegmental neurons in Parkinson's disease (Javoy-Aqid & Aqid, 1980; Maingay et al., 2006); these tegmental neurons do connect to many parts of the cortex (Oades & Halliday, 1987), and Parkinson's disease signs and symptoms may derive from loss of dopaminergic innervation of the cortex as well as of the basal ganglia. We must also consider the possibility of adaptive (and maladaptive) changes in cortical circuits that arise secondarily because of basal ganglion pathology. Such cortical adaptations would presumably arise through alterations in basal ganglia inputs to the thalamus, with consequent changes in thalamocortical influences (the detailed nature of which is not easy to picture).

Dopamine receptors fall into two main classes, D1-like and D2-like, with the former class including D1 and D5 receptors, and the latter class including D2, D3, and D4 receptors (Werkman et al., 2006). Having a molecular classification is an invaluable start, allowing (in principle, at least) the design of receptor-specific agonists and antagonists, and the development of transgenically modified mice (Waddington et al., 2005).

Dopaminergic terminals are found in the cortex: in the rat, terminals have been reported in visual and retrosplenial cortex, as well as anterior cingulate, in layers I to III, with a columnar organization suggested (Berger et al., 1985). In the human brain (Gaspar et al., 1989; Smiley et al., 1992), there is dopaminergic innervation of all cortical areas, especially layer I, a layer rich in interneurons and the apical tufts of large layer V pyramidal cells. The densest innervation was in the agranular (i.e., containing a minimal layer IV) cortex: primary and secondary motor areas (which would receive input from basal ganglia and cerebellum, via the ventral thalamus), as well as anterior cingulate and insula; this input was to be found in all cortical layers. There was a less dense innervation of granular cortex, in layers I, V, and VI (i.e., skipping superficial pyramidal layers and layer IV). Dopaminergic synapses were symmetrical and asymmetrical, mainly the former, and were to be found on dendritic spines and shafts, as well as probably on synaptic terminals.

The detailed study of dopamine actions, on neurons and synapses, has mostly involved the prefrontal cortex, motivated in large part by four related issues: (1) the large dopaminergic innervation of that region, combined with (2) some basic observations on involvement of prefrontal cortex in working memory (Goldman-Rakic, 1998), on (3) disruptions in working memory in schizophrenia (Goldman-Rakic, 1999; Tan et al., 2007; see Chapter 6), as well as attention-deficit disorders (Arnsten & Li, 2005), and dementing illnesses with frontal lobe pathology (Robbins, 1996); and on (4) the use of dopamine-blocking drugs in schizophrenia (see Chapter 6). Thus, in 1971 (Fuster & Alexander, 1971), it was discovered that some cells in monkey prefrontal cortex, but not other cells, increase their firing rates during a "delayed response" test, and maintain an increased firing rate throughout the delay period, thereby providing evidence that prefrontal cortex is somehow important for working memory. Fuster and Alexander did not prove that there was specific meaning in *which* neurons increased their firing rates, but such evidence was provided later (Vaadia et al., 1995), and indeed it was shown that correlations in firing times, between particular neurons, could develop rapidly even without alterations in firing rates. [*Explanation and slight digression*: In such a test, with "delayed match to sample" being one variety of the test, an animal or human is shown a stimulus, and then enters a waiting or delay period; after the delay, another stimulus is shown, and—depending on the relationship between the two stimuli, one or another response is to be produced. This type of test has been invaluable in studies of working memory, and in the encoding of stimulus properties in the firing rates of selected sub-populations of neurons: "cell assemblies"; such studies complement other experiments demonstrating encoding of stimulus properties in oscillation synchrony and other aspects of collective neuronal activities; it should be noted, however, that some cortical regions—prefrontal, inferior temporal (Miyashita, 1988)—have been studied with delayed response tests; others (e.g., visual cortex) has been studied in terms of stimulus-induced oscillatory responses; and still others (motor cortex, somatosensory) have been studied both ways.]

It is essential that the reader understand the historical context in which dopamine actions on prefrontal cortex are so often studied—the studies are generally not in terms of oscillations, but of activated states (although, of course, activated states might coexist with oscillations). Not only is it clear that prefrontal actions of dopamine could be clinically relevant, in disorders affecting cognition, but in addition studies are often organized around a hypothesis—either stated explicitly or not—as to the nature of the underlying physiological issue. Again, we must digress a bit to explain this hypothesis. The idea may be paraphrased this way: a "memory" is postulated to correspond to selection of a defined subset of neurons, and the corresponding exclusion of other neurons. One theoretical model, often referred to as the "Hopfield model" (Hopfield, 1982; Hopfield & Tank, 1986), constructively demonstrates how such a selection might take place, and be implemented in terms of high firing rates for the "selected" neurons, and low firing rates for the rest of the neurons. The "memory" is conceptually identified with the set of neurons firing at high rates. [In variations of this model, the selected neurons—constituting a memory or a "percept"—oscillate synchronously in phase, without necessarily have higher firing rates, while other neurons oscillate at different frequencies or phases, or do not oscillate at all—however, there is no comparably elegant model as to how the selection of neurons is to occur (there are models of the synchronization itself (Traub et al., 1996c), but that is a different matter.)] Not only is the idea of "memory as subset of neurons" based on theoretical models, the idea is also rooted in experimental observations on the slow oscillation of sleep, discovered by Mircea Steriade and colleagues (Steriade et al., 1993a,b,d). (Whether the idea is *correctly* rooted in such experimental observations is, however, another matter.) As we have mentioned previously, during the slow oscillation, cortical neurons—both excitatory and inhibitory—alternate between depolarized so-called "upstates" and hyperpolarized so-called "downstates," with virtually all neurons participating. Entry into the upstate requires glutamatergic neurotransmission, originally claimed to result from α-amino-3-hydroxyl-5-methyl-4-isoxazole-propionate (AMPA) receptor activation (Sanchez-Vives & McCormick, 2000), but actually dependent on kainate receptor activation (Cunningham et al., 2006b), at least in the entorhinal cortex in vitro. Conceptually, there is some affinity between (1) the idea of recurrent excitation bringing neurons collectively into an activated state, even as interneurons are simultaneously active, with (2) the idea of selected high-firing-rate neurons in the Hopfield model; and some researchers hypothesize that working memory involves processes analogous to the slow oscillation. There are, however, some important caveats that cannot be ignored: (1) a memory involves, hypothetically, a *subset* of neurons to be activated, whereas the slow oscillation involves almost all the neurons; (2) the slow oscillation occurs during slow-wave sleep, a brain state characterized not only by decreased mobility and responsiveness, but also by *absence of memory*; (3) it has not been proved that the glutamatergic activation leading to an upstate actually results from recurrent excitatory synapses (as required in Hopfield-type models), as opposed to glutamate derived from another source, say astrocytic release (Montana et al., 2006).

Whether working memory is essentially involved with selection of pyramidal cells to be activated or no, this is the leading hypothesis in the field, and dopamine is often studied from the point of view of how the compound can bias the properties of neuronal networks, in such a way that selected activation can be favored. An excellent review (Seamans & Yang, 2004) illustrates this point; that review, as well, discusses the many methodological complexities involved in the analysis of dopamine actions on neuron membranes and synapses in vitro: the history-dependence of its actions, the importance of the concentration, and other experimental details. Perhaps the experimental problem is this: neuronal circuits are most often studied in terms of neuronal membrane potentials, or $[Ca^{2+}]$ signals, while some manipulation is performed. In the present context, at least part of the manipulation involves application of dopamine itself, or a receptor subtype agonist or antagonist.

Unfortunately, neuronal membrane potentials, and [Ca^{2+}] signals, are rather "downstream" from dopamine actions, which are exerted through concentration changes, within subcellular compartments, in a number of moieties such as cyclic AMP, inositol triphosphate (IP_3), and any number of proteins in various phosphorylation states—and these concentration changes themselves are not directly observable in the experiment, certainly not on the time scales with which neural activities are measured. Thus, there is ample opportunity for these "hidden variables" to move in directions not anticipated by, or determinable by, the experimenter.

Dopamine has actions on *intrinsic membrane currents* and *synaptic interactions*, and there is direct and indirect evidence as well that dopamine affects *gap junctions*. Presumably related to these more basic actions, dopamine also has effects on network oscillations in vitro. Let us consider these in turn, with the proviso that some of the data are conflicting:

• Dopamine has been reported to diminish peak Na^+ currents in hippocampal pyramidal neurons (Few et al., 2007); consistent with this, dopamine at concentrations of 0.1 to 20 µM had effects on rat prefrontal regular-spiking neurons which were consistent with a reduction of Na^+ currents, including suppression of slow inward currents and intrinsic membrane oscillations, and *decreasing* the firing rate for a given current injection (Geijo-Barrientos & Pastore, 1995). In contrast, in a different study, the properties of fast Na^+-dependent action potentials in rat layer 5 pyramidal cells (rat prelimbic prefrontal cortex) were not affected by dopamine (Gulledge & Stuart, 2003). In a still different study, however, Yang and Seamans (1996) reported *excitatory* effects of dopamine (via D1 receptor actions) in rat layer V to VI prefrontal cortex neurons, for example a lowered firing threshold, effects attributed to a prolongation of persistent Na^+ current and reduction of the dendrotoxin-sensitive "D" type of K^+ current (both of which are present in the axons of pyramidal cells; see later); as well, high-threshold dendritic Ca^{2+} spikes were suppressed by dopamine. The next effects of these actions would be to suppress burst-firing induced by dendritic depolarization, while rendering the cell prone to fire higher-frequency trains of single action potentials. Henze et al. (2000) likewise reported increased rates of regular single-action-potential firing in layer III cells of primate prefrontal cortex. But what is one to make of the seemingly contradictory data: increased, decreased, or unaffected regular firing? It may be that details of the in vitro experimental conditions are critical in determining the balance of diverse signal pathways activated by applied dopamine—and this leaves us uncertain how to extrapolate the data to clinically relevant in vivo situations.

• Dopamine receptors have been reported, for prefrontal cortex, on presynaptic glutamate-releasing terminals onto principal cells (D1: Paspalas & Goldman-Rakic, 2005); and dopamine increases the excitability of fast-spiking interneurons, by a D1 effect (Gorelova et al., 2002; Kröner et al., 2007). There is a postsynaptically mediated dopamine (D1) enhancement of N-methyl-D-aspartate

(NMDA) responses in prefrontal pyramidal neurons (Chen et al., 2004; Wang & O'Donnell, 2001). Glutamate release is inhibited at a subpopulation of corticostriatal presynaptic terminals, via D2 receptors (Bamford et al., 2004). In the hippocampus, via D1 receptor–mediated actions (probably presynaptic), dopamine increases glutamatergic transmission at some synapses (mossy fiber terminals onto CA3 pyramidal cells, Kobayashi & Suzuki, 2007), and decreases transmission at others (onto subicular pyramidal cells, Behr et al., 2000). Again, interpretation of these data in terms of effects on brain oscillations is not (yet) straightforward.

• A highly unusual action of dopamine, at least partly mediated through D2 receptors, has been described in the pars compacta of the substantia nigra, although not yet in cortex: the reduction of I_h, the hyperpolarization-activated cation current (Vandecasteele et al., 2008).

Dopamine and Gap Junctions

Cortical neuronal oscillations depend on neuronal intrinsic properties, synaptic interactions, and—at least in many instances—gap junctions. It is therefore necessary for us to consider dopamine effects on gap junctions, or at least on dye-coupling, a parameter often easier to measure experimentally than gap junction-mediated electrical coupling itself—and far easier to examine than gap junction ultrastructure. The subject is complex, as dopamine exerts opposite effects in different kinds of cell, via distinct receptor subtypes and distinct signaling pathways. We list a few examples: electrical coupling has been shown directly, along with dye-coupling, in GABAergic striatal neurons (Venance et al., 2004); dye coupling is reduced between such neurons by dopamine acting on D2 receptors, probably by elevation of cyclic AMP (Onn & Grace, 1994). [Cyclic AMP was previously shown to *increase* coupling of rat hepatocytes (Saez et al., 1986). In the case of retinal AII amacrine cells, dopamine increases cyclic AMP and uncouples the cells, but it acts through D1 receptors (Hampson et al., 1992).] In developing neocortex, dopamine uncouples pyramidal cells, acting via *both* D1 and D2 receptors (Rörig et al., 1993). A negative result concerns the lack of effect of D1 receptors on electrical coupling between fast-spiking interneurons (Towers & Hestrin, 2008); we are not aware of data concerning D2 effects on coupling between interneurons.

If systemically administered L-dopa (converted to dopamine in the brain) exerts some of its useful effects via modification of gap junctions, then interpretation of the mechanisms requires that we know which gap junctions are the relevant ones. As we shall hypothesize, following the work of Peter Brown and others, that it is *neocortical* oscillations that most matter for Parkinsonian motor symptoms (and that are suppressed by systemic dopamine, see later), then it follows that neocortical gap junctions are the ones that matter—and we would expect that dopamine would tend to block them. Consistent with this notion, as we shall see in later chapters, many in vitro neocortical oscillations at frequencies recorded in Parkinson's disease cortex [e.g., beta (15–30 Hz)], are also suppressed by gap junction blockade.

Experimental Models of Parkinson's Disease

A basic question in Parkinson's disease research is this: how is it that dopamine depletion in the brain actually leads to so many of the signs and symptoms of the disease? In vitro and molecular data are necessary to address this question, but even more crucially, in vivo "systems" data are required: one needs to be able to observe the akinesia and rigidity and make measurements of what the brain is doing during such states. Some of the "systems" data derive from patients undergoing surgery, or in the wake of surgery, either ablative or for implantation of a deep brain stimulating (DBS) device; this situation is somewhat analogous to invasive recording in epilepsy patients: although the goal in epilepsy surgery is usually to localize the origin of seizures, whereas the goal in movement disorder surgery is to guide an instrument or electrode to a desired therapeutic location. Other "systems" data, however, derive from experimental models of Parkinson's disease, of which we mention two: first is the use of injection of 6-hydroxydopamine (6-OHDA), which is taken up by catecholaminergic neurons, and can be made relatively specific for dopaminergic neurons by injecting the compound into pars compacta of substantia nigra (e.g., of a rat), and by pretreating the animal with the antidepressant desipramine and the monoamine oxidase (MAO) inhibitor pargyline (Mendez & Finn, 1975). The second model, called the MPTP model, has an interesting history, and is described in more detail (Burns et al., 1983, 1985):

1-Methyl-4-phenyl-1,2,3,6,-tetrahydropyridine (MPTP) is an impurity that can appear in the improperly conducted (and often illegal) synthesis of MPPP [1-methyl-4-phenyl-4-propionoxypiperidine, a synthetic opiate related to meperidine (Demerol)]. After systemic administration, MPTP enters the brain and is converted by glial MAO-B to 1-methyl-4-phenylpyridinium (MPP+), which is in turn taken up by presynaptic terminals of dopaminergic neurons. MPP+ interferes with cellular respiration (electron transport in mitochondria), causing cell death, particularly in the pars compacta of substantia nigra (with its dopaminergic neurons), but also in the ventral tegmental area (VTA) —to a lesser extent—as well (Gnanalingham et al., 1993; Jackson-Lewis et al., 1995). At the cellular level, dopaminergic input to the basal ganglia is significantly reduced, and one expects it to be reduced, at least somewhat, in limbic, frontal and other cortical regions as well. Clinically, affected individuals develop severe Parkinsonism, with rigidity, akinesia, postural instability, and sometimes tremor; nonhuman primates are susceptible to MPTP as well, also developing a Parkinsonian syndrome, while rodents are less susceptible (but rodents are still used in certain experimental studies). Burns et al. (1985)

have pointed out that naturally occurring Parkinson's disease in humans is associated with loss of norepinephrine neurons (e.g., in locus coeruleus) as well as dopaminergic neurons, whereas only the latter are affected in the MPTP syndrome. MPTP-induced Parkinsonism, in both humans and non-human primates, responds clinically to L-dopa, as does the naturally occurring human disorder.

The Subthalamic Nucleus

One of the most effective nonpharmacological treatments of Parkinson's disease consists of chronic high-frequency stimulation, via an implanted device, of one or both subthalamic nuclei: a so-called deep-brain stimulation (DBS) technique. The corresponding implanted electrodes have been used to provide important pathophysiological data, and the treatment itself may work through its effects on brain oscillations, so we shall need to discuss subthalamic DBS. First, however, it is appropriate to say something about the subthalamic nuclei themselves.

The subthalamic nucleus consists of glutamatergic cells with rather complex intrinsic properties (Kass & Mintz, 2006), which are able to fire in at least four different "modes": a silent hyperpolarized mode, a depolarized plateau, tonic firing of single spikes, and rhythmic slow bursts. Currents that have been described in these neurons include (in addition to action potential-generating currents) these: persistent Na^+ (Beurrier et al., 2000); multiple types of Ca^{2+} currents, although not P-type, and with T-channels—as is the case for nucleus reticularis thalami neurons—probably located in dendrites (Beurrier et al., 1999; Song et al., 2000); Kv3.1-like K^+ channels (Wigmore & Lacey, 2000), probably necessary for high-frequency firing; an inward rectifier (Magariño-Ascone et al., 2002); and I_{CAN} (Ca^{2+}-activated nonspecific cation current; Zhu et al., 2004), which may be involved in plateau potentials and in bursting. Dopamine has effects on the intrinsic properties of subthalamic neurons: it depolarizes them and increases their firing rates, probably by reducing a K^+ current, through a D2-like receptor action (Zhu et al., 2002).

It is not clear if subthalamic neurons interact with each other directly, either through chemical synapses or through gap junctions: there was no evidence of collective (i.e., synchronized) neuronal activity in the subthalamus in vitro, in sagittal midbrain slices, even with activation by bath-applied muscarine, ACPD (trans-1-aminocyclopentyl - 1,3 - dicarboxylic acid) (a metabotropic glutamate receptor agonist), NMDA, dopamine, or bicuculline (Wilson et al., 2004); of course, it is always possible that some yet untested agent might do this. In vivo, gamma oscillations at 46 to 70 Hz have been recorded in subthalamus, whose power is increased by D2-like dopamine receptor activation (Brown et al., 2002); presumably—although this cannot be asserted with absolute certainty—these (and other) subthalamic oscillations either (1) arise in other structures (such as cerebral cortex) and are projected to the subthalamus or (2) they arise through interaction between subthalamus and another connected brain region. The latter interaction-based mechanism has been shown to be possible in organotypic slice cultures, including globus pallidus externa and subthalamic nucleus, at frequencies of 0.4 to 1.8 Hz (Plenz & Kitai, 1999). However, it is not clear if such mechanisms apply in vivo, and the frequencies involved are much slower than those that seem relevant in Parkinson's disease.

Let us next consider synaptic connections to and from structures outside the subthalamic nucleus. Subthalamic axons project into the globus pallidus (both internal and external segments), and into the pars reticulata of substantia nigra: all structures whose principal neurons are GABAergic; and sometimes the same axon can project to more than one of these structures (Sato et al., 2000). GABAergic inputs into the subthalamic nucleus derive from the globus pallidus (so that globus pallidus and subthalamic nucleus are reciprocally interconnected), and from the ventral pallidum (Bell et al., 1995). There is also an extensive glutamatergic input to the subthalamus from the cerebral cortex, including prefrontal, motor, and somatosensory cortices; in the case of the somatosensory cortex, the cells of origin have been shown to lie in layer Vb (Canteras et al., 1988; Kolomiets et al., 2001).

Synaptic projections to the subthalamus have been studied with electrical stimulation as well as by tracing axons (Bevan et al., 2002a, 2007; Li S. et al., 2007; Magill et al., 2004). Two important issues stand out from these studies: first, synaptic inhibition of subthalamic neurons cannot be interpreted simply in terms of decreases in firing rates. This is the case because phasic synaptic inhibition can be followed by rebound generation of action potentials in subthalamic neurons, another example of a general principle that has been noted at, for example, nucleus reticularis thalami/thalamocortical relay cell connections, and at cerebellar Purkinje cell/deep cerebellar nuclei connections. Second, electrical stimulation of the subthalamic nucleus not only generates synaptic potentials in globus pallidus and substantia nigra reticulata; it also produces antidromic stimulation of the cortex (Li S. et al., 2007). In particular, stimulation of subthalamic nucleus at typical deep brain stimulating frequencies, around 130 Hz, induces cortical oscillations at similar frequencies, and—as well—suppresses cortical beta oscillations (see later). We shall have more to say in later chapters about the interplay between very fast (>70 Hz) oscillations and beta in the cortex: suffice it to say here that both can be generated in populations of layer V pyramidal cells through gap junction-mediated interactions. We also must note that beta-2 (20–30 Hz) oscillations—in motor, supplementary motor, and somatosensory cortices—are associated with holding still (i.e., akinesia), and conceivably an increase in such beta-2 could be relevant clinically for the pathologically enhanced akinesia in Parkinson's disease. It is to this and related topics that we turn next.

Cortical and Basal Ganglion Oscillations in Parkinson's Disease, and Human and Experimental Models

One may speak of two general classes of oscillations measurable in Parkinson's disease. First are the oscillations observable in a patient (or at least many patients) by the clinician: resting tremor (typically at 4–5 Hz), and "cogwheeling," the tremor superimposed on rigidity, felt when passively moving a joint, at 7.5 to 9 Hz (Findley et al., 1981). Second are those oscillations that can be recorded from the brain, at frequencies higher than tremor frequency, whose relation to clinical phenomenology is indirect. The most prominent of these are beta (15–30 Hz) and very fast (>70 Hz) oscillations.

The mechanisms of Parkinsonian tremor remain unclear (Rivlin-Etzion et al., 2006). It seems likely that the thalamus is involved, as extracellular recordings, obtained in patients having surgical procedures, have identified thalamic neurons, often bursting ones (as would occur in thalamocortical relay cells when hyperpolarized), whose firing is phase-locked to tremor (Lenz et al., 1994; Zirh et al., 1998). Thalamocortical relay cells in the motor thalamus [ventrolateral (VL) and ventral intermediate (Vim)] project to layers I and III of motor and supplementary motor cortex (and also somatosensory cortex), and are able to elicit EPSPs in layer V pyramidal cells (Deschênes et al., 1979; Donoghue & Ebner, 1981; Kawaguchi et al., 1983); correspondingly, microstimulation of VL in the cat can elicit contraction of a single limb muscle (Asanuma & Hunsperger, 1975), making it plausible that coordinated network activity within the motor thalamus might express itself as rhythmic movement. Magnetoencephalographic recordings indicate the presence of oscillating signals, at tremor frequency, in the diencephalon (i.e., including thalamus)—but also in widespread cortical regions (Schnitzler et al., 2006; Volkmann et al., 1996)—making it hard to disentangle cause and effect. Whether the tremor-generating circuits are intrinsic to motor thalamus or not, destructive lesions in Vim, and high-frequency stimulation there, are both able to reduce or abolish Parkinsonian tremor (Benabid et al., 1987).

Let us consider next those oscillations recorded from the brain that may be related to akinesia. The Parkinsonian state, in the naturally occurring human disorder and in experimental models, is associated with a general increase in correlations between firing times of neurons, and of neuronal rhythmicity, in many regions of the cerebral cortex and basal ganglia; there are, in addition, abnormal burst discharges in primary motor cortex neurons, not, however, associated with movement (Goldberg et al., 2002; Marsden et al., 2001; Nini et al., 1995; for reviews, see Bevan et al., 2002b; Brown 2003, 2006). Two types of oscillatory synchrony are of special interest, as they are clinically/experimentally striking, and as both have plausible cellular mechanisms established by in vitro and modeling studies (see later chapters): First, beta (15–30 Hz) oscillations are prominent in the Parkinsonian state (patient and experimental), in cortex and subthalamic nucleus, with coherence between the two (Brown et al., 2001; Marsden et al., 2001; Weinberger et al., 2006; Williams et al., 2002; Wingeier et al., 2006). Figure 5.1 shows an example of enhanced cortical–subthalamic beta oscillations from an experimental Parkinson model, the 6-OHDA-treated rat. [Beta occurs in the globus pallidus as well, in this rat model of Parkinsonism (Mallet et al., 2008b).] Treatment with L-dopa or with subthalamic stimulation suppresses the beta oscillations, and unmasks the normally occurring subthalamic gamma at 46 to 70 Hz (Brown et al., 2001, 2002); further, the *extent* of motor-performance-improvement, induced by subthalamic stimulation, is correlated with the *extent* by which beta oscillations are reduced acutely by the stimulation (Kühn et al., 2008). Enhanced beta oscillations, in cortex and subthalamic nucleus, do not result *acutely* from dopamine depletion, but rather take some days to develop in an experimental preparation (Mallet et al., 2008a)—apparently,

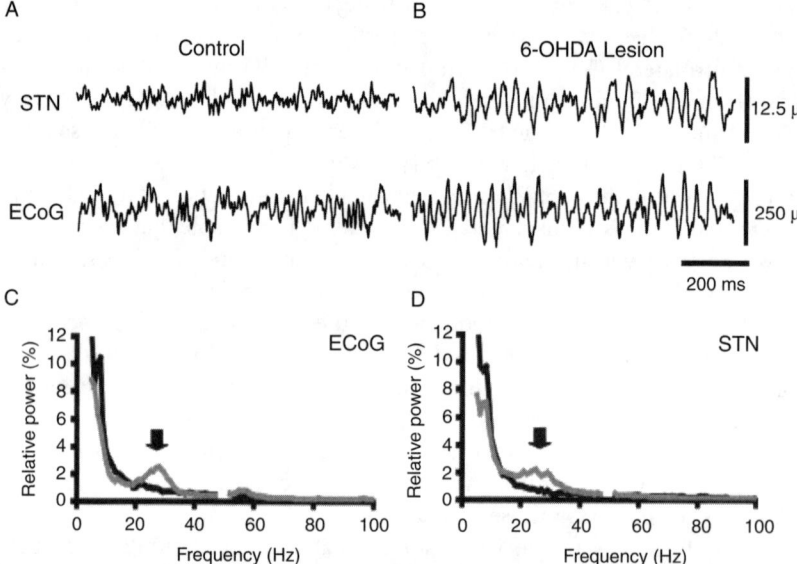

Figure 5.1 Unilateral destruction of brainstem dopaminergic neurons (in the awake rat) leads to coherent beta (22–32 Hz) oscillations in frontal cortex and ipsilateral subthalamic nucleus. **A:** Local field potentials from STN (subthalamic nucleus) and ipsilateral electrocorticogram (ECoG) from control rat. **B:** Likewise after unilateral injection of 6-OHDA (6-hydroxydopamine) into median forebrain bundle, in the presence of desipramine and pargyline, to destroy dopaminergic neurons unilaterally. Note the prominent oscillations. **C:** Power spectra of ECoG signals (black, control; gray, 6-OHDA lesion); note the beta peak after the lesion. **D:** Power spectra of STN field potentials (black, control; gray, 6-OHDA lesion). Again note the beta peak. Coherence between frontal cortex and STN is demonstrated in the original paper.

(Reproduced from Sharott et al., 2005 with permission.)

a trophic effect is involved. These observations are of particular interest, as beta-2 oscillations (20–30 Hz) occur as a normal phenomenon in motor, supplementary motor, and somatosensory cortices, during expectant immobility (see Chapter 11); and because oscillations at this frequency occur in somatosensory cortex slices, in layer V pyramidal cells, via coupling mediated by gap junctions (see Chapter 11)—a type of cellular interaction likely to be modifiable by dopamine. It is plausible to hypothesize that dopamine deficiency leads to enhanced electrical coupling between layer V cortical pyramidal cells, and thereby to excessive beta oscillations; and that these cortical beta oscillations are projected to the subthalamic nucleus via the numerous cortical fibers thereto, thus explaining the coherence of the oscillations between the two structures. What is not clear, however, from this line of thought, is the mechanism by which cortical beta oscillations actually lead to akinesia.

In addition to enhancing beta, it appears that the Parkinsonian state is associated with suppression of a subthalamic approximately 300-Hz oscillation, that is either present normally, or at least appears after therapeutically successful treatment with L-dopa (Foffani et al., 2003). Figure 5.2 shows an example of this very fast oscillatory activity in the subthalamic nucleus. The functional significance of the approximately 300-Hz oscillation is not clear, nor is it

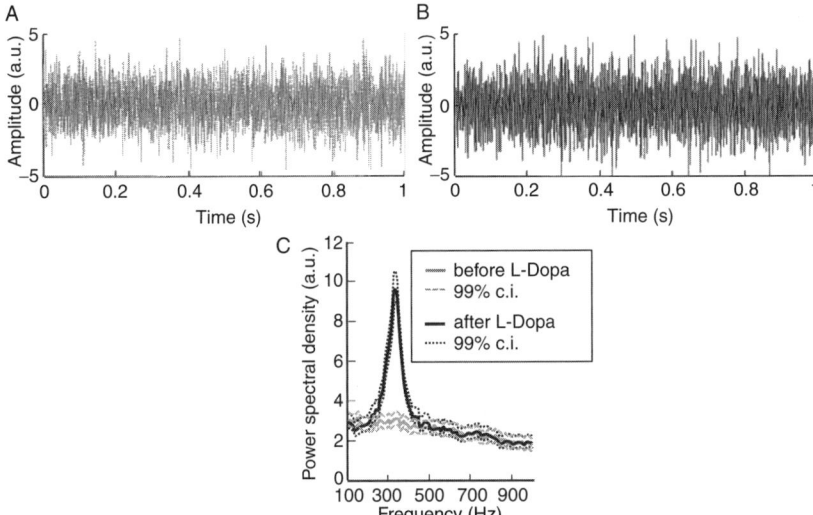

Figure 5.2 300 Hz subthalamic nucleus oscillations are enhanced by L-DOPA in a patient with Parkinson's disease. **A:** Local field potential in subthalamic nucleus (STN) of a patient with Parkinson's disease. The activity has a relatively flat power spectrum between 100 and 1000 Hz (gray line in **C**). **B:** STN local field potential in the same patient, 30–40 minutes after oral fast-acting levodopa (100–200 mg). There is a pronounced spectral peak at ~300 Hz (dark line in **C**).

(Reproduced from Foffani et al., 2003 with permission.)

known how the oscillation is generated. The only examples of network oscillations at comparable frequencies, where the cellular mechanisms have been worked out, involve electrical coupling between principal cell axons (Draguhn et al., 1998; Traub et al., 1999b).

Deep Brain Stimulation and Ablative Treatments of Parkinson's Disease, and Their Effects on Brain Oscillations

We have mentioned thalamotomy as a treatment for Parkinsonian tremor; unfortunately, this modality is not effective against the bradykinesia/akinesia that are generally far more debilitating for the patient than is the tremor. Fortunately, however, first small destructive lesions, and then chronic "deep brain" stimulation (DBS), of either the globus pallidus or of the subthalamus, have proven to be effective treatments for the akinesia (as well as rigidity and tremor), and serve as well to decrease L-dopa requirements, and hence to decrease associated dyskinesias and other side effects of dopamine (Benabid et al., 2001; Krack et al., 2003; Kumar et al., 2000; Limousin et al., 1995, 1997, 1998). Stimulation frequencies of 130 to 185 Hz are commonly used. As mentioned, subthalamic stimulation may work via antidromic effects on the cortex (Li S. et al., 2007), and, indeed, some initial studies indicate that direct stimulation of motor cortex itself, using epidural electrodes, can also be effective, in at least some patients (Cioni et al., 2007). A side effect of subthalamic stimulation, particularly when the stimulus amplitude is too large, consists of abnormal involuntary movements such as hemiballism (a type of rapid throwing movement), a complication as well of destructive lesions in the subthalamus (Hammond et al., 1979; Limousin et al., 1995).

In summary, beta and very fast oscillations are altered in the cortex and subthalamic nucleus of Parkinson's disease patients and in Parkinsonian animal models. The beta oscillations are probably generated within the cortex, and they may be an enhanced version of a normal signal associated with (and perhaps causing) immobility. Pharmacological and stimulation treatments reduce cortical/subthalamic beta as they produce clinical improvement. Beta oscillations occur in in vitro models (e.g., Roopun et al., 2006; Yamawaki et al., 2008; Chapter 11), in a manner dependent on gap junctions, and with properties suggesting that similar oscillations could occur in vivo. [Note that the in vitro models of Roopun et al. (2006) and Yamawaki et al. (2008) both use mature tissue: gap-junction dependent beta oscillations are not confined to immature cortex (Dupont et al., 2006).] As dopamine affects gap junctions, it is possible that dopamine depletion (associated with the Parkinsonian state) produces some of its deleterious effects through alterations of gap junction–dependent brain oscillations. Perhaps, when these matters are more clearly understood, a more precisely targeting treatment can be found than dopamine itself, and its agonists.

6

Schizophrenia

With thanks to Jeremy Coplan, M.D.

Schizophrenia is diagnosed by its clinical presentation and evolution; it is particularly important to recognize a clear sensorium in the patient and to exclude psychotic states induced by drugs and by neurological disorders, especially disorders that may have specific treatments (such as encephalitis). No particular laboratory test—including imaging, CSF analysis, EEG, and molecular genetic studies—clinches the diagnosis. Schizophrenic-like disorders may, however, occur on a background of inherited disease, such as velocardiofacial/ DiGeorge syndrome (Gothelf, 2007; van Amelsvoort et al., 2004) or metachromatic leukodystrophy (Gieselmann, 2008; Hyde et al., 1992). Psychotic states induced by particular drugs, although transient, constitute a means to examine the pathophysiology of the disease—especially as the drugs can be administered to animals as well as to human subjects, and can be applied to in vitro brain preparations. One of these drugs, ketamine, produces striking effects on neuronal oscillations, and this observation provides one of the motivations for considering schizophrenia in this book.

Schizophrenia, then, is defined predominantly by how patients behave, particularly (but not exclusively) by what they say and write, and how they seem to think. (Of course, there are other manifestations as well, in motor behavior, interpersonal relations, tendency to suicide, and many other regimes.) We reproduce here two examples of the language and thought of schizophrenic patients, taken from McKenna's 1997 monograph, which in turn depended on original sources (Kraepelin, 1913b; Neale & Oltmanns, 1980).

Example 1

A patient made notes on what his "voices" said to him. His internal speech, directed to the voices, is in parentheses.

(What is your real object?)

"We wish to kill you. You have offended divine providence. Our object is morally irrelevant. M. must laugh at you. Our object is your cleansing. But Absalom! We love and hate you. Our object is terrible establishment of women's regiment. We are silly."

(Are you human beings or spirits?)

"We are human beings, old topswine. Oh that needs an insane patience! I will show you my last aims. We weep about you. You have been very prudent. We are climbing up Ararat. Now then, little spirits! Little folk, brownies! You are fundamentally insanely deep!"

(Are you near?)

"No, far away. What shall we do contrary to your interests? No, in the middle of your head."

Example 2

The following is an illustration of thought disorder (McKenna, 1997):

Interviewer: Have you been nervous or tense lately?

Patient: No, I got a head of lettuce.

Interviewer: You got a head of lettuce? I don't understand. Tell me about lettuce.

Patient: Well, lettuce is a transformation of a dead cougar that suffered a relapse on the lion's toe. And he swallowed the lion and something happened. The . . . see, the . . . Gloria and Tommy, they're two heads and they're not whales. But they escaped with herds of vomit, and things like that.

Schizophrenia is also associated with a decline in many aspects of cognitive functions (Andreasen, 1999). Perceptual disorganization (Uhlhaas & Mishara, 2007; Uhlhaas & Silverstein, 2005: Uhlhaas et al., 2006;) and working memory deficits (Goldman-Rakic, 1994; Frith & Dolan, 1996) figure prominently. Both of these symptoms are, to some extent, quantifiable at a psychophysical and neurophysiological level, facilitating studies that have revealed a link between schizophrenic psychopathology and brain rhythm generation (for reviews, see Uhlhaas et al., 2008; Whittington et al., 2008). The above examples of hallucinations and thought disorder in patients are complemented by examples we reproduce below from Uhlhaas and Mishara (2007), revealing more specific aspects of disrupted Gestalt processes: limitations in perception of a whole sensory object through combination of a set of smaller, simpler features:

Example 3 (Chapman, 1966)

Everything I see is split up. It's like a photograph that's torn to bits and put together again. If somebody moves or speaks quickly, everything I see disappears quickly and I have to put it together.

Example 4 (Matussek, 1952)

I may look at the garden, but I don't see it as I normally do. I can only concentrate on details. For instance, I can lose myself in looking at a bird on a branch, but then I don't see anything else.

For depictions of schizophrenia in biography and film, the reader is referred to Sylvia Nasar's biography of the mathematician John Nash, *A Beautiful Mind* (1998, Simon & Schuster, New York); and to Ingmar Bergman's 1961 film *Through a Glass Darkly* (*Såsom i en spegel*).

General Features of the Disease, with Historical Considerations

In the 19th century, the distinction was not yet clear between diseases, all of which cause severe mental dysfunction, but that might or might not be associated with distinctive neuropathology. For example, only early in the 20th century did Alois Alzheimer (1864–1915), using Nissl and Bielschowsky staining techniques, define the characteristic neuropathology of the dementing disease now named for him: neuronal loss, senile (or amyloid) plaques, and neurofibrillary tangles (Alzheimer, 1907, 1911; Graeber et al., 1997). As a second example, consider another common cause of psychiatric disability (and death) in the 19th century, one of the forms of late neurosyphilis, discovered by Antoine-Laurent-Jesse Bayle (1799–1858), and named general paresis of the insane (GPI), or general paralysis, by Louis-Florentin Calmeil in 1826 (Brown, 1994). Hideyo Noguchi (1876–1928) is credited with proving that GPI was indeed a form of syphilis, by isolation of the spirochete from the brain of a patient

(Noguchi & Moore, 1913). Neurosyphilis is associated with clear pathological changes in the brain and/or spinal cord, both macroscopic and microscopic (Blackwood et al., 1967).

Two psychiatrists, the German Emil Kraepelin (1856–1926, a sometime associate of Alzheimer), and the Swiss Eugen Bleuler (1857–1939) are credited with the description and delineation of schizophrenia, a psychiatric disorder without clear neuropathology—at least not clear at the time, and relatively subtle even with modern methods (Bleuler, 1911; Géraud, 2007; Hippius & Müller, 2008; Kohl, 1999; Kraepelin, 1913a,b; Tölle, 2008). Besides separating schizophrenia from what are now considered neurological diseases (Alzheimer's disease, neurosyphilis, and the like), a major achievement for Kraepelin and Bleuler was the separation of schizophrenia from the "affective psychoses," such as manic–depressive disorder—although, even now, an intermediate psychiatric illness, schizoaffective disorder, is recognized. The name "schizophrenia" is due to Bleuler; Kraepelin called the disease "dementia praecox," emphasizing what he considered to be a usual course of progressive mental disability. "Schizophrenia" derives from the Greek, meaning "splitting of the mind"— not in the sense of split personality (generally a manifestation of hysteria), but rather in what was considered a fragmentation of the psyche, a concept difficult to nail down precisely.

According to McKenna (1997), the cardinal clinical manifestations of schizophrenia include *delusions*, that may be grandiose or of persecution; abnormal perceptions, including auditory *hallucinations* in the form of internal voices speaking to the patient, perhaps commenting on him or her (see Example 1); "*formal thought disorder*," in which the patient's trains of thought are difficult or impossible to follow (see Example 2); *motor, volitional, and behavioral disorders*, including catatonia (immobility and mutism, perhaps with automatic following of commands), mannerisms and stereotypic repeated behaviors; and *emotional disorders*, including flattened or inappropriate affect, and withdrawal. A striking type of symptom is the feeling that one's thoughts are being broadcast, or, in contrast, thoughts are inserted into one's own mind from some external source, and by implausible means—and said thoughts and influences may be experienced as controlling one's actions and sensations. Schizophrenia has acute and chronic presentations and courses. The pioneers, Kraepelin and Bleuler, delineated four types—paranoid (with suspicious delusions), hebephrenic (with prominent thought disorder), catatonic, and simple (in which a progressive loss of mental function occurs). These types are less used now than previously: researchers are more likely to speak of "positive" symptoms, such as hallucinations, and "negative" symptoms, such as apathy, muteness, and social withdrawal. Liddle (1987) has written of three basic syndromes, which he calls *psychomotor poverty* (diminished speech and movement, emotional blunting), *disorganization* (inappropriate affect, empty speech, and formal thought disorder), and *reality distortion* (delusions and hallucinations).

The Human Burden of Schizophrenia

Estimates of the prevalence of schizophrenia vary, but one careful review of worldwide data arrived at a point prevalence of 0.4%, similar to the prevalence of epilepsy (Bhugra, 2005; Saha et al., 2005). The clinical picture is surprisingly consistent across different cultures (Jablensky, 1987).

The disease most often (certainly not always) strikes in adolescence and young adulthood, and can be relapsing/remitting or progressive—characteristics shared with multiple sclerosis. (And, also like in multiple sclerosis, complete or near-complete recovery can sometimes occur.) Schizophrenia, once it strikes, is likely to be lifelong. It imposes suffering on patients and family alike. The disease impairs marriage and other human relationships, and employability. It increases the likelihood of homelessness, suicide, and premature death from medical conditions, perhaps because patients do not properly care for themselves (Mueser & Jeste, 2008). Cigarette smoking is more common among schizophrenics than in the general population, possibly a result of psychiatric self-medication (Bidzan, 2007); increased smoking rates also contribute to the medical morbidity. Members of the public are likely to judge schizophrenics frightening, odd, or generally to be avoided. Patients often require extensive support, both psychiatric for the illness itself, and social or institutional for the superimposed difficulties in coping with life.

Effective—up to a point—pharmacotherapy became possible in the 1950s with the introduction of chlorpromazine (Thorazine, Largactil) (Labhardt, 1954) and related phenothiazines, followed by haloperidol (Haldol) and its relatives. These drugs are blockers of dopamine D2/D3 receptors (Strange, 2008); perhaps that provides a clue as to schizophrenia pathogenesis (see later), but, not surprisingly, these antipsychotic drugs regularly produce Parkinsonism as a side effect (see Chapter 5) (Haddad & Dursun, 2008), as well as other quite troublesome basal ganglia syndromes, both acute and chronic: such as akathisia (motor restlessness) and tardive dyskinesia [consisting of involuntary movements of the face, mouth, and tongue, which may persist after the offending drug is discontinued (Soares-Weiser & Fernandez, 2007)]. Predisposition to the sometimes-fatal neuroleptic malignant syndrome is also a risk (Bhanushali & Tuite, 2004). There is some consensus that antipsychotic medications, particularly the older ones, are more effective against positive symptoms (hallucinations, delusions, and thought disorder) than against negative symptoms ("blunted affect, anhedonia, asociality, inability to initiate and carry out complex tasks to completion") (Risch, 1996); however, the literature on this crucial issue is not consistent (Axelsson & Ohman, 1987; Breier et al., 1987; McLaren et al., 1992).

About 20 years after introduction of chlorpromazine treatment for schizophrenia, reports of a new drug, clozapine, began to appear (Faltus et al., 1973). Clozapine was the first of a new series of agents, perhaps misleadingly labeled as "atypical antipsychotics," with less consistent blockade of D2 receptors, and more affinity for 5-hydroxytryptamine (5-HT, serotonin) receptor subtypes (5-HT$_{2A}$ and 5-HT$_{2C}$, and in a few instances, 5-HT$_{1A}$) (Nikam & Awasthi, 2008); possibly because of the different features of receptor binding (the older antipsychotics are virtually devoid of serotonin receptor binding), the atypical antipsychotic agents have greater efficacy against negative symptoms than do the older antipsychotic drugs. Other members of the drug family include risperidone and olanzapine. Unfortunately, Parkinsonism, tardive dyskinesia, and other "extrapyramidal" side effects still occur with the atypical drugs, as do a whole range of medical (i.e., nonpsychiatric) side effects, including diabetes, weight gain, and (in the case of clozapine) a sometimes fatal agranulocytosis (Correll & Schenk, 2008; Haddad & Sharma, 2007). In addition, the specific cognitive aspects of schizophrenia respond poorly to antipsychotic therapy. Indeed, some antipsychotic agents can make cognitive decline worse. Deficits in spatial working memory are exacerbated by some antipsychotics (Reilly et al., 2006, 2007). These detrimental effects have been variously linked to drug action on dopamine D1 receptors and muscarinic cholinoceptors in the prefrontal cortex (Mori et al., 2004), with effects being highly selective for the specific antipsychotic administered (McGurk et al., 2005).

In part because of the side effects, and because of incomplete amelioration of symptoms—but not entirely explained by these factors—noncompliance with prescribed medications is a real problem for those caring for schizophrenic patients (Sharif, 2008; Voruganti et al., 2008). Perhaps the disease itself contributes to noncompliance; perhaps the lessening of schizophrenic symptoms is counterbalanced by other unpleasant mental states. What we must conclude is this: drug treatment has greatly diminished the human burden of this disease, as is true for epilepsy and Parkinson's disease—but the burden is very far from eliminated.

Schizophrenia Is a Genetic Disorder, in Large Part

Diseases that are mysterious and that seem to be incurable attract no end of explanations. In the early 20th century, Sigmund Freud believed that psychological mechanisms somehow underlay schizophrenia, and treatment might be possible with the "gold of psychoanalysis." In the 1950s, the anthropologist Gregory Bateson came up with the now discredited idea of the "schizophrenogenic mother" (Hartwell, 1996).

A solid scientific approach to schizophrenia probably must start with the incontrovertible genetic findings, for example, oft-repeated studies of identical twins: the concordance rate is about 50% (Tsuang, 2000), far above chance, although not the 100% concordance to be expected with complete penetrance in a disorder determined by a single allele or matched pair of alleles. The "heritability in liability to schizophrenia" has been estimated at 81%, with a 95% confidence interval of 73% to 90% (Sullivan et al., 2003). While the remaining risk of developing the disease, not accounted for by genotype, has

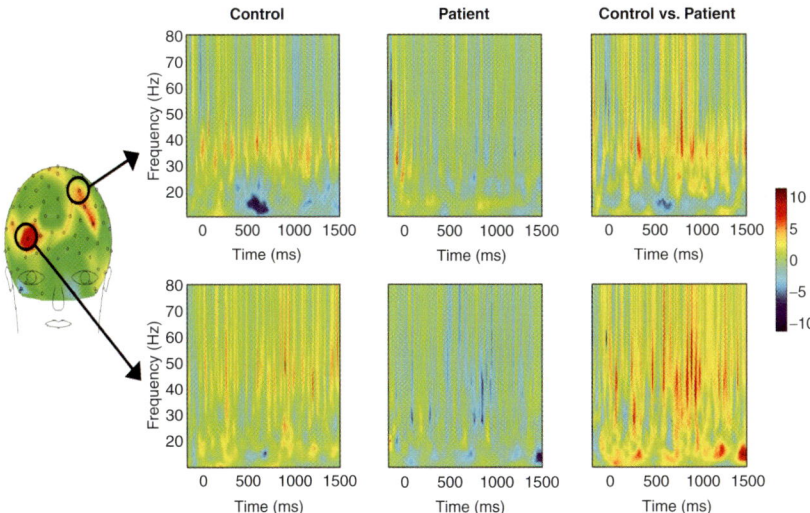

Figure 6.3 Nonschizophrenic people modulate the amplitude of induced frontal cortex gamma activity, according to visual task difficulty; schizophrenic (and schizoaffective) patients do not modulate frontal gamma this way—and also make more errors on the harder task. The task was visual: first a square was shown, either green or red; after a delay, an arrow was shown either left-pointing or right-pointing. If the square had been green ("easy task"), the subject was to respond with the hand to which the arrow pointed. If the square had been red, the subject was to respond with the hand opposite to where the arrow pointed ("hard task"). The two rows of data are time-frequency plots of EEG activity during the delay period, for left frontal (top row), and right frontal (bottom row)—note the gamma oscillations. In the left and center columns, "hot" colors encode a high degree of gamma power modulation by task difficulty. In the right column, "hot" colors encode greater modulation in the control subjects than the patients. In summary: control subjects modulate frontal gamma power (in the delay period), with task difficulty, to a greater extent than do schizophrenic and schizoaffective subjects.

(From Cho et al., 2006 with permission.)

been attributed to undetermined environmental stresses (drug use, concomitant illness, other unknown factors), it is at least theoretically possible that the nonheritable component is attributable to some random event, within the brain of the affected individual. An analogy would be to the situation of mutations in DNA repair enzymes, which place some individuals at risk for cancer. For an individual with such a mutation actually to develop cancer, an error in DNA replication must first occur that is not repaired properly, but that also has the appropriate properties to initiate a propagating series of cellular abnormalities that eventually shows itself as a cancer. It remains to be proven whether a corresponding sequence of events, determined solely by cellular biology, can operate in a psychiatric disorder.

Knowing that schizophrenia has a genetic basis, at least in large part, is one thing; determining what that basis actually is turns out to be another matter entirely (Burmeister et al., 2008). The genetics are certainly nowhere as straightforward as for "classical" genetic disorders such as phenylketonuria, sickle cell disease, or Huntington's disease. The most compelling data suggest that any one patient might possess one or more abnormalities (including small deletions or duplications) in any of a number of distinct genes, where many of the genes affected involve signaling pathways (Walsh et al., 2008). Another study (Stefansson et al., 2008) found three deletions (of unknown functional significance) that were associated with schizophrenia. Some specific proteins that have been mentioned as possible risk factors include *dysbindin*, *neuregulin 1*, *DAOA*, *COMT* (catechol-O-methyl transferase), and *DISC1* (Ross et al., 2006). In such a view, any two schizophrenia patients might or might not have similar genetic abnormalities—similar in that the same gene is affected. These data suggest to us that there must be one or more "common functional pathways" between the genetic abnormalities and the psychiatric presentation. At least some of these proposed common functional pathways appear to involve brain oscillations.

Schizophrenia is associated with diminished reproductive fitness, but occurs in the population at a relatively high (and presumably stable) rate; one study has suggested that this anomalous situation is possible because relatives of schizophrenics reproduce at higher rates than the general population (Avila et al., 2001).

One of the major themes of this book concerns the role of gap junctions is oscillations. For schizophrenia, there does not appear to be compelling evidence for an inherited abnormality of a CNS gap junction protein, at least yet. Aleksic et al. (2007) did not find a significant association between schizophrenia (in the Japanese population) and single nucleotide polymorphisms involving the genes for connexin-36 or pannexin-2 (connexin-36 is a major neuronal gap junction protein, although not the only one). Meyer et al. (2002) also reported negative results for connexin-36. Ni et al. (2007) reported that connexin-50 might play a role in susceptibility to schizophrenia; however, connexin-50 is found in the eye, especially the lens (White et al., 1992), so that the mechanism of any causal connection between the connexin abnormality and mental disorder is unclear. As there is compelling evidence for disturbed brain oscillations in schizophrenia (see later), the issue of gap junction genetics obviously requires further study.

Anatomical and Cellular Abnormalities

Alfred Meyer, in his chapter "Psychoses of Obscure Pathology" in Greenfield's *Neuropathology* (Blackwood et al., 1967), begins by reviewing the anatomical work on schizophrenia and other psychoses, up to the mid-1920s. His comment: "The histological changes in the brain described by these earlier workers were rather slender." The findings that Meyer reviewed up to the time of his writing were characterized by a certain indefiniteness, often complicated by lack of proper controls. Since 1967, reproducible data have accrued in two types of macroscopic brain pathology—although the significance of the findings is far from clear. First, the whole brain of schizophrenics is a few percent less massive than controls, there is reduction in volume of cortical structures with thinning of the cerebral cortex itself, and lateral ventricles are larger (Lawrie & Abukmeil, 1998; Nesvåg et al., 2008); the overall cortical gyral pattern is probably not affected, however (Noga et al., 1996).

The next set of findings involves a magnetic resonance imaging (MRI) technique called diffusion tensor imaging (DTI), used to study white matter tracts (Friedman et al., 2008; Kyriakopoulos et al., 2008). [A discussion on "diffusion tensor": Classical diffusion, as studied by Fick and Einstein, is "isotropic"—the same in all directions—and uniform, and is described by the equation $\partial C/\partial t = D \nabla^2 C$, where C is concentration, and D is the scalar diffusion coefficient. In nonisotropic diffusion ("anisotropic," i.e., different in different directions), the evolution of concentration in a small volume depends not on C per se, but on the gradient of C, $\nabla \cdot C$, and on a 3×3 matrix operating on this gradient: the matrix is called the diffusion tensor, and the components of the tensor depend on position. In DTI, the diffusion tensor for water diffusion is estimated from MRI signals. The anisotropy of water diffusion in brain white matter arises because of the presence of long thin axons—along which water diffuses readily—that are surrounded by lipid-rich myelin, through which water diffuses poorly.] Kyriakopoulos et al. (2008) reviewed a large number (40) of DTI studies, comparing schizophrenic subjects with controls. A common finding is reduced anisotropy in one or another white matter region, such as subcortical whiter matter, corpus callosum, or cerebellar peduncles. In the absence of direct morphological data, however, the possible functional significance of such abnormalities is not clear. It is not even known if there is pathology in axons themselves, myelin, or in the statistical properties of the way that the axons are organized in space.

Another type of abnormality exists in schizophrenia, at the cellular level, requiring for its demonstration molecular techniques that were (obviously) not available to the schizophrenia pioneers. This abnormality has been particularly analyzed by David Lewis and colleagues, focusing on dorsolateral prefrontal cortex—an area exhibiting functional abnormalities in schizophrenia, and also (as discussed in Chapters 5) a key brain region for working memory (see also later). The findings, based on autopsy studies, consist of a series of quantified changes in a number of molecular markers for inhibitory interneuron function (see Lewis et al., 2005 and Gonzalez-Burgos & Lewis, 2008) for reviews). The primary deficit appears to be a selective reduction in the enzyme glutamic acid decarboxylase 67 (GAD67), an enzyme critical for synthesis of GABA (Akbarian et al., 1995; Mirnics et al., 2000). Interestingly, another isoform of this GABA synthesizing enzyme, GAD65, appears unaffected in schizophrenia (Guidotti et al., 2000). It is GAD67 that appears to be critical for maintaining normal levels of the inhibitory transmitter gamma-aminobutyric acid (GABA): GAD65 gene knockout causes a large decrease in brain GABA levels in animal models (Asada et al., 1997). Deficit in the expression levels for GAD67 would therefore be expected to have a blanket disruptive effect on all GABAergic inhibition in the brain of schizophrenic patients.

This pattern of GAD changes is seen qualitatively in other brain regions, besides dorsolateral prefrontal cortex, such as anterior cingulate cortex (Ohnuma et al., 1999), superior temporal gyrus (Impagnatiello et al., 1998), and entorhinal cortex in animal models of schizophrenia (Cunningham et al., 2006a), suggesting that disrupted GABAergic inhibition in schizophrenia may be a common feature in neocortex (Heckers et al., 2002). Complementary studies on other markers for inhibitory interneuron function strongly suggest, however, a peculiar specificity of deficit, involving only a subset of cortical interneurons, regardless of neocortical region. In dorsolateral prefrontal cortex, the overall decrease in GAD67 levels was associated with an overt neuron-specific decrease in about 30% of GABAergic interneurons, while the remaining GABAergic neurons had normal levels of the enzyme (Volk et al., 2000). The changes were most marked in superficial layers, and were paralleled by changes in the GABA uptake transporter GAT1. In addition, decreases in the neuron-specific expression levels of the calcium binding protein parvalbumin (PV), a marker for a subset of GABAergic interneurons, parallels that of GAD67 (Hashimoto et al., 2003); and changes in levels of $GABA_A$ receptor subtypes and of co-released peptides also appear to be neuron-specific (Hashimoto et al., 2008). Further neuronal specificity in the deficit is suggested by the observation in this study, that only a subset of PV-immunopositive interneurons lacked detectable levels of GAD67. Identifying this subset of interneurons is the subject of much interest in the field of schizophrenia research. Currently, the focus is on chandelier neurons in neocortex, a PV-immunopositive interneuron subtype that controls pyramidal cell output (Woo et al., 1998), and selectivity for deficits in interneurons that have a large complement of N-methyl-D-aspartate (NMDA) receptors has also been reported (Cunningham et al., 2006a; Woo et al., 2004).

While the nature of interneuron subtypes specifically affected in schizophrenia is subject to much attention, it is also not clear what primary pathophysiological processes may underlie the robust postmortem changes described briefly in the preceding text. One promising line of inquiry centers on the activity of the tropomyosin-related kinase B (TrkB) receptor. Activation of this receptor promotes development of PV-immunopositive GABAergic neurons during the so-called critical period and, specifically, the enzyme GAD67 that is so closely associated with cellular pathology in schizophrenia (Yamada et al., 2002). TrkB mRNA is reduced in schizophrenia (Weickert et al., 2005a), and mice with reduced TrkB displayed significantly diminished GAD67 and PV mRNA expression (Hashimoto et al., 2005), in a cell-type specific manner that shows remarkable similarity to that seen in postmortem samples from schizophrenia patients (Volk et al., 2000).

There has been considerable speculation about the functional consequences of the PV and GABA receptor abnormalities, based on extrapolations from model systems. For example, in $PV^{-/-}$ knockout mice, IPSCs evoked at high (gamma, ~30 Hz) frequencies are increased relative to wild-type (Vreugdenhil et al., 2003), suggesting a partial compensatory role with respect to the major deficits in markers for GABA synthesis (Gonzalez-Burgos & Lewis, 2008; Roopun et al., 2008a). Unfortunately, one can imagine any number of possibilities, and—in the absence of electrophysiological evidence—a cautious attitude is advisable. How might such evidence be ethically obtained in humans? The answer may depend on difficult questions, such as to what extent schizophrenia is a contraindication to epilepsy surgery in a patient who happens to have both conditions, and who would be a surgical candidate in the absence of the psychiatric disease. If tissue could be obtained from a human patient, following proper ethical procedures of course, it might be possible to determine what functional consequences there are of interneuron abnormalities. In the meantime, we must depend on electrophysiological studies of animal model systems (Fig. 6.1, and also later).

A particular reason why the data on PV^+ interneurons (Lewis et al., 2005) are of interest concerns the observations that dopamine D2 and NMDA NR2A receptors (both receptor types being implicated in schizophrenia) may play a trophic role, in maintaining the viability of these interneurons, in cultured neurons (Kinney et al., 2006), and also in vivo—in rodents and in humans (Morris et al., 2005; Porter et al., 1999; Wang et al., 2007; Woo et al., 2004, 2008). It is not clear how this effect is realized, but a role for control of redox state (via NADPH oxidase) and NMDA-receptor initiated developmental plastic changes mediated by TrkB signaling (see earlier) have been implicated (Behrens et al., 2007b; Y. Liu et al., 2007).

Our next topic develops further the relationship between neurotransmitter receptors and schizophrenia.

Transmitters and Receptors

As we noted earlier, the early antipsychotic drugs (phenothiazines, butyrophenones) bound to dopamine D2 receptors, and—to some extent, at least—later

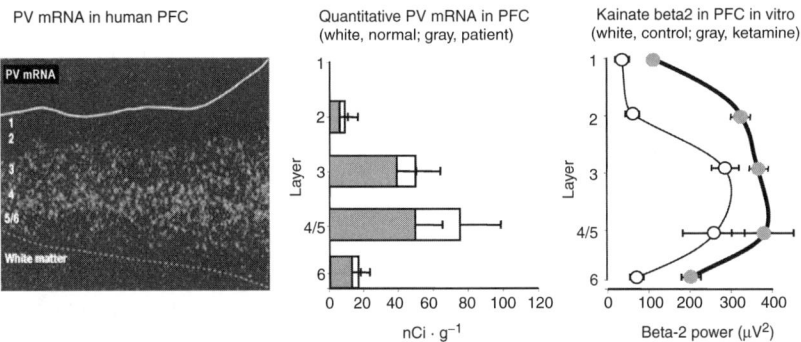

PV mRNA in human PFC

Quantitative PV mRNA in PFC
(white, normal; gray, patient)

Kainate beta2 in PFC in vitro
(white, control; gray, ketamine)

Figure 6.1 Parvalbumin mRNA (messenger RNA) is reduced in layers 3, 4, and 5 of human dorsolateral prefrontal cortex (PFC), in schizophrenia. Beta-2 (20–30 Hz) oscillations occur in these same layers in rat prefrontal cortex in vitro, in response to kainate; and ketamine (which induces psychotic symptomatology in human subjects) enhances beta-2 in these layers. Left panel shows autoradiographically determined mRNA in control human prefrontal cortex. The density of somata of parvalbumin-positive interneurons (such as basket cells and axo–axonic or chandelier cells) would follow this distribution. Middle panel shows summary data on parvalbumin distribution from a series of control individuals, and of schizophrenic (including some schizoaffective) individuals. Note the relative loss of parvalbumin in layers 3, 4, and 5 in the schizophrenic patients. The right panel shows the power of kainate-induced beta-2 oscillations in rat prefrontal cortex in vitro, in baseline conditions (white circles), and with the further addition of ketamine (filled gray circles). Note that ketamine enhances beta-2 power in layers 3, 4, and 5.

[Composite figure from Roopun et al. (2008). Left and middle panels were taken from Lewis, Hashimoto and Volk (2005), using data in Hashimoto et al. (2003). All reproduced with permission.]

antipsychotic drugs do as well. The original observation of a strong correlation between binding kinetics at dopamine receptors, and effective clinical concentrations of drugs, in part led to one of the most enduring hypotheses of schizophrenia pathology (Bunney et al., 1991; Grace, 1991)—the dopamine hypothesis. [There has since been further refinement of the hypothesis that some abnormality in the dopamine system actually underlies the disease, or at least the positive symptoms (Seeman, 2006; Seeman et al., 2005a, 2005b, 2006).] Of further note, certain drugs of abuse, such as cocaine and amphetamine and its derivatives, block the dopamine transporter DAT (Iversen, 2006); and these drugs are capable of causing positive psychotic symptoms in nonschizophrenic individuals (Featherstone et al., 2007; Sherer, 1988). To our knowledge, there still are no in vitro network experiments that develop the hypothesis at a functional, brain-dynamic level. It is worth noting, however, the strong association between dopamine receptor activity and the development and recruitment of fast-spiking (PV-immunopositive) interneurons, neurons that are known to be critical for many types of faster cortical EEG rhythms of cognitive relevance.

Thus, dopamine is capable of directly exciting PV-immunopositive inter-neurons via the D1 subtype of receptor, particularly in prefrontal cortex (Trantham-Davidson et al., 2008). In addition, this receptor subtype is also responsible for boosting glutamatergic synaptic excitation of fast-spiking interneurons acutely (Paspalas & Goldman-Rakic, 2005), but can also facili-tate long-term depression of glutamatergic synapses during patterned input (Otani et al., 1998). Dopamine is also important for maintaining normal endophenotype for PV-immunopositive neurons: lesion of dopaminergic pathways in adult rodents (in zona incerta) reduces PV-immunopositive cell numbers, without reducing the total number of cells (Heise & Mitrofanis, 2005). In contrast, increasing exposure to dopamine enhanced the number of PV-immunopositive neurons in frontal cortex (Ross & Porter, 2002), an effect that was mediated specifically by D2 receptors, in a manner dependent on coactivation of NMDA receptors (Porter et al., 1999).

Let us now consider NMDA receptors in more detail, specifically, the hypothesis that hypofunction of NMDA receptors is a key component in schizophrenia pathophysiology (Greene, 2001; Lee et al., 2003; Lisman et al., 2008). Before discussing the link between schizophrenia and NMDA receptors, however, we must first discuss the receptors themselves. They are, after all, rather complicated.

NMDA receptors (Davies et al., 1979) form one of three classes of ionotropic glutamate receptors: in response to extracellular glutamate binding to the mem-brane receptor protein, there is a change in receptor configuration, and a trans-membrane flow of ionic current, typically inward (depolarizing). NMDA receptors are distinguished from α-amino-3-hydroxyl-5-methyl-4-isoxazole-propionate (AMPA) and kainate types of glutamate receptors, not only in the molecular composition of the subunits that assemble to form the receptors, but in a number of special properties: the relatively long duration of NMDA currents (hundreds of milliseconds), the voltage-dependence of channel gating, permea-bility of the channels to Ca^{2+} ions (Goldberg et al., 2003; Jahr & Stevens, 1993), and sensitivity to extracellular cations and small amino acids. NMDA receptors are critical in the development of the nervous system (Gu et al., 1989; Sato & Momose-Sato, 2008), in long-term potentiation (Harris et al., 1984), in hip-pocampus-dependent memory operations (Nakazawa et al., 2002, 2003), in the operation of central pattern generating circuits (Morgado-Valle & Feldman, 2007), and in epileptogenesis, at least in certain experimental models including low $[Mg^{2+}]_o$ (Avoli et al., 1987, 1991; Traub et al., 1994a), and $GABA_A$ receptor blockade (Traub et al., 1993). In addition, NMDA receptors are presumed to contribute to "information processing" operations in the brain, even when plas-ticity is not involved; however, the role of NMDA receptors in normal circuit operations is surprisingly ill-defined. It is known that NMDA receptors are key for certain types of network oscillations (Gillies et al., 2002). Network modelers sometimes assume that any slow synaptic inward current must be due to NMDA receptors—a dangerous assumption, given that ionotropic kainate receptors can induce synaptic currents lasting tens of ms (Cunningham et al., 2006b).

As is true for most ligand-gated channels, particularly including AMPA receptors, NMDA receptors are assembled from a repertoire of subunits, whereby (1) the choice of subunits constituting a given receptor macromolecule determines its gating properties, and b) different cell types are likely to use various combinations of subunits. In particular—again as for AMPA receptors (Geiger et al., 1995)—combinations of NMDA subunits are different between pyramidal cells and interneurons (Monyer et al., 1992, 1994), with NR2C and NR2D more common in interneurons; in addition, some cells (e.g., layer 4 spiny stellate neurons) can individually express at least two different types of NMDA receptor (Fleidervish et al., 1998). Spiny stellate neurons are glutamatergic but, like GABAergic interneurons, they also contain the NR2C subunit (Binshtok et al., 2006). [Spiny stellate cells are located in layer 4, hence are found rarely in "agranular" cortex—including motor and premotor cortex—in which layer 4 is poorly developed; but spiny stellate cells are found in prefrontal and temporal neocortices, regions of interest for schizophrenia.] Pyramidal cells in neocortex can also apparently express, individually, multiple types of NMDA receptor (Kumar & Huguenard, 2003). Further, receptor subunit composition depends on developmental age (Flint et al., 1997), and this—along with any number of other factors—may be important in determining the age of disease onset.

Most, but not all, of the functional, biophysical data on NMDA receptor gating properties derives from studies of pyramidal cells; while the properties of these receptors on interneurons (rather than on pyramidal cells) may be more relevant for schizophrenia.

In pyramidal cells, NMDA receptor conductance is regulated by transmembrane voltage and by external Mg^{2+} concentration (Jahr & Stevens, 1990a; Nowak et al., 1984), in such a way that higher external Mg^{2+} concentrations tend to block the channel, but so that membrane depolarization can overcome the block. Very roughly: in physiological conditions, the NMDA-receptor/bound-glutamate complex behaves like a voltage-gated channel, opening with depolarization; but in low magnesium conditions, the channel acts as a purely ligand-gated channel with relatively slow kinetics, including an opening time constant of about 10 ms, and closing time constants of tens of milliseconds (for the fast time constant) and >200 ms (for the slow time constant, present in interneurons) (Martina et al., 2003): all of these time constants are much slower than for AMPA receptors. Interneurons may, however, behave differently with respect to the voltage-dependence of NMDA currents, given that interneurons tend to express NR2C: NR2C diminishes the Mg^{2+}-dependence of channel gating (Kuner & Schoepfer, 1996; Sobolevsky et al., 2007).

Another interesting property of NMDA receptors is the potentiation of glutamate-induced currents by binding of small amino acids (glycine, D-alanine, D-serine) to the so-called "glycine site" on the receptor—a different site, obviously, than the glycine receptor responsible for one type of synaptic inhibition (Johnson & Ascher, 1987; Kleckner & Dingledine, 1988; Martina et al., 2003).

The amino acid composition of cerebrospinal fluid (CSF) is altered in certain medical conditions such as hepatic encephalopathy (Shi & Chang, 1984), but apparently not in schizophrenia (Fuchs et al., 2008). Nevertheless, D-serine has been shown to produce beneficial effects in schizophrenia, when used as "add-on" therapy with certain antipsychotic drugs (risperidone and olanzapine, but not clozapine) (Heresco-Levy et al., 2005). Whether the beneficial effect of serine comes about because of actions on NMDA receptors—and, if so, which NMDA receptors—remains to be determined.

When the NMDA receptor blocker MK801 is administered to awake rats, behavioral abnormalities are observed, consisting of "behavioral stereotypy" (in the following: ambulation, freezing, turning, grooming, sniffing up or down, mouth movements, jaw tremor, bed digging, head wagging, and rearing); and also impaired working memory, as measured by alternating arm entry in a four-armed "plus maze"; in addition, prefrontal cortex "regular-spiking" neurons (which are expected to be mostly pyramidal cells) were found to fire at increased rates (Jackson et al., 2004). (Unfortunately, cortical oscillations were not recorded.) These in vivo data are consistent with the idea that NMDA receptor blockade, to an extent causing abnormal behaviors possibly relevant to human psychiatric disorders, predominantly reflects diminished excitability of *interneurons* and not of pyramidal cells: if the diminished excitability were predominantly of pyramidal cells, one would have expected *decreased* firing rates of regular-spiking neurons with receptor blockade, rather than the observed *increased* firing rates.

Pinault (2008) did study the effects of ketamine and MK801 on cortical oscillations in vivo, examining gamma oscillations (30–80 Hz, exclusive of possibly artifactual 50 Hz) in behaving rat frontoparietal cortex, after drug doses sufficient to induce ataxia (but not anesthesia). Such doses increased gamma amplitude by more than 2-fold, and increased mean frequency by about 10 Hz. Apomorphine and *d*-amphetamine (drugs activating dopamine receptors), at concentrations inducing hyperactivity and behavioral stereotypy, also enhanced gamma oscillations, but not to the same extent as the NMDA receptor antagonists.

With this background in mind, on the properties and effects of NMDA receptors, let us now consider how it is that clinicians have related schizophrenia to NMDA receptors. In large part, this is due to the fact that the compounds *phencyclidine* (PCP, a drug of abuse), and *ketamine* (formerly used as an anesthetic in humans, still used as a veterinary anesthetic), both cause, or enhance, positive symptoms of schizophrenia—for example, delusions and hallucinations—in normal volunteers and in schizophrenic volunteers (Krystal et al., 1994; Lahti et al., 2001). In addition, mice with a 95% reduction of NR1 (an essential subunit of the NMDA receptor) exhibit stereotypic behaviors, and abnormal social interactions, that are improved with antipsychotic drugs used to treat schizophrenia: haloperidol and clozapine (Mohn et al., 1999). Although both ketamine and phencyclidine bind to dopamine D2 receptors, as well as to NMDA receptors (Seeman et al., 2005a), it is generally assumed that the most relevant psychotogenic effects of these agents are, in fact, on NMDA receptors, and probably on interneuron NMDA receptors. Clearly, many critical details remain to be further elucidated. Later in this chapter, we will consider specific effects of ketamine on brain oscillations in vitro, and discuss why these effects might be clinically relevant.

Focus on Brain Areas: Prefrontal Cortex and Medial Temporal Lobe

McKenna (1997), drawing on Schneider's 1958 monograph, lists these "first rank" symptoms of schizophrenia: "audible thoughts; voices heard arguing; voices heard commenting on one's actions; the experience of influences playing on the body; thought withdrawal and other interferences with thought; diffusion of thought; delusional perception; and feelings, impulses, and volitional acts that are experienced by the patient as the work or influence of others." McKenna (1997) also quotes the World Health Organization 1979 study (WHO, 1979), listing the 15 most frequent symptoms, 2 years after acute presentation of schizophrenia: lack of insight, flatness of affect, poor rapport, inadequate description, suspiciousness, unwillingness to cooperate, apathy, ideas of reference, delusions of reference, restricted speech, emotional withdrawal, delusions of persecution, hypochondriacal complaints, auditory hallucinations, and verbal hallucinations. Conspicuous by its absence from these lists is mention of *working memory*, at least explicit mention. Nevertheless, the schizophrenia research community devotes a great deal of attention to disorders of working memory, and problems in the brain regions necessary for working memory (including prefrontal and temporal cortex), based on the hypothesis that working memory is somehow central to the disease (Barch, 2005; Goldman-Rakic, 1994; Piskulic et al., 2007) (Fig. 6.2). Why should this be?

Part of the answer may derive from the existence of an experimental paradigm that has been extremely productive in studying working memory: delayed match-to-sample, and related tasks, in awake behaving animals (usually monkeys), occasionally in humans; with simultaneous recording of EEG, field potentials, or single and multiple neuronal units—all from cortical regions found to be essential in these types of tasks (parietal, temporal, prefrontal cortices). These sorts of study were pioneered by Joaquin M. Fuster (born 1930), developed by other investigators such as Richard Andersen and Yasushi Miyashita, and now employed in many laboratories. Such studies (some described in Chapter 5; see also Fig. 6.2) have provided insight into certain aspects of "cognitive" processes, and therefore ought to be (one hopes) applicable in psychiatric disorders: they give, as do neuronal oscillations, a possible experimental handle on a disease mechanism. In addition, however, there is a body of neuropsychology, EEG, and functional imaging evidence, from schizophrenic patients, that points to involvement in the disease of

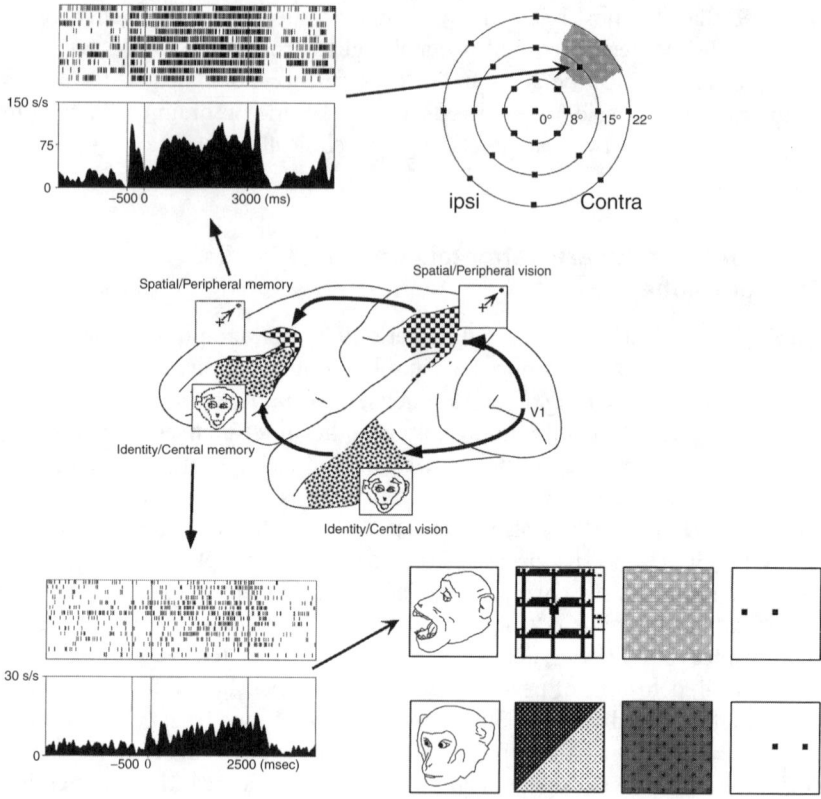

Figure 6.2 Cortical regions involved in working memory, as studied in macaque monkeys with visual delayed match-to-sample and oculomotor delayed response paradigms. Prefrontal cortex (on the left) is of interest in schizophrenia, based on neuropsychological tests in patients (and analogous animal experiments), measurements of brain oscillations, and functional imaging data. The diagram shows primary visual cortex (V1) on the extreme right. The magnocellular visual pathway diverges superiorly toward the parietal lobe (checkerboard pattern), which processes the location and movement-direction of objects, used to guide eye movements, reaching, grasping, and the like; and the diagram shows the parvocellular pathway diverging toward the inferior temporal lobe—this pathway processes form and object-classification of visual images, including faces. The two pathways are sometimes called "where" and "what," respectively. The visual pathways in turn converge on the lateral prefrontal cortex (PFC), with the "where" pathway more dorsal, and the "what" pathway more lateral. The left insets show neuronal firing data (multiunit raster plots above, firing rates below) for working memory tasks: a stimulus is presented briefly, remembered during the delay period (0–3000 ms), and then another stimulus is presented. Depending on the relative properties of the two stimuli, the monkey must make a saccadic eye movement in some particular direction (above), or a visual pattern identification (below).

(From Goldman-Rakic, 1999 with permission.)

dorsolateral prefrontal cortex (DLPFC)—a region important for working memory. We review some of that evidence here.

Three large reviews have concluded that there is a "broadly based cognitive impairment" in schizophrenia (Heinrichs & Zakzanis, 1998; see also Kebir & Tabbane, 2008), and that memory in particular is impaired: more so recall than recognition memory (Aleman et al., 1999). Lee and Park (2005) have suggested that "encoding and/or [the] early part of maintenance may be problematic," in explaining what they considered to be a significant problem with working memory in schizophrenics. According to Ranganath et al. (2008), schizophrenics have trouble with memory tasks involving relations between items, rather than just items per se; the memory deficit is not the same as in, say, Alzheimer's disease, in which items to recall simply vanish within minutes [not surprisingly, working memory is diminished in Alzheimer's disease (MacDonald et al., 2001)]. Ranganath et al. (2008) also describe a difficulty in applying one or another strategy to the particular cognitive task at hand—a sort of difficulty one would associate with prefrontal cortex.

Functional imaging (e.g., with fMRI) has also been extensively applied in psychiatry; this subject has been critically reviewed by Gur et al. (2007: 927), who note, "The breadth of approaches has precluded the establishment of a functional imaging phenotype of schizophrenia." Nevertheless, one should note the meta-analysis of 18 functional imaging studies, undertaken by Achim and Lepage (2005). During retrieval memory tasks, Achim and Lepage list the following brain regions as being more activated in control subjects than in schizophrenics: left inferior prefrontal cortex, left middle frontal gyrus, left medial frontal gyrus, right subgenual region, left precentral gyrus, bilateral thalamus, left anterior hippocampus, right fusiform gyrus, and bilateral cerebellum. Control subjects had *less* activation than schizophrenics in the right anterior medial temporal lobe. While the involvement of the cerebellum in memory seems rather mysterious, these data are broadly supportive of the schizophrenia field's attentiveness to prefrontal cortex and medial temporal lobe.

Yoon et al. (2008) administered a delayed matching task, that they call the "AX" task, while performing fMRI. In the AX task, the subject is shown a letter for 0.5 seconds; after a 3.5-second delay, a second letter is shown. The subject responds with an index finger button push if, and only if, the first letter is A and the second X; otherwise, there is to be a middle finger push. Schizophrenics perform less well than controls on this task, especially when the second letter is indeed X, but the first one is something other than A. During the AX task, there is less activation of left DLPFC in schizophrenics than in controls.

In the next sections, we consider brain oscillation data, in patients and in in vitro model systems. Most of the data (in accord with the preceding discussion) are from frontal cortex and medial temporal lobe.

Abnormal Brain Oscillations in Patients

Howard et al. (2003) studied how cortical gamma oscillation power varied with the difficulty (and duration) of a "Sternberg task." In this task, a series of images, say letters, is presented, one after the other. After a pause (typically of some seconds), a test image is presented, and the subject must decide if this test image was in the original series. The length of the initial series of images can be varied as a parameter. The two subjects for Howard et al. were epilepsy patients, who were not psychotic; the patients each had subdural grids of electrodes placed and, in one case, a strip of subdural electrodes as well—this was part of the preoperative evaluation of the patients (see Chapter 4). Howard et al. found that neocortical gamma oscillation power kept increasing as the series of presented images continued; once the series was terminated, gamma power started to fall off. In some sense, then, under these conditions at least, the "amount" of gamma depends on "information processing demands."

Cho et al. (2006) performed a somewhat analogous experiment (Fig. 6.3), comparing control and schizophrenic (and schizoaffective) subjects—although not with the Sternberg test; instead, difficulty of the task was determined by whether the subject uses a hand on the same side as an arrowpoint (easy) or on the opposite side (hard). The signals recorded were scalp EEG, rather than subdural. The finding was that control subjects modulate gamma activity with task difficulty, somewhat along the lines of the Howard et al. 2003 study, but schizophrenic subjects have much less modulation. In addition, a connection between gamma rhythms and schizophrenia has now been made by many researchers (e.g., K.H. Lee et al., 2003). Changes in auditory and visual steady-state gamma rhythm generation have been seen (Clementz et al., 2004; Light et al., 2006), as well as decreased induced-gamma power and event-related frontotemporal/parietal coherence (Bucci et al., 2007); and numerous studies on visual Gestalt stimuli in schizophrenics have revealed significant deficits in measures of global cortical synchronization at gamma frequencies (Spencer et al., 2003; Uhlhaas et al., 2006). Similar deficits in gamma rhythmogenesis, associated with early auditory processing, have also been reported (Ford et al., 2007; Kwon et al., 1999). Working memory processes, associated with gamma rhythms, are also seen to be disrupted in schizophrenia (Haenschel et al., 2007), with fundamental differences in response to GABA-modulating drugs, when comparing schizophrenics with controls in target discrimination (Menzies et al., 2007).

Of course, to interpret these various observations in cellular terms, one needs to know more about how the gamma oscillations are generated, and modulated, in vivo. We shall show later on how in vitro data can, at least, suggest hypotheses concerning this issue.

Uhlhaas et al. (2006) used a different task than the one in Figure 6.3, although one that was still visual: subjects had to determine if an image was a face or not; again, the recorded signals were scalp EEG. Here (Fig. 6.4), schizophrenic subjects tended to exhibit less phase synchrony than control subjects,

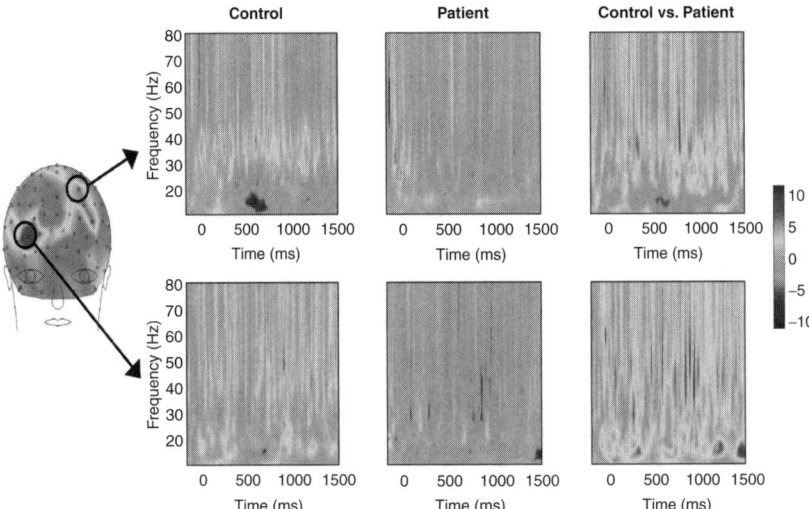

Figure 6.3 Nonschizophrenic people modulate the amplitude of induced frontal cortex gamma activity, according to visual task difficulty; schizophrenic (and schizoaffective) patients do not modulate frontal gamma this way—and also make more errors on the harder task. The task was visual: first a square was shown, either green or red; after a delay, an arrow was shown either left-pointing or right-pointing. If the square had been green ("easy task"), the subject was to respond with the hand to which the arrow pointed. If the square had been red, the subject was to respond with the hand opposite to where the arrow pointed ("hard task"). The two rows of data are time-frequency plots of EEG activity during the delay period, for left frontal (top row), and right frontal (bottom row)—note the gamma oscillations. In the left and center columns, "hot" colors encode a high degree of gamma power modulation by task difficulty. In the right column, "hot" colors encode greater modulation in the control subjects than the patients. In summary: control subjects modulate frontal gamma power (in the delay period), with task difficulty, to a greater extent than do schizophrenic and schizoaffective subjects. Please see color insert.

(From Cho et al., 2006 with permission.)

when comparing signals in pairs of EEG electrodes—in the frequency range 15 to 30 Hz. Oscillations in this frequency range, upper beta-1 and beta-2, when occurring in sensorimotor cortical regions, have been associated with immobility and preparation for movement (see Chapter 5). These brief bursts of beta activity, before directed motor task performance, are, however, greatly attenuated in schizophrenic patients (Ford et al., 2007). We should note that there are several forms of beta rhythm, generated by different local brain circuits, with different patterns of activation. Each is associated with its own cellular and network mechanisms (Roopun et al., 2006; Traub et al., 1999a; Whittington et al., 1997b; see Part III of the book). In our opinion, then, the interpretation of these data must await a better understanding of

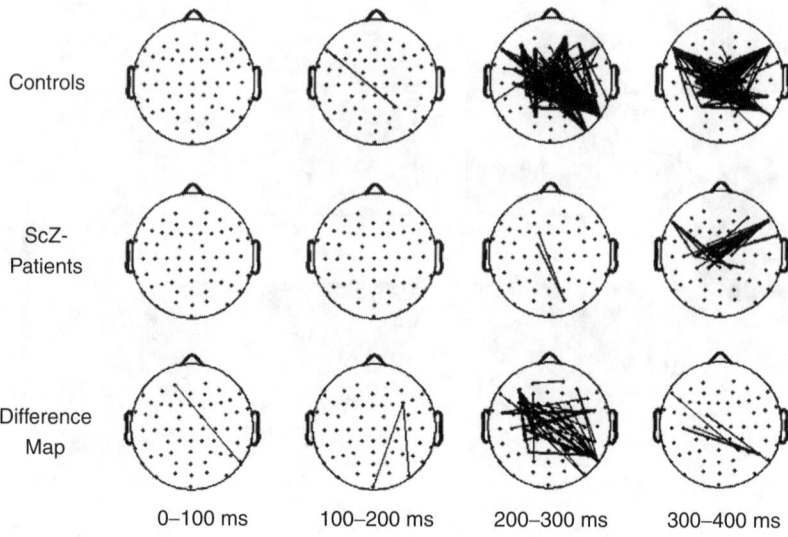

Controls

ScZ-Patients

Difference Map

0–100 ms   100–200 ms   200–300 ms   300–400 ms

Figure 6.4 Nonschizophrenic people develop significant EEG phase synchrony, at beta frequencies (15–30 Hz), across much of the brain, during a visual task; schizophrenic subjects have far less phase synchrony. Subjects were briefly shown a black-and-white image (without gray), that was either a face or not a face; the subjects were to determine if the image was a face or not. EEG electrodes were in place. The figure illustrates pooled data from the condition where a face was shown, and EEG phase synchrony (15–30 Hz) measured between pairs of electrodes, during four time intervals after stimulus presentation (0–100 ms, 100–200 ms, etc.) A line was drawn between two electrode sites if there was significant phase synchrony between those sites, during the corresponding timer interval. For the period 200–400 ms poststimulus, control subjects (top row) showed far more phase synchrony than did the schizophrenia patients (middle row).
(From Uhlhaas, Linden, Singer, et al., 2006 with permission.)

how the particular beta oscillations—induced by the task in question— are actually generated.

Is it plausible that disrupted cortical oscillations, of any sort whatsoever, could be causative of cognitive difficulties? This is the next question that we shall consider.

Experimental Studies Suggesting a Possible Role for Abnormal Oscillations

The NMDA Hypothesis

We have discussed in the preceding text the rather tentative evidence for the hypothesis that, in schizophrenia, there is insufficient activation (for whatever reason) of NMDA receptors on interneurons, or at least certain types of interneurons; and we have mentioned the rather puzzling observation of Pinault (2008) that ketamine actually *increases* gamma oscillation power, at least in part of the cortex (frontoparietal). In the following, we make the case that if gamma oscillations are actually disrupted—somewhere in the cortex or hippocampus—then such disruption might plausibly be related to a behavioral abnormality. Finally, we show that (at least in vitro) the effects of ketamine on brain oscillations are region specific, and probably layer specific as well.

The plausibility argument, for relating gamma oscillations to behavior, is based on the effects of genetic manipulation of AMPA receptors. Before describing the genetic manipulation in detail, we must first note that AMPA receptors, like NMDA receptors, are (1) built from subunits (GluR-A, GluR-B, GluR-C, GluR-D; alternatively, GluR-1, . . . ,GluR-4); that (2) the subunit composition is different in different sorts of cells (with GluR-B present in pyramidal cells and virtually absent in interneurons, while GluR-A and GluR-D are major components for interneurons); and (3) channel kinetics, calcium permeability, and other channel properties are determined by subunit composition, as well as nuclear RNA editing and alternative splicing (Burnashev et al., 1992, 1995; Geiger et al., 1995; Jonas et al., 1994; Keinänen et al., 1990; Mosbacher et al., 1994). To elaborate on this, a critical observation on hippocampal and cortical excitatory postsynaptic potentials (EPSPs), which have a major effect on oscillation properties, is this: that EPSPs in pyramidal cells tend to be rather broad, but EPSPs in fast-spiking interneurons have an abrupt onset and rapid decay kinetics (Geiger et al., 1997; Gulyás et al., 1993; Miles, 1990; Miles & Wong, 1986; see also Part II)—and this difference in kinetics is a result of the fact that GluR-B tends to be expressed in pyramidal cells, and GluR-D in interneurons.

Genetic manipulation of AMPA receptor subunits (Fuchs et al., 2001) provided a critical test of our model of long-range synchronization of induced gamma oscillations, the interneuron doublet model (Traub et al., 1996c; Traub et al., 1996c). The doublet model involves the encoding of phase differences, between oscillating sites, by small differences in timing of action potential pairs in fast-spiking interneurons, as analyzed mathematically by Ermentrout and Kopell (1998). For this mechanism to work, interneuron EPSPs must be precise, with kinetics on the order of 1 to 2 ms. This notion was tested by Fuchs et al. (2001), who introduced GluR-B into fast-spiking interneurons, where they are not normally expressed, thereby prolonging the EPSPs. In vitro–induced gamma oscillations in slices from the mutant mouse were indistinguishable from wild type oscillations, when the oscillations were evoked at a single site; but when the oscillations were evoked at two interconnected sites, synchrony occurred in the wild-type, but not in the GluR-B knock-in, consistent with prediction. Behavioral studies, however, were not conducted in the GluR-B knock-in mouse.

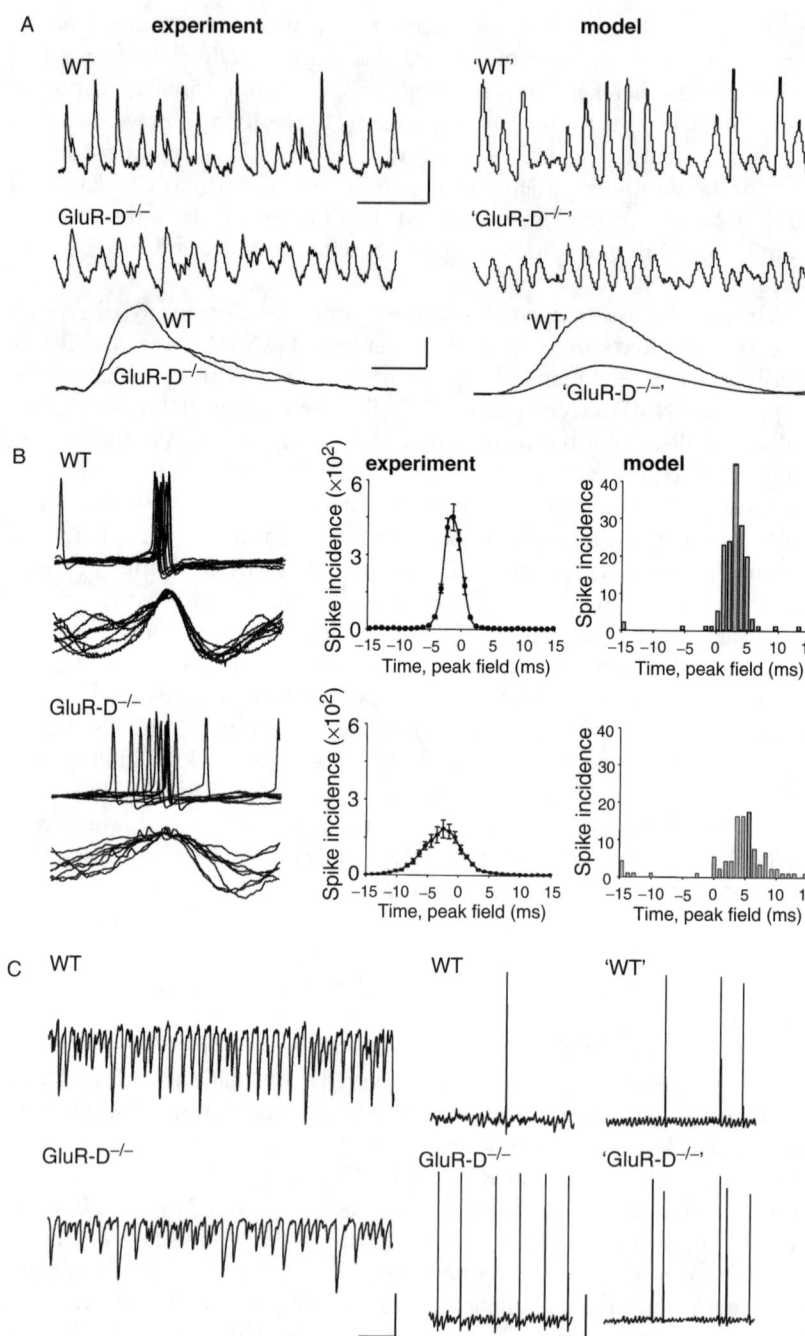

Figure 6.5 In a knockout mouse affecting (predominantly) parvalbumin-positive (PV+) fast-spiking interneurons, in vitro hippocampal gamma oscillations have reduced power: a result of loss of precision of spike-timing in the affected interneurons. The knockout (GluR-D$^{-/-}$) was of the AMPA receptor subunit GluR-D (GluR4), which is expressed almost exclusively in PV$^+$ fast-spiking (FS) interneurons. Unitary EPSCs in these interneurons, in the KO, have reduced amplitude and slightly slower kinetics. Traces from wild-type (WT) animals shown in black, from the KO shown in red. Experimental gamma oscillations were induced with bath-applied kainate. Model traces are from a network model of gamma oscillations (described in a later chapter), in which the amplitude and time course of unitary EPSCs in specific interneuron types can be varied as parameters. **A:** During network oscillations, compound EPSPs in interneurons are reduced in the KO. This is a consequence of smaller unitary EPSCs, and of reduced synchrony of firing. Average compound EPSPs are shown in the lower traces. **B:** During network oscillations, the timing of interneuron spiking is less precise in the KO. To measure this, spike times in FS interneurons were correlated with the maximum of the local field. Ten superimposed traces are shown on the left, spike histograms on the right. **C:** Because of dispersed interneuron spike times, IPSPs in pyramidal cells, during network oscillations, are smaller and more erratic in the KO. That, in turn, leads to more pyramidal cell action potentials in the KO. Trains of pyramidal cell IPSPs are shown on the left, and trains of action potentials on the right, first experimental data, then model data. [Interestingly, blockade of NMDA receptors in vivo, at doses that interfere with working memory (see Fig. 6.6), also lead to increased firing rates of pyramidal cells in behaving rat prefrontal cortex (Jackson et al., 2004).]

[From Fuchs et al. (2007). For data on the power spectra of the oscillations in WT and KO, see Fig. 4 of the original paper. All reproduced with permission.]

An AMPA Receptor Subunit Knockout Mouse Indirectly Supports the NMDA Hypothesis

In a second study, however, the kinetics of excitatory postsynaptic conductances (EPSCs) in fast-spiking interneurons were altered by different techniques (Fuchs et al., 2007). One of these techniques consisted of knockout of the AMPA receptor subunit GluR-D, which confers large-amplitude and fast kinetics to EPSCs. Some of the cellular effects are shown in Figure 6.5, and some of the behavioral effects in Figure 6.6. In the GluR-D knockout, in vitro oscillations were examined with a different experimental paradigm than was used in the GluR-B knock-in: tetanically evoked oscillations in the GluR-B knock-in (analogous to sensory stimulation in vivo), versus persistent, chemically induced, oscillations in the GluR-D knockout, analogous to spontaneously occurring hippocampal gamma in the "theta state" (Buzsáki, 2002; Fisahn et al., 1998).

As Figure 6.5 demonstrates (and the reader is referred as well to the original paper), compound EPSPs are reduced in interneurons during gamma oscillations evoked by bath-applied kainate (unitary EPSCs are, of course, reduced as well, although this is not shown in the figure) (Fig. 6.5A). Data in this figure are from in vitro brain slices ("experiment") and from a network model of persistent gamma oscillations ("model"; see Traub et al., 2000, 2003b, 2003c). As a result of the smaller (and somewhat slower) interneuron

EPSCs, the timing of interneuron action potentials becomes dispersed in the knockout (Fig. 6.5B), leading to reduced gamma power (as shown in the original paper). Finally, the dispersed—and reduced—interneuron firing leads to smaller inhibitory postsynaptic potentials (IPSPs) in pyramidal cells (Fig. 6.5C), and more frequent pyramidal cell action potentials, recalling the increased firing rates seen in vivo, after ketamine, by Jackson et al. (2004).

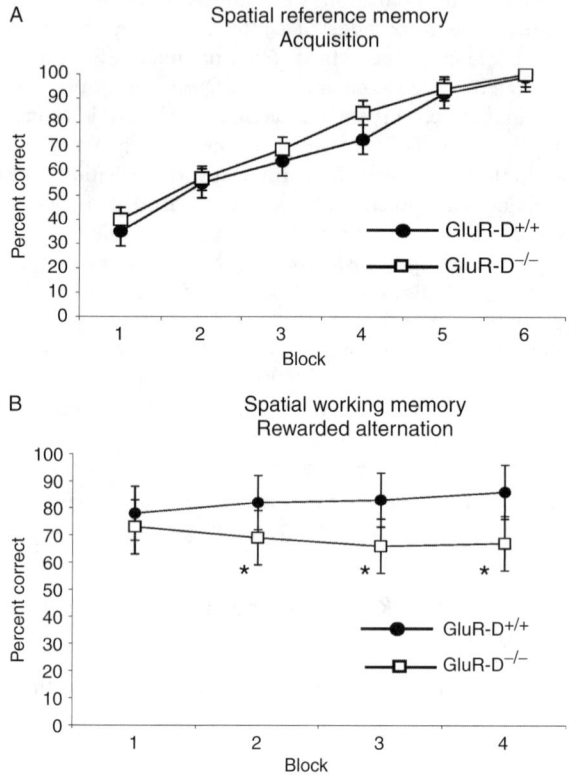

Figure 6.6 In the GluR-D⁻/⁻ KO mouse (see Fig. 6.5), working memory is abnormal, although spatial reference memory is normal. Thus, a genotype affecting synaptic excitation of PV⁺ FS interneurons produces both an abnormal oscillation phenotype, and an abnormal working memory phenotype; however, the mechanistic relationship between the two phenotypic abnormalities remains to be elucidated. **A:** Spatial reference memory is normal in the GluR-D⁻/⁻ KO mouse. In this task, the mouse was place in the center of a Y-maze, and had to learn that a particular branch of the maze contained a drop of milk at the end of it. The abscissa, "block", refers to a group of trials. **B:** Spatial working memory is impaired in the GluR-D⁻/⁻ KO mouse. In this task, the mouse entered either the left or right arm of a T-maze, whichever arm was not obstructed by the experimenter. At 15 seconds later, the mouse was to enter the *opposite* arm for a drop of milk (the obstruction having been removed). This requires the mouse to remember where it went the first time (or where the obstruction was), for a period of seconds. (From Fuchs et al., 2007 with permission.)

Figure 6.6 shows that GluR-D knockout mice are also impaired, at least somewhat, on spatial working memory tasks, although not on spatial reference memory tasks.

Can one infer from these data that the altered oscillations *cause* the behavioral deficit? We would argue that the data do not—at least not yet—allow such an inference. The data do show, however, that there is a plausible connection between a genetic abnormality, altered oscillations, and behavior. The path from the genetic manipulation to cellular abnormality, and on to disrupted oscillations, can be explained in mechanistic terms, using a testable simulation model; however, the path to altered behavior is far more difficult to analyze.

We shall now consider the effects ketamine has on fast oscillations in brain slices. How it is that ketamine exerts its effects is not entirely clear. The point to be made here is that ketamine effects are region specific.

Ketamine Effects on Cortical Oscillations In Vitro

Figure 6.7 shows that ketamine significantly attenuates gamma oscillations (produced by bath-applied kainate) in the superficial layers of the medial entorhinal cortex (EC) in vitro, but not in the deep layers, or in the CA3 region of the hippocampus. As the original paper documents (Cunningham et al., 2006a), the disrupted superficial EC gamma oscillations are accompanied by smaller IPSPs in glutamatergic EC stellate neurons, and increased firing rates in these cells (again recall Jackson et al., 2004). Gamma oscillations occur normally in vivo, in the entorhinal cortex of awake, behaving rat (Chrobak & Buzsáki, 1998), and this activity in superficial medial entorhinal cortex is expected to be projected along output pathways: to the dentate gyrus, CA1 and CA3, subiculum, perirhinal cortex, and several cortical areas (e.g., medial frontal) (Baks-Te Bulte et al., 2005; Insausti et al., 1997; van Groen et al., 2003). We can speculate—but it remains unproven—that the temporal disorganization of spike-timing in EC, induced by ketamine, makes the output signals appear "nonsensical" to downstream structures.

Figure 6.8 provides a type of survey of ketamine effects on kainate-induced cortical oscillations in the rat, in vitro: both on gamma (30–80 Hz, here), and beta-2 (20–30 Hz). [Note: in at least some cortical regions, such as somatosensory cortex, gamma and beta-2 are generated in distinct cortical layers, with gamma superficial, and beta-2 deep (Roopun et al., 2006; and see also Part III of this book).] The data shown in Figure 6.8 confirm the attenuation by ketamine of gamma in medial entorhinal cortex (Fig. 6.7), but also show that gamma is *enhanced* by ketamine in primary auditory cortex in vitro—perhaps consistent with the in vivo data, cited previously, of Pinault (2008). In addition, ketamine enhances beta-2 oscillations in secondary somatosensory cortex and prelimbic cortex—an effect tentatively attributed to disinhibition of pyramidal neurons, but with the mechanism awaiting further investigation.

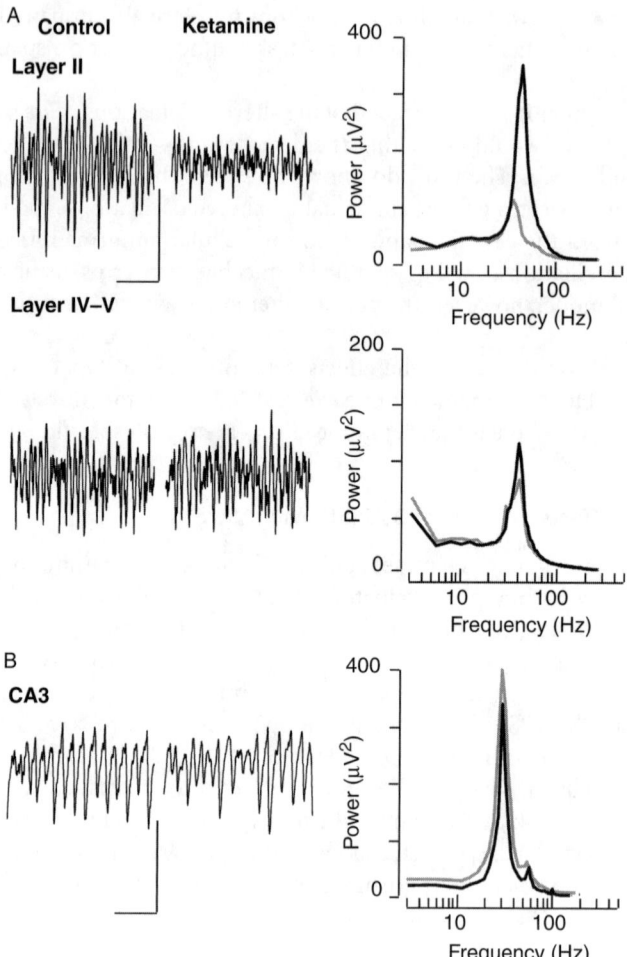

Figure 6.7 Ketamine is a noncompetitive antagonist of NMDA receptors; administered to human subjects, it produces effects similar to both positive and negative symptoms of schizophrenia (Krystal et al., 1994). Ketamine in vitro (20 μM) attenuates gamma oscillations (induced by 400 nM kainate) in the superficial layers of rat and mouse entorhinal cortex, but not in the deep layers or in the CA3 region of the hippocampus. [At least some fast-spiking interneurons in superficial entorhinal cortex are strongly excited by NMDA receptors (Jones & Buhl, 1993).] The figure shows field potential oscillations in control conditions (kainate alone), and kainate + ketamine. Corresponding power spectra are on the right (control, black; ketamine, dashed). **A:** Ketamine attenuates gamma oscillations in entorhinal cortex layer II (superficial), but not in layers IV–V (deep). **B:** Ketamine has no effects on kainate-induced gamma oscillations in hippocampal CA3 region. Calibrations: 50 μV, 100 ms.

(From Cunningham et al., 2006 with permission.)

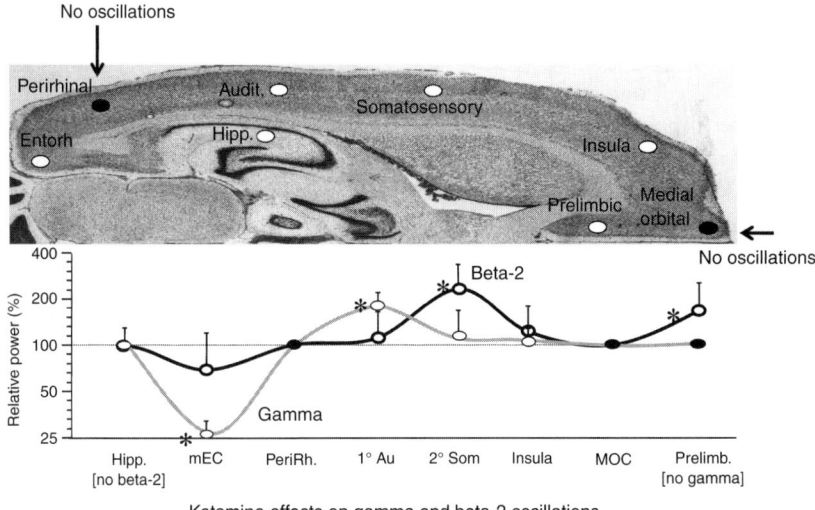

Ketamine effects on gamma and beta-2 oscillations
at various cortical sites

Figure 6.8 Topography of ketamine effects on gamma (30–80 Hz) and beta-2 (20–29 Hz) oscillations at various cortical sites (rat, in vitro). Ketamine reduces gamma in medial entorhinal cortex (mEC), and increases gamma in primary auditory cortex (1° Au); ketamine increases beta2 in secondary somatosensory cortex (2° Som) and prelimbic cortex. [The figure does not illustrate the cortical layers at which oscillations were found and ketamine exerted its effects.] Schizophrenic-like symptoms induced by ketamine can not be explained by a global decrease or increase of gamma/beta oscillations. Oscillations were induced by kainate, 100–400 nM, and ketamine was bath-applied at 10–20 μM. Ketamine effects on oscillation power are shown in the graph below, gray line for gamma, black line for beta-2. Approximate recording sites are shown in the section of rat brain above. Kainate did not induce any oscillations in perirhinal cortex or medial orbital cortex; kainate induced no beta-2 in the hippocampus, and no gamma in prelimbic cortex. *Hipp*, hippocampus; *mEC*, medial entorhinal cortex; *PeriRh*, perirhinal cortex; *1° Au*, primary auditory cortex; *2°*, secondary somatosensory cortex; *MOC*, medial orbital cortex; *Prelimb.*, prelimbic cortex. Asterisks mark sites where ketamine effects reached statistical significance.

(From Roopun et al., 2008 with permission.)

Comment on Nicotine

We have mentioned earlier that the rate of cigarette smoking among schizophrenics is much higher than among the general population. It is interesting that nicotine has been shown to enhance the excitability of axons (Kawai et al., 2007); and nicotine is a potent inducer of gamma oscillations in the cerebellum (Middleton et al., 2008). Whether nicotine enhances cortical gamma oscillations, and thereby has a beneficial effect on schizophrenia, remains an open question.

In Vitro Theta Oscillations also Depend on NMDA Receptors, Probably on Interneurons

While we have concentrated in this chapter on beta-2 and gamma oscillations, it may be important to note that in vitro theta oscillations (at least in certain experimental models) also depend on NMDA receptors. Specifically, Gillies et al. (2002) studied theta oscillations in the CA1 region of hippocampal slices; theta was induced either by (2) a combination of metabotropic glutamate receptor activation with blockade of AMPA/kainate receptors (with the drugs DHPG and NBQX); or (2) with muscarinic activation using carbachol. In the former case, block of NMDA receptors virtually abolished theta; in the latter case, block of NMDA receptors attenuated the power of theta oscillations, and slowed the frequency of the remaining oscillations. As EPSCs in pyramidal cells bathed in carbachol are quite small (Fisahn et al., 1998), it is quite possible that the relevant site of NMDA receptor blockade lies on interneurons. If this is so, this particular in vitro model of theta may be relevant to the NMDA hypothesis for schizophrenia. One expects that disruption of limbic theta in vivo would be associated with cognitive deficits (Buzsáki, 2002).

Summary and Hypothesis

Based on the data presented in this chapter, we offer a tentative synthesis, that attempts to draw connections between cortical oscillations and at least some aspects of schizophrenia (by no means all!). Our synthesis has three components. Thus, in Figure 6.8, we note that ketamine can, depending on cortical area:

    a. Diminish gamma oscillations
    b. Enhance gamma oscillations
    c. Enhance beta-2 oscillations

What might the significance of these alterations be for schizophrenia?

We propose, for (a), that significant diminution of persistent gamma oscillations in a brain area (as in medial entorhinal cortex, Figs. 6.7 and 6.8) is causally associated with a loss of function in that area. A plausibility argument, but not definitive proof, for such a notion comes from the GluR-D KO mouse (Figs. 6.5 and 6.6), in which the mouse phenotype combines diminished hippocampal gamma oscillations (in vitro) with abnormalities in spatial working memory in vivo; but, we must emphasize that a causal relationship between the disrupted oscillations and the disrupted behavior remains to be proved. For the entorhinal cortex specifically, we note that gamma oscillations do indeed occur there in vivo, at least in the rat (Chrobak & Buzsáki, 1998); that a smaller left entorhinal cortex, and entorhinal cytoarchitectural abnormalities, have been reported in schizophrenia (Arnold, 2000; Prasad et al., 2004); and even unilateral entorhinal ablation has a detrimental effect on spatial learning in rats

(Kopniczky et al., 2006)—hardly a surprising finding, in view of the "grid cells" described in entorhinal cortex (Hafting et al., 2005). We again emphasize, however, that (to our knowledge) experimental data still do not exist, which show a causal relationship between reduction of medial entorhinal cortex oscillations—specifically—and impaired behavioral function.

Next, we propose for (b) that sufficient *enhancement* of gamma oscillations in a sensory area (such as primary auditory cortex; see Fig. 6.8) predisposes that area to the generation of hallucinations in the respective modality. (And note, of course, the prevalence of auditory hallucinations in schizophrenia.) This notion is consistent with two observations in patients: first, the case report of continuous gamma oscillations in somatosensory cortex in a patient with somatic hallucinations (Baldeweg et al., 1998); and second, the conclusion of Shergill et al. (2001), who studied a 36-year-old schizophrenic man via fMRI: "Hallucinations in a given modality seem to involve [cortical] areas that normally process sensory information in that modality." Both sorts of observations could bear repeating.

Third, we propose for (c) that enhanced beta-2 oscillations in secondary somatosensory cortex (see Fig. 6.8) may—as is the case for Parkinson's disease (Chapter 5)—be associated with poverty of movement, one of the negative symptoms of schizophrenia. We need, however, here to consider carefully the multiplicity of circuit mechanisms that are capable of generating beta rhythms. The enhanced persistent beta-2 oscillations, seen in secondary somatosensory rat cortex under ketamine, may represent quite different network processes than are involved in the diminished beta activity bursts, seen in schizophrenics during motor task anticipation (see earlier and Ford et al., 2007).

The cellular physiology of in vitro oscillations in prefrontal cortex, where so much basic research on schizophrenia has focused, demands further study. Particularly helpful would be more understanding of the cellular mechanisms of gamma and beta oscillations in human prefrontal cortex, removed during surgery (obviously for conditions other than schizophrenia itself) and studied in vitro.

Is it possible to offer a synthesis of prevailing transmitter and cellular theories of schizophrenia, encompassing the roles of parvalbumin-positive interneurons, NMDA receptors, and dopamine D2 receptors? Most likely it is reckless to attempt such a synthesis, but we shall try: thus, we suggest that neither interneuron loss, nor NMDA receptor block, nor D2 overactivity is in any way primary; but that each plays a role in a set of chains (so far understood only vaguely) that leads to either imprecise oscillations or unregulated oscillations. The human burden of schizophrenia urgently demands practical approaches, yet it seems that much basic research lies before us, a situation not unlike that for cancer.

In part III of this book, we enter into much greater detail concerning the issues of just how gamma and beta-2 oscillations are produced in cortex, and how it is that drugs and cellular abnormalities can alter these oscillations. It provides, we hope, an entry into the necessary basic research.

7

Cerebellar Ataxia

With thanks to Drs. Thomas Knöpfel and Steven J. Middleton

. . .there is no difference between the symptoms of ataxia that
result from lesions of afferent and efferent cerebellar pathways
(e.g., those of cerebellar peduncles or within the brainstem) and
those that result from the cerebellum itself.
—Timman & Diener, 2003

Signs and Symptoms of Cerebellar Pathology

As for any brain region, clues to the function of the cerebellum derive from a
number of sources: clinical observation of the effects of pathology; anatomy;
electrical stimulation; experimental lesions; recording electrical, magnetic,
BOLD and other sorts of signals; study of naturally occurring and genetically
engineered mutant animals; in vitro slice experiments; and many other
methods of well. [The reader is referred also to a textbook clinical descriptions
of cerebellar disorders, such as Chapter 5 of Victor and Ropper (2001) and
Chapter 17 of Goetz (2007).]

Unfortunately, the cerebellum (and its input and output pathways) is
affected by a large variety of disease processes, of which a few may be mentioned here: a significant fraction of multiple sclerosis patients experience
cerebellar symptoms; tumors, both primary and metastatic; so-called remote
effects of cancer (Henson & Urich, 1982); cerebrovascular disease (stroke); a
number of hereditary progressive degenerative conditions; by the abuse of
alcohol. The principal cell type of the cerebellum, the Purkinje cell, is susceptible to anoxia and other biochemical and metabolic stresses; of course, being
neurons, these cells are not replaced, once lost. Multiple sclerosis is a common
neurological disease [prevalence ~0.9–1.0 cases per 1000 population (Hirtz
et al., 2007; Turabelidze et al., 2008)], with ataxia and tremor in greater than
50% in one study (Alusi et al., 2001b), and up to 80% of patients experiencing
ataxia at some point in their course (Mills et al., 2008). Cerebellar ataxia and
tremor are, in almost all cases, extremely difficult to treat. Ataxia and tremor

can be so disabling as to make a patient bedfast, unable to speak or swallow, and completely dependent on caregivers; stereotaxic surgery has been of benefit in some patients (Alusi et al., 2001a).

The phenomenology of cerebellar disorders has been described exhaustively; the anatomy and cellular physiology of the cerebellum have been researched extensively, by some of the field's most eminent practitioners (Chambers & Sprague, 1995; Eccles et al., 1964, 1966). But the underlying pathophysiology remains largely mysterious.

Historically, a study of immense importance was that of the British neurologist Gordon Holmes (1876–1966), undertaken at forward military hospitals on the Western front of World War I (Holmes, 1917; McDonald, 2007; van Gijn, 2007), and following the pioneering work of Joseph Babinski (1857–1932); Holmes observed the effects of "gunshot" (including shrapnel and fragment) injuries to the cerebellum of soldiers—of course, without benefit of modern imaging methods to precisely localize pathology; but with keen clinical observation, quantitation and permanent recording of motor behavior with improvised instruments, skull radiographs, and (on occasion) pathological study.

The major signs and symptoms of focal cerebellar injury, described and analyzed by Holmes, are listed next. Holmes' observations have been confirmed, but also refined, over the decades. So called pan-cerebellar and vermal syndromes, as well as exotic diseases such as kuru (Gajdusek & Zigas, 1957; Gajdusek et al., 1967), produce yet additional signs and symptoms.

• "Ataxia," a lack of coordination of movement that cannot be attributed to weakness, or to impaired sensory (including proprioceptive) input. Cerebellar ataxia typically is not worsened by eye closure, unlike "sensory ataxia," which is so enhanced (because vision partly compensates for loss of proprioception). Ataxia is notable in learned complex movements, of the limbs and of articulation; but gait and eye movements can also be ataxic.

• Tremor, especially with movement, such as reaching or grasping ("kinetic" or "intention" tremor), but possibly also with holding a posture. (Of course, not all tremors are due to disease of the cerebellum or its pathways.)

• Decomposition of movement. Most movements in humans involve the smooth interaction of flexions and extensions (and sometimes rotations) across several joints, both within a given limb (e.g., shoulder, elbow, and wrist), and between limbs (swinging the arms and legs in proper relative phases during walking). This type of interaction may fail with cerebellar disease, to be replaced not by paralysis, but by a less elegant performance, so that one joint carries out its action, then the next, then the next: "decomposition."

• Impaired rapid alternating movements. This is a sign that may be elicited during neurological examination, without the patient necessarily being aware of it. Dancers and musicians are aware of such difficulties. Rapid alternating movements are tested in the distal limbs by, for example, having the patient alternately strike the ball of the foot, or the heel, onto the floor; or using the thumb tip to tap onto the fingertips of the same hand in repetitive sequences.

The test can be altered to apply to other parts of the body as well, such as the tongue. With cerebellar disorders, it becomes difficult or impossible to maintain a quick and steady rhythm of such movements, and to keep the amplitude constant.

- Decreased muscle tone, without paralysis.
- "Pendular" deep tendon reflexes, in which, say, after a knee jerk is elicited, the foot swings to and fro.
- Walking into, or away from, a fixed obstacle, during turning. After a chair is placed in the middle of the floor, a normal person can walk rapidly in small circles around the chair, without hitting it or moving further away from it. With a lesion in a cerebellar hemisphere, a patient may walk in circles of smaller and smaller radius (eventually walking into the chair), or larger and larger circles.
- Other disorders of equilibrium and balance, in which a patient may be unable to sit or stand, with exhibiting of postural sway, slow tremor, and need for support; or, in less severe cases, showing an inability to walk on a narrow track ("broad-based" gait), or to walk heel-to-toe ("tandem" walking).
- Nystagmus, a type of involuntary movement of the eyes, most often (but not necessarily) involving conjugate movements of both eyes together. [Some types of nystagmus are normal—e.g., "optokinetic nystagmus," elicited by moving a tape with stripes (or rotating a striped cylindrical drum) in front of a person: the eyes follow a given stripe as it moves, until this is no longer possible, then jump back so as to follow the next stripe. Most sorts of nystagmus are, however, not normal.]

Thus, the cerebellum is not sufficient by itself to produce movements, but it is necessary for movements—especially complicated ones, involving many joints or muscles—to be executed smoothly and in coordinated fashion; and it is particularly required when sensory input (visual, proprioceptive, or vestibular, alone or in combination) is helping to guide the movement. The movements in question may be of the limbs, trunk, head, mouth, or eyes.

How is it that the cerebellum performs, or at least contributes to, its functions, and, in particular, can one give an account in cellular terms? Despite many landmark studies in anatomy and both systems and cellular neurophysiology (and we must mention especially the classic 1967 monograph of Eccles, Ito, and Szentágothai), there is no clear answer. In this chapter, we argue that cerebellar oscillations at least provide some clues and suggest research directions.

Structures Intimately Associated with the Cerebellum

While the discussion that follows concentrates on the cerebellar cortex (Latin *cortex* = "bark"), it is important to keep in mind that the cerebellum is intimately interconnected with a number of other important paired (i.e., existing on the left and right sides of the brain) structures, including the inferior olive in the medulla, the red nucleus in the midbrain (which consists of two subregions having distinct connectivity), the vestibular nuclei in the medulla (four on each side), and—perhaps most critically the deep cerebellar nuclei. In humans, there are four named cerebellar nuclei on each side—globose, fastigial, emboliform, and dentate; although, by analogy with the simpler anatomy of lower mammals, the fastigial and emboliform are often lumped together as the interpositus nucleus (Fig. 7.1). Gordon Holmes (1917) recognized what is now well known, that when cerebellar pathology impinged on the deep nuclei, then clinical symptoms were likely to be more severe and more long-lasting than when not. Detailed discussion of these "fraternal" structures is beyond the scope of this book, but we mention here the prominent existence of gap junctions in the inferior olive (Bourrat & Sotelo, 1983; King et al., 1975; Llinás et al., 1974; Weickert et al., 2005b) and vestibular nuclei (Korn et al., 1973), and the marked tendency of the inferior olive to generate its own oscillations

Figure 7.1 Section of a part of a human cerebellum, showing cerebellar cortex, white matter, and deep cerebellar nuclei.

(From Voogd, 2003, after Stilling, 1864, with permission.)

(Llinás & Yarom, 1981a,b, 1986), whose synchrony depends on gap junctions (Blenkinsop & Lang, 2006; Leznik & Llinás, 2005; Long et al., 2002). Further considerations on the inferior olive appear in the text that follows.

Overview of the Cerebellum

Like the cerebrum, the cerebellum has a folded, convoluted cortex (Fig. 7.1), containing neurons and locally ramifying and extending axons—but the folia are thinner and shallower than in the cerebral cortex. The folia predominantly run transversely (i.e., sideways). Again as for the cerebrum, beneath the cortex lies white matter, consisting mainly of myelinated axons; but there is a contrast: the *cerebral* subcortical white matter contains fibers that run varying distances from cortical region to cortical region, with some of these distances being quite long (e.g., from parietal lobe to frontal lobe). In the *cerebellar* cortex, such long-distance subcortical white matter pathways appear to be absent: the cerebellar cortex is a far more "modularized" piece of apparatus than the cerebral cortex—a point to which we must return later, as the apparent "modularization" may be deceptive. The cerebellar white matter contains afferents to, and efferents from, the cerebellar cortex.

Contained in the white matter on each side, the cerebellum has four *deep cerebellar nuclei* (Fig. 7.1), named dentate, fastigial, globose and emboliform—with the globose and emboliform lumped together into the interpositus nucleus in "lower" animals. These nuclei play a role vaguely analogous to the thalamus for the cerebral cortex, but with critical differences here as well: (1) Sensory afferent fibers do not proceed directly to the cerebral cortex, but instead most of them synapse on thalamic relay cells, and the relay cells in turn project axons to the cerebral cortex—whereas afferents to the cerebellar cortex proceed directly there, sending collaterals to the deep cerebellar nuclei. (2) While the cerebral cortex has a significant projection to the thalamus, both to the relay cells and to the nucleus reticularis GABAergic cells, the *most prominent output tracts* (e.g., the corticospinal tract, the corticopontine pathway, the corticostriate projection) bypass the thalamus. In the case of the cerebellum, however, the great majority of output fibers from the cerebellar cortex synapse within the deep cerebellar nuclei.

The surface topography of the cerebellum has been cartographically described in some detail (see, e.g., Palay & Chan-Palay, 1974). A much simpler classification, used by clinicians, suffices here: the most critical structures are the flocculus and nodulus (forming the *flocculo-nodular lobe* or *vestibulocerebellum*); the central vermis ("worm"); the anterior lobe; and the cerebellar hemispheres. The anterior lobe and most of the vermis together form the "spinocerebellum," receiving somatosensory inputs from the spinal cord; the cerebellar hemispheres and a small portion of the vermis form the pontocerebellum, or neocerebellum, with major inputs from the pontine nuclei (Dietrichs, 2008). The division of the cerebellum into these three major divisions has anatomical significance, in terms of input and output pathways, and functional significance for behavior as well; but the cellular architecture and intrinsic connectivity of the cerebellar cortex is strikingly uniform. [The cerebellar uniformity can again be contrasted with the cerebral cortex, which is divided into dozens of cytoarchitectonic regions (Brodmann, 1909), and where gross structural differences are often easily apparent—e.g., the relative absence of layer 4 in motor cortex, or the subdivisions of layer 4 in primary visual cortex.]

Intrinsic Organization of the Cerebellar Cortex: Cell Types, Synaptic Organization, Purkinje Cell Behavior

The basic cellular elements of the cerebellar cortex were delineated by the great Ramón y Cajal (Fig. 7.2), although some new cell types have been discovered since (see later). We list here the elements shown in Figure 7.2, starting with what Cajal drew, then elaborating with some physiological data:

• *Purkinje cells* (Fig. 7.2a), the principal neuron of the cerebellar cortex, receiving two distinct types of excitatory afferent (to be described), as well as GABAergic input from other cerebellar neurons. Purkinje cells themselves are GABAergic (inhibitory)—another major contrast with the cerebral cortex, whose major output neurons (deep layer pyramidal cells) are glutamatergic and excitatory. The Purkinje cell has an extensive dendritic arborization, shown in Figure 7.2, with a proximal smooth (nonspiny) portion, and innumerable spiny branchlets that receive, in total, more than 100,000 excitatory synapses [although many or most of these synapses are probably "silent" (Isope & Barbour, 2002; Jörntell & Hansel, 2006)]. The dendrites lie roughly in a plane, and the dendritic planes of nearby Purkinje cells are nearly parallel, and orthogonal to the axis of the local folium. Purkinje axons (*o* in Fig. 7.2) descend through the white matter to one or another deep nucleus, but also emit collaterals to nearby Purkinje cells: two such collaterals are drawn in Figure 7.2. There are an estimated 340,000 Purkinje cells in rat cerebellum. Their somata lie in a single layer, the "Purkinje cell layer." Ghosts of numerous Purkinje cells are drawn in Figure 7.2, allowing one to picture this layer. (The region inhabited by the Purkinje cell dendrites is called the "molecular layer"; and the region just below the Purkinje somata, through which the Purkinje axons pass, is called the "granular layer," for reasons that will become apparent.)

• The electrophysiological properties of Purkinje cells are extremely rich (Llinás & Sugimori 1980a,b, and many dozens of succeeding papers.) While we shall return to this subject in Chapter 10, when considering the detailed cellular mechanisms of cerebellar oscillations, suffice it to say that Purkinje cells can generate (1) fast Na^+ spikes, that initiate in the axon and propagate forward as well as into the soma (but not much past the proximal dendrites, in the retrograde direction), (2) dendritic Ca^{2+} spikes, which influence spike

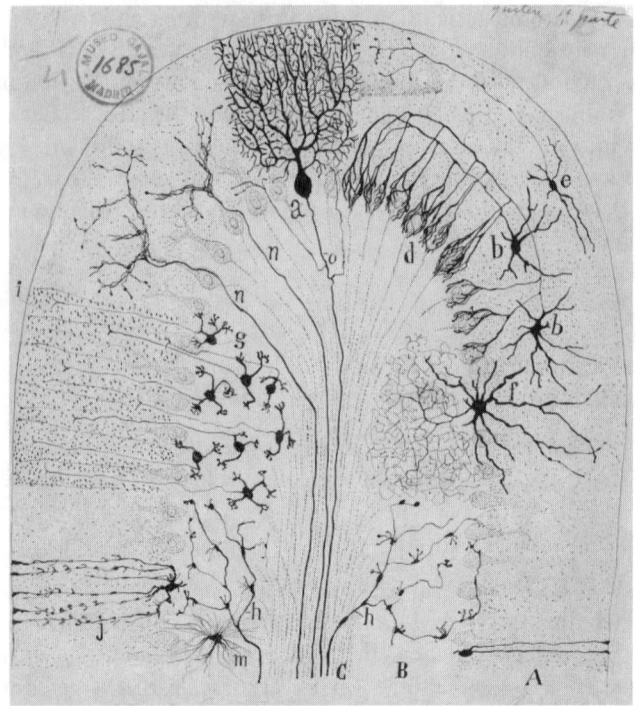

Figure 7.2   Semischematic composite drawing of the cells and fibers in mammalian cerebellar cortex by Ramón y Cajal. Structures include: *a*, Purkinje cell, with *o*, axon; *b*, basket cell, with *d*, baskets around soma and proximal axon of Purkinje cells; *e*, stellate cell; *f*, Golgi cell; *g*, granule cells (the many dots around the axons of the granule cells represent cut parallel fibers that run orthogonally to the section shown); *h*, mossy fibers; *n*, climbing fibers; *j* and *m*, glia.

[From Sotelo (2003) with permission from the Cajal Institute (original deposited at the Cajal Institute CSIC, Madrid).]

output, synaptic integration, and synaptic plasticity, (3) a variety of other forms of electrogenesis.

• *Climbing fibers* (Fig. 7.2n), excitatory axons originating from neurons in the inferior olivary complex of the medulla, which send collaterals into the deep nuclei, branch in the white matter, and which each contact several Purkinje cells—but so that a Purkinje cell receives input from only one climbing fiber. [The olive in turn receives inputs from multiple sources, including the cerebellum itself, the periphery, motor cortex, and the caudate nuclei (Sedgwick & Williams, 1967).] A climbing fiber has about 500 release sites, primarily on smooth dendrites, with quantal amplitude approximately 0.5 nS (Silver et al., 1998), so that mean excitatory postsynaptic conductance (EPSC) amplitude would be a remarkable 250 nS. [As a contrast, pyramidal cell-to-interneuron synaptic connections may use only *one* release site (Gulyás et al., 1993), although this one site may still be enough to initiate an action potential.] Needless to say, climbing fiber-induced excitatory postsynaptic potentials (EPSPs) in Purkinje cells are large and long, not only because of the huge synaptic conductance [mediated by α-amino-3-hydroxyl-5-methyl-4-isoxazole-propionate (AMPA)/kainate receptors], but also because of augmentation by dendritic Ca^{2+} spikes. Several high-frequency fast action potentials occur during the climbing fiber response—the so-called "complex spike"—with the first action potential being most reliably conducted along the axon, and a mean of two spikes being propagated per response; in contrast, "simple" spikes—those not elicited by climbing fibers—propagate with approximately 90% reliability (Khaliq & Raman, 2005). Climbing fiber responses (complex spikes) occur at about 1 Hz, on average, in vivo (Armstrong & Rawson, 1979); and they are followed by a brief pause in Purkinje cell firing. [In contrast, the other kind of Purkinje cell response—so-called "simple" (or individual) spikes—occur at dozens of Hz in vivo.]

• *Mossy fibers* (Fig. 7.2h), the most numerous cerebellar afferents. These excitatory axons originate from many sources, including, but not limited to, Clarke's column in the spinal cord (carrying proprioceptive and other somatosensory information), the pontine nuclei (in turn receiving cortical inputs), the vestibular nuclei, and the reticular formation. The designation "mossy" derives from the small blobs that are apparent on these axons.

• *Granule cells* (Fig. 7.2g) receive synaptic excitation from mossy fibers, as well as synaptic inhibition from Golgi interneurons. Granule cells are extremely numerous (estimated 92,000,000 in the rat, ~275 times more numerous than Purkinje cells (Harvey & Napper, 1988), and small (soma diameter 5–8 μm), with 3–5 stubby dendrites that each receive only one excitatory terminal (Jörntell & Ekerot, 2006; Palay & Chan-Palay, 1974). Granule cell dendrites, mossy fiber terminals, and Golgi axon terminals form structures called *glomeruli* (from the Latin meaning "little ball").

• *Parallel fibers* (Fig. 7.2i) are formed as the ascending axons of granule cells (some of which are visible in Fig. 7.2) bifurcating in the molecular layer into two branches. The parallel fibers run *along* folia, and Figure 7.2 depicts (schematically) a section orthogonal to a folium: hence, the parallel fibers are cut in cross section, and Fig. 7.2 appropriately shows them as a collection of dots (*i*). The reader is referred to the Eccles et al. (1967) for some beautiful colored three-dimensional drawings that show the parallel fibers as actual parallel fibers (versions of these drawings have appeared in textbooks ever since). The parallel fibers are unmyelinated axons, and form excitatory synapses on the spines of Purkinje cells, among other places (Fig. 7.3).

• *Golgi neurons* (Fig. 7.2f) are GABAergic interneurons having a dendrite extending up into the molecular layer, and an axon ramifying in the granular layer.

• *Basket cells* (Fig. 7.2b) are GABAergic interneurons with somata lying in proximal stratum moleculare, with axons forming "baskets" (d) around Purkinje cell somata and proximal axons. Some of the axon terminals of basket cells are interconnected by specialized junctions—septate junctions—of unknown function (Gobel, 1971). The collection of interconnected terminals around the Purkinje axon initial segment is called a pinceau [French: (paint)brush]. Korn and Axelrad (1980) have suggested that the pinceau inhibits the initial segment by a "field effect," that is, effects on transmembrane potential caused by current flow in the extracellular space—in addition to (but sooner than) effects produced by GABA release. Basket cells receive synaptic inhibition from Purkinje cell collaterals and from stellate cells (Bishop et al., 1993).

• *Stellate cells* (Fig. 7.2e) are GABAergic interneurons with somata lying in stratum moleculare, whose axons provide synaptic inhibition to Purkinje cell dendrites. Molecular layer interneurons, including both basket cells and stellate cells, are dye-coupled with one another and also electrically coupled (Mann-Metzer & Yarom, 1999); gap junctions have been demonstrated in this region of the cerebellum with ultrastructure (Sotelo & Llinás, 1972).

• Further information on the synaptic organization of the cerebellum can be gleaned from Figure 7.3, reproduced from a review by Masao Ito (2006). Figure 7.3 is, of course, highly schematic, but highlights the following:

• Parallel fibers excite not only Purkinje cell dendrites, but also the dendrites of Golgi, stellate, and basket cells (GO, SC, BC).

• Additional cell types have been discovered in the cerebellum since Ramón y Cajal. Examples in the figure are the *Lugaro cell* (LC), a GABAergic cell excited by serotonergic afferents (SR) derived from the raphe nuclei; and the *unipolar brush cell*, an excitatory interneuron found mostly in the vestibulocerebellum (Kalinichenko & Okhotin, 2005) [and in the dorsal cochlear nucleus in the brain stem, a nucleus that has a cellular organization reminiscent of the cerebellar cortex (Diño et al., 2008)].

• A pathway exists from the deep cerebellar nuclei to the inferior olive, the origin of the climbing fibers (the *nucleo-olivary pathway*, N-O)

• The deep cerebellar nuclei are another source of mossy fibers (*nucleocortical afferents*, N-C).

• The red nuclei receive input from the deep nuclei, and in turn send output to the inferior olive (R-O).

The Lack of Recurrent Synaptic Excitation Within the Cerebellar Cortex

We come now to what is perhaps the most striking difference between cerebellar and cerebral cortices, so far as circuit organization goes: most of the synapses within the cerebral cortex are excitatory synapses lying on excitatory neurons. Thus, with suppression of synaptic inhibition, synchronized epileptiform discharges will develop (Chapter 4). In the cerebellar cortex, however, at least outside the vestibulocerebellum (which contains unipolar brush cells), the only excitatory cells—the granule cells—lie in the afferent stream of mossy fibers, and serve to distribute this stream along the parallel fibers. All of the

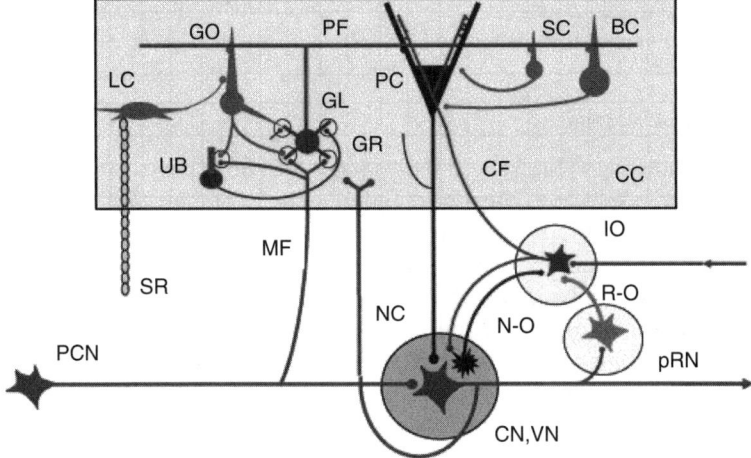

Figure 7.3 Schematic of synaptic circuitry within the cerebellar cortex (represented by what is in the tan box, CC = cerebellar cortical "microzone"), and afferents to and efferents from the cerebellar cortex. *BC*, GABAergic basket cell, in stratum moleculare; *CF*, climbing fiber, originating in the inferior olive, and contacting soma and smooth dendrites of Purkinje cell, with collaterals to cerebellar nuclei; *CN*, *VN*, cerebellar and vestibular nuclei, which receive collaterals of cerebellar afferents and also receive contacts from Purkinje cell axons; *GL*, glomerulus; *GR*, granule cells, giving rise to *PF*, parallel fibers; *GO*, Golgi cell; *IO*, inferior olive; *LC*, Lugaro cell; *MF*, mossy fiber, originating from any of a number of sources, synapsing in the glomeruli, with collaterals to cerebellar nuclei; *N-C*, nucleo-cortical mossy fiber afferent; *N-O*, nucleo-olivary connection (inhibitory); *PC*, Purkinje cell; *PCN*, "precerebellar" neuron, e.g. in pontine nuclei, or in Clarke's column in lower spinal cord; *PF*, parallel fiber, originating from granule cells, and making excitatory synapses on Purkinje cell dendrites, and on Golgi, stellate and basket cells; *pRN*, parvicellular portion of red nucleus, which receives a part of the cerebellar output, and provides some of the input to the inferior olive (*R-O*, rubro-olivary excitatory connection); *SC*, stellate cell; *SR*, serotonergic fiber, contacting a Lugaro cell; *UB*, unipolar brush cell.

(From Ito, 2006 with permission.)

other cerebellar cortical neurons are inhibitory neurons, and all of the recurrent synaptic connections are GABAergic. So far as we know, no collective activity in the cerebellar cortex has been described, which consists of synchronized burst discharges resembling a cerebral cortical interictal burst or electrographic seizure.

Nevertheless, the absence of recurrent synaptic excitation in the cerebellar cortex does not mean that it is impossible for action potentials in one neuron to induce action potentials in another neuron—such induction is possible (in principle) across gap junctions. Whether the appropriate gap junctions actually exist is a matter for experiment to determine. We shall return to this topic in due course.

Cerebellar Afferents

Cerebellar afferents (climbing and mossy fibers), and efferents, are somatotopically (topographically) organized, but in a more complex fashion than for, say, primary somatosensory cortex (Garwicz et al., 1998). In the ensuing, we follow the review of Manni and Petrosini (2004). The somatotopic arrangements of mossy and climbing fiber inputs correspond, at least roughly, and both are most readily demonstrated in the anterior lobe (spinocerebellum) by stimulation of the periphery—with, for example, tactile stimulation, muscle stretch, and joint movements (Adrian, 1943). Even here, however, somatotopy is most clearly defined in anesthetized animals, and is more diffuse without anesthesia (Dow, 1939; Combs, 1954). Manni and Petrosini (2004) propose that there may be two superimposed sensory representations, one diffuse, one precise. In addition, throughout the cerebellum, representation of body regions on the cerebellar surface is not continuous; rather, regions that are contiguous on the body surface map to regions that are not contiguous on the cerebellar surface—and each body region maps to a number of spots on the cerebellar surface. This whole arrangement has been called "fractured somatotopy" (Shambes et al., 1978).

There is spatial organization within the deep nuclei as well as for the cerebellar cortex (Asanuma et al., 1983); and there is a more global set of inter-relations between regions of deep nuclei, inferior olive, and parasagittal strips of cerebellar cortex—these are referred to as "complexes."

The neocerebellum has a massive extent in humans, and a correspondingly massive mossy fiber input from the cerebral cortex, relayed through the pontine nuclei. In the cat, all cortical regions—not just motor regions—project to the pontine nuclei (Chiba, 1980). [In this sense, the cerebellum is not purely a motor organ—although it is not easy to show this clinically, at the bedside.] Corticopontine fibers are often collaterals of corticospinal or corticobulbar axons; and each axon entering the pontine nuclei is likely to contact several neurons (Bolstad et al., 2007).

Cerebellar Efferents and the Deep Cerebellar Nuclei

The entire output of the cerebellar cortex is carried by Purkinje cell axons. Where does it go? Some Purkinje axons project directly to the vestibular nuclei or reticular formation, but the preponderant cerebellar output is relayed in the deep cerebellar nuclei, which contain their own patterns of organization (Asanuma et al., 1983). Projection cells in the deep nuclei are both glutamatergic and GABAergic; there are smaller GABAergic cells as well (Uusisaari et al., 2008). There seems to be relatively little recurrent synaptic connectivity within the deep nuclei (Uusisaari & Knöpfel, 2008); gap junctions have not been ruled out. Most of the long-range output fibers proceed in the superior cerebellar peduncle (in which they cross to the opposite side), and synapse in the red nucleus or in the motor (or "cerebellar") thalamus (Thach, 1987). As Figure 7.4 outlines, there is considerable organization in the mapping from cerebellar cortex, through particular deep nuclei, on to selected thalamic nuclei, and there to cortex, especially (but not exclusively) "motor" regions—primary

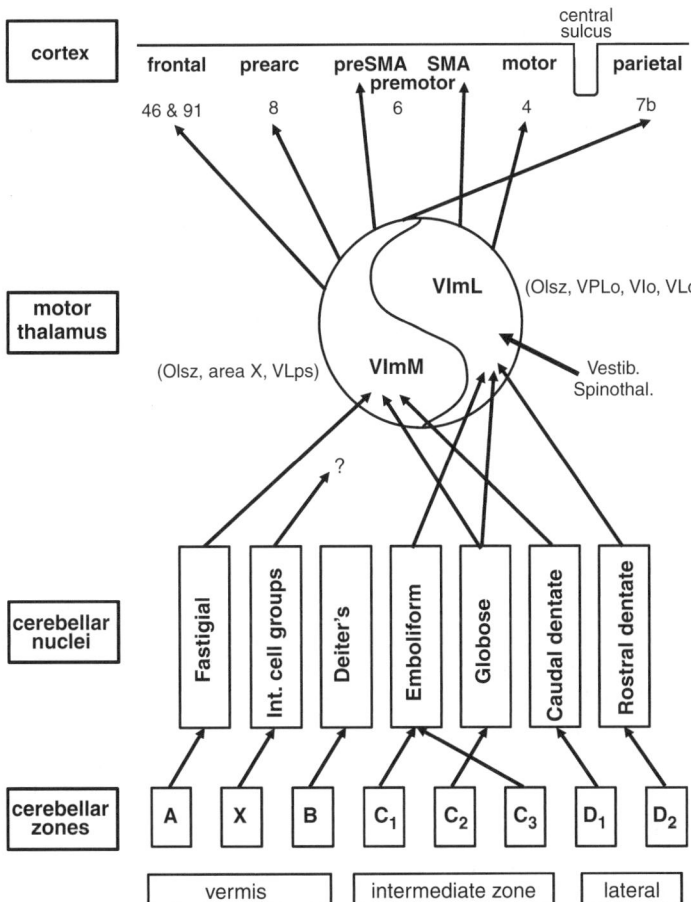

Figure 7.4 Projections—in primates—from portions of the cerebellar cortex ("zones") to portions of the cerebral cortex (labeled by Brodmann areas), through particular deep cerebellar nuclei, and through particular regions of the motor (cerebellar) thalamus. *Mot*, primary motor cortex; *Olsz*, Olszewski; *Pariet*, parietal lobe; *Prearc*, prearcuate area; *Premot*, premotor area; *SMA*, supplementary motor area; *VLc*, caudal ventrolateral nucleus; *VLo*, oral ventrolateral nucleus; *VLps*, posterior ventrolateral nucleus; *VPLo*, oral lateral ventral posterior nucleus. (From Voogd, 2003 with permission.)

and supplementary motor areas, and premotor cortex. This organization takes on relevance when we come to consider coherent oscillations that involve both cerebellum and cerebral cortex.

Recall that Purkinje neurons are GABAergic and induce inhibitory postsynaptic potentials (IPSPs) in target neurons. Nevertheless, at least in some preparations (but see later), deep cerebellar nuclear neurons can exhibit rebound firing after a hyperpolarizing pulse (Llinás & Mühlethaler, 1988), so that—in principle at least—deep nuclear cells might be excited (after a delay) by Purkinje cell activity. In addition, experimental stimulation of deep nuclei is capable of producing movement, possibly employing a pathway through the red nucleus (Asanuma & Hunsperger, 1975; Giuffrida et al., 1982); the more usual effects of stimulation of the cerebellum are, however, interference with the performance of an ongoing movement. A recent paper (Alviña et al., 2008) suggests that rebound firing, in deep cerebellar nuclear neurons, is actually rather unusual (but not impossible) in vivo, and is most likely to occur under unphysiological conditions, in which membrane potential is brought below the $GABA_A$ reversal potential. All of these observations are logically compatible: deep nuclear cells can provoke movement if made to fire; suppressing deep nuclear activity by Purkinje cell activation would suppress movement; but perhaps, under special conditions, a minority of nuclear neurons exhibit delayed excitation after cerebellar cortical activity.

Some Theories of Cerebellar Function: Problems of Time and of "Binding" Within the Motor System

Inferring nervous system function from structure alone has proved to be a daunting task even for towering figures like Ramón y Cajal and Lorente de Nó; with the provocatively titled 1967 monograph of Eccles et al., however, it began to look as though enough combined physiological/anatomical knowledge on the cerebellum was available, so that useful theories on function could be offered. David Marr (1945–1980) offered one of the best-known models in his 1969 *Journal of Physiology* paper. Marr proposed several ideas: that inferior olivary neurons (hence individual climbing fibers) correspond to a "cerebral instruction for an elemental movement"; that firing of a Purkinje cell is able to induce such a movement; that Purkinje cell discharge can also be induced by an appropriate pattern, or combination, of parallel fiber inputs; and that a climbing fiber input acts—so to speak—as a teacher of the Purkinje cell, so that the Purkinje cell eventually will fire in response to each appropriate parallel-fiber pattern, without the help of simultaneous climbing fiber input: the teaching takes place by synaptic potentiation (what we would now call long-term potentiation (LTP)). James Albus (born 1935) independently developed a somewhat similar theory (Albus, 1971), again emphasizing a role for inferior olivary neurons/climbing fibers in motor learning, teaching Purkinje cells by modifying parallel fiber synaptic strengths. There were two key differences from the Marr theory: now, inferior olivary neurons generate error signals, between intended and actual motion, rather than forming simply part of a large repertoire of possible movements; and parallel fiber EPSPs were predicted by Albus to be *depressed* by conjunction with climbing fiber inputs, rather than *potentiated*. Nevertheless, both theories—collectively now designated the Marr–Albus theory—similarly emphasized the selection of patterns of parallel fiber inputs, and the importance of climbing fiber inputs in learning. An enormous boost was given to the Marr–Albus corpus by the experimental demonstration of long-term depression of parallel fiber inputs by coincident complex spikes (Ito, 1982), with the (as subsequently shown) necessary entry of Ca^{2+} into Purkinje-cell dendrites (Konnerth et al., 1992). The study of cerebellar long-term depression (LTD) has indeed become a large field in its own right (Ito, 2006).

One ethological problem in the motor-learning hypothesis, however, is this: why is it that many animals have a prominent cerebellum, yet do not obviously need to engage in motor learning? Frogs do not learn to play the piano or dance the tango. Is it not possible that the cerebellum is engaged in motor performance, even without learning, and that complex spikes have some other functional role besides teaching?

Valentino Braitenberg (born 1926) proposed, in 1961, a rather different idea, which might be paraphrased this way: if the cerebellum helps to provide fine control of movement, it must have some way of measuring time, or at least computing time delays. Braitenberg proposed that such computation could be accomplished through the conduction times along "beams" of parallel fibers, which run long distances (millimeters at least) along the cerebellar cortex, and which contact the planar dendritic fields of Purkinje cells one by one. Or, to put it another way, two Purkinje cells that lie along a beam, separated by distance Δx, ought to be excited by a "volley" of parallel fibers with *defined* temporal separation Δt (Freeman & Nicholson, 1970). While this notion may be valid, the recent data of Heck et al. (2007) cast doubt on whether one should look at the cerebellum as centered around "parallel fibers as delay lines." Thus, what Heck et al. (2007) showed was this: they recorded from pairs of Purkinje cells in awake head-restrained rats that performed reaching movements, to gather food pellets. The two Purkinje cells might be "on beam" (lying along the same beam of parallel fibers), or "off beam". When the cells were on-beam, phase-locking of simple spike firing occurred between the two Purkinje neurons, at statistically significant levels, independent of firing rate, but induced by the movement (i.e., rather than being present all the time). While the phase-locking sometimes exhibited time delays consistent with parallel fiber conduction delays (i.e., consistent with Braitenberg's 1961 hypothesis), by far the most common temporal interaction was *synchrony* of firing. In this sense, the cerebellar cortex behaves, during a learned motor activity, rather analogously to the cerebral cortex (at least parts of it) during appropriate sensory stimulation—there is synchrony. But synchrony in the cerebral cortex is associated with oscillations.

The time is clearly approaching when we must consider oscillations in the cerebellar cortex, also.

First, however, we raise another analogy with oscillations in the cerebral cortex. Locally synchronized oscillations, evoked by sensory input, may provide for binding between distinct neocortical regions, by virtue of inter-regional synchrony (Chapter 3). Further, motor cortical oscillations are phase locked to muscle electromyographic (EMG) oscillations (Chapter 3). Do similar principles apply to the cerebellum? One imagines that this might be the case, because the cerebellum is involved in coordination of different parts of the body—and different parts of the body are represented in different parts of the cerebellum: these different cerebellar parts must surely interact with each other somehow. Do they oscillate together, as in the cerebral cortex? And, if so, how is long-range synchrony achieved, given the absence of recurrent synaptic excitation?

In the next sections, we consider two types of cerebellar oscillations: very fast oscillations (> ~70 Hz, "VFO"), and oscillations at lower frequencies: gamma down to theta (Hartmann & Bower, 1998). The latter oscillations occur during physiological conditions, and may be related to cerebellar function. VFO, however, is probably pathological in the cerebellum; however, pathological or not, VFO offers vital clues to circuit organization. These clues are introduced here, but expanded upon in a later chapter.

Network Oscillations in the Cerebellum and Connected Systems: VFO

Cerebellar very fast oscillations (VFO) were described by Adrian in 1935 [see Isope et al. (2002) for a reproduction of one of Adrian's figures]. De Solages et al. (2008) observed what may be the same phenomenon, in the cerebellum of both anesthetized and unanesthetized mice. Field potential oscillations at approximately 200 Hz were maximal in the Purkinje cell layer, having there amplitudes of tens to hundreds of microvolts; the firing of Purkinje cells was synchronized between pairs of cells, and to the field potential, and synchrony was present for cells firing both at high and low rates (reminiscent of Heck et al., 2007). Unlike the findings of Heck et al., however, the VFO of de Solages was coherent over hundreds of micrometers, either "on-beam" or "off-beam." The in vivo VFO did not require afferent input, as it persisted after section of the cerebellar peduncles (containing all of the input pathways to the cerebellum), and it likewise persisted after pharmacological blockade of AMPA receptors. Blockade of $GABA_A$ receptors with picrotoxin, however, suppressed the VFO. [One cannot infer from this, however, that such an experiment proves involvement of phasic IPSPs in generating the VFO: $GABA_A$ receptor blockade also suppresses VFO that appears to be nonsynaptically generated (Traub et al., 2003b), probably because of blockade of axonal $GABA_A$ receptors (Stasheff et al., 1993a,b).] Unfortunately, gap junction pharmacology was not addressed. De Solages et al. (2008) accounted for their VFO with a network model in which the primary means of interaction between cells was the system of Purkinje cell collaterals, which communicate via GABAergic synapses on other Purkinje cells; the period of the VFO was determined by the latency and rise time of IPSC, as well as the rapid decay kinetics of IPSCs.

Guy Chéron, Laurent Servais, and their colleagues in Belgium have investigated a remarkable collection of genetically modified mice, all of which are ataxic, and all of which exhibit cerebellar VFO, occurring in short epochs (Cheron et al., 2004, 2005a,b, 2008: Servais & Cheron, 2005; Servais et al., 2005). The genetic modifications fall into two general categories: (1) knockout of one or another calcium-binding protein, or double knockout of a pair of such proteins (e.g. calbindin, parvalbumin) and (2) loss of the maternal copy of the *ube3a* gene, coding for ubiquitin ligase. The latter mouse constitutes an animal model of a devastating human disorder, Angelman syndrome (Lalande et al., 1999; Williams, 2005). (Angelman syndrome consists of mental retardation with the lack of speech acquisition as the major problem, with relative preservation of social interactions, and a generally cheerful disposition; ataxia and epileptic seizures can also occur. Of course, the genetically modified mice can not exhibit the speech disorder.)

As Figure 7.5 shows, cerebellar VFO in the mice studied by the Belgian group (which we shall call "mutant VFO") can be particularly robust and

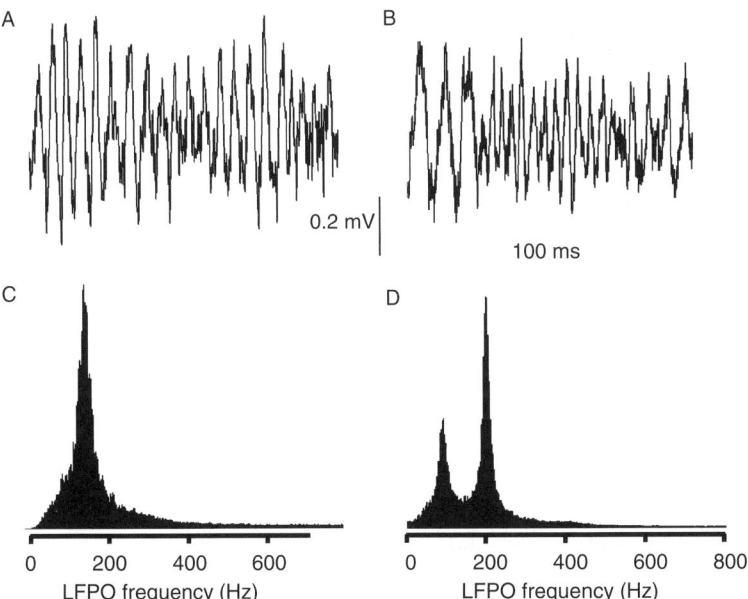

Figure 7.5 Very fast oscillations (VFOs) in the cerebellar cortex (vermis) of ataxic genetically modified mice. Field potentials are shown above, power spectra (for 60 s of data) below. **A, C**: Oscillations from parvalbumin knockout, Pv$^{-/-}$; **B, D**: Oscillations from parvalbumin-calbindin D-28K double knockout, Pv$^{-/-}$ Cb$^{-/-}$. Note the dual spectral peaks in the double knockout.

(From Cheron et al., 2008, originally in Servais et al., 2005. Reproduced with permission.)

large, with amplitudes approaching 0.5 mV. Interestingly, in some of the mutated animals, two distinct frequency peaks appear (Fig. 7.5B,D). Mutant VFO is also of maximal amplitude in the Purkinje cell layer, as is VFO in wild-type animals, and it is similarly suppressed by GABA-receptor blockade [and also, strangely enough, by N-methyl-D-aspartate (NMDA) receptor blockade]. In addition, the Belgian group observed that locally applied carbenoxolone, a gap junction blocker, would suppress mutant VFO (Cheron et al., 2005b; Servais et al., 2005;).

From the available data, then, there seem to be two possibilities: (1) mutant VFO is simply an exaggerated form of VFO that occurs in the cerebellar cortex or (2) there are two distinct types of VFO: a slower one at frequencies of approximately 160 Hz and less (as in the parvalbumin knockout in Fig. 7.5A,C; and in

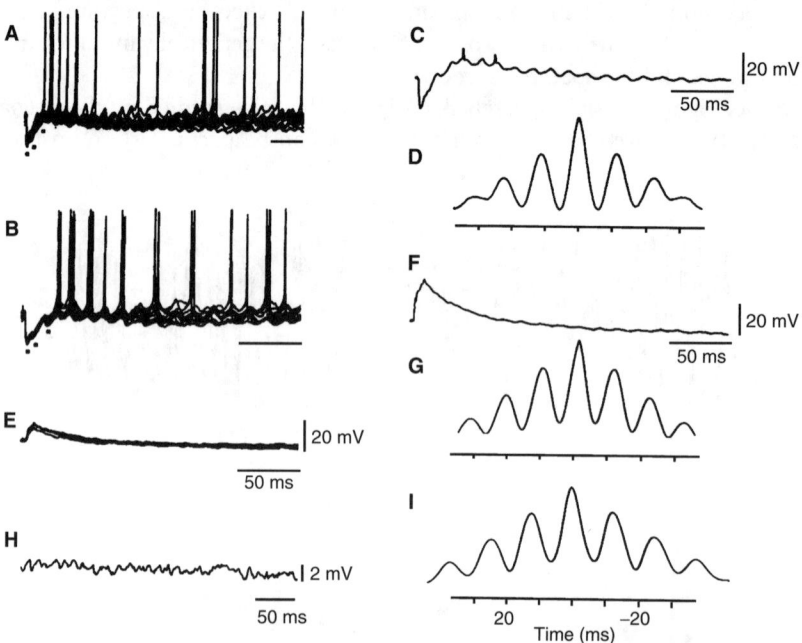

Figure 7.6 Very fast rhythmic IPSPs in deep cerebellar nuclear neurons, in the brain stem–cerebellar preparation of the guinea pig. It is expected that cerebellar cortical VFO (as in Fig. 7.5) would induce similar (but faster) activity in these cells. **A:** A shock induces an IPSP followed by rebound firing with rhythmic IPSPs (multiple traces superimposed). **B:** Similar data at faster sweep speed. **C:** Rhythmic IPSPs are still present in averaged traces. **D:** Autocorrelation of data from C shows frequency of ~83 Hz. **E:** The induced IPSP is reversed by hyperpolarization **F:** Rhythmic synaptic potentials continue to be present when the initial IPSP is reversed, **G**, at about the same frequency as before. **H:** Rhythmic synaptic potentials ocurring spontaneously. **I:** Autocorrelation of data in H.

(From Llinás and Mühlethaler, 1988 with permission.)

the double knockout in Fig. 7.5B,D); and a faster one at approximately 200 Hz (as in the double knockout in Fig. 7.5B,D; and in nonmutant animals). This issue remains to be resolved.

What might be the effects of cerebellar VFO on downstream structures, the deep cerebellar nuclei? Some clue to the answer to this question derives from the data in Figure 7.6, which shows recordings from deep cerebellar neurons in the isolated brainstem-cerebellar preparation, following a shock to the white matter (Llinás & Mühlethaler, 1988): one observes sustained (hundreds of milliseconds; compare Figs. 7.5 and 7.7) rhythmical IPSPs, in this case at approximately 83 Hz. Unfortunately, simultaneous field potentials were not recorded in the cerebellar cortex; these would have perhaps helped to confirm that the IPSPs represent a network phenomenon (produced by ensembles of Purkinje cells), rather than the output of a single Purkinje cell.

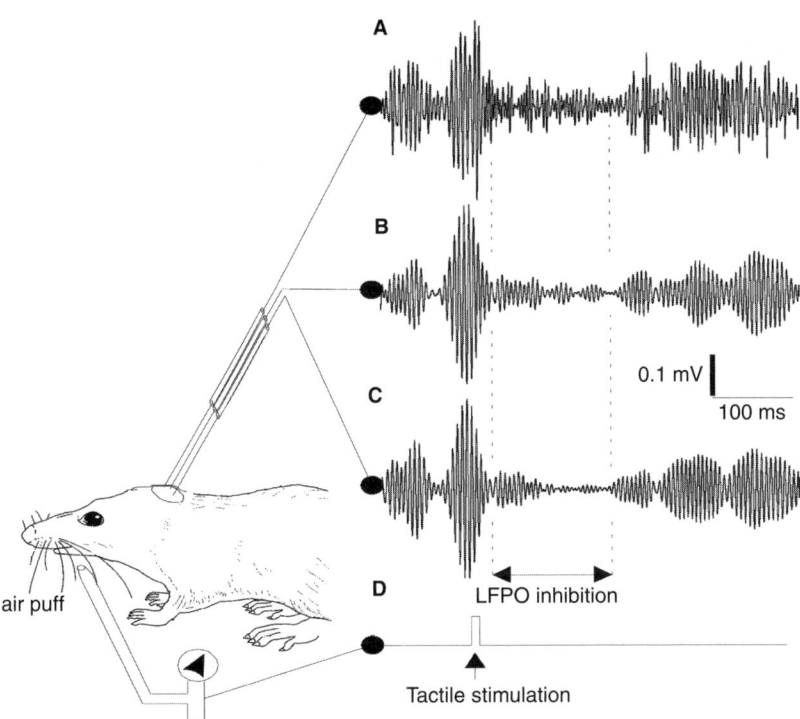

Figure 7.7   Cerebellar cortical VFO in a different ataxic mouse (modeling Angelman Syndrome) is temporarily suppressed by sensory input. The mice had a maternal null mutation in the gene Ube3a (Ube3a m−/p+). Field oscillations were recorded from lobules 9a, 9b, and Crus Ia, and low-pass filtered (<200 Hz), then averaged, using as trigger the first wave after the tactile stimulus. The stimulus was a brief air puff to the whisker region. Data in **A**, **B**, **C** were recorded from electrodes 250 μm apart.

(From Cheron et al., 2005b with permission.)

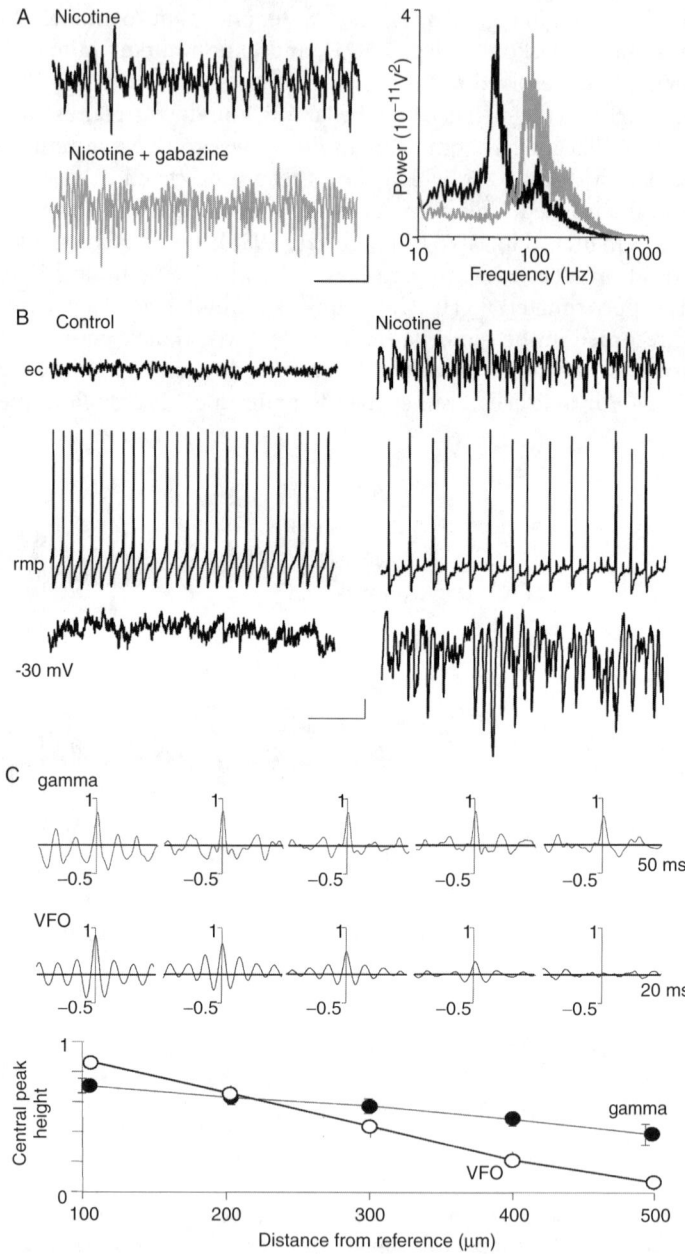

Figure 7.8 Both gamma (~40 Hz) and VFO (~100 Hz) network oscillations can be induced pharmacologically in rat cerebellar slices. **A**: Oscillating field potentials (Purkinje cell layer) in Crus I/Crus II. Nicotine (10 µM) induces gamma oscillations; addition of gabazine (1 µM, to block GABA$_A$ receptors) converts the gamma to very fast oscillations, VFO, here about 100 Hz. Power spectra on the right. Scale bars: 0.1 mV, 100 ms. **B**: With the development of nicotine-induced gamma oscillations (right traces), Purkinje cell firing rates *drop*, spikelets occur, and there are large rhythmic IPSPs in the Purkinje cells. Traces are: field (ec), intracellular at resting membrane potential (rmp), intracellular when holding at -30 mV (to highlight the IPSPs). Scale bars: 100 ms and 50 µV, 10 mV, 2 mV. **C**: Oscillations were recorded simultaneously with two electrodes in the Purkinje cell layer, at various separations, and cross-correlations were calculated (from 2 s epochs of pooled data)—shown in the upper traces for gamma and for VFO. Gamma oscillations remain highly correlated over at least 500- µm separations.

(From Middleton et al., 2008 with permission.)

The size of the initial IPSP, however, suggests that multiple Purkinje cells must be participating in this response. With deep cerebellar nuclear cells entrained to an autonomous rhythm, presumably they are not available to perform their normal computations, and this state is related to the ataxia in the mutant mice discussed above; even if this notion is true, however, it does not account for the presence of cerebellar VFO in putatively normal mice. One reasonable hypothesis would be as follows: the VFO that is normally present in the cerebellar cortex is not sufficient to prevent firing in deep cerebellar nuclear cells; but when cortical VFO becomes excessive (e.g., in the mutant ataxic mice), then nuclear firing is indeed suppressed, producing effects comparable to cooling of a deep nucleus: interference with movement (Conrad & Brooks, 1974; Tsujimoto et al., 1993).

One remarkable feature of mutant VFO is that it is suppressible for 100 ms or so by brief sensory input, in alert mice (Fig. 7.7). Thus, even as the de Solages et al. (2008) data suggest that cerebellar afferents are not required to generate VFO, still, cerebellar afferents can influence VFO, at least in the mutant mice. Presumably, one or more types of cerebellar interneurons are responsible for the VFO suppression, but this has not yet been proven directly.

While detailed consideration of oscillation mechanisms are deferred until later chapters, we would like to introduce here the fact that cerebellar VFO can be induced in vitro, in slices of both rat and also human biopsy tissue (Fig. 7.8 and Middleton et al. (2008)). At least in certain regions of cerebellar cortex, nicotine induces approximately 40 Hz gamma oscillations, intermixed with VFO (Fig. 7.8A), while subsequent blockade of GABA$_A$ receptors (in this case with gabazine) suppresses the gamma and enhances VFO. [One's first conjecture is that this in vitro VFO *has to be* different than the in vivo VFO described earlier, as in vitro VFO is *enhanced* by GABA$_A$ receptor block, and in vivo VFO is *suppressed* thereby. But this conjecture is very possibly wrong: in vivo VFO may require axonal excitation provided by GABA (Traub et al., 2003b),

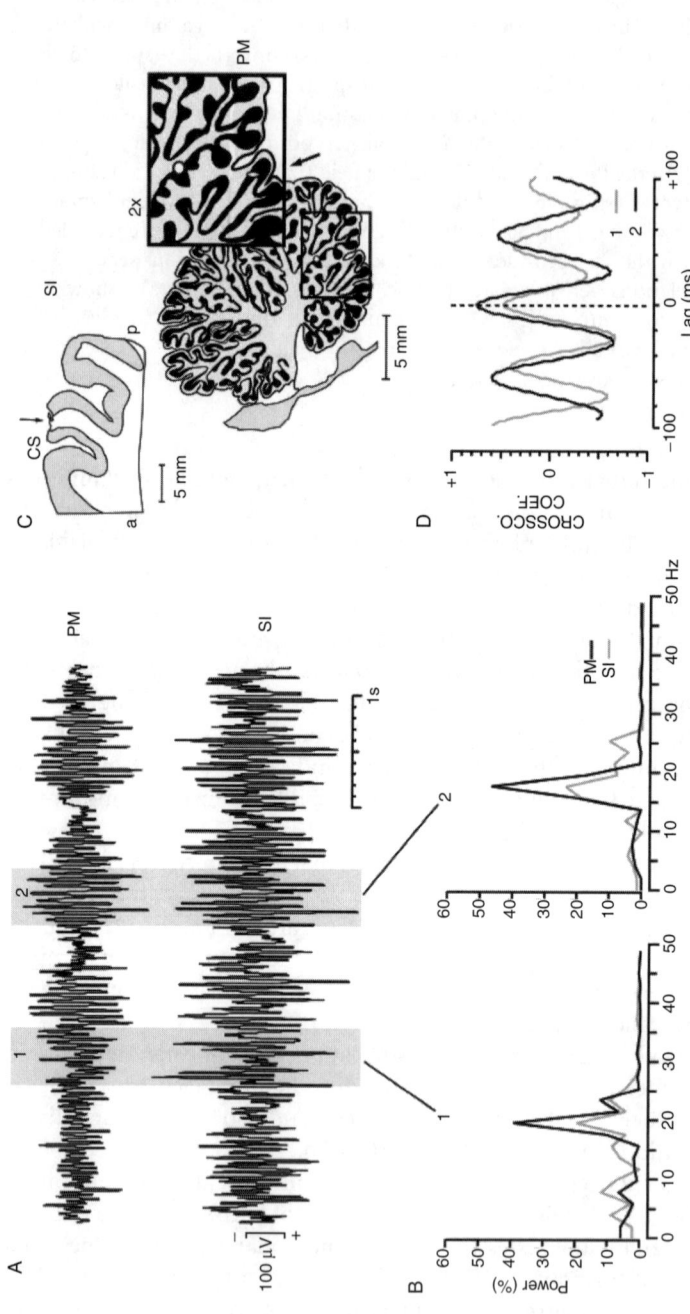

Figure 7.9 Beta oscillations can synchronize tightly, between primary somatosensory cortex (*SI*) and cerebellar cortex (paramedian lobule, *PM*): awake resting monkey. **A**: Field potential oscillations, recorded simultaneously, from PM and SI. Power spectra from two epochs (1, 2) are shown in **B**, with peaks 15-20 Hz. **C**: Details of recording sites (*CS*, central sulcus). **D**: Cross-correlations of the field potential data, from epochs 1 and 2.

(From Courtemanche and Lamarre, 2005 with permission.)

and so be suppressed by picrotoxin; whereas nicotine is also known to excite axons (Kawai et al., 2007), so that in the slice conditions, i.e., with nicotine present, GABA$_A$ receptors may not be required for VFO to occur.]

Strikingly, as gamma oscillations develop in the cerebellar slice, simple spike frequencies in Purkinje cells actually drop, and spikelets appear (Fig. 7.8B); the appearance of spikelets indicates axonal origin for the oscillation, and both experimental and modeling data support this idea (Middleton et al., 2008; Traub et al., 2008; see also later in this monograph). As for mutant VFO in the cerebellum, in vitro cerebellar VFO is suppressed by gap junction blockade (Middleton et al., 2008); and, like in vivo VFO (de Solange et al., 2008), it is coherent over hundreds of micrometers (Fig. 7.8C).

From a clinical perspective, one would very much like to know if cerebellar VFO occurs in humans as they go about their normal activities, and if it is different in people who are ataxic, as opposed to those not ataxic. The reader will recall that intracranial, even intracerebral, recordings are sometimes obtained in epileptic patients as part of their preoperative evaluation (Lüders, 2008); and that such recordings provide information on many types of cerebral oscillations. There are no clinical indications for comparable types of recordings to be obtained in patients with cerebellar disorders. Noninvasive techniques, such as magnetoencephalography (Ioannides & Fenwick, 2005), will need to be refined for the detection of cerebellar oscillations in humans.

Network Oscillations in the Cerebellum and Connected Systems: Beta and Gamma

Although VFO is coherent within cerebellar cortex itself (over hundreds of microns at least), and may be coherent with activity in the deep nuclei (Fig. 7.8), it is not known if VFO is coherent between cerebellum and noncerebellar structures. It is the case, however, that lower frequency oscillations do have such coherence properties, with Figures 7.9 and 7.10 providing two examples. Figure 7.9 demonstrates coherent beta-frequency oscillations, recorded in the awake resting monkey, in the cerebellar cortex (paramedian lobule) and primary somatosensory neocortex. Cross-correlations (Fig. 7.9D) indicate synchronization with an accuracy of a few milliseconds. Figure 7.10 shows data (also recorded from the awake monkey) indicating phase-locking of beta-2/gamma oscillations, involving a deep cerebellar nucleus (interpositus), and several shoulder and arm muscles. Whatever functions such synchrony and phase-locking may serve, widespread and structured oscillations throughout the motor system seem to be a fact of life, demanding explanation.

The interactions between cerebellar cortex and neocortex are polysynaptic. In one direction, we have: Purkinje cells → deep cerebellar nuclei (inhibitory) → motor thalamus → neocortex. In the reverse direction, we have: layer 5 pyramidal cells → pontine nuclei → cerebellar granule cells → Purkinje cells; there will be collaterals from the pontine nuclei to the deep cerebellar nuclei. Connections

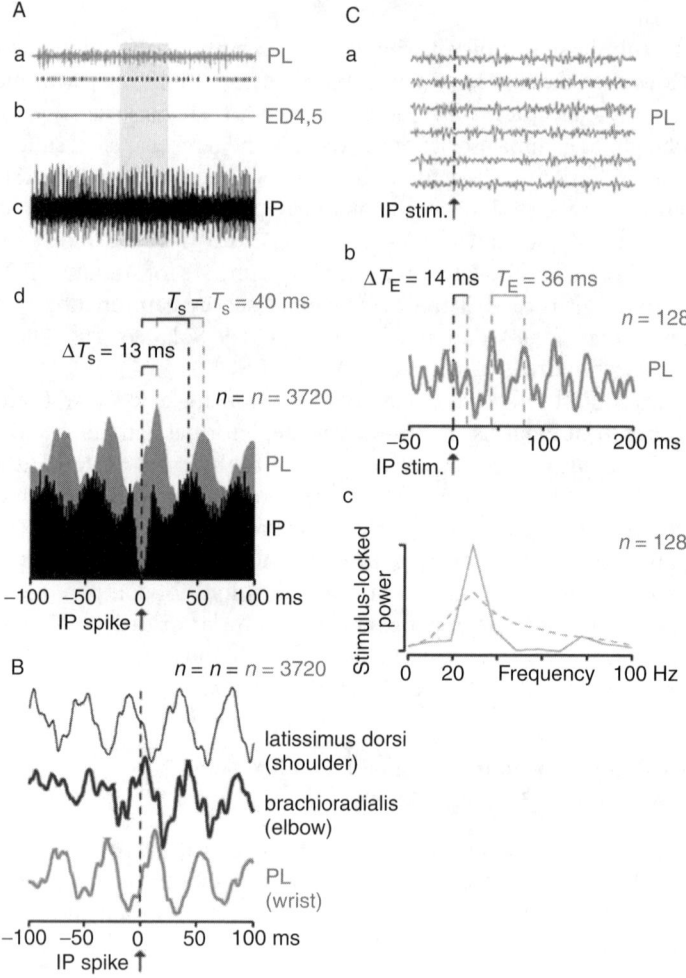

wrist flexion hold

Figure 7.10 Oscillating neurons (at beta frequency) in a deep cerebellar nucleus (interpositus) phase-lock to EMG activity, during a voluntary motor task: awake behaving monkey. In the case illustrated, the monkey was flexing its wrist against an elastic load. **A:** EMG from palmaris longus (**A:** PL), an extensor digitorum (**B:** ED4,5), and unit recording from an interpositus (deep cerebellar nuclear) neuron (c, IP), with raster plot of this cell below a. Beta-2 activity, ~25 Hz, occurs in the neuron during the shaded epoch. (d) shows the autocorrelogram of the IP neuron discharges (black, central peak removed), and averaged and smoothed PL EMG data, triggered by the IP neuron. T_S shows the oscillation period, ~40 ms, and ΔT_S the phase lag, 13 ms. **B:** Spike-triggered (using IP neuron) EMG signals from three different muscles (EMG data were rectified and smoothed). Note the beta-2 oscillations. **C:** Small extracellular stimuli were given in the interpositus nucleus (IP stim). a shows raw EMG data from palmaris longus in six trials; b shows stimulus-triggered average data (rectified and smoothed), with period $T_E = 36$ ms (28 Hz), and lag ΔT_E 14 ms; c shows power spectrum of average rectified and smooth EMG; dashed line = 95% confidence interval.

(From Aumann and Fetz, 2004 with permission.)

between spinal cord and cerebellum are also polysynaptic. How oscillations are organized, and synchronized, in such a large and heterogeneous system—this is a difficult, perhaps intractable, problem with our present level of understanding.

We do wish to point out, however, that the synaptic delays may not be as long as "classical" chemical synaptic delays: there is suggestive evidence that the synapses between deep cerebellar nuclear cells and motor thalamic cells may be "mixed," that is, contain gap junction–mediated electrical coupling, in addition to transmitter release sites. In the present context, the suggestive evidence consists of intracellular recordings, made by Igor Timofeev and Mircea Steriade, in ventrolateral (VL) thalamic neurons during (and between) stimulations to cerebellar afferents (Fig. 7.11). These recordings demonstrate spikelets, or so-called fast prepotentials, that do not have the appearance of usual EPSPs. Electrical coupling at the synaptic connections could act to couple diverse brain regions with shorter synaptic delays than is typically the case.

Conclusion

In this chapter, we have commented on the devastating consequences of cerebellar ataxia in at least some patients, including a subset of patients with

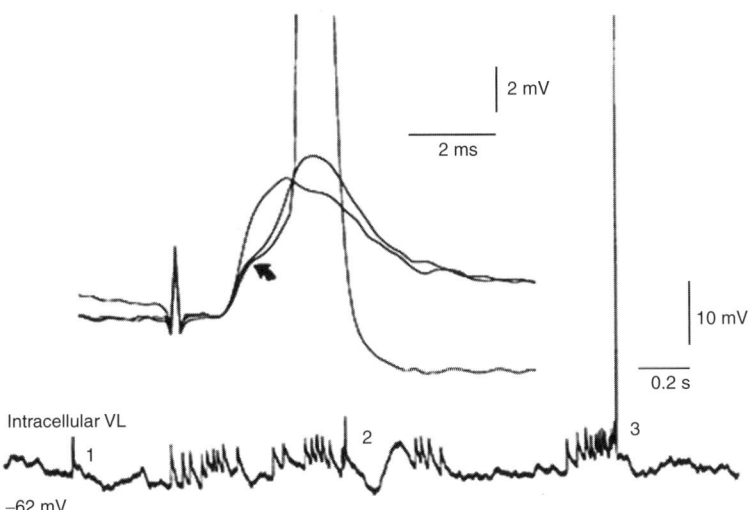

Figure 7.11 Stimulation of the superior cerebellar peduncle (brachium conjunctivum) induces spikelets in VL thalamic neurons, suggesting the possibility of mixed synapses at this connection. Ketamine/xylazine anesthetized cat. Upper traces show details of three different stimulations, marked 1, 2, 3. Compare with Hamzei-Sichani et al. (in prep.).

(From Timofeev and Steriade, 1997 with permission.)

multiple sclerosis; and we have pointed out the intractability of this clinical problem. We have shown that beta and gamma oscillations are organized (phase-locked, sometimes at zero phase, i.e., synchronized) throughout much of the motor system (see also Courtemanche & Lamarre, 2005; Kelly & Strick, 2003; Marsden et al., 2000; Soteropoulos & Baker, 2006;); and we have shown that VFO occurs normally in the cerebellar cortex, but may be enhanced in certain genetically modified mice that happen to be ataxic—suggesting, if not proving, a causal relationship. Finally, we have introduced an in vitro model of cerebellar cortical oscillations, both for gamma frequencies and for VFO.

In a later chapter, we discuss the evidence that both gamma and VFO in the cerebellum require gap junctions. Here, we cite some clinical data that imply a possible link between the present chapter, and the basic scientific data that are to come: there is a group of inherited disorders associated with transient episodes of ataxia (Jen et al., 2007). Some of these so-called primary episodic ataxias, particularly type 2, are responsive to orally administered acetazolamide—a drug that causes metabolic acidosis, and that is expected to reduce gap junction conductances (Spray et al., 1981).

This observation, and the data on so-called mutant VFO, lead us then to an hypothesis: that gap junctions are essential for a repertoire of oscillations that the motor system requires for its normal functions; but that when the conductance of selected gap junctions becomes too large, some of the oscillations change their properties (e.g. by involving too many neurons, or continuing autonomously and independent of afferent inputs)—and the altered oscillation properties in turn lead to functional abnormalities.

PART II

Basic Properties of Single Neurons
and Gap Junctions

8

Cortical Neurons and Their Models

The study of the shapes and intrinsic properties of the vastly many types of neurons, and of the synaptic relations between neurons, is one of the most beautiful of scientific endeavors (Llinás, 1988; Somogyi et al., 1998). To review such a vast subject here is not possible. What we attempt to do is to cover selected topics on the properties of single neurons, and their chemical synaptic connections, that are of special salience to network oscillations. Even so, we must be selective. The subject of gap junctions—electrical synaptic connections whose activity appears to underlie a great many persistent network rhythms—is covered in Chapter 9.

From a physical aspect, what are the essential issues to consider in a monograph such as this? They are: in what ways can individual neurons oscillate on their own? And, what effects can neurons exert on one another that act to promote network oscillations, even when the individual cells are not oscillatory? We shall consider examples of each of these issues. The goal is to illustrate how one attempts to capture the properties of neurons, in such a way as to be able to model them mathematically, and to model networks as well. The larger purpose is to provide physical and biological intuition, and to provide predictions that help to guide experiments. For some of us, there are aesthetic and philosophical goals as well.

Single neurons need to be considered from many points of view, all at once. There are the *"passive properties,"* that is, those aspects of electrical behavior that exist independently of voltage- and Ca^{2+}-dependent membrane channels, but that depend on shape, on the physical properties of the intracellular contents, and on the membrane capacitance and electrical "leak" (Rall, 1962).

Next (but successfully and profoundly investigated by Hodgkin and Huxley *before* Rall's cable theory was worked out) are the physiological principles governing *action potentials*—first, conventional fast action potentials (Hodgkin & Huxley, 1952), and later, many other sorts of Ca^{2+}-mediated action potentials. Some of the principles involved in action potential generation are these: that membrane conductances exist that are ion selective; that can produce transmembrane currents that are inward or outward; that have different time courses; that can be characterized by "state variables" governed by differential equations involving time, membrane potential, and the state variables themselves; and which can involve multiple states having faster or slower kinetics (that is to say, channels can have rapid activation/deactivation and slower inactivation/recovery). Further, membrane conductances influence each other through the common medium of the local membrane potential: cooperativity exists even at the level of small membrane patches. The study of action potentials—which generally involves characterizing the kinetics of two or more ionic conductances, at a macroscopic, multichannel level—in turn depends on a field of study that evolved decades after the original Hodgkin–Huxley experiments and theory: the study of *individual channels*, isolated by physical and pharmacological methods, using patch techniques (Armstrong & Hille, 1998; Sakmann & Neher, 1984). The channels themselves can now be analyzed in expression systems other than the neurons where the channels normally reside, and can be modified with all manner of molecular and biochemical techniques. Next, one can study *firing patterns*, either under natural in vivo conditions, or in response to a synaptic stimulus or injected current pulse. Finally, one can attempt to integrate various of these kinds of data into a mathematical model (Dodge & Cooley, 1973; Traub & Llinás, 1977).

In the case of neocortical neurons, action potential firing patterns, in response to steady current pulses, form the basis of one descriptive classification of neuronal types. This classification is extremely useful, but before proceeding to illustrate it, let us mention two caution-inducing, important points. First, the firing pattern of a given neuron is not something unalterably fixed throughout the lifetime of that neuron: the firing pattern may be modifiable by phosphorylating a membrane channel, by intense firing activity, or even by simply altering membrane potential. For an excellent review of the lability of cortical principal cell firing patterns, see Steriade (2003). It should also be noted that firing patterns in individual neurons may vary enormously in different subcompartments, with axonal, somatic and dendritic regions producing markedly differing 'outputs' in response to identical electrical inputs. Second, there are many other sorts of classification schemes, which—in the case of interneurons—can become baroquely complicated indeed. To expand on this, neuronal taxonomy depends on the location and shape of the cell body, the nearby and farther destinations of axonal collaterals, branching patterns of dendrites, presence or absence of spines, the repertoire of calcium-binding proteins, and peptides co-released with the main transmitter—to name some of the possibilities. Such is biology.

These issues relating to classification need not overly complicate matters if one considers some basic building blocks of individual neurons—their passive properties, type and distribution of active conductances—as defining their functional signature. Assembling a model neuron from these building blocks can generate many of the complex activity patterns as emergent properties, as illustrated in the examples in this chapter. Although these models (Fig. 8.1) may contain only about a dozen different membrane conductances, and the number of compartments is not especially large—61 in the case of

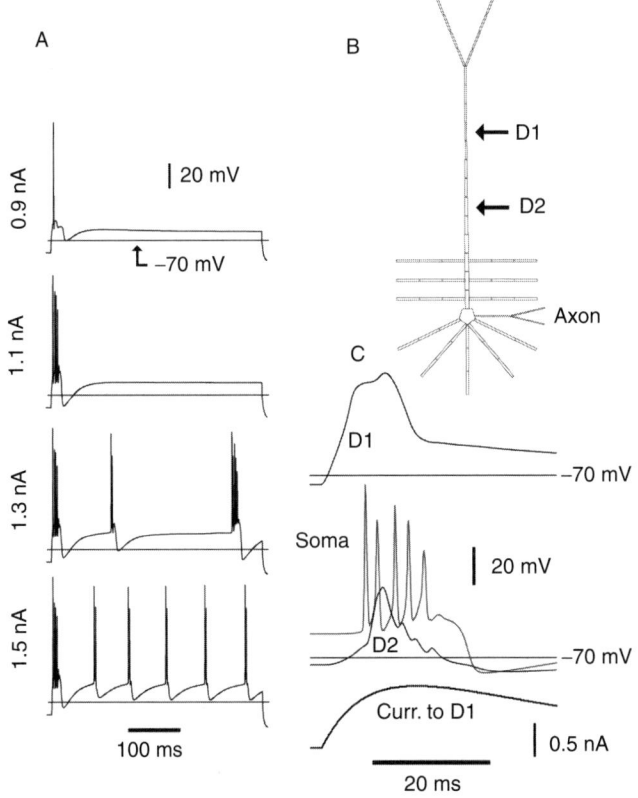

Figure 8.1 Multicompartment model of an intrinsically bursting tufted layer 5 pyramidal neuron. **A:** Intrinsic bursts in response to sufficiently large somatic current pulses. **B:** Schematic structure of the model, with soma, tufted apical dendrite, basal and oblique dendrites, and axon. The various compartments are endowed with voltage- and Ca^{2+}-gated conductances [multiple types of g_{Na}, g_{Ca}, g_K, and an anomolous rectifier (h-current)]. **C:** Somatic burst produced by current injection into the distal apical shaft. Note the Na$^+$- and Ca^{2+}-mediated dendritic electrogenesis. Variations of this model are used in a number of network simulations illustrated in this book.

(From Traub et al., 2005 with permission.)

layer 5 (LV) pyramids (a far smaller number than would be needed to capture the cell morphology with anatomical realism)—it suffices to have a schematic representation of axon, soma, and different sorts of dendrites, with approximately the right membrane surface area; to use reasonable kinetics for the conductances; and to allow for the possibility that membrane conductances need not be distributed with uniform density and voltage thresholds (Colbert & Johnston, 1996; Colbert & Pan, 2002; Kole et al., 2008). Models following these general principles are available for a number of other neocortical cell types, including superficial and deep regular spiking pyramids, fast-spiking (FS) and low-threshold-spiking (LTS) interneurons, fast rhythmic bursting (FRB) neurons, and spiny stellate cells (Cunningham et al., 2004b; Traub et al., 2005a).

A practical consequence of being able to capture much of the intrinsic electrophysiology in relatively simple models is this: it becomes possible to simulate large networks of neurons on a parallel computer. To simulate a population of 15,000 cells, with multiple functional subclasses, is readily possible. Such simulations are, as we shall see, an essential tool in understanding oscillations.

Firing Patterns of Some Cortical Interneurons

Figure 8.2 illustrates two firing patterns that are characteristic of interneurons: FS and LTS. Interneurons may also demonstrate regular spiking and bursting but we will not consider these types in any detail here. FS cells have narrow spikes, can fire rapidly (as their name would imply), and have little or no accommodation (i.e., slowing of the frequency with repetitive firing). LTS cells (Deuchars & Thomson, 1995; Kawaguchi & Kubota, 1993) have somewhat broader spikes, fire an initial burst of spikes when depolarized from a hyperpolarized resting potential, and do not fire rapidly. Although the classification of interneurons is a contentious and intricate subject (Ascoli et al., 2008; Burkhalter, 2008; Markram et al., 2004; Miles, 2000), it is helpful to have a tentative scheme for linking firing patterns to other features of interneurons. For this, we follow conventional notions: FS firing patterns are most often, but not exclusively, found in parvalbumin-positive interneurons, which include many (but not all) basket cells and axo–axonic (chandelier) cells—the sorts of interneurons which, respectively, contact the soma and proximal dendrites of pyramidal cells, and the axon initial segments of pyramidal cells (and basket cells also contact each other, as well as various other sorts of interneuron). We might note, however, that identified hippocampal axo–axonic cells can fire doublets, in addition to rapid single spikes (Buhl et al., 1994) and some *non*-proximal targeting interneurons also show fast spiking behavior—for example, trilaminar cells and bistratified cells [of which only a proportion are parvalbumin-positive (Gloveli et al., 2005a)]. In addition, some proximal targeting interneurons do not have FS firing patterns: "Ivy" interneurons in

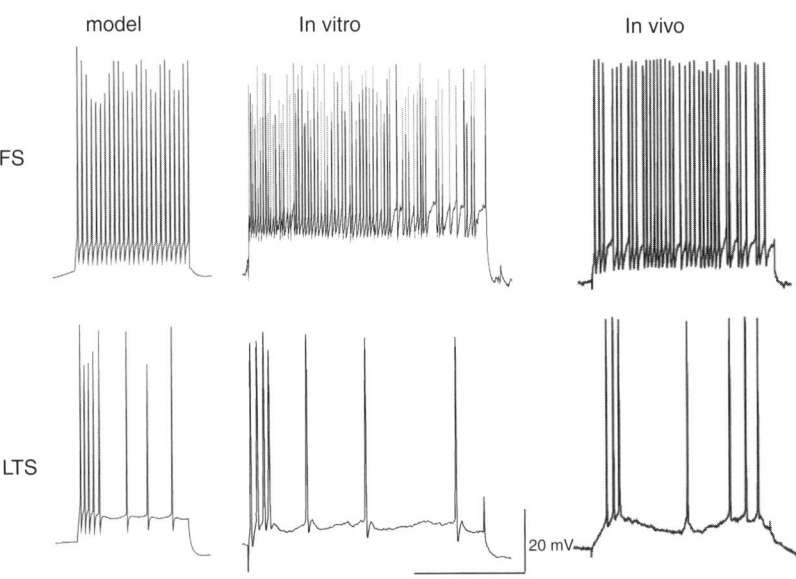

Figure 8.2 Firing patterns associated with some (but not all) interneurons. "FS" refers to fast-spiking: in response to strong enough depolarization, such cells fire at high rates with little accommodation. Basket cells are usually fast-spiking. "LTS" refers to low-threshold spiking: If the cell is depolarized with a pulse, from a hyperpolarizing resting potential, there is an initial burst of action potentials, followed by a sparser pattern of action potentials.
(From Traub et al., 2005 with permission.)

hippocampus have firing patterns which closely resemble pyramidal cells and not FS interneurons (Fuentealba et al., 2008).

In contrast to FS cells, LTS interneurons are not parvalbumin positive, but are likely to stain for calbindin$_{D28K}$, to have vertically oriented (rather than horizontally oriented) axons, and to contact dendritic shafts of pyramidal cells (Deuchars & Thomson, 1995; Kawaguchi & Kubota, 1993). Xiang et al. (2002) found that the decay time constants for IPSCs produced in pyramidal cells by LTS cells were about the same as for inhibitory postsynaptic conductance (IPSCs) produced by FS cells, although the LTS-induced IPSCs had a somewhat slower onset. In contrast, Martinotti cells (a type of inhibitory cell in neocortex, Fig. 8.3) provide dendritic inhibition to pyramidal cells that has considerably slower kinetics than that provided by FS cells (Silberberg & Markram, 2007).

Given this complexity, and the observation that firing patterns of interneurons do not correlate well with molecular markers for subtypes, we must be cautious in relating specific interneuron subtypes to specific brain rhythms. Sometimes the evidence is reasonably unequivocal: in the hippocampal and the neocortical slice, during gamma oscillations induced by kainate,

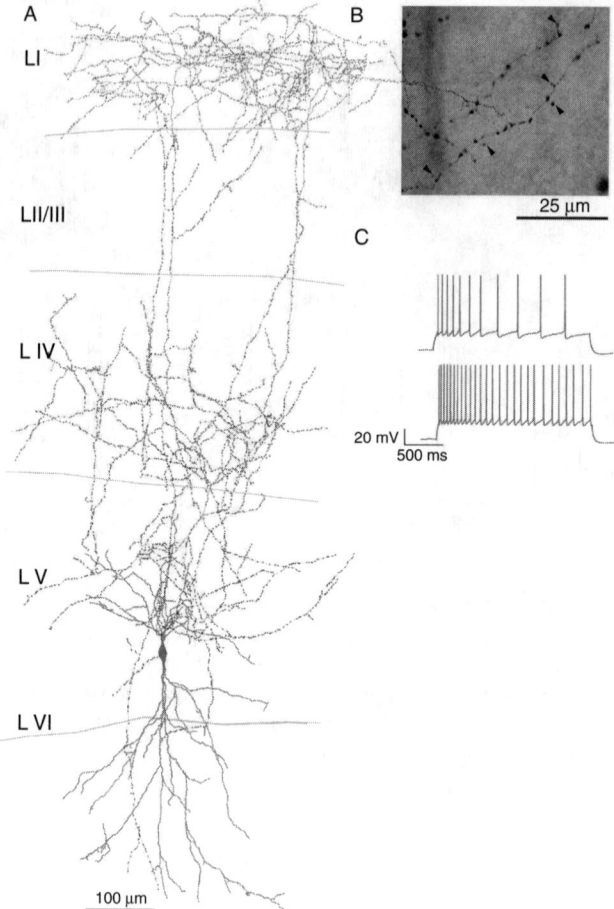

Figure 8.3 An interneuron which fires accommodating trains of action potential, a pattern different than for FS, LTS, and neurogliaform cells. **A:** morphology of layer 5 Martinotti cell (soma and dendrites, red; axon, blue). Note the extensive ramifications in layers 4 and 1. **B:** Magnified view of spiny boutons on the axon (arrows). **C:** Accommodating trains of action potentials, in response to current pulses.
(From Wang et al., 2004 with permission.)

basket cells fire on most gamma cycles, and are presumed to be critically involved in the actual generation of the oscillation (Gloveli et al., 2005a; Traub et al., 2005a)—both the ability to sustain firing at approximately 40 Hz, and also the anatomical location of the output synapses, are likely to be important features of the basket cells. But even with the relatively well studied gamma rhythm we cannot categorically state that basket cell morphology and physiology are paramount. For example, in hippocampus, intense, gamma frequency spiking is also seen in bistratified and trilaminar cells (Gloveli et al., 2005a). Further, it is becoming clear that not all basket cells play a role in this hippocampal gamma rhythm. In hippocampus, basket cells immunopositive for cholecystokinin (CCK) do not appear to generate gamma frequency outputs, and in neocortex some basket-like cells have intrinsic conductances which strongly drive their output at theta, not gamma, frequencies (Blatow et al., 2003).

Cortical Interneurons Differ in the Way They Respond to Oscillation-Inducing Neuromodulators

One of the most effective tools for studying oscillations in vitro is to apply a neuromodulatory compound to the bath; such application may then be capable of eliciting—in appropriately prepared brain slices—a network oscillation that lasts many hours (Fisahn et al., 1998). To work out the mechanisms by which the oscillations are generated, it is, of course, necessary to know what the modulator actually does to each relevant cell type, and what it does to the interactions between cells. Such information is not as complete as we would like. Figure 8.4 (from Bacci et al., 2005) provides an example of the kind of data available for cortex: what does acetylcholine do to different sorts of interneurons, in this case in layer 5 of rat sensorimotor cortex? [Carbachol, which has both muscarinic and nicotinic actions, would be expected to exert similar effects to acetylcholine, although lasting longer than acetylcholine, because of carbachol's resistance to hydrolysis.] As Figure 8.4 shows, acetylcholine has opposite effects in layer 5, on LTS interneurons (depolarized), compared with FS interneurons (hyperpolarized). Further, these effects are mediated, respectively, by different types of receptor, nicotinic in the first case, and muscarinic in the second. Similar dual actions of cholinergic receptors have been reported previously, for example in the thalamus: cholinergic activation hyperpolarizes nucleus reticularis cells (McCormick & Prince, 1986), but depolarizes thalamocortical relay cells (Zhu & Uhlrich, 1998); in this case, however, both actions are muscarinic.

Although the preceding examples suggest interneuron subtypes may be readily characterized by their responses to neuromodulators, this is unfortunately not always the case. Parra et al. (1998) demonstrated no fewer than 25 different response combinations to neuromodulators in a heterogeneous population of hippocampal interneurons. Kawaguchi (1997) has also reported rather complex effects of cholinergic activation on different interneuron types. Using frontal cortex, cholinergic modulation was found to affect only regular and burst-firing interneurons, and not fast-spiking parvalbumin-containing interneurons. These data are in contrast to the studies above taken from somatosensory cortex. Thus, we need to consider not only the subtype of interneuron but also its anatomical location. Given this complexity, identifying precisely which type or types of inhibitory neuron—*if any*—mediate specific

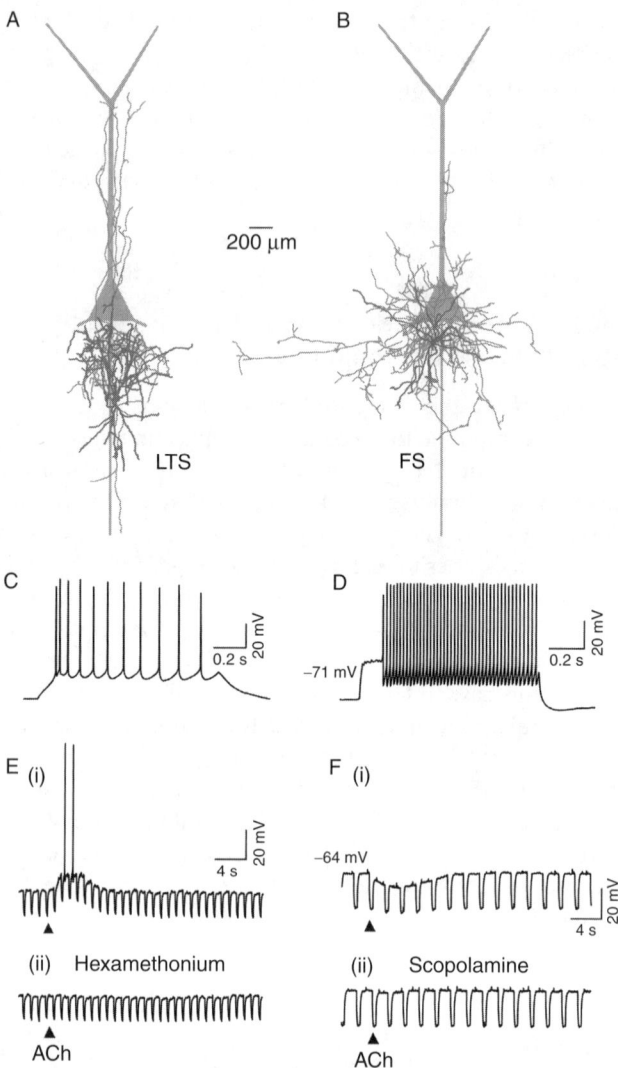

Figure 8.4 Differential morphology, connectivity, firing patterns and sensitivity to neuromodulators in neocortical interneurons. Data from layer 5 of rat sensorimotor cortex in vitro. Upper panels show a schematic tufted layer 5 pyramidal neuron in gray; the soma and dendrites of an interneuron in red (low-threshold spiking, or LTS, in **A**; fast-spiking, or FS, in **B**); and the respective interneuron axons in blue. Note how the axon of the LTS cell wraps around the apical and basal dendrites, while the axon of the FS cell is mostly confined to the perisomatic regions. **C, D:** Firing patterns of the LTS and FS cell, in response to a depolarizing current pulse. **E, F:** A puff of acetylcholine (ACH) depolarizes the LTS cell, but hyperpolarizes the FS cell. The depolarization of the LTS cell is mediated by nicotinic receptors, as it is blocked by hexamethonium; the hyperpolarization of the FS cell is mediated by muscarinic receptors, as it is blocked by scopolamine.

(From Bacci et al., 2005 with permission.)

frequencies of network rhythm is likely to be critical, in understanding underlying pathology where particular brain rhythms are detrimentally affected. We consider this in more detail in later chapters but, for now, move on to firing properties in excitatory neurons.

Fast Rhythmic Bursting in Pyramidal Cells

Figure 8.5 illustrates two types of firing pattern that are seen in cortical principal neurons: "FRB" or fast rhythmic bursting, also called "chattering" (Gray & McCormick, 1996; Steriade et al., 1998b; see also Chapter 3, Fig. 3.5); and "RS" or regular spiking (Connors et al., 1982; McCormick et al., 1982; Nowak et al., 2003). FRB behavior is unusual; it is found mostly in pyramidal neurons in superficial layers, but also in pyramidal and nonpyramidal neurons in other cortical layers, and, indeed, also in the thalamus (Steriade et al., 1993e). Regular spiking is the most common firing pattern seen in cortical pyramidal cells and spiny stellate cells; it is also found in some interneurons. Figure 8.5 illustrates simulation data, as well as intracellular recordings from in vitro and in vivo preparations, to emphasize the similarity.

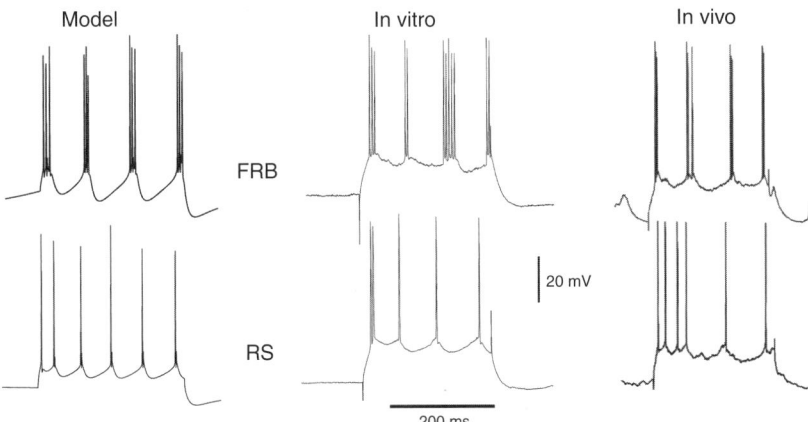

Figure 8.5 Different types of excitatory neurons have distinctive firing patterns, when stimulated with depolarizing current pulses. The figure shows two different firing patterns: fast rhythmic bursting ("FRB," also called "chattering"), and regular spiking (RS). FRB neurons can fire short bursts of high-frequency action potentials, with an interburst frequency in the beta/gamma range (see also Chapter 3, Fig. 3.5). RS neurons fire at high rates initially, but then the firing rate slows and becomes more regular ("adaptation" or "accommodation"). Firing patterns are shown for simulated multicompartment neurons ("model"), for neurons recorded in neocortical slices ("in vitro") and from the neocortex in vivo. Most, but not all, neurons with FRB and RS firing patterns are principal (excitatory) cells.
(From Traub et al., 2005 with permission.)

With fast rhythmic bursting, there are two time scales regulating action potential frequencies: within-burst frequencies are several hundred Hertz; and the bursts themselves recur at frequencies from 10–20 Hz to greater than 50 Hz. Early studies of in vivo sensory-induced gamma oscillations (e.g., Gray, 1994) provided extracellular recording evidence suggesting the existence of chattering, or fast rhythmic bursting, and FRB has long been suspected as important in the generation of gamma oscillations, because of the intrinsic gamma period exhibited by these neurons; however, as shown in Chapter 3, the participation of FRB neurons in visually induced gamma oscillations is rather complex. There is, nevertheless, evidence (to be discussed in Chapter 12) that FRB neurons are important in the generation of persistent gamma oscillations.

During regular spiking, as Figure 8.5 shows, the neuron generates single action potentials, which, however, tend to slow in frequency over time, before reaching steady-state. There may also be a short burst of action potentials at the onset of a current pulse.

The period between fast rhythmic bursts, and between regular spikes, is largely determined by K^+ currents, especially the "M-current," which exhibits an activation/deactivation time constant in the tens of milliseconds, at membrane potentials near the resting potential. This particular current is highly relevant to many brain oscillations, including beta2 and VFO, so let us dwell on it briefly. The name "M" stands for "muscarinic," because the underlying conductance is reduced by cholinergic muscarinic receptors (Adams & Brown, 1982; Adams et al., 1982; Brown & Adams, 1980); the pathway is now known to depend on phospholipase C activation (Suh & Hille, 2002). Functionally, the M current mediates so-called medium duration afterhyperpolarizations (AHPs), and regulates the tendency of neurons to generate bursts of action potentials. The current also most likely has a direct regulatory effect on the excitability of axons, because the conductance is located, at least in part, on the proximal axon and on nodes of Ranvier (Cooper et al., 2001; Devaux et al., 2004; Pan et al., 2006). The M-conductance is mediated, at least to a significant extent, by heteromeric KCNQ2/3 potassium channels (Prole & Marrion, 2004). These channels tend to close in the presence of acidic conditions (Prole et al., 2003) which, in the case of axons, would also be expected to close nearby gap junctions.

So which mechanisms underlie fast rhythmic bursting (chattering)? X-J Wang (1993, 1999) proposed that FRB behavior could be explained in this way: that—analogously to the spatial separation of Ca^{2+} currents and spike generation used to account for pyramidal cell intrinsic bursts (see later; Pinsky & Rinzel, 1994; Traub et al., 1991)—there was also a spatial separation of a slower inward current, and of spike-generating currents, in FRB cells. For FRB cells, however, the slower inward current was postulated to be a persistent (slowly inactivating) Na^+ current (French et al., 1990; Kay et al., 1998; Llinás & Sugimori, 1980a). Such a slow inward current, interacting with fast Na^+ and K^+ channels could, in principle explain the bursts themselves, especially if the slow inward current was separated (hence independently controllable by a second, slower, K^+ current) from the action potential currents. The slower K^+ current was postulated to regulate the period between bursts. X-J Wang located the persistent Na^+ current in the dendrites of his model neuron.

As Figure 8.6 demonstrates, Brumberg et al. (2000), using pharmacological methods either to increase, or to suppress, persistent Na^+ conductances, showed that such conductances indeed were essential for FRB. Brumberg et al. recorded from pyramidal cells in superficial layers of ferret visual cortex. Brumberg et al. also proved experimentally that Ca^{2+} conductances were *not* needed for FRB, nor were Ca^{2+}-dependent conductances. Subsequently, we (Traub et al., 2003a) developed a multicompartment model of a layer 2/3 pyramidal cell, that could switch back and forth between regular spiking and FRB, depending on a balance between persistent g_{Na} and a particular Ca^{2+}-dependent conductance, carried by BK channels (Wanner et al., 1999); this conductance is also voltage-dependent, has rapid kinetics, and contributes to spike repolarization (Shao et al., 1999) and fast AHPs. Experiments with the scorpion toxin, iberiotoxin, a specific blocker of BK channels, showed that reduction of BK currents did, indeed, favor FRB behavior. In our model of layer 2/3 pyramidal

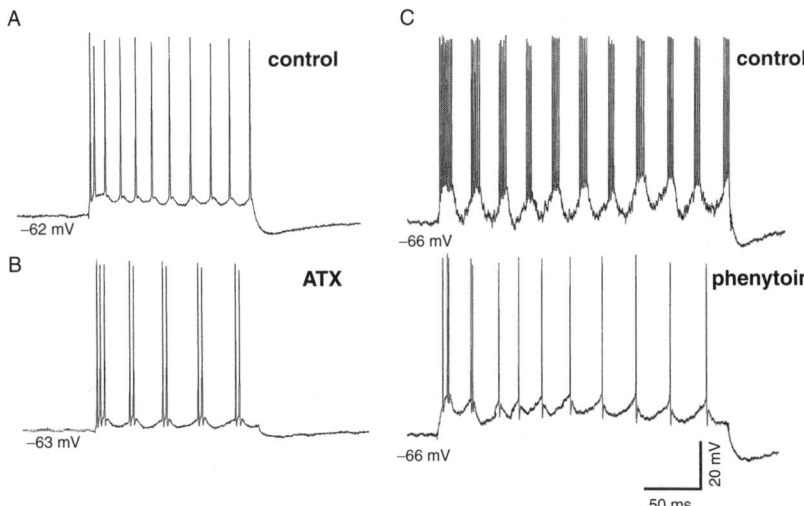

Figure 8.6 Persistent Na^+ current contributes to FRB (fast rhythmic bursting, chattering) behavior in layer 2/3 cortical neurons. **A:** Regular spiking behavior in response to 0.5 nA current, ferret superficial visual cortex in vitro. **B:** Addition of a sea anemone toxin (Alsen, 1983), which decreases inactivation of g_{Na}, unmasks FRB behavior. **C:** Phenytoin (which preferentially blocks persistent g_{Na}) converts FRB behavior to regular spiking, again in response to a 0.5 nA current pulse (ferret superficial visual cortex).

(From Brumberg et al., 2000 with permission.)

cells, we located a portion of both persistent Na^+ conductance, and of the BK channel conductance, in the axon (Astman et al., 2006; Wanner et al., 1999); it appears that the axon may be the critical site for the generation of fast rhythmic bursts.

Ca^{2+}-Dependent Intrinsic Bursting: An Alternate Burst-Mode Distinct from FRB Behavior

Let us now direct our attention to pyramidal neurons, the most common class of cell morphology to be found in the cortex, concentrating on large layer 5 pyramidal cells—probably the most intensively studied cell type in the brain. Many of these layer 5 pyramids generate intrinsic bursts—although it is important to note, following the lead of M. Steriade (2001), that the proportion of cells (in vivo) which exhibit intrinsic bursting depends on the behavioral state of the animal, and presumably reflects background synaptic activity, membrane potential, and neuromodulatory effects: 15% in the intact cortex of an anesthetized animal, 30% to 40% in small isolated cortical slabs, but less than 5% in awake behaving animals.

Figure 8.7 illustrates the shape of three sorts of pyramidal neurons, as visualized after horseradish peroxidase fills, and corresponding firing patterns in response to a current pulse: a smaller regular-spiking pyramid such as is commonly found in superficial cortical layers, and containing a small dendritic tuft in layer 1; a layer 5 regularly spiking pyramid with a relatively small apical dendrite, that does not reach layer 1; and an intrinsically bursting layer 5 pyramid, with a large apical dendrite which ascends to layer 1, and branches there into a tuft. In response to a sustained current pulse, this type of intrinsic bursting may continue rhythmically, or may switch to regular spiking. Some details of the shape of the intrinsic burst are interesting as well, being common to intrinsic bursts in hippocampal CA3 pyramidal cells: the fast AHP after the first action potential (but not later ones), and the declining amplitude and increasing width of spikes, as the burst progresses—this in contrast to bursts in fast rhythmic bursting cells, where fluctuations in somatic spike amplitude are smaller and irregular (compare with Fig. 8.6). [In addition, the *axon* of intrinsically bursting cells does not display attenuation of the successive spikes in a burst (Williams & Stuart, 1999).] As we shall see, and also in a manner similar to hippocampal pyramids, layer 5 intrinsic bursts depend in part on dendritic calcium electrogenesis (Williams & Stuart, 1999; Wong & Prince, 1981—but see also Jensen et al., 1996).

Intrinsic bursts are especially prominent during epileptiform events, where they are amplified by recurrent synaptic excitation, and can recur at frequencies as high as 20 Hz or more (Chapter 4; see also Chapter 13). That, of course, is a pathological situation, so that one must ask: why do intrinsic bursts occur in the normal brain? There are data to support two related functions: first, one naturally would suppose that a burst of action potentials serves to amplify the postsynaptic effects of the bursting neuron. This may be true,

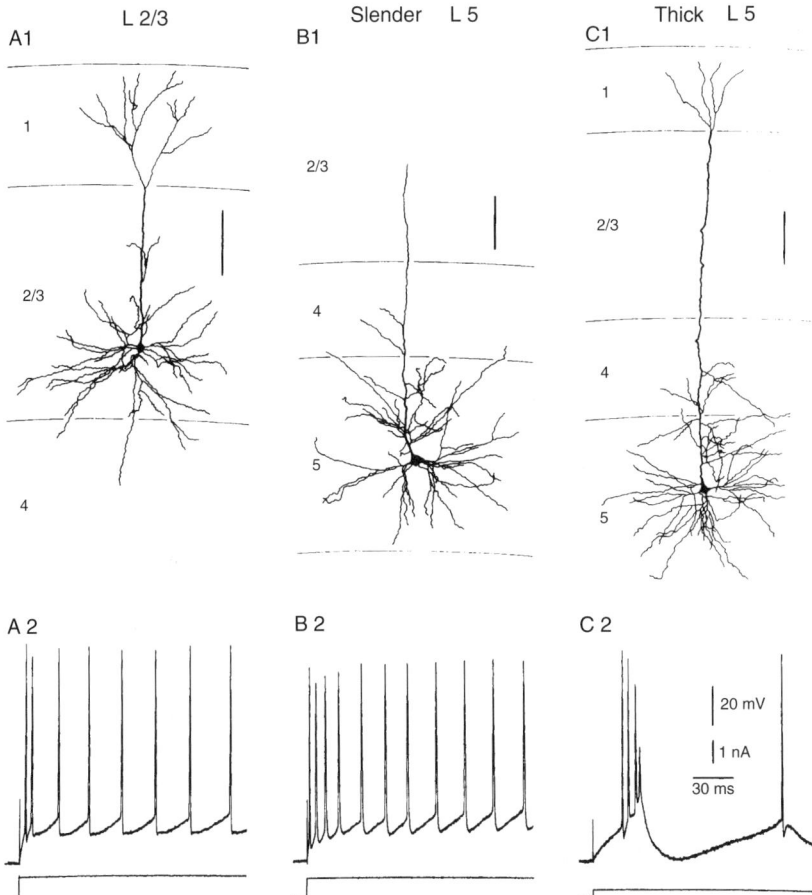

Figure 8.7 Regular-spiking and intrinsic bursting in pyramidal neurons of rat visual cortex in vitro. Drawings of the neuronal morphology were based on fills of the cells (recorded with sharp microelectrodes) with horseradish peroxidase. For each cell, soma, basal and oblique dendrites, and the ascending apical dendrite can be discerned. For the layer 2/3 and the "thick" (i.e., having a thick apical dendrite) layer 5 cells, the apical dendrite ascends to layer 1 and branches into a tuft. The thin apical dendrite of the cell in **B** does not reach layer 1, and does not possess a tuft. The neurons exhibit the typical firing patterns associated with the morphology and location: regular spiking for layer 2/3 and thin-dendrite layer 5 pyramids, and intrinsic bursting for tufted thick-dendrite layer 5 pyramids (but there are exceptions to this rule). Vertical spatial scale bars are 100 μm.

(From Mason & Larkman, 1990 with permission.)

but is more intricate than first appears: as studied in simultaneous recordings of synaptically connected layer 5 pyramids, excitatory postsynaptic potentials (EPSPs) elicited by successive spikes in a presynaptic burst exhibit depression— that is, they decline in amplitude, one by one; however, if the postsynaptic cell is depolarized a bit, then one or more intrinsic currents (probably a persistent g_{Na}) act to compensate for the depression (Williams & Stuart, 1999). If the postsynaptic target is an interneuron then bursting appears to be essential to activate certain types of inhibition—outputs from Martinotti cells are optimized only by facilitation of trains of synaptic inputs as generated by presynaptic cell burst discharges (Silberberg & Markram, 2007). Second, intrinsic bursts can induce Ca^{2+} concentration increases in the basal dendrites of the postsynaptic cell, and these resulting intracellular $[Ca^{2+}]$ increases favor spike-time–dependent synaptic plasticity (Kampa et al., 2006).

Let us now consider some of the critical details of apical dendritic electro-genesis, in layer 5 pyramidal cells, as investigated so extensively by Bert Sakmann and his school, using the dual techniques of simultaneously recording two or more sites of the same neuron, and of correlated optical imaging mea-surements. Figure 8.8 (upper and middle traces) shows an intrinsic burst—simultaneously recorded in the soma and apical dendrite—that has been evoked by synaptic stimulation in layer 2/3. Both the soma and the dendrite exhibit depolarizing events on two different time scales. The *faster* time scale corresponds to Na^+ spikes that are initiated in the axon (Gulledge & Stuart, 2003; Palmer & Stuart, 2006; Stuart et al., 1997; Williams & Stuart, 1999), and then propagate into the soma and dendrites, broadening as they do. The *slower* time scale corresponds, at least in part, to dendritic calcium electrogenesis, that leads to a slow depolarizing envelope, which is larger in the dendrite than the soma—and which causes the dendritic fast spikes to *increase* in amplitude even as the somatic fast spikes are *decreasing* in amplitude. That calcium elec-trogenesis is a major contributor to the slow depolarization is demonstrated by the lower traces that show suppression of this depolarization with cad-mium, an ion that blocks calcium channels. (Note that cadmium will block Ca^{2+}-dependent transmitter release, so that the burst in the lower traces could not be evoked with synaptic stimulation; a somatic current pulse was used instead.)

As discussed in the preceding text, the issue of Ca^{2+} electrogenesis, with respect to intrinsic bursting, involves the apical dendrites of tufted layer 5 pyramidal cells. The dendrites of the layer 2/3 pyramids, being smaller, have not been studied as extensively as the dendrites of layer 5 cells; nevertheless, data do exist for superficial pyramidal neurons, concerning the backpropaga-tion of perisomatically generated Na^+ spikes, and the apical dendritic genera-tion of Ca^{2+} spikes (e.g., Larkum et al., 2007; Svoboda et al., 1999). (Figure 5A of the Larkum et al. study incidentally also shows that apical dendritic spikes—this time in layer 2/3 pyramids—do not elicit somatic spikelets.) The putative dendritic recordings of Amitai et al. (1993) have revealed attenuated back-propagated Na^+ spikes, as well as tetrodotoxin-resistant (hence presumably Ca^{2+}-mediated) slow spikes. Figure 8.9 illustrates similar phenomenology, and emphasizes that dendritic Ca^{2+} spikes can occur rhythmically, as shown many years ago in Purkinje cell (Llinás & Sugimori, 1980b) and hippocampal pyramidal cell dendrites (Traub et al., 1993). In the simulation data shown in Figure 8.9, as expected, the fast spikes are initiated in the axon, and the slow

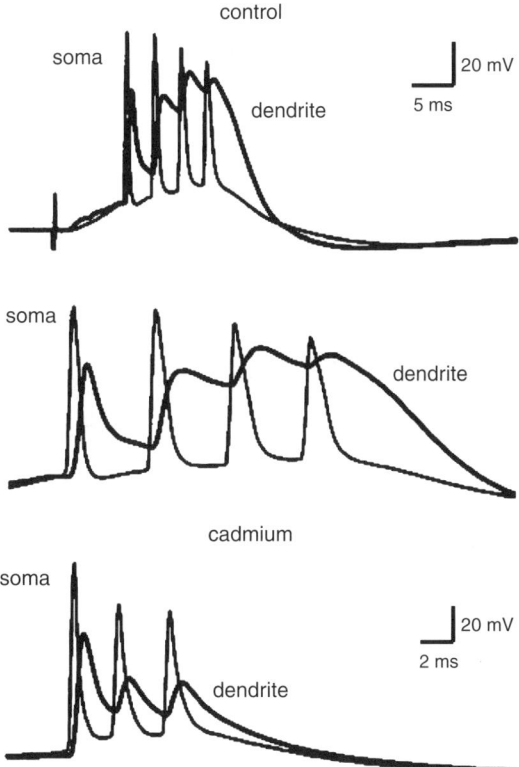

Figure 8.8 Dendritic bursts in pyramidal cells are associated with calcium electrogenesis. Simultaneous recordings were made from apical dendrites and somata of tufted layer 5 pyramidal cells, in rat somatosensory cortex in vitro. The top and middle recordings show the same data at different time scales: synaptic excitation was evoked in layer 2/3, leading to a burst of action potentials at the soma. Each somatic action potential "backpropagates" to elicit a delayed and broadened action potential in the dendrite (325 μm from the soma). The dendritic spikes increase in amplitude, even as the somatic spikes decrease, probably because of a slow depolarizing envelope in the dendrites. This latter slow envelope is probably caused by inward Ca^{2+} currents, because it is blocked by cadmium (200 μM, lower traces, dendritic recording 390 μm from the soma, burst evoked by somatic current pulse of 0.5 nA, 200 ms). With Ca^{2+} currents blocked, the dendritic spikes decrease with time, instead of increasing. Note that there is still a residual slow depolarizing envelope at the soma, perhaps resulting from a persistent Na^+ current.

(From Stuart, Schiller, & Sakmann, 1997 with permission.)

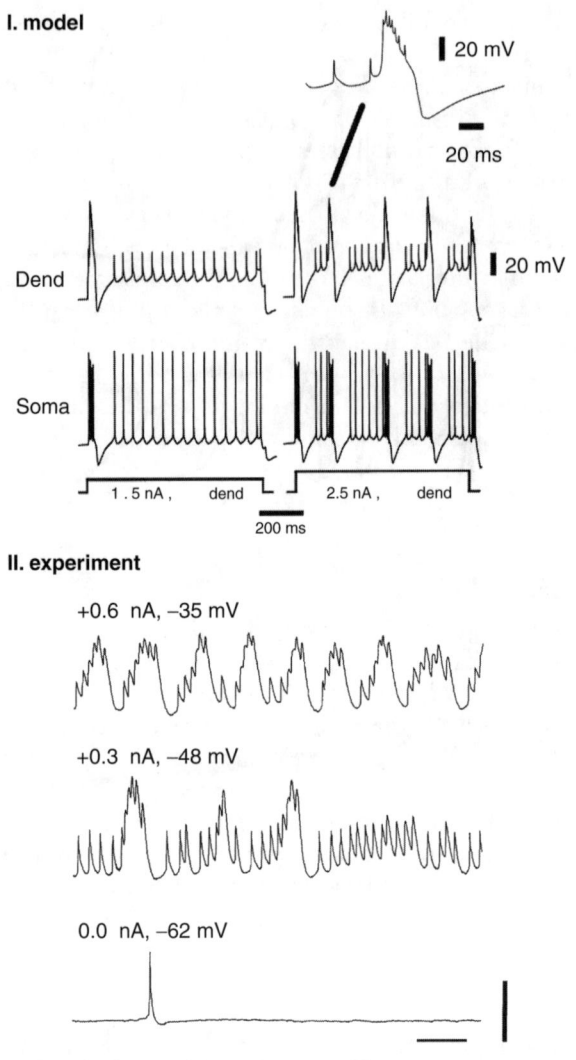

Figure 8.9 Dendritic electrogenesis in layer 2/3. I. Simulations of a multicompartment model of a layer 2/3 regular spiking pyramidal cell. 1.5 nA current pulse to the apical dendrite elicits (after an initial burst) regular spiking at the soma, with spikes backpropagated to the apical dendrite. 2.5 nA dendritic current pulse elicits a mixture of backpropagated Na$^+$ spikes and locally generated, broad, Ca^{2+} spikes. II: injection of sufficiently large current pulses into putative layer 2/3 pyramidal cell dendrite elicits a similar mixture of small fast spikes, and larger broad spikes. Scale bars 120 ms, 20 mV.

(From Traub et al., 2003 with permission.)

Ca^{2+}-mediated spikes are initiated in the apical dendrites. Dendritic electrogenesis, involving fast sodium and slow calcium spiking, may therefore be an ubiquitous phenomenon in cortical pyramidal cells.

Strikingly Different Patterns of Electrogenesis can Occur in the Dendrites vis-à-vis the Soma

The contrast between mechanisms underlying fast rhythmic bursting (predominantly axo–somatic conductances) and Ca^{2+}-dependent bursting (dendritic conductances) illustrates clearly the phenomenon of compartmentalization of activity patterns in principal cells. One of the remarkable features of many neurons, which first became apparent in recordings of Purkinje cells (Llinás & Nicholson, 1971), is that the electrical behavior of one cell region can look quite different from the behavior in a different region. Given that tufted layer 5 pyramidal cells can have an apical dendrite approaching 1 mm in length, such cells are natural candidates for the demonstration of regional specialization. And indeed it is so, as Figure 8.10 demonstrates: the apical tuft can generate large Ca^{2+} spikes, independent of the soma. (The apical voltage transients are known to be Ca^{2+} spikes because they are blocked by cadmium, and are associated with intracellular $[Ca^{2+}]$ increases; the $\sim$10 ms time course is also consistent.)

What is the significance of such dendritic electrogenesis for oscillations? There are at least three reasons to consider dendrites. First, dendritic bursts occur during seizures (Miles et al., 1984). Second, the beta-2 cortical oscillation is generated by intrinsically bursting neurons in layer 5, so a contribution of the dendrites to cell behavior must be considered (Roopun et al., 2006)— although it remains uncertain if a dendritic contribution is essential for the population behavior, and the apical tufts themselves can not be critical: the beta-2 oscillation persists after amputation of the distal apical dendrites. The presence of dendritic calcium spikes and fast, small sodium-dependent spikes in dendrites may have significance in shaping pyramidal cell involvement in gamma and theta network rhythms, however. In hippocampus, pyramidal cells show calcium spiking phase-locked to theta rhythms in vivo (Ylinen et al., 1995b). Direct depolarization of neocortical pyramidal cell dendrites generates a mixture of fast sodium spikes nested within larger, slow calcium spiking (Fig. 8.9; Traub et al., 2003a). Autocorrelation analysis showed a robust theta frequency component to the calcium spikes and a gamma frequency component to the sodium spikes. Thus, within the dendrites of single neocortical pyramids, intrinsic conductances combine to mimic some of the more common features of persistent cortical network rhythms (theta/gamma nesting, Fig. 8.9). Finally, if there is regional specialization between soma and dendrites, we must also be prepared for regional specialization between soma and axon (see later)—and much of the "action" in generating oscillations takes place in and between axons.

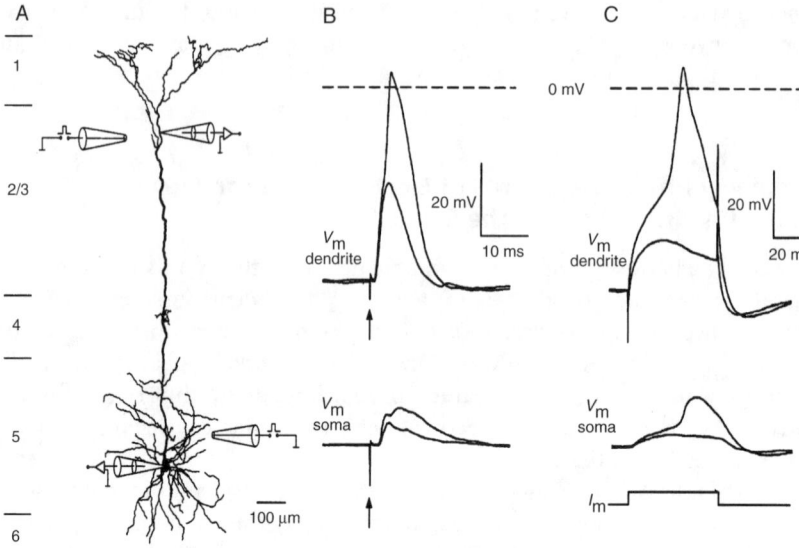

Figure 8.10 Ca^{2+} spikes can be generated in the apical tuft of larger layer 5 pyramidal cells. **A:** Biocytin-filled tufted pyramid, with cortical layers marked on the left, as well as sites of patch recording (soma and apical tuft, 920 μm from the soma) and extracellular stimulation. **B:** Dendritic Ca^{2+} spikes, evoked at two intensities of distal dendritic extracellular stimulation. [These are known to be Ca^{2+} spikes because subsequent experiments showed them to be associated with intracellular [Ca^{2+}] transients, and to be blocked by cadmium.] **C:** Similar dendritic calcium spike evoked by local dendritic current injection. Ca^{2+} spikes have been recorded in the dendrites of a number of neuronal types, including hippocampal pyramidal neurons (Wong et al., 1979) and cerebellar Purkinje cells (Llinás & Nicholson, 1971; Llinás & Sugimori, 1980a,b).

(From Schiller, Schiller, Stuart, & Sakmann, 1997 with permission.)

The Firing Properties of Axons are also not the Same as for Somata

Very fast brain oscillations depend for their generation on physical interactions between the axons of principal (excitatory) neurons; certain gamma and beta oscillations depend on such interactions as well (see Chapters 11 and 12). Both the intrinsic properties of axons, and the means by which axons communicate with each other, are evidently vital to our story. Communication will be considered in Chapter 9. Here, we consider certain aspects of axonal physiology. The study of mammalian axons is, of course, technically difficult, partly because of the small size, and partly because of the myelin that ensheathes large portions of many of them. Following our usual pattern, let us first outline some of the history. The list below will be recognized to be sketchy, but touches on some of the essential points.

1. Adrian et al. (1931) found a means to record from single axons in frog peripheral nerves.
2. In the late 1940s and early 1950s, Hodgkin and Huxley performed their classical experiments on, and analysis of, the generation and propagation of action potentials in the isolated squid giant axon—a preparation hundreds of μm in diameter (e.g., Hodgkin & Huxley, 1952). It was possible to "space clamp" the preparation (i.e., to minimize longitudinal intracellular currents, and effectively isolate the transmembrane currents), by passing a tiny wire through the core of the axon. The local action potential could be accounted for by a linear circuit (membrane capacitance and leak conductance), interacting with two membrane nonlinear circuit elements: g_{Na} and g_K. The time-dependent evolution of the nonlinear circuit elements could be accounted for, phenomenologically, by a set of differential equations describing how each conductance evolved over time, with kinetics depending on transmembrane potential. In a stunning technical tour de force, now largely taken for granted, Hodgkin and Huxley showed that their theory of the locally generated action potential, in space-clamp conditions, could account for the propagation velocity in non-space-clamp conditions.
3. Frankenhaeuser and colleagues (e.g., Dodge & Frankenhaeuser, 1958) extended Hodgkin and Huxley's techniques and ideas to the vertebrate, studying the frog node of Ranvier.
4. Kocsis and Waxman (1982) were able to perform intracellular recordings from mammalian myelinated axons, in the dorsal columns of the spinal cord. It became apparent that axons have a large variety of ionic conductances, and not just the two major ones studied by Hodgkin and Huxley (Black et al., 1990).
5. Rasminsky, Sears, Bostock, and their collaborators, using mainly extracellular measurements, analyzed spike conduction in myelinated and experimentally demyelinated fibers, as well as the generation of ectopic action potentials in diseased axons. These studies have implications for multiple sclerosis and pain syndromes (Baker & Bostock, 1992; Rasminsky & Sears, 1972; Scholz et al., 1993; Schwarz et al., 2006).
6. Patch clamp recordings, starting in the early 1990s, became possible from the proximal axons of mammalian pyramidal cells and Purkinje cells, as well as from selected larger sorts of presynaptic terminals— hippocampal mossy fiber terminals, and Calyx of Held terminals (Borst & Sakmann, 1998; Clark et al., 2005; Colbert & Johnston, 1996; Forsythe, 1994; Geiger & Jonas, 2000; Kole et al., 2007, 2008; Schmitz et al., 2001; Stuart & Häusser, 1994).
7. Raastad and Shepherd (2003) were able to record extracellularly from single axons of CA3 pyramidal cells that were visualized, using uptake

of the fluorescent dye diI. They were able to show that the refractory times of these axons ranged from about 2.5 to about 10 ms at 32°C.

David A. McCormick and his colleagues have recently developed a preparation which allows even more detailed analysis of the properties of the axons of layer 5 pyramidal cells: in neocortical slices, in prefrontal cortex, a bleb could sometimes be identified, at the cut end of an axon at the slice surface. In fortunate circumstances, it was possible to record directly from the axonal bleb and also simultaneously from the soma of the large pyramidal neuron giving rise to the cut axon. Prefrontal cortex is a favorable region for this type of experiment, because in that region myelination does not occur until greater than 100 μm from the soma (see later); layer 5 pyramids in other cortical regions, such as somatosensory cortex, myelinate at closer to 40 μm from the soma (Palmer & Stuart, 2006). We discuss some of the data from this bleb preparation, shown in Figures 8.11 to 8.13. An essential point is that axons contain a number of membrane conductances besides those involved directly in the generation and repolarization of action potentials, that is, besides the conductances involved in the original Hodgkin–Huxley description. The power of the axonal bleb recording technique is illustrated first in relation to an issue involving epilepsy.

From the earliest intracellular studies of epileptiform interictal and ictal events (Goldensohn & Purpura, 1963), it was recognized that the somata of pyramidal cells could become so depolarized as to be unable to generate action potentials. This state was known as "depolarization block," and was presumed to reflect voltage-dependent inactivation of Na^+ channels. In accord with this terminology, the large depolarizations during epileptiform events were called "depolarization shifts" (Prince, 1968). Interictal bursts arise through collective mutual excitation between pyramidal cells (Traub & Wong, 1982); such excitation is mediated synaptically (and is now known also to involve non-chemical-synaptic interactions), so that—in both the synaptic and the nonsynaptic cases—the axons will be critical in bringing about collective behavior. Thus, it became a matter of some interest to know how the axon was behaving during an epileptiform burst vis-á-vis the soma: there was no a priori reason to assume that the axon also entered depolarization block, given that the large depolarization at the soma was primarily (not exclusively) driven by excitatory synaptic currents (Johnston & Brown, 1981), and the axon would be relatively isolated from such currents. This is the case, of course, because excitatory synapses on pyramidal cells lie on spines, and the overwhelming majority of spines lie on the dendrites [there are, however, rare exceptions (Kosaka, 1980)].

Figure 8.11 shows simultaneous axonal and somatic recordings from a layer 5 pyramidal cell, during an epileptiform event in vitro, produced by bath-applied picrotoxin (Shu et al., 2007a, 2007b). The data in Figure 8.11 confirm the suspicion that the axon—at least far enough away from the soma—does not undergo a depolarization shift in the way the soma itself

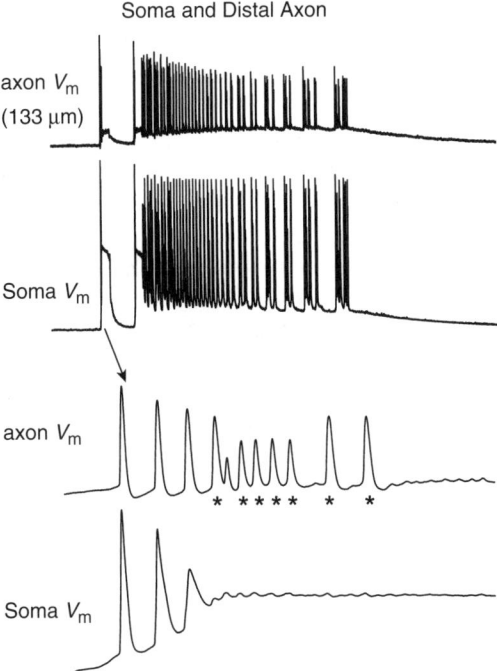

Soma and Distal Axon

axon V_m
(133 µm)

Soma V_m

axon V_m

* * * * *   *   *

Soma V_m

Figure 8.11 The axon of a cortical neuron can fire repetitively (presumably driving downstream neurons), when the soma exhibits "depolarization block". The recording was taken from a layer 5 pyramidal cell in ferret prefrontal cortex in vitro, with the GABA$_A$ receptor blocker picrotoxin (50 µM) in the bath to promote epileptiform discharges. The axon (133 µm from the soma) and soma were recorded simultaneously. The upper traces show an epileptiform burst lasting several seconds; the lower traces, on an expanded time scale, show the soma in "depolarization block" after three action potentials, while the axon continues to fire.
(From Shu, Duque et al., 2007 with permission.)

does; and, indeed, the axon fires repetitive spikes at high frequency, in a way expected to propagate to the presynaptic terminals (and perhaps to other axons—see later). These data are important not only in themselves, but also serve to emphasize that different portions of a neuron can do somewhat different things: soma vs. dendrites, and soma vs. axon.

Under most physiological conditions, action potentials are initiated in the proximal axon, in both pyramidal cells (Palmer & Stuart, 2006; Stuart et al., 1997) and Purkinje cells (Stuart & Häusser, 1994). The proximal axon is of interest for two additional—related—reasons. First, the axon initial segment of pyramidal cells is the site of inhibitory synapses, whose presynaptic elements belong to a specialized type of interneuron, the axo–axonic or chandelier cell (Buhl et al., 1994; Freund et al., 1983; Somogyi et al., 1982); and—at least under certain conditions—it is possible that these synapses can paradoxically be exciting (Szabadics et al., 2006). Second, dye-coupling data (Schmitz et al., 2001) suggest the existence of gap junctions between the initial segments of different axons; in the case of CA1 pyramidal neurons, this dye coupling was found up to 150 μm from the soma, in the unmyelinated portion of the axon. Given the importance that axonal gap junctions appear to have, in the generation of network oscillations, it is important to know where myelination first occurs in pyramidal cells. [One would not expect gap junctions to occur between the myelinated portions of axons, although gap junctions between nodes of Ranvier are possible in principle.] Figure 8.12 provides data substantiating a statement made earlier, that in prefrontal cortex layer 5, pyramidal cell axons may not myelinate for hundreds of micrometers. Even in other parts of the cortex, myelination does not occur for tens of micrometers, for layer 5 pyramidal cells, in principle allowing the possibility of close contacts between neighboring axons. It would be most helpful to have comparable data for superficial pyramidal cells and for layer 4 spiny stellate cells.

We have already mentioned above how axons contain persistent (slowly inactivating) Na$^+$ channels, and also the "M" type of K$^+$ channels (although neither of these channel types is specifically confined to axons). The axon bleb

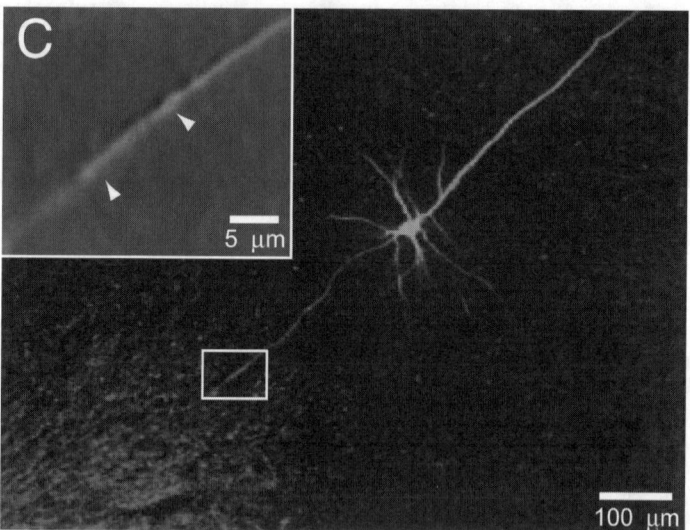

Figure 8.12 In layer 5 pyramidal neurons of ferret prefrontal cortex, myelination can begin hundreds of micrometers from the soma. The neuron was filled with biocytin, which fluoresces green in this image. After fixation, the specimen was stained with an antibody to myelin basic protein, and then developed with Alexa-488 conjugated to a secondary antibody: thus, myelin appears red in the image. In this cell, myelin is first observed about 200 μm from the soma. Please see color insert. (From Shu, Duque et al., 2007 with permission.)

recording technique reveals (Fig. 8.13) still another class of channels in axons, channels that are blocked by a mamba snake venom toxin called α-dendrotoxin: this group includes Kv1.1, 1.2 and 1.6, and according to Harvey (1997), it is inactivating forms of K⁺ conductances mediated by these channels that are most blocked by the toxin (although the data in Fig. 8.13 seem not completely in accord with that notion; see later).

In view of Figure 8.11, what Figure 8.13A shows is rather surprising. Thus, Figure 8.11 demonstrated that a cell soma could go into depolarization block, while the axon fired repetitive action potentials; note, however, that in such a

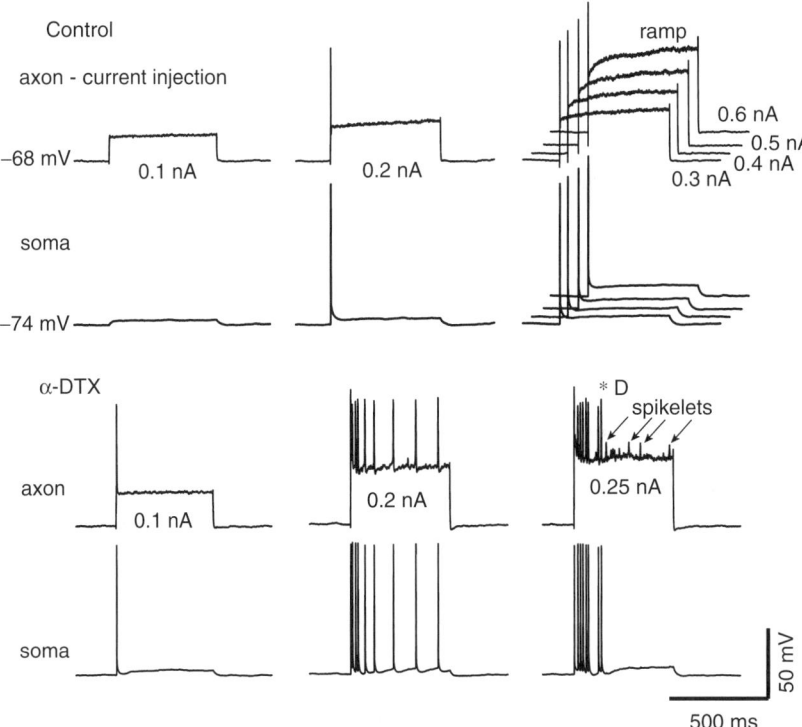

Figure 8.13 In layer 5 pyramidal cells of rat prefrontal cortex, strong axonal depolarization leads to suppression of action potential firing; block of selected K⁺ channels with α-dendrotoxin allows the axon to fire bursts of action potentials. Recordings were made simultaneously from the soma and axon, 110 μm from the soma, and depolarizing current pulses and ramps were injected into the axon. **A:** Native conditions; at most, one axonal spike is generated. **B:** After application of 100 nM α-dendrotoxin, the axon is able to fire bursts of action potentials, and to generate spikelets. [α-dendrotoxin is derived from the venom of mamba snakes, and blocks Kv1.1, Kv1.2, and Kv1.6 types of K⁺ channels, preferentially the inactivating components (Harvey, 1997; Harvey & Robertson, 2004).]

(From Shu, Yu et al., 2007 with permission.)

situation, the axon is depolarized by currents that have to flow longitudinally, along the thin axon, from the depolarized soma—so that the axon itself is not strongly depolarized. In Figure 8.13A, in contrast, we see what happens when a depolarizing current is injected *directly* into the axon: there is a single action potential followed by depolarization block. The depolarization block in this case seems not to be caused (at least for the most part) by Na^+ channel inactivation, but rather because the depolarization is opening K^+ channels that shunt the membrane. The reason for thinking this is shown in Figure 8.13B: in the presence of the toxin (and provided current injection is large enough), strong depolarization still occurs, indeed even more depolarization than in baseline conditions—but there is now, nevertheless, repetitive firing of the axon. The lesson here, for the study of oscillations mediated by networks of axons, is that the state of axonal K^+ channels is likely to be critical for network behavior.

Spikelets do not Originate in the Apical Dendrite in Layer 5 Pyramidal Cells

The simultaneous dendrite/soma recording technique, illustrated in Figure 8.8, also provides useful information on a fundamental question in the study of oscillations: a number of oscillations, including gamma (Cunningham et al., 2004a,b; Fisahn et al., 2004; Chapter 12), beta-2 (Roopun et al., 2006; Chapter 11), and very fast oscillations (Chapter 10), are associated with spikelets in single neurons. Spikelets resemble miniature action potentials. An early study (Spencer & Kandel, 1961) examined these events—there called "fast prepotentials"—in hippocampal pyramidal neurons in vivo, and postulated an origin from spikes generated in the apical dendrite, that were electrotonically transmitted to the soma. Schmitz et al. (2001) showed, however, that it was possible to elicit somatic spikelets (in hippocampal pyramidal cells) by axonal stimulation, with the spikelet actually conducted actively along the axon, and with the involvement of one or more gap junctions along the pathway (from stimulating electrode to recorded soma), in at least some instances (Chapter 9). In a word, these latter data indicated an *axonal* origin for spikelets, not dendritic. Those of our network models in which spikelets occur—and that are extremely successful in accounting for experimental data—postulate an axonal origin for spikelets. The question of the origin of spikelets is, therefore, critical for us.

The data illustrated in Figure 8.14A suggest that it is extremely improbable that somatic spikelets are caused by apical dendritic spikes, in layer 5 pyramidal cells. (Fig. 8.14B shows spikelets in a layer 5 pyramidal cell during the course of a beta-2 oscillation: these spikelets are much sharper than the somatic potential in Fig. 8.14A). Gasparini et al. (2004) have provided somatic/dendritic recordings from CA1 pyramidal cells that are similar in appearance to Figure 8.14A: hippocampal pyramidal cell spikelets are also unlikely to

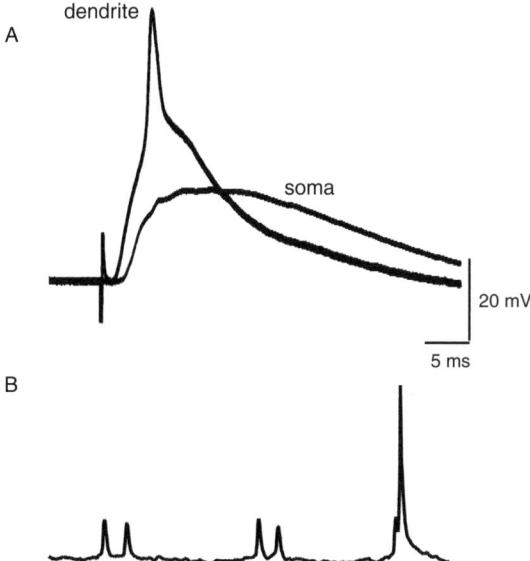

Figure 8.14 Apical dendritic action potentials evoke slow, smooth, potentials at the soma, rather than spikelets. **A:** Simultaneous patch-clamp recordings from soma and apical dendrite (440 μm from the soma) of a tufted layer 5 pyramidal cell. A dendritic spike was evoked by synaptic stimulation in layer 2/3. A subthreshold electrotonically filtered somatic potential was produced, lasting over 15 ms. **B:** Spikelets and a presumed antidromic action potential recorded in a layer 5 intrinsically bursting neuron, during a beta-2 network oscillation induced by kainate (see also Chapter 11). The spikelets here have about the same amplitude as the somatic potential in A, but last only about 1 ms.
(**A** from Stuart, Schiller & Sakmann, 1997; B from Roopun et al., 2006. All reproduced with permission.)

originate in the apical dendrite. Spikes in the basal dendrites of layer 5 pyramids (which are not easy to elicit) can, however, produce spikelets of rapid onset, although the repolarization phase is still much slower than for spikelets that occur during network oscillations (Nevian et al., 2007). The possibility of basal dendritic electrogenesis must at least be considered during network oscillations (Memmesheimer & Timme, 2006). It is also conceivable that at least some spikelets originate from mixed synaptic contacts onto pyramidal neurons (i.e. synapses containing both classical chemical components and also nearby gap junctions).

GABA Receptor-Mediated Inhibitory Synaptic Responses

Since many brain oscillations—particularly gamma oscillations—are regulated by gamma-aminobutyric acid (GABA)-receptor–mediated synaptic inhibition, it is well to consider some of the principles which govern this inhibition. Here, we list the following basics:

1. The effects GABA exerts on the postsynaptic neuron are critically determined by whether the receptor (to which the GABA must bind) is $GABA_A$ or $GABA_B$ in type (Ulrich & Bettler, 2007).
2. $GABA_A$ receptors are primarily permeable to Cl^- under physiological conditions, and $GABA_B$ to K^+; consequently, $GABA_B$-mediated currents will have a lower reversal potential.
3. $GABA_A$ receptors are ionotropic: GABA binds to the receptor/channel, changing the conformation of the latter, and current flows directly the channel. In contrast, $GABA_B$ receptors are metabotropic. GABA binding leads to a G-protein–mediated series of biochemical steps, which eventually cause a separate set of channels to open. Hence, $GABA_B$ receptor–mediated currents have a relatively slow onset, and they last a long time (often hundreds of milliseconds or more).
4. GABA receptors have different properties, determined by the composition of subunits (see also later).
5. Almost all $GABA_B$ receptors are extrasynaptic, and are activated by GABA "spillover." Some $GABA_A$ receptors are also extrasynaptic, but most are localized at synaptic specializations.
6. $GABA_A$ receptor–mediated currents have different time courses in different cells; but even in the same kind of cell (say pyramidal cells), there are faster and slower duration $GABA_A$ receptor-mediated currents, as first studied in the hippocampus (Banks et al., 1998; Banks & Pearce, 2000; Pearce, 1993).
7. Within the neocortex, a particular interneuron type—the neurogliaform cell (Fig. 8.15)—can elicit (even via a single action potential) both slow $GABA_A$ and $GABA_B$ responses (Szabadics et al., 2007; Tamás et al., 2003). Fast-spiking interneurons, in contrast, elicit classical rapid $GABA_A$ responses (Somogyi et al., 1998; Tamás et al., 1997). Neurogliaform cells have an interesting firing pattern, so-called delayed spiking (Fig. 8.15, see also Kawaguchi, 1997).

Synaptic Responses to a Given Neurotransmitter Depend on the Postsynaptic Cell Type

We have mentioned earlier how α-amino-3-hydroxyl-5-methyl-4-isoxazole-propionate (AMPA) receptor-mediated EPSPs/EPSCs are faster in fast-spiking interneurons than in pyramidal cells (Miles, 1990; Miles & Wong, 1986). Here, we shall provide further details on this important issue.

AMPA receptors are built out of four subunits, variously called GluR-A, . . . , GluR-D, or GluR1, . . . ,GluR4 (Hollmann & Heinemann, 1994; Seeburg, 1993;). The properties of neuronal glutamate receptors can be studied in relative

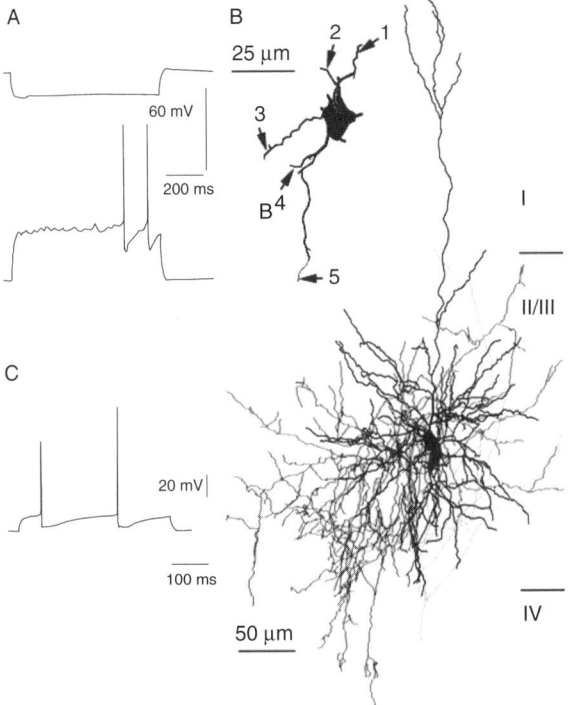

Figure 8.15 Neurogliaform interneurons exhibit delayed firing, and produce slow IPSPs in pyramidal cells. **A:** Response of an identified neurogliaform cell to hyperpolarizing current pulse (above), and small depolarizing current pulse (below). Note the delayed onset of firing. **B:** Morphology of layer 2/3 neurogliaform cell and a synaptically connected layer 2/3 pyramidal cell. The soma and dendrites of the neurogliaform cell are blue, and its axon red; the soma and dendrites of the pyramidal cell are black, and its axon green. (The original figure shows synaptic potentials produced at this connection.) Inset shows sites of ultrastructurally identified synapses. **C:** Delayed firing in a multicompartmental model neurogliaform cell, with a high density of transient K+ conductance in the soma and dendrites.

(**A** and **B** are from Tamás et al.,(2003); **C** is unpublished data of R.D. Traub and Miles A. Whittington. All reproduced with permission.)

isolation, using patch techniques, and receptor properties then compared with the repertoire of subunits expressed in the neuron from which the analyzed receptor(s) was isolated—subunit expression is estimated by single-cell polymerase chain reaction (PCR) techniques (Jonas & Monyer, 1999). In this way, it has been confirmed that AMPA receptor-mediated EPSCs on interneurons are indeed extremely rapid (submillisecond rise and decay times at 34°C) (Geiger et al., 1997); and that AMPA receptors in FS interneurons and pyramidal cells have systematically different properties—including Ca^{2+} permeability in interneurons but not pyramidal cells, faster kinetics in interneurons, and susceptibility to polyamine block in interneurons (Hollmann et al., 1991; Jonas et al., 1994; Koh et al., 1995a,b). Finally, these systematic differences in properties correlate with AMPA receptor subunit composition: GluR-B is expressed in pyramidal cells, and confers Ca^{2+} impermeability and relatively *slow* kinetics; GluR-B is absent in FS interneurons—this, together with the expression of GluR-D in FS interneurons confers *fast* kinetics to AMPA receptor excitatory postsynaptic conductances (EPSCs; Fuchs et al., 2007; Geiger et al., 1995). Interestingly, when GluR-B expression is induced in FS interneurons, by transgenic techniques, this produces slowing of AMPA receptor kinetics (Fuchs et al., 2001).

That AMPA receptor kinetics in FS interneurons are so rapid has several consequences for brain oscillations. We have mentioned one consequence already: fast (and large) EPSCs in interneurons allow interneuron spike doublets to enhance two-site synchronization in the tetanic gamma oscillation model (Traub et al., 1996c, 1999c), and synchronization is degraded when interneuron EPSCs are slowed (Fuchs et al., 2001). In addition, as we shall expand upon in Chapter 12, rapid interneuron EPSCs allow for an interplay between gamma and very fast oscillations in the persistent gamma model.

$GABA_A$ receptor-mediated IPSCs also appear to be faster in FS interneurons than in principal cells (pyramidal cells and hippocampal dentate granule cells). Bartos et al. (2001) reported at decay time constant for IPSCs in parvalbumin-positive dentate gyrus FS cells (visualized in a mouse, in which EGFP [green fluorescent protein] was expressed under control of the parvalbumin promoter) of approximately 2.5 ms at 32°C, and a decay time constant of approximately 5 ms in nearby dentate granule cells. In a subsequent study (Bartos et al., 2002), IPSC decay was fit with two exponentials, with the majority of the IPSC due to the first, rapidly decaying component. Example fits were, for interneurons, $\tau 1$ of 1.0 ms and $\tau 2$ of 10 ms for dentate gyrus interneurons; and for dentate granule cells, $\tau 1$ of 1.8 ms and $\tau 2$ of 8.5 ms. Bartos et al. (2002) argued that the rapid time course of interneuron IPSCs, along with the relatively depolarized reversal potential of these IPSCs (to produce "shunting inhibition") was critically important for the generation of network gamma oscillations by mutual inhibition, when synaptic excitation was suppressed; they did not, however, provide experimental oscillation data to test experimental predictions. Significantly, mutual synaptic inhibition between parvalbumin-positive interneurons is not necessary for hippocampal gamma oscillations in vivo; hippocampal gamma appears normal (in field potential recordings) in a genetically modified mouse, in which synaptic inhibition onto parvalbumin-positive interneurons has been removed (Wulff et al., 2009).

The relation between $GABA_A$ receptor kinetics and subunit composition is, to the best of our knowledge, not clear. This is a consequence of the large number of possible subunit combinations, and the existence within a given cell type of different phasic IPSC kinetics, and of both phasic and tonic IPSCs (Mody & Pearce, 2004).

We must not forget that there are differences between principal neurons and FS interneurons, not only in synaptic currents and in cell shape, but also in the properties of intrinsic membrane conductances, particularly the Na^+ and K^+ channels that are responsible for action potentials (Martina & Jonas, 1997; Martina et al., 1998).

Kainate Receptors

Bath application of kainate constitutes a particularly robust method for the induction of oscillations in cortical circuits, when used in concentrations of 1 μM or less ("nanomolar kainate")—concentrations far below those used in studies of excitotoxicity [typically ~10 mM in vitro (Kristensen et al., 2001)]. And indeed, network oscillations are produced at kainate concentrations near the threshold for evoking kainate responses at all, at least in cultured hippocampal neurons [~500 nM (Wilding & Huettner, 1997)]. This is a suitable point to elaborate on a few of the details concerning kainate receptors, details that are relevant to understanding the mechanisms by which kainate facilitates network oscillations.

We have mentioned above the fact that N-methyl-D-aspartate (NMDA) and AMPA types of glutamate receptors are constituted of different types of subunits, with receptor properties depending on which types of subunits have co-assembled. The same principles apply to kainate receptors, whose subunits are designated GluR5, GluR6, GluR7, KA1, and KA2 (Bleakman & Lodge, 1998; Hollman & Heinemann, 1994; Wisden & Seeburg, 1993). GluR5 and GluR6 can exist in different forms, depending on RNA editing (Bernard & Khrestchatisky, 1994).

The situation with kainate receptors is complicated, for a number of reasons: the stoichiometry of subunit assembly is unknown; native receptors and recombinantly expressed receptors can have different properties; kainate receptor-induced currents even within a given preparation can have distinctive properties (depending on how the currents are activated, see later); and there are both ionotropic and metabotropic actions of kainate receptors (Rodríguez-Moreno & Lerma, 2007). Kainate receptors are located at presynaptic terminals, where they contribute to regulation of both GABA and glutamate in rather complicated ways (Lerma, 2006; Pinheiro & Mulle, 2006). Kainate receptors are also located on somatic/dendritic membranes of both principal cells and interneurons, where they contribute to ionotropic EPSCs (Frerking et al., 1998; see later), as well as to the metabotropic regulation of AHP currents through GluR6 (Melyan et al., 2004; Fisahn et al., 2005); and, kainate receptors are probably located on axons, at least of interneurons, where there is evidence that kainate receptor activation enhances axonal excitability in some unknown fashion (Fisahn et al., 2004; Semyanov & Kullman, 2001). Kainate receptors are also found, somewhat intriguingly, on astrocytes (Usowicz et al., 1989).

We here concentrate on kainate-induced membrane currents, as (1) metabotropic kainate actions on AHP currents have been shown to be unnecessary for the generation of kainate oscillations (Fisahn et al., 2004); (2) the significance of kainate regulation of synaptic transmission for oscillations remains to be defined; (3) while direct kainate actions on axons (and gap junctions) could be of immense importance for oscillations, the evidence for such actions is indirect and of unclear mechanism; (4) it seems likely that the well-studied cellular depolarizations induced by kainate, in both pyramidal cells and interneurons, are important for oscillations (Fisahn et al., 2004). For example, in vivo, fast oscillations occur during the depolarizing phases of the slow (<1 Hz) oscillation (Grenier et al., 2001, 2003; Steriade et al., 1995). Thus, it seems that postsynaptic, ionotropic actions of kainate, consisting of cellular depolarizations, are key factors in generating oscillations. The receptor molecules KA1 and KA2 bind to kainate, but do not appear to mediate transmembrane currents by themselves; we shall not discuss them further. Most of the literature on membrane currents induced by kainate concerns GluR5 and GluR6, and we therefore limit our considerations to those subunits.

The issues then, are these: what do individual kainate EPSCs look like, when evoked by an extracellular stimulus or by stimulating a single presynaptic neuron? And further, how can we extrapolate from the individual EPSCs, so as to deduce the effects of bath-applied kainate, as used in oscillation protocols? A main consideration in this question concerns the extent to which kainate receptors desensitize at the concentrations used in oscillation experiments (typically <1 µM).

Figure 8.16 shows data taken from Ali (2003), consisting of EPSCs evoked by an extracellular shock, in a layer 5 pyramidal cell, and in a cortical fast-spiking interneuron. The kainate receptor-induced EPSC was isolated pharmacologically, by blockade of AMPA receptors (with GYKI53655) and NMDA receptors (with D-AP5). The kainate EPSC so produced is considerably smaller than the control EPSC, and has slower onset and decay kinetics. In contrast, it was found in the same study that kainate EPSCs, produced by activation of single presynaptic neurons, usually had kinetics similar to control EPSCs. By whatever means the slower EPSCs are produced (whether through extrasynaptic receptors, or through synaptic receptors present at synapses not activated in the experimental protocols used), it seems likely that bath applied kainate would activate all kainate receptors, and hence generate slow depolarizations; and indeed this is observed experimentally, in pyramidal cells and interneurons in the hippocampal slice (Fisahn et al., 2004; Frerking et al., 1998). These slow depolarizations seem to be produced by activation of kainate receptors containing GluR6 (Fisahn et al., 2004). [Note also that the threshold concentration for activating GluR6 receptors is much lower than the threshold for activating GluR5 receptors (Wilding & Huettner, 1997; Sommer et al., 1992). Note as well that the roles of particular kainate

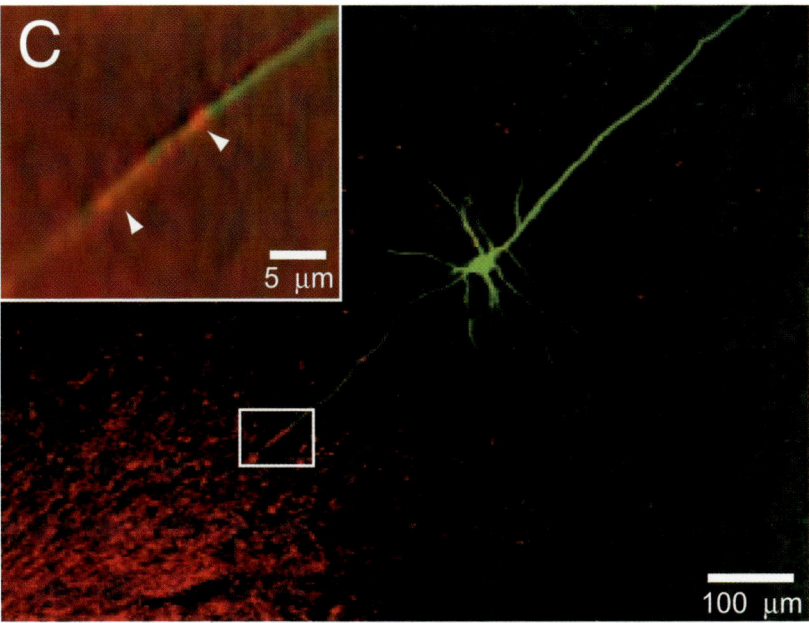

Figure 8.12 In layer 5 pyramidal neurons of ferret prefrontal cortex, myelination can begin hundreds of micrometers from the soma. The neuron was filled with biocytin, which fluoresces green in this image. After fixation, the specimen was stained with an antibody to myelin basic protein, and then developed with Alexa-488 conjugated to a secondary antibody: thus, myelin appears red in the image. In this cell, myelin is first observed about 200 μm from the soma.

(From Shu, Duque et al., 2007 with permission.)

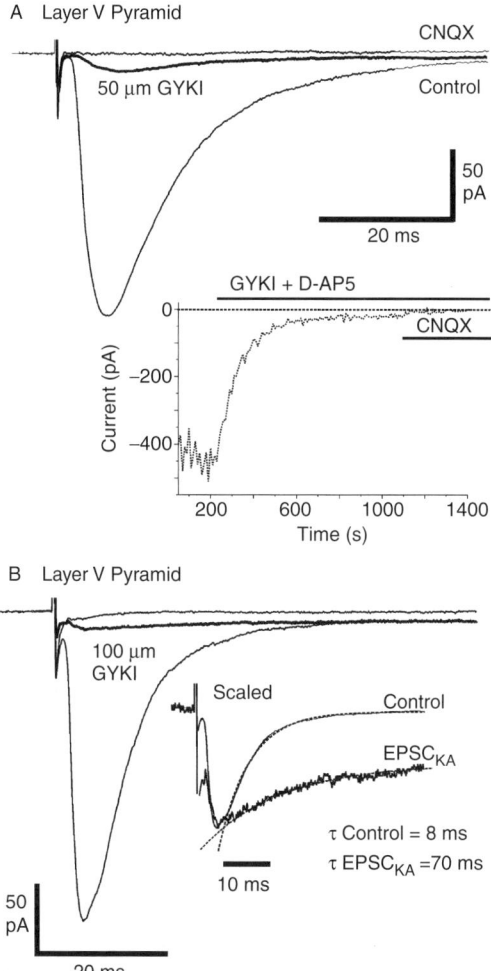

Figure 8.16 Kainate receptor–mediated EPSCs in cortical neurons in vitro, evoked by a single shock, are smaller and slower than AMPA receptor–mediated EPSCs. [Kainate receptor–mediated EPSCS evoked by stimulating a single presynaptic cortical neuron, however, usually have fast kinetics, for unclear reasons.] The largest EPSCs shown here were evoked in control solution, and contain AMPA, kainate and perhaps NMDA receptor mediated components. Kainate EPSCs were isolated by addition of 100 μM GYKI53655 (to block AMPA receptors), and 100 μM AP5 to block NMDA receptors. The uppermost potentials—with virtually no synaptic current —were recorded after the further addition of 10 μM CNQX, which blocks both kainate receptors (as well as AMPA receptors); this serves as confirmation that the middle traces do correspond to kainate receptor stimulation. **A:** Recordings in a layer 5 pyramidal cell. The evoked EPSC attributable to kainate receptors – EPSC$_{KA}$ - is the middle one (thick line). **B:** Similar recordings from an FS interneuron. The inset shows an expanded version of EPSC$_{KA}$ along with the response when AMPA, kainate, and NMDA receptors are all blocked. Again, note the prolonged time course of EPSC$_{KA}$.

(Modified from Ali, 2003 with permission.)

receptors—GluR5 vs. GluR6—may possibly be different in neocortex vis-à-vis the hippocampus.]

What about desensitization? Here, it has been shown that kainate currents, particularly those mediated by GluR6, desensitize only partly, with a persistent current that is about 30% of the initial peak current (Wilding & Huettner, 1997; Paternain et al., 1998). We would thus expect that bath-applied kainate would lead to a persistent depolarization in principal cells and interneurons, just as the oscillations themselves persist.

While GluR6 is absolutely required for kainate gamma oscillations in the hippocampal CA3 region in vitro (Fisahn et al., 2004), that does not mean GluR5 plays no role. The role, however, seems to be regulatory. When GluR5 is knocked out in mice, the resulting slices oscillate at lower kainate concentrations than do slices from control mice; and slices from the knockout are more prone to generate epileptiform bursts in kainate, than are slices from control animals.

ATP-Gated K+-Channels

ATP-gated K^+ channels turn out, surprisingly, to be important in regulating the cortical slow oscillation, at least in vitro; the channels also regulate network oscillations in pools of hypoglossal motorneurons (Sharifullina et al., 2005). The slow oscillation is the setting for many oscillations in vivo, so this is a proper place to provide some information on the channels (Seino & Miki, 2003). ATP-gated K^+ channels are of clinical relevance in at least two ways: *first*, the channels are present in pancreatic beta cells: their closure leads to membrane depolarization, Ca^{2+} entry through voltage-gated channels, and subsequent insulin release. It turns out that beta cell channels can be closed (thereby promoting insulin release), in a manner that is independent of intracellular ATP levels (whereby ATP rises tend to close the channels): this can be accomplished through binding of the channels to the sulfonylurea compound tolbutamide. Tolbutamide is absorbed from the intestine, and so it is used as an oral agent in the treatment of those diabetic patients who have sufficiently many beta cells left to supply useful amounts of insulin; patients with too few pancreatic beta cells need to be administered exogenous insulin. Another drug, diazoxide, opens the channels, thus suppressing insulin release. Orally administered diazoxide is therefore used to treat certain patients with hypoglycemia resulting from excessive insulin release. Second, ATP-gated K^+ channels are present in vascular smooth muscle cells; closing the channels tends to depolarize these cells, increasing vascular tone and elevating blood pressure. Opening the channels relaxes vascular smooth muscle and reduces blood pressure. Intravenous diazoxide is thus used to treat hypertensive crisis.

ATP-gated K^+ channels are also present in brain (Allen & Brown, 2004; Ohno- Jiang & Haddad, 1997; Shosaku & Yamamoto, 1992); one proposed function is to protect neurons from hypoxic/ischemic injury (Sun et al., 2007): ATP depletion in potentially injurious conditions will be depleted, causing channel opening, membrane hyperpolarization, and thus reduced action potential firing and reduced transmitter release—both of these processes being sources of metabolic demand on neurons.

The channels are formed by assembly of four inner K^+ channel subunits to form the conducting pore, and four outer proteins that contain the sulfonylurea binding sites (Liss & Roeper, 2001; Seino & Miki, 2003). The K^+ channel subunits (typically Kir6.2, for neurons) belong to a family of K^+ channels called "inward rectifiers": they pass current into the cell more readily than they pass current outward. Given the fact that the K^+ reversal potential is typically below reversal potential, this means that inwardly rectifying K^+ channels can generally only be open at hyperpolarized membrane potentials. The presence of of the sulfonylurea binding sites in brain ATP-gated K^+ channels means that tolbutamide and diazoxide can be used as pharmacological tools in the study of brain oscillations in vitro (Cunningham et al., 2006b).

9

Gap Junctions and the Notion of Electrical Coupling Between Axons

Many, although not all, brain oscillations depend on gap junctions, which can be visualized as a collection of small tunnels between cell interiors. Even when a type of brain oscillation does not absolutely require gap junctions, it might still occur more robustly when gap junctions are open. Much of this book documents the importance of gap junctions for oscillations; here, let us take that notion as a given, and consider the structure, function, and distribution of gap junctions themselves in the nervous system. We concentrate particularly on the evidence for gap junctions between principal neurons, including pyramidal neurons, hippocampal dentate gyrus cells, and cerebellar Purkinje cells. We will also introduce some counterintuitive ideas (to be developed further later): for example, that even though gap junctions are structurally small tunnels, it does not follow that cells in a gap-junctionally connected network must be behaving similarly; or that the behavior of the individual cells, when the gap junctions are shut, will look anything like the behavior when the junctions are open.

It is fair to say that chemical synaptic transmission (Coombs et al., 1955a,b) dominates neuronal network thinking. Indeed, the McCulloch–Pitts model (1943) postulates a form of interaction between "neurons" that is loosely based on chemical synapses—and this model predates the definitive physiological experiments that defined how chemical synapses actually work. The reader is also undoubtedly familiar with the controversy, dating back to the late 19th century, in which the Ramón y Cajal school postulated separable, individually defined nerve cells, and the Camillo Golgi school postulated a neuronal syncytium (López-Muñoz et al., 2006). Ramón y Cajal and Golgi shared the

1906 Nobel Prize for Physiology or Medicine; yet the Cajal school became dominant as electron microscopy confirmed the discrete, separable nature of neurons, and both the structural details and also the overwhelming numbers of chemical synapses; and, there have been an astronomical number of published studies on synaptic function. The dominant place of synaptic physiology in all domains of Neuroscience has become firmly and forever ingrained. But, as we shall see, chemical synapses do not make the whole story; and further, a rather small number of strategically placed gap junctions can (and do) have striking and profound effects on network activities. Perhaps ironically, gap junctions and chemical synapses often occur immediately adjacent to each other ("mixed" synapses) (Martin & Pilar, 1963; Rash et al., 1996).

Before proceeding, it is necessary to clarify terminology and basic concepts. The Reader is familiar with the distinction between "*membrane channels*"—concrete objects made out of protein—and "*membrane conductances*," mathematical/physical descriptions of electrical circuit behaviors, that involve ionic current flows across cell membranes. The physical properties of "membrane conductances" are determined by the structure, amino acid sequence, and biophysics of the "membrane channels," as well as on details in the environment and modifications of the channel proteins (e.g., by ionic conductance gradients, or by which particular channel amino acids have been phosphorylated). Nevertheless, we must be able to distinguish conceptually between the channels (the molecules) and the conductances (the physical circuit behavior). In an analogous fashion, in this chapter, we must distinguish conceptually between "*gap junctions*"—ensembles of gap junction channels, each of which is assembled out of protein molecules—and between "*electrotonic coupling*" (alternatively, "*electrical coupling*")—one of the physical behaviors, conferred by gap junctions onto abutting pairs of cells. (And we must also recognize that electrotonic coupling might, in principle, be produced by structures other than gap junctions; and, that gap junctions confer physical behaviors, on cell pairs, other than electrotonic coupling.) In each of the two cases, a concrete object confers a physical behavior, but historically the physical behavior was discovered first, by one set of experimental techniques, and the structural and molecular bases were discovered later, via somewhat different techniques (Bennett et al., 1963; Revel & Karnovsky, 1967; Robertson, 1963).

In the world of living things, gap junctions do not occur in plants (whose cells have walls that would make gap junction formation impossible), nor, of course, do they occur in single-celled creatures such as bacteria and protozoa. [This latter statement is not as vacuous as it sounds, as—in animals—gap junctions may interconnect different domains of a single cell (e.g. a Schwann cell).] In any case, gap junctions are confined to multi-celled animals. They occur in extremely primitive animals (e.g., phylum Cnidaria, which includes coral animals, jellyfish, and sea anemones), in ourselves, and probably in all other creatures of intermediate complexity: all types of vertebrates, arthropods, molluscs, earthworms, nematodes, and many other sorts of animal (Andreuccetti et al., 1987; Hadley et al., 1983; Hanna et al., 1978; Moss et al., 2005). Within a given type of creature, gap junctions occur very early in development (in our case, even prior to fertilization, as gap junctions interconnect oocytes with granulosa cells); and, indeed, gap junctions are crucial for the embryo to develop properly (Elias & Kriegstein, 2008; Warner et al., 1984). And, even though in many tissues (including the nervous system), gap junctions may decline in number as development proceeds, nevertheless, in adult vertebrates, gap junctions continue to be found almost everywhere: the uterus, the liver, the heart, the lens of the eye, the brain, and so forth. Gap junctions in the heart are crucial to life: it is only through them that a cardiac action potential can conduct from myocyte to myocyte, so that an organized pattern of muscle contraction (and effective cardiac pumping) is possible. Gap junctions are also present between nerve cells, where they allow for a type of neuron-neuron signaling that complements chemical synapses [itself a form of neuronal signaling present in extremely primitive creatures, like sea anemones (Westfall et al., 2002)].

Gap junction channels have an aqueous interior (see also later), through which ionic currents can flow—the physical basis of electrotonic coupling between pairs of cells. (An exception to this statement consists, however, of Ca^{2+} ions which usually close gap junction channels.) In addition, however, other molecular species (generally of molecular weight up to ~1,000) can also pass through the channel. This allows smaller fluorescent dyes (or other molecules that may be visible microscopically, after appropriate chemical processing) to travel from cell to cell, a process known as "dye-coupling." In addition, passage of small molecules, such as inositol triphosphate (IP3), through gap junctions is likely to be of physiological relevance.

How are gap junction channels put together? First, Nature starts with the protein molecules. In invertebrates, the proteins belong to a family called the innexins (Phelan, 2005). In the case of vertebrates, the proteins are chosen from either of two families: the connexins [of which there are more than 20 types (Söhl et al., 2005)], or the pannexins; these latter have molecular homology with the innexins (Panchin et al., 2000). Three sorts of pannexins are known, two of them in the nervous system (Barbe et al., 2006; Bruzzone et al., 2003, 2005; Litvin et al., 2006; Ray et al., 2005; Vogt et al., 2005). Connexins are each identified with a number that is roughly proportional to the molecular weight; for example, there is connexin 36 (molecular mass ~36 kD), the major mammalian neuronal connexin. Each of these gap junction proteins has a number of cytoplasmic and transmembrane domains, with a significant part of the protein located on the outside of the cell.

To visualize the next step in gap junction channel construction, it helps to know that the channel has a sixfold rotational symmetry. Each of the two cell membranes, which will be linked by the channel, contains a protein hexamer; in the case of connexins, the hexamer is called a *connexon*. Before association of two connexons (one in each membrane), a membrane-located connexon forms what is called a *hemichannel*, a potential channel from the cell interior to the extracellular space (we say "potential" because the hemichannel is likely to be closed by Ca^{2+} ions in the extracellular milieu). Finally, two hemichannels (one in each membrane) "lock" together to form the channel. We shall consider further details of this structure below.

The term "gap junction" is usually used to refer to an assembly of contiguous, packed, gap junction channels: anywhere from just a few of them to many thousands. The channels are likely to be packed tightly together, most often into a hexagonal array, although not always (Larsen, 1977; Revel & Karnovsky, 1967). The assembly as a whole will have a characteristic shape: usually a disk or plaque, but possibly a ribbon or a network-like ("reticular") structure (Kamasawa et al., 2006; Larsen, 1977). There may even be yet another level of scaling, as a given pair of cells may be interconnected by several gap junctions (see later).

Other types of intercellular junctions occur as well, particularly in epithelia (two-dimensional layers of cells) that form a boundary between one space and another. Examples are tight junctions and septate junctions. Tight junctions serve to seal off regions of extracellular space; in tight junctions, membranes of separate cells are closely apposed—the membranes appear to fuse, although they do not actually do so (Gumbiner, 1987). The function of most sorts of specialized junctions is, however, not precisely known.

Next, we outline some of the methods by which gap junction structure and function have been studied. These methods can be divided into *physiological* and *functional*; *structural* and *imaging*; and *molecular*. Of course, any actual scientific study is likely to combine techniques from each of these categories.

Physiological analyses would include the following: first, pairs of putatively connected cells can be recorded simultaneously with electrodes, while slow voltage deflections are induced in one cell, and the corresponding deflection in the other cell measured (this is called "DC coupling"); alternatively, action potentials can be induced in one cell and the resulting voltage transients ("spikelets") recorded in the other. These studies provide definitive evidence of electrotonic coupling, and suggestive evidence for the existence of gap junctions. Sometimes, however, it is not feasible to record from two cells simultaneously. In such cases, it may be possible to record from a single cell, and to stimulate (extracellularly) the axons of nearby homologous cells, or stimulate an afferent pathway that impinges on the recorded cell. Characteristic potentials—spikelets (also called, in this context, "short latency depolarizations")—may be evoked, and their occurrence suggests the existence of a gap junction somewhere along the path between the stimulus and the recorded response. Physiological measurements of this kind can be combined with experimental manipulations that enhance or diminish gap junction conductances (see later); and, at least in vitro, the experiments can also be combined with manipulations that block chemical synapses, so that evoked responses can more readily be attributed to gap junctions rather than to synaptic potentials. In addition, synaptic responses have amplitudes that depend on membrane potential, because the reversal potential for synaptic responses tends to remain fixed, so that altering the membrane potential will change the "driving force" of the synaptic current; in contrast, current flowing through gap junctions does not depend on a reversal potential, so that the gap-junction-mediated responses tend to be less dependent on membrane potential. (There are, however, exceptions to this rule, if the gap junction is rectifying, i.e., passes current in only one direction, or if the gap junction conductance is voltage-dependent; or if the gap junction response requires recruitment of voltage-dependent conductances.)

Structural and imaging techniques include the following: direct microscopic visualization of the gap junction, in appropriately fixed and prepared tissue sections; and the demonstration of dye coupling (suggestive of gap junctions; Kanno & Loewenstein, 1964). Dye coupling experiments can be performed acutely in vitro (or in lightly fixed tissue), with fluorescent dyes (e.g., Lucifer yellow), using light microscopy and possibly confocal imaging methods; or, dye coupling can be performed by injecting biocytin or neurobiotin into a cell and then performing post-hoc visualization (again, usually with a light microscope). Direct visualization of gap junctions is not, however, possible with the light microscope; electron microscopy ("ultrastructure") is required. Electron microscopy can be carried out on ultrathin sections of tissue—transmission electron microscopy (TEM); it can also be performed on freeze-fractured tissue that is treated with a lipid-removing compound, and then coated with a thin layer of carbon, or other material. Freeze-fracture imaging can be combined with gold particle-complexed antibodies that bind to gap junction proteins ("FRIL", or freeze-fracture immunogold replica labeling).

Molecular methods for studying gap junctions are possible because many gap junction proteins have been cloned and sequenced. Therefore, it is possible to breed mice in which one (or more) of the proteins has either been genetically modified, or knocked out. The resulting mice can be examined for behavioral phenotypes, EEG abnormalities, and in vitro and morphological studies also performed. Molecular techniques also allow the recombinant expression of gap junction proteins in diverse cell types that can be examined in vitro. Typically, the recombinant proteins assemble to form functional hemichannels; or, in the case of oocytes that are brought into sufficiently close physical proximity, functional gap junctions—of defined chemical composition—can be "manufactured" (Bruzzone et al., 2005; Loewenstein, 1981; Loewenstein et al., 1978).

Finally, we should mention a more indirect method of studying gap junctions: the analysis of network phenomena that depend on the junctions. In this case, the relation between the intercellular coupling—between pairs of neurons—produced by individual gap junctions, and a network phenomenon (such as an oscillation), is an indirect one. A model is needed to relate the cell-cell interactions to the population activity. Of course, this same concept applies to every network phenomenon in the brain, including those produced solely by chemical synaptic interactions: one always needs a model, in order to relate the properties of individual cells and synapses to a population behavior.

Structure of a Gap Junction Channel

High-resolution x-ray crystallography would, in principle, provide structural data useful in understanding the biophysics of gap junction channels. Most membrane proteins are, however, unstable in solution, and do not lend themselves to crystallization (Engel & Gaub, 2008); these proteins are sometimes studied in expression systems (i.e., at high densities in cell membranes) as "two-dimensional crystals," wherein structure can be analyzed at relatively low resolution with electron beams. Figure 9.1 illustrates data of this sort for a modified cardiac gap junction protein, which has assembled to form a gap junction channel in an expression system. Several features are of note in this image, including the sixfold radial symmetry, the tunnel-like appearance of the pore, and the interdigitating "fingers" of the two connexons in the extracellular space ("E" in Fig. 9.1a). These latter interdigitations allow one to

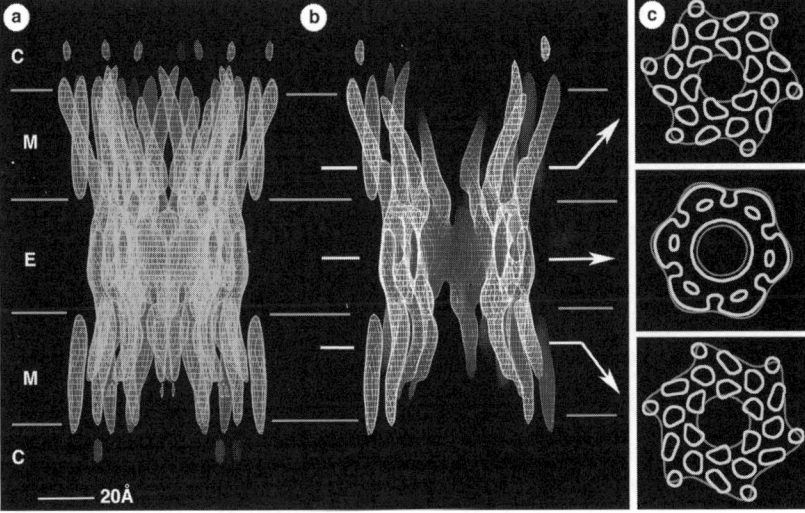

Figure 9.1　Electron crystallographic structure of a recombinant cardiac gap junction (modified connexin-43), expressed in baby hamster kidney cells. ~7.5 Å resolution in the membrane plane, 21 Å in the vertical plane. *C*, cytoplasm; *M*, membrane (with lipid bilayer); *E*, extracellular space. **A:** Side view of the junction. Note the interdigitations of the proteins extending from the two cells, and the separation of the cell membranes (the "gap," in this case about 2 nm). **B:** A view through the channel proper. Red asterisk indicates narrowest part of the pore. **C:** Cross-sectional views at the levels shown by the arrows. Note the 6-fold symmetry, corresponding to the assembly of 6 connexin molecules to form each of the two connexons that constitute the channel. Please see color insert.

(From Unger et al., 1999 with permission.)

understand where the "gap" in the gap junction comes from, and one reason a gap junction has a different ultrastructural appearance than a tight junction: the connexin polypeptide chains, which extend well into the extracellular space, binding together by electrostatic forces sterically, produce a space between the membranes of several nm (tens of angstroms).

This is an appropriate point to mention some of the experimental (and natural) manipulations that alter gap junction conductances. Let us begin with some caveats: the biophysical/biochemical means by which gap junctions are modified are, in general, not properly understood. Nor are there yet compounds that close the channels both completely and with the specificity that is possible with ligand-gated receptor/channels. In situ, gap junction conductances are modified by cyclic nucleotides [e.g., cyclic AMP (Saez et al., 1986)], which themselves may be modified by neurotransmitters/modulators like acetylcholine, serotonin (Rörig & Sutor, 1996) or dopamine; however, different gap junctions may respond to a particular transmitter, or cyclic nucleotide, by increasing or decreasing its conductance (Gladwell & Jefferys, 2001; Hampson et al., 1992; Moss et al., 2005; Onn & Grace, 1994; Pereda et al., 1992; Perez Velazquez et al., 1997; Piccolino et al., 1984; Saez et al., 1986).

Gap junction conductance is also (in general) decreased by intracellular acidification and increased by alkalinization (Arellano et al., 1990; Church & Baimbridge, 1991; Gutnick & Lobel-Yaakov, 1983; Spray et al., 1981; Turin & Warner, 1977)—a fact of great experimental importance, and probably significant clinically as well, for example in the initiation of epileptic seizures (see later on). Please note González-Nieto et al., 2008, for a possible counter-example to this rule.

Gap junctions also tend to be exquisitely sensitive to (and close in response to) Ca^{2+} ions (Arellano et al., 1990; Bennett & Verselis, 1992), a property possibly related to pH sensitivity, but possibly distinct from it; such a sensitivity to Ca^{2+} ions is experimentally convenient: a low-calcium bathing medium can be used to block synaptic transmission, while at the same time enhancing gap junction conductances, thereby unmasking, or amplifying, gap-junction-dependent circuit behaviors.

In terms of exogenous chemical agents that modify gap junction conductances, modafinil has been reported to open the channels (Urbano et al., 2007). The following compounds (not an exhaustive list) are used experimentally to decrease gap junction conductances: octanol and certain other alcohols (Peracchia, 1991); the gas anesthetic halothane and some of its congeners (Peracchia, 1991); carbenoxolone, a derivative of 18-glycerrhytinic acid (an alkaloid found in the roots of the licorice plant) (Davidson & Baumgarten, 1988); and the antimalarial drug mefloquine (Cruikshank et al., 2004). Interestingly, both carbenoxolone (in Europe) and mefloquine are administered to humans for clinical purposes; it is not known if brain concentrations after oral clinical use are sufficient to affect gap junctions. Eating large quantities of licorice (>0.25 kg) has been reported to cause transient visual loss (Dobbins & Saul, 2000), an effect that may be due to blockade of retinal gap junctions (which are extremely numerous and important (Söhl et al., 2005)).

What Sorts of Neurons Are Coupled by Gap Junctions, in Mammals?

This is a hard question to answer rigorously, because data are more compelling for some sorts of neurons than others; and because there may be species differences. There are certainly also major changes in gap junction distribution with development, with mostly a decline with age occurring rather than the reverse—one always has to consider developmental age in judging a scientific study. Finally, trauma (and perhaps other types of neurological lesion) can induce the formation of gap junctions, a topic to which we return later.

Adult mammalian brain regions and cell types, for which gap junctions (and/or electrotonic coupling or dye coupling) have been reported, include (but are not limited to): the retina (Söhl et al., 2005); basal ganglia (Onn & Grace, 1994); the cerebral cortex and hippocampus (Katsumaru et al., 1988; Sloper & Powell, 1978a,b; see also later); nucleus reticularis thalami (Landisman et al., 2002); thalamocortical relay cells (Hughes et al., 2002a); interneurons in the molecular layer of the cerebellum (Mann-Metzer & Yarom, 1999; Sotelo & Llinás, 1972); the inferior olive (Bourrat & Sotelo, 1983; Llinás et al., 1974; Sotelo et al., 1974); the mesencephalic nucleus of the 5th cranial nerve (Baker & Llinás, 1971); and the lateral vestibular nucleus (Korn et al., 1973). Were we to include lower vertebrates, such as frogs and fish, and consider the spinal cord as well as the brain, the list would become even more extensive, and would include, in particular, spinal motorneurons (see later). Of these examples, two locations of gap junctions have been extensively studied for their involvement in network oscillations: between the dendrites of *inferior olivary neurons*, a type of glutamatergic cell that gives rise to the climbing fibers in the cerebellum (Chapter 7); and between the dendrites of a variety of *GABAergic neurons* in the cortex, hippocampus, and thalamus. While these oscillations are not, for the most part, of central importance for this book, we briefly discuss here some of the connections between the gap junctions and the oscillations.

The Inferior Olive

The cells in this medullary structure give rise to the cerebellar climbing fibers. Inferior olivary neurons are remarkable in two ways: they have intrinsic subthreshold oscillatory properties, dependent on a low-threshold-activated Ca^{2+} conductance (Llinás & Yarom, 1981a,b); and they are interconnected by gap junctions containing (at least in part) connexin36 (Bourrat & Sotelo, 1983;

Weickert et al., 2005b). Studies of the inferior olive, in vitro and in vivo, with pharmacological blockade of gap junctions and with connexin36 knockout and other molecular techniques, suggest that the gap junctions act to phase-lock the cellular oscillators, and produce a temporally coordinated output; the uncoupled neurons are individually oscillators, but oscillate out of phase with each other when gap junctions are blocked (Blenkinsop & Lang, 2006; Leznik & Llinás, 2005; Long et al., 2002): the coordinated output is probably important in motor control (Welsh & Llinás, 1997). (This is one type of network oscillation, therefore, where modeling with a coupled oscillator theory makes sense; in general, however, other theoretical approaches are necessary.) The functional significance of gap junction-induced olivary synchrony is, however, more subtle than expected. One approach to analyzing this function has been study of the tremor induced by the alkaloid harmaline (Lamarre et al., 1971); harmaline hyperpolarizes olivary neurons and enhances collective oscillations (through effects on de-inactivation of g_{Ca}), at approximately 10 Hz (Llinás &Yarom, 1986). Surprisingly, harmaline tremor is not altered (in vivo) in a connexin-36 knockout mouse (Long et al., 2002). Even so, intraperitoneal injection of gap junction blocking compounds (in mice) suppressed harmaline tremor (Martin & Handforth, 2006); and the coherence of harmaline tremor, as assessed with multisite intramuscular recordings, was reduced when connexin-36 was suppressed with a lentivirus (Placantonakis et al., 2004). Perhaps compensations, via upregulation of other gap junction proteins (like connexin-45; Li et al., 2008), occur when connexin-36 is knocked out—an important issue, as we shall see, when connexin-36 knockout mice are used to study other types of oscillations.

Gap Junctions and Interneuron Network Oscillations

There are a number of experimental oscillation models, at theta to gamma frequencies, in which pharmacologically isolated populations of GABAergic neurons generate network oscillations (Blatow et al., 2003; Deans et al., 2001; Traub et al., 1996b; Long et al., 2004; Whittington et al., 1995). Hippocampal and neocortical interneuron network oscillations require mutual GABAergic synaptic inhibition; while nucleus reticularis thalami oscillations do not. In the case of hippocampal gamma (Traub et al., 2001), a neocortical approximately 3 to 5 Hz oscillation generated by low-threshold spiking (LTS) interneurons (Deans et al., 2001), and nucleus reticularis approximately 10 Hz oscillations, gap junctions either enhance synchrony, or are required for synchronous activity to occur at all. In addition, gap junctions [probably between interneurons (Traub et al., 2003c)] enhance the power of persistent gamma oscillations in vitro (Hormuzdi et al., 2001). For these reasons, consideration of interneuron gap junctions is necessary.

Interneuron gap junctions have been studied with ultrastructure, immunocytochemistry, dye coupling (in the cerebellum), paired intracellular recordings, via network oscillations, and with molecular techniques. Some of the principal findings are these:

1. Interneuron gap junctions are most commonly located on dendrites, but sometimes on somata. They may occur adjacent to dendrodendritic chemical release sites. Ultrastructural data are most extensive for gap junctions between the dendrites of parvalbumin positive interneurons; with such neurons, one neuron may couple to dozens of others. Coupled cortical interneurons need not be located in the same cortical column (Fukuda & Kosaka, 2000, 2003; Fukuda et al., 2006; Katsumaru et al., 1988; Kosaka 1983a,c; Sloper & Powell, 1978b).

2. Interneuron gap junctions are constituted, at least in large part, of connexin-36 (Deans et al., 2001; Fukuda et al., 2006; Hormuzdi et al., 2001; Landisman et al., 2002; Rash et al., 2000; Venance et al., 2000;). Transfected connexin-36 channels have a low voltage sensitivity, allow dye coupling, and have a unitary conductance of 10 to 15 pS (Srinivas et al., 1999).

3. Dual interneuronal intracellular recordings, of nearby cells, demonstrate DC coupling and on occasion spike synchronization, as well as spikes in one cell correlating with spikelets in the other. Functional electrical coupling is most common between interneurons having homologous firing pattern (e.g., FS cells or LTS cells), but heterologous coupling is possible, especially when one of the interneurons is a neurogliaform cell; on occasion, an interneuron can even couple to a spiny stellate cell (i.e. an excitatory neuron) (Beierlein et al., 2000; Galarreta & Hestrin, 1999, 2002; Gibson et al., 1999, 2005; Mann-Metzer & Yarom, 1999; Simon et al., 2005; Venance et al., 2000; Zsiros & Maccaferri, 2008).

It remains to be seen whether gap junctions between local GABAergic interneurons play a truly critical (as opposed to modulatory) role for the generation of in vivo brain oscillations. (Certainly, given how numerous they are, interneuron gap junctions must surely play some sort of important functional role, but this role may not be related to oscillations.) In contrast, many sorts of in vivo brain oscillations do appear to depend on gap junctions between principal neurons. This conclusion depends rather strongly on in vitro data, however, so we must take particular pains to argue that the in vitro oscillations really do capture what happens in vivo. Further, by an unfortunate irony, the ultrastructural data, demonstrating gap junctions between principal neurons, are far less extensive than is the case for interneuronal gap junctions—probably because there are simply many fewer gap junctions between principal neurons than between interneurons. In this instance, we must convince the reader that the *number* of gap junctions does not, by itself, predict importance for oscillations. Why this is true depends on understanding oscillation mechanisms as well as gap junctions in and of themselves, and we will elaborate on those mechanisms in succeeding chapters. First, in the next sections, we will review the data indicating that gap junctions and electrical coupling do exist between principal neurons, and review data indicating that at least some of these gap junctions lie on axons—a site where they can exert much more powerful functional effects than do dendritic gap junctions.

Morphological Evidence for Gap Junctions Between Axons

The notion that gap junctions should exist between principal cell axons dates back over 10 years, and was originally based on electrophysiological data (Draguhn et al., 1998). There are, however, precedents for the idea, in the form of suggestive ultrastructure in mammals, and in the form of data from non-mammalian species.

Thus, Figure 9.2 is an ultrastructural image, taken from Kosaka (1983b) of "some sort of junction" (to quote the author) between two obliquely cut

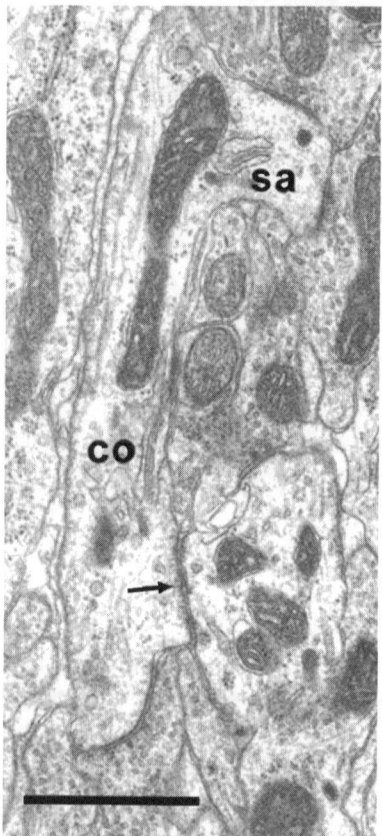

Figure 9.2 Two obliquely cut axon initial segments (of dentate granule cells), with a junction between them (arrow), in thin-section transmission EM image. This junction could be a reticular gap junction (Kamasawa et al., 2006), although it is impossible to be certain (John E. Rash, personal communication). *co*, cisternal organelle; *sa*, spine apparatus. There are also chemical synapses on the upper initial segment. Scale bar, 1 μm.

(From Kosaka, 1983 with permission.)

axon initial segments, of hippocampal dentate granule cells. Although it is impossible to know whether the junction is actually a gap junction, that is at least possible (John E. Rash, personal communication; Kamasawa et al., 2006). Figure 9.3, taken from Shiosaka et al. (1989), shows what appears to be a gap junction between a presynaptic terminal and an axonal shaft, in adult rat hippocampus (stratum oriens, where many principal cell axons run, although other axons are present there as well).

Gap junctions exist between presynaptic terminals and axonal initial segments, in the medullary pacemaker nucleus of weakly electric fish (Fig. 9.4)—a structure whose function, interestingly enough, is to generate very fast oscillations, approximately 1 kHz (Dye, 1991; Dye & Heiligenberg, 1987); the very fast oscillations generated in the nucleus are projected to spinal electromotor neurons which, interestingly, are also electrically coupled (Bennett et al., 1963). [The medullary pacemaker nucleus may, however, act as a collection of coupled neuronal oscillators (Moortgat et al., 2000a,b), so that the network mechanisms operative there are not the same as for the generation of very fast oscillations in mammalian brain.]

Figure 9.5 shows transmission electron microscopic evidence for gap junctions between mossy fiber axons (the unmyelinated axons of glutamatergic

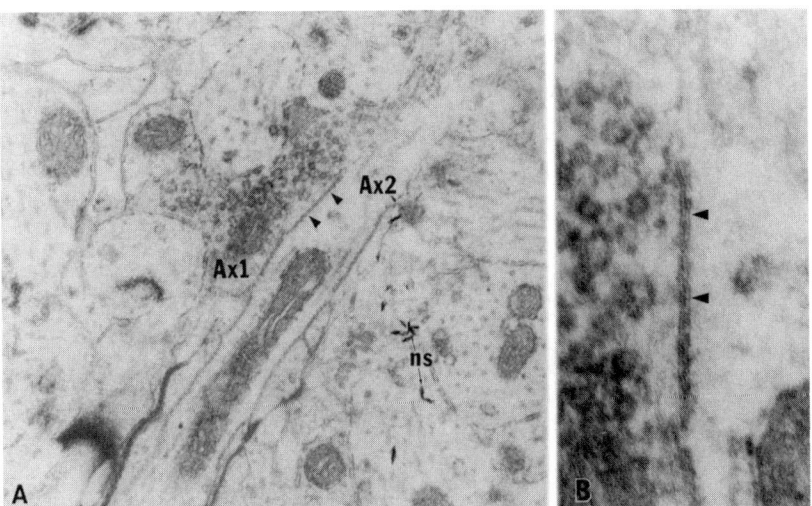

Figure 9.3 Possible axo-axonal gap junction in adult rat hippocampus. The tissue was prepared for immunolabeling with antibodies to a 27-kD rat liver gap junction protein. An axon (Ax2) was myelinated and running in stratum oriens. Shortly after losing its myelin it was contacted (arrowheads) by an immunolabeled terminal (Ax1). **A:** Magnification 72,000. Ns, counterstaining contamination. **B:** Magnification, 138,000 of the gap junction-like structure, 14–16 nm in length.

(From Shiosaka et al., 1989 with permission.)

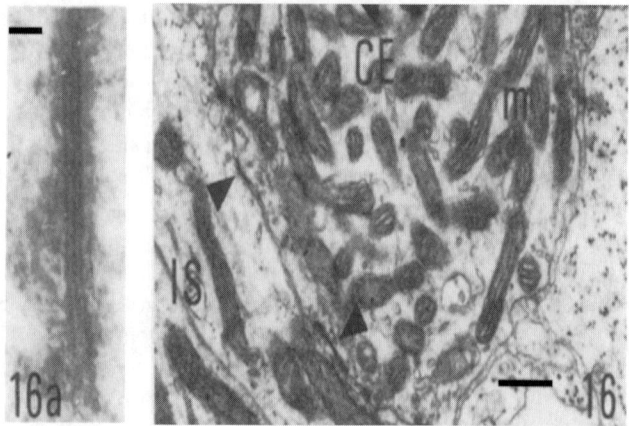

Figure 9.4   Gap junctions (arrowheads) on an axon initial segment of a neuron in the medullary pacemaker nucleus of a weakly electric fish. Scale bar = 500 nm. *IS*, axon initial segment; *gl*, glial processes; *CE*, club ending (analogous to a presynaptic terminal); *A*, myelinated axon preterminal segment; *m*, mitochondria. Inset shows one of the gap junctions at higher power, scale bar = 35 nm.

(From Elekes and Szabo, 1985 with permission.)

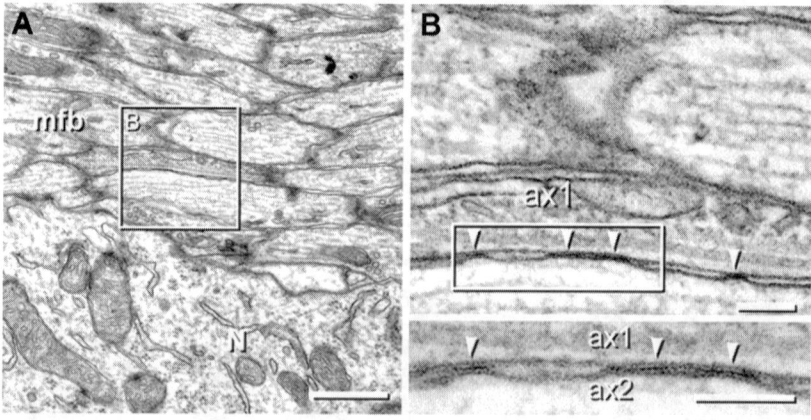

Figure 9.5   Thin section transmission EM of gap junctions between hippocampal mossy fiber axons (adult rat). **A:** Longitudinal section through mossy fiber bundle (mfb) in CA3 stratum lucidum (just on the apical dendritic side of stratum pyramidale), where thin, unmyelinated, mossy fibers run and contact giant spines of CA3 pyramidal cells. *N*, pyramidal neuron cell body. Scale bar = 500 nm. **B:** Higher power view of boxed region in **A**. gap junctions identified by arrowheads. Inset below shows still higher power view of boxed region, with gap junctions between 2 axons (ax1, ax2). Scale bars = 100 nm.

(From Hamzei-Sichani et al., 2007 with permission.)

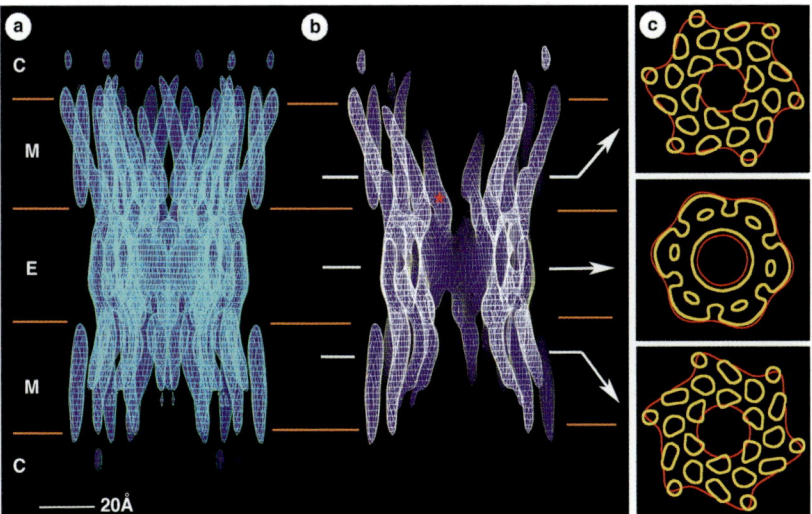

Figure 9.1 Electron crystallographic structure of a recombinant cardiac gap junction (modified connexin-43), expressed in baby hamster kidney cells. ~7.5 Å resolution in the membrane plane, 21 Å in the vertical plane. C, cytoplasm; M, membrane (with lipid bilayer); E, extracellular space. **A:** Side view of the junction. Note the interdigitations of the proteins extending from the two cells, and the separation of the cell membranes (the "gap," in this case about 2 nm). **B:** A view through the channel proper. Red asterisk indicates narrowest part of the pore. **C:** Cross-sectional views at the levels shown by the arrows. Note the 6-fold symmetry, corresponding to the assembly of 6 connexin molecules to form each of the two connexons that constitute the channel.

(From Unger et al., 1999 with permission.)

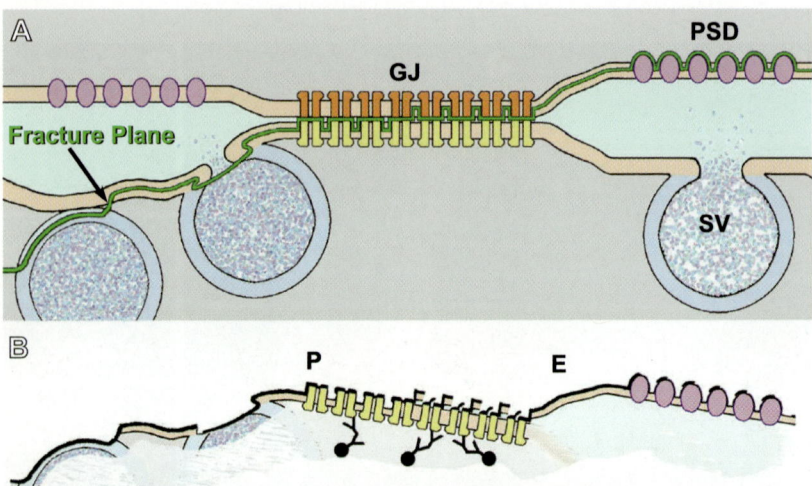

Figure 9.6 Principles of freeze-fracture replica immunogold labeling (FRIL), as applied to gap junctions. **A:** Schematic side view of a mixed synapse, with a gap junction (GJ) in the middle and chemical release sites on either side. The membranes are colored tan, the extracellular space light green. Gap junction connexons on one side of the junction are yellow, the other side orange. The impending fracture plane is indicated by the dark green line: Note that this fracture plane passes through the lipid bilayer of portions of the top and bottom cells, but separates connexons symmetrically. *SV*, synaptic vesicle; *PSD*, postsynaptic density of IMP (intramembrane particles), for example glutamate receptors. **B:** Schematic of immunogold labeling of gap junction protein. "P" and "E" label respective fracture faces. Gap junction proteins are labeled first with rabbit anti-connexin-36 antibody; these antibodies are then identified by labeling them with goat anti-rabbit IgG, to which molecules have been complexed 10 nM gold beads (black discs). The beads can be visualized with the electron microscope; their presence indicates nearby connexin-36.

(From Nagy et al., 2004 with permission.)

hippocampal dentate granule cells); these axons are small (hundreds of nanometers in diameter) and run in tight bundles. The putative gap junctions in Figure 9.5 are only tens of nanometers long, but in the case illustrated, the junctions occur several in a row. It is possible, therefore, that the total number of interconnecting gap junction channels is similar to interneuron junctions, which have been reported to be approximately 220 nm long (Fukuda & Kosaka, 2003).

The most compelling evidence that gap junctions exist on mammalian excitatory axons derives from freeze–fracture replica immunogold labeling (FRIL), the principles of which are illustrated in Figure 9.6. Figure 9.7 shows stereoscopic images of a gap junction on a hippocampal mossy fiber, that has been specifically immunolabeled with two different antibodies to connexin-36. The gap junction contains connexin-36; therefore, although other (unlabeled)

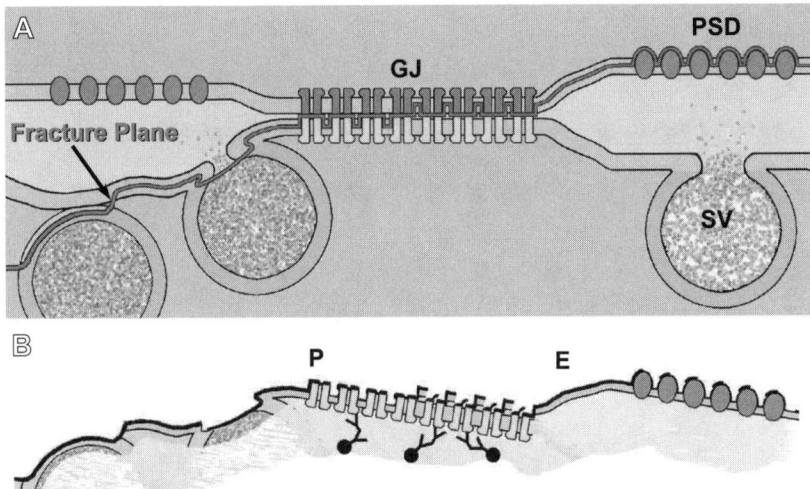

Figure 9.6 Principles of freeze-fracture replica immunogold labeling (FRIL), as applied to gap junctions. **A:** Schematic side view of a mixed synapse, with a gap junction (GJ) in the middle and chemical release sites on either side. The membranes are colored tan, the extracellular space light green. Gap junction connexons on one side of the junction are yellow, the other side orange. The impending fracture plane is indicated by the dark green line: Note that this fracture plane passes through the lipid bilayer of portions of the top and bottom cells, but separates connexons symmetrically. *SV*, synaptic vesicle; *PSD*, postsynaptic density of IMP (intramembrane particles), for example glutamate receptors. **B:** Schematic of immunogold labeling of gap junction protein. "P" and "E" label respective fracture faces. Gap junction proteins are labeled first with rabbit anti-connexin-36 antibody; these antibodies are then identified by labeling them with goat anti-rabbit IgG, to which molecules have been complexed 10 nM gold beads (black discs). The beads can be visualized with the electron microscope; their presence indicates nearby connexin-36. Please see color insert.
(From Nagy et al., 2004 with permission.)

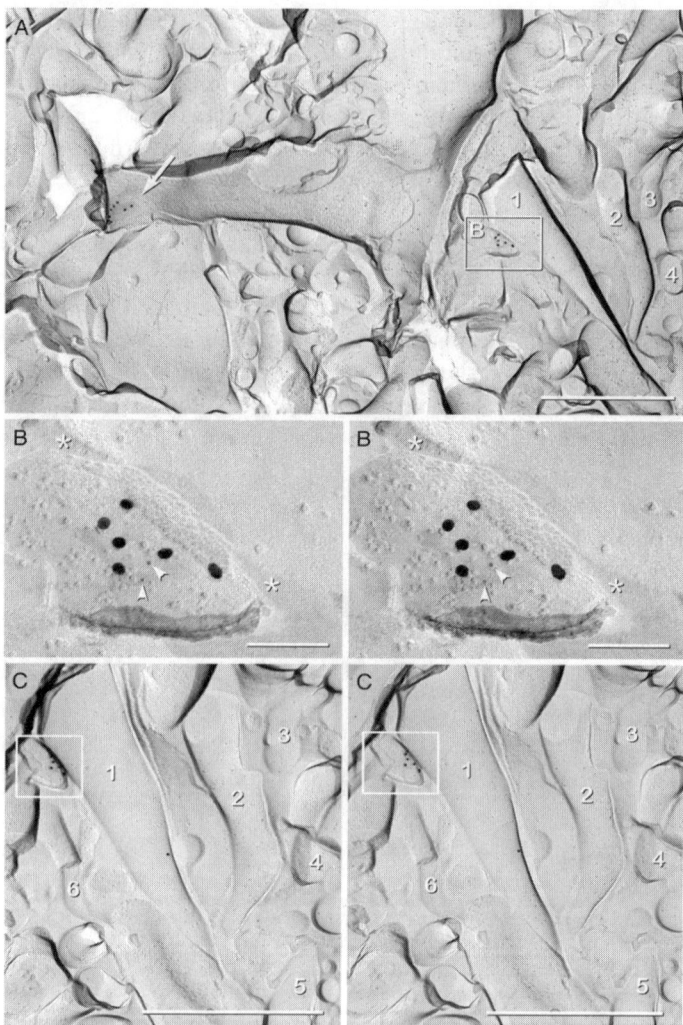

Figure 9.7 FRIL demonstration of a gap junction on a hippocampal mossy fiber axon (CA3c stratum lucidum, 150-g rat). **A:** Freeze–fracture replica showing two gap junctions labeled with two different gold bead–complexed antibodies identifying connexin-36; the beads are are 18 nm and 6 nm. The left gap junction (arrow) lies on a dendrite. The right one (box B) lies on mossy fiber axon 1 (other axons in the image are numbered as well). Scale bar = 1 μm. **B:** Dual stereoscopic images of the gap junction in box **B** above. Arrowheads mark the 6-nm gold beads. The small gap junction plaque has been shaded in red. It contains about 100 connexons. Asterisks mark the extracellular space, which is ~3 nm at the gap junction site. Scale bar = 100 nm. **C:** Stereoscopic view with the replica tilted, and mossy fiber axons numbered. Scale bar = 1 μm.

(From Hamzei-Sichani et al., 2007 with permission.)

connexins could be present as well. The junction is about 100 nm long, is associated with an extracellular gap of approximately 3 nm, and contains an estimated 100 connexons. If the channels were all made of connexin-36, the total gap junction conductance—for the illustrated plaque—would then be about 1.0 to 1.5 nS, given the unitary channel conductance of 10 to 15 pS. If additional plaques exist (which is not possible to determine from this image), the total gap junction conductance would, of course, be correspondingly larger.

We shall now consider dye-coupling evidence. Dye coupling was described between principal cells in hippocampus and neocortex beginning almost 30 years ago (Andrew et al., 1982; Connors et al., 1983, 1984; Gutnick & Lobel-Yaakov, 1983; Gutnick & Prince, 1981; Gutnick et al., 1985; MacVicar & Dudek, 1980, 1982; MacVicar et al., 1982). Figure 9.8 illustrates two particularly instructive examples. The cell pair in Figure 9.8A is fluorescing with Lucifer Yellow, after one of the cells (probably the lower one) was injected in vivo. The lower cell is a large pyramidal cell with spiny dendrites (inset), intrinsically bursting membrane properties, and a recipient of synaptic input from the thalamus. Unfortunately, one cannot determine from this image (or indeed from virtually any of the published dye coupling images) which portions of the neurons were actually coupled—somata are obviously excluded, because the somata are physically separated, but were dendrites or axons coupled? In the case of the two CA1 pyramidal cells in Figure 9.8B, however, it was possible to identify the site of dye coupling, and to show—by light microscopic criteria—that the coupling site was between axons. In four cases where this identification was possible, the coupling site was approximately 50 to 150 μm from the soma, and presumably proximal to the beginning of the myelin sheath.

To how many different cells does a given cell (say, a pyramidal cell) couple, on average? This depends on at least two parameters. The first parameter is developmental age, with extensive coupling (up to 80 cells fluorescing after injecting a single neuron) possible early on (Connors et al., 1983; Peinado et al., 1993). Another parameter is pH, with alkalinization greatly increasing the extent of dye coupling [consistent with dye coupling being mediated by gap junctions (Spray et al., 1981)]. The relation between pH and dye coupling was investigated extensively by Church and Baimbridge (1991) in rat hippocampal CA1 pyramidal cells, in vitro, using Lucifer Yellow. At pH 7.4, one cell was coupled to 0.62 others, on average. This is below the so-called percolation limit (in which one cell couples to one other, on average), the limit at which collective behavior becomes possible (as we discuss in Chapter 10). At pH 7.9, however, one cell coupled to 2.25 others, on average, well above the percolation limit—but still well below the density of recurrent excitatory synaptic connections in CA1: Deuchars and Thomson (1996) estimated that the synaptic coupling probability—in one direction or the other—was about 1% for CA1 pyramidal cells. Thus, in a population of a few thousand cells, one cell might contact dozens of others. Nevertheless, it is vitally important to

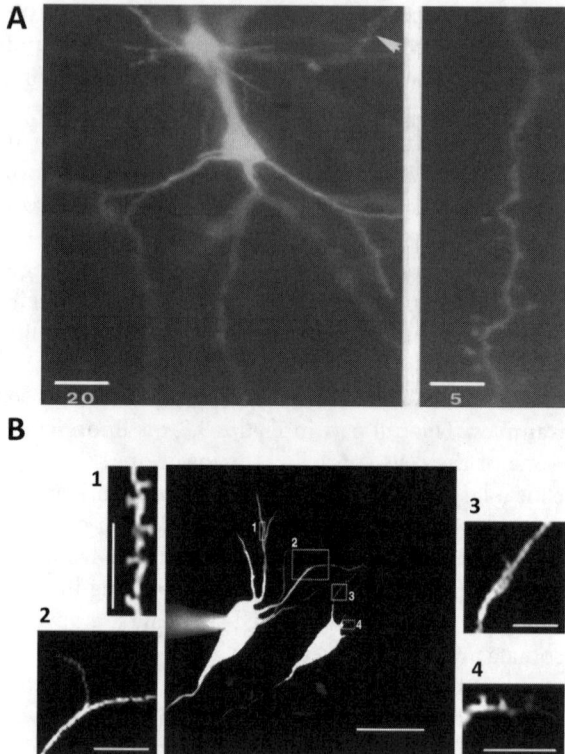

Figure 9.8 Dye-coupling of pyramidal neurons, suggestive of gap junctional coupling. In the hippocampal case, the coupled structures are axons (by light microscopic criteria). **A:** Area 5 of cat neocortex in vivo, depth ~0.5 mm. An intrinsically bursting cell (probably the lower pyramid) was filled with Lucifer yellow, which diffused into a second neuron. The injected cell exhibited short-latency EPSPs after stimulation of two thalamic nuclei (lateral posterior and centrolaminar). Inset shows spiny dendrite of the lower neuron. Scale bars in μm. **B:** Hippocampal CA1 region in vitro. The left cell (note the pipette) was filled with rhodamine-123, while up to 80 confocal images were captured every 15–30 s. Both the injected cell and the coupled cell are pyramidal cells; note the spiny dendrites (B1 and B4). It was possible to identify the site of coupling: The coupled structures (B2 and B3) lack spines and have collaterals emerging at approximately right angles, and thus the coupled structures appear to be axons. Scale bars: 50 μm in center; 5 μm in 1 and 4; 10 μm in 2 and 3.

(**A** from Steriade et al., 1993a, their Fig. 10D; **B** from Schmitz et al., 2001. All reproduced with permission.)

understand a much underappreciated mathematical fact: even if each cell contacts very few others, a syncytium will exist, provided the connectivity is random—that is why collective behavior is possible. Dye injection does not reveal the entire syncytium when each cell contacts—*directly*—only a few others, because dye concentration drops as the compound crosses from cell to cell; and below a critical concentration of dye, fluorescence cannot be distinguished from the background. Nevertheless, the syncytium will exist, functionally.

Other parameters that influence dye coupling are neurotransmitters/ modulators (Hampson et al., 1992; Onn & Grace, 1994; Perez Velazquez et al., 1997), the presence of gap junction blockers such as halothane (Peinado et al., 1993), and extracellular Ca^{2+} (which tends to reduce coupling) (Perez Velaquez et al., 1994). In other words, dye coupling is influenced by the same experimental parameters that modulate electrotonic coupling.

Dye Coupling Between Heterogeneous Cell Types

In invertebrate central pattern generating circuits, electrotonic coupling exists between different cell types (Marder, 1984), and we have mentioned how electrotonic coupling can exist between interneurons and spiny stellate cells in

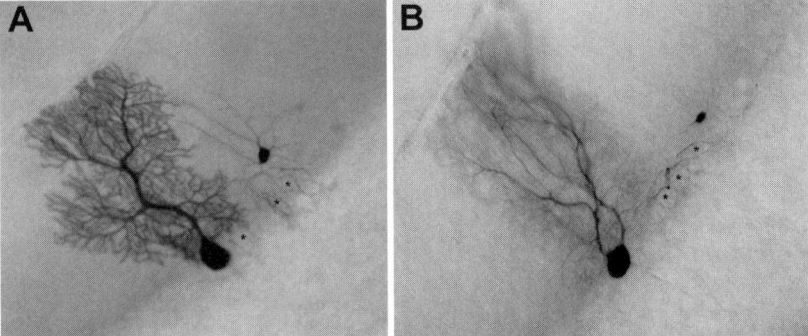

Figure 9.9 Biocytin coupling of mouse Purkinje neurons to cerebellar basket neurons, and to another Purkinje cell. Mouse, in vitro. **A:** Impalement of a Purkinje cell (left) leads to filling of a basket cell (right). Asterisks mark "ghosts" of Purkinje cell somata, surrounded by baskets. (Compare the neurobiotin coupling of a rat hippocampal dentate gyrus cell to a putative axo–axonic interneuron, shown in Haring et al. (1997). **B:** Injection of a different Purkinje cell also leads to coupling of a basket cell, and to a second Purkinje cell in addition. These data raise several questions: (1) Szentágothai (1965) estimated that at least 50 basket cells contact a given Purkinje cell; how then is it that, in these images, only one basket cell fills? (2) Does coupling take place between the septate-like junctions that have been shown to exist between basket cell axon terminals, and between these terminals and Purkinje cell axon initial segments (Gobel, 1971; Sotelo & Llinás, 1972)? It has been proposed that septate junctions could mediate intercellular communication (Gilula et al., 1970). (3) Alternatively, is the coupling between nearby (but yet to be demonstrated) gap junctions? Gap junctions can exist adjacent to septate junctions in invertebrates (Baldwin & Hakim, 1981; Caveney & Podgorski, 1975).
(From Middleton et al., 2008 with permission.)

neocortex (Venance et al., 2000). Interestingly, dye coupling in the mammal has been described between heterologous cells in mammalian brain. In two such examples, the coupling involved interneurons that synaptically (i.e. via chemical GABAergic synapses) contacted axon initial segments. Thus, Haring et al. (1997) showed dye coupling between a dentate granule cell and an axo–axonic interneuron. Figure 9.9 shows two examples of a cerebellar Purkinje cell being dye coupled to a basket cell, an interneuron that synaptically contacts the perisomatic region (including axon initial segment) of Purkinje cells. In Figure 9.9B, a second Purkinje cell is dimly fluorescing. This phenomenon is important, because it implies the possibility that principal neurons might—in principle—be electrically coupled to each other indirectly, through the presynaptic terminals of a heterologous neuron.

For the coupling shown in Figure 9.9, it is not at all clear what the structural basis for the dye passage is. Gobel (1971) and Sotelo and Llinás (1972) described septate-like junctions (1) between Purkinje cell axon initial segments and basket cell terminals, and (2) between pairs of basket cell terminals. Further, a transverse section through one of these septate-like junctions showed a honeycomb appearance (Sotelo & Llinás, 1972), perhaps capable of passing dyes—although that has not been proven. In some invertebrate tissues, gap junctions are found close to septate junctions (Baldwin & Hakim, 1981), so that it is conceivable that gap junctions (hitherto not visualized) exist on basket cell terminals in the cerebellum.

Biological Precedent for Spike Transduction Across an Axonal Gap Junction

As we shall see later on, one means by which very fast oscillations can be generated involves the passage of action potentials, from axon to axon, across a gap junction. To the best of our knowledge, such a phenomenon has never been demonstrated by direct simultaneous recordings of two isolated axons, in a mammalian preparation; spike transduction has been shown with dual somatic recordings (see later) in mammalian cells, but without identification of the anatomical site of electrical coupling. Spikes crossing from axon to axon have, however, been demonstrated in an invertebrate preparation (Fig. 9.10), in one of the very earliest physiological studies of electrical coupling (Furshpan & Potter, 1959). [The junction mediating the electrical coupling was studied structurally only later (Hanna et al., 1978).] A schematic of the preparation— part of the isolated nerve cord of a crayfish—is shown at the top of Figure 9.10. Furshpan and Potter demonstrated that a current-induced action potential in the lateral giant fiber can induce—following the direction of normal orthodromic transmission—an action potential in the giant motor fiber, with a latency of a fraction of a millisecond (lower part of Fig. 9.10). Interestingly, the junction studied by Furshpan and Potter is rectifying (i.e. passes current in only one direction, unlike, say, connexin-36 gap junction channels); thus,

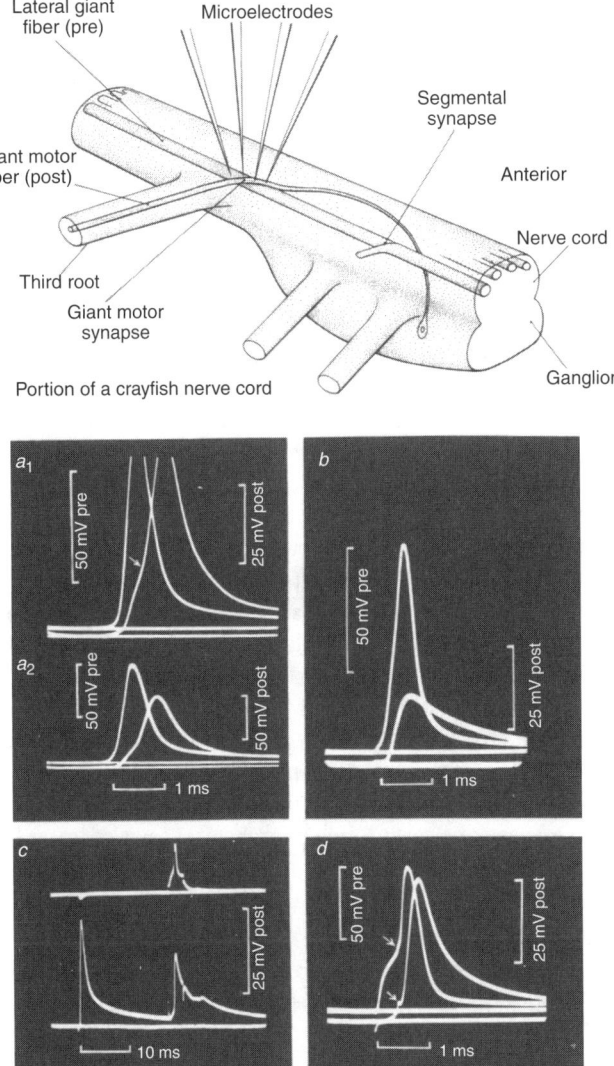

Figure 9.10 Action potential crossing from one axon to another. Crayfish nerve cord in vitro. Upper diagram illustrates the preparation, consisting of an isolated portion of the nerve cord. Pairs of intracellular electrodes were inserted into each of two axons: a lateral giant fiber, and a giant motor fiber. These axons are now known to be coupled by a structural gap junction (Hanna et al., 1978), which allows current to pass when the lateral giant fiber is depolarized relative to the giant motor fiber (reflecting the normal physiological direction of information flow). Lower portion shows an induced action potential in the lateral giant fiber (the first spike in each case), inducing an action potential in the coupled axon, at short latency (<0.5 ms). Note the inflection on the rising phase of the coupled spike (e.g., arrow in a_1); the inflection reflects charging of the "postsynaptic" membrane, followed by active spike initiation. [Action potentials have also been shown to percolate through a network of spinal electromotor neurons in fish (Bennett et al., 1963).]

(From Furshpan and Potter, 1959 with permission.)

if an ectopic spike were to arise in the giant motor fiber, it would not propagate retrograde into the lateral giant fiber. A signaling mechanism like this would seem to be suitable for a case where one axon must excite a small number of others, but would not work if one axon had to excite many other fibers (there would be too much current drained through the gap junctions); nor can a gap junction produce "inhibition," reversing the polarity of the effect of a spike in one cell on a second cell. Perhaps these considerations have something to do with why chemical synapses evolved, and Nature did not simply stick with gap junctions for signaling between neurons. Of course, chemical synaptic transmission allows for many other properties not available to gap junctional coupling: for example, postsynaptic effects that last much longer than the presynaptic action potential, and receptor desensitization.

There is also indirect evidence of spikes propagating from neuron to neuron, in a vertebrate, specifically in spinal electromotor neurons of an electric fish: a spike in one neuron could be precisely correlated with spikes in another neuron 3 mm away (Bennett et al., 1963).

We now turn to the subject of electrophysiological evidence for electrical coupling between mammalian principal neurons, specifically at axonal sites. The hypothesis that electrotonic coupling with such properties should exist grew out of studies on approximately 200 Hz oscillations—so-called "ripples"—in hippocampal slices. By way of historical background, hippocampal ripples occur in vivo, superimposed on so-called physiological sharp waves. The latter are a normal, transient, EEG phenomenon in the hippocampus and other limbic structures, which occur during certain forms of anesthesia, and are most characteristic of certain behavioral states (awake immobility, slow-wave sleep). They were first discovered by György Buzsáki in the 1980s (Buzsáki, 1986); he and his colleagues also first characterized the superimposition of ripples on the sharp waves in vivo (Buzsáki et al., 1992). Subsequently, an in vitro model of sharp wave-ripple complexes in mouse hippocampal slices was developed (Behrens et al., 2005, 2007a; Maier et al., 2002, 2003; Nimmrich et al., 2005).

Approximately 200-Hz Ripples Under Conditions of Blocked Synaptic Transmission

It is a remarkable fact that the sharp wave, and the superimposed ripple, can be dissociated. The slower component of the sharp wave, in CA1, lasting tens of milliseonds, corresponds to chemical synaptic currents, induced at least in part by synaptic inputs from CA3 (Chrobak & Buzsáki, 1996; Ylinen et al., 1995a). On the other hand, ripples can be induced in vitro, without sharp waves, under conditions where chemical synaptic transmission is blocked (Fig. 9.11; Draguhn et al., 1998). Indeed, lowering extracellular Ca^{2+} concentration not only blocks synaptic transmission, but it *enhances* the very fast oscillations, effects that could be mediated by increases in membrane excitability (Frankenhaeuser & Hodgkin, 1957), as well as by opening gap

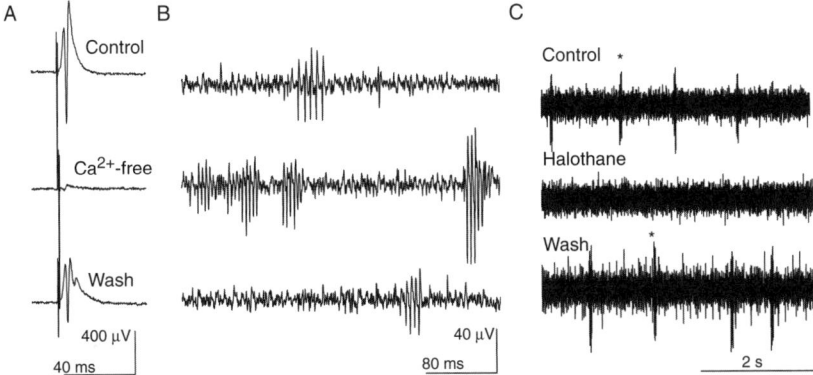

Figure 9.11 ~200 Hz ripples, in the hippocampal slice (CA1), are enhanced in Ca^{2+}-free media (that blocks chemical synaptic transmission), and are suppressed by the gap junction blocker halothane. These data suggest that the ripples depend on gap junctions, rather than chemical synapses, and that the requisite gap junctions are not pannexins [which are Ca^{2+}-insensitive (Bruzzone et al., 2005)]. **A:** Extracellular synaptic field potentials are reversibly blocked by Ca^{2+}-free media. **B:** ~200 ripples occur in extracellular recordings, in stratum pyramidale, in control media (topmost and bottommost traces); the ripples are larger and longer-lasting in Ca^{2+}-free media (middle trace). **C:** Ripples are reversibly suppressed by 5 mM halothane (note the longer time scale in *c*).

(From Draguhn et al., 1998 with permission.)

junction channels. That gap junctions are required for the very fast oscilla-tions (VFO) is shown by the ripple-depressing effects of treatments that reduce gap junction conductances (e.g., halothane, as in Figure 9.11c, octanol, carbenoxolone, and acidification); conversely, alkalinization correspondingly enhanced the VFO. The in vitro VFO has maximal amplitude in stratum pyra-midale and just below it (i.e., toward stratum oriens), as is the case in vivo.

The problem—and it really is a problem—is to understand how (and which) gap junctions are essential to VFO. The fact that VFO can be detected as recurring population spikes suggests that it is synchronized pyramidal cell—rather than interneuron—activities which lead to oscillating potential signals. Draguhn et al. (1998) were not able record from interneurons whose somatic action potentials correlated with the rippling field; however, they were able to record from a pyramidal cell whose activity correlated precisely with the field (Fig. 9.12). Interestingly, the recorded pyramidal cell produced either of two types of depolarizations that correlated with the population spikes: (1) full action potentials, that appeared to be antidromic, because they were inflected, or had a shoulder or notch at the onset (such as would be pro-duced by an axonal spike, that induces a somatic spike after a slight delay); or (2) spikelets (Fig. 9.12). These data immediately tell us that the cells—or at

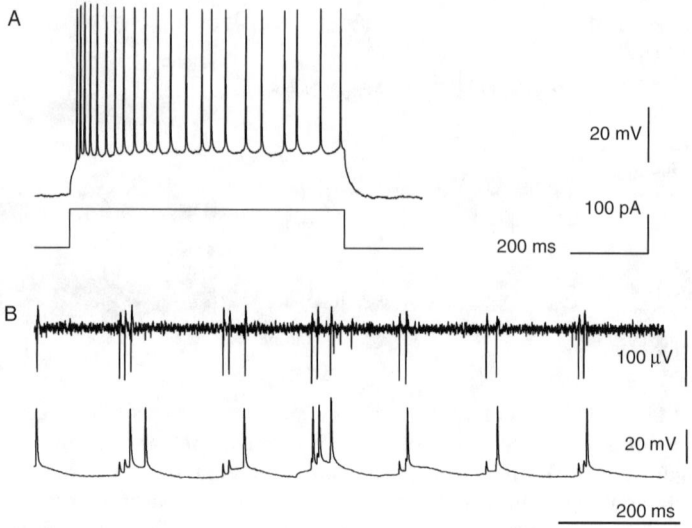

Figure 9.12 During ~200 Hz ripples in CA1 in vitro (Ca²⁺-free medium), intracellular spikelets are temporally correlated with population spikes; this suggests that each spikelet is induced by firing in one or more electrotonically coupled neurons. (The firing could in principle be in the axon, soma, or both, of the coupled neuron or neurons.) **A:** Regular-spiking behavior, in response to an injected current pulse, in a visually identified pyramidal cell (whole cell patch recording). **B:** Rippling field potential (above) was recorded a few tens of micrometers away from the recorded cell soma. Occasionally a full action potential occurs that is not correlated with a population spike, indicating that the rippling does not simply reflect unit activity; however, each spikelet is correlated with a population spike.

(From Draguhn et al., 1998 with permission.)

least the somata—are not intrinsic 200-Hz oscillators, whose firing times are synchronized by electrical coupling. The data do not, however, rule out the hypothesis that the *axons* are intrinsic oscillators (possibly firing in bursts at 200 Hz), that are coupled together by gap junctions. Elimination of that hypothesis, and support for an alternative hypothesis, depends on the observation that the frequency of network activity, and of spikelets, is tunable with gap junction conductance; that notion is developed in Chapter 10.

Without clear ultrastructural guidance (as was the case in 1998), how could one guess which compartments of the pyramidal neurons might be electrically coupled? An initial attempt to answer this question depended on simulations (Draguhn et al., 1998). This attempt illustrated a simple physical principle involving gap junctions (Bennett, 1966): that they tend to act as low-pass filters (with exceptions, however). To appreciate this idea, think of the gap junction as a resistor joining two neurons, one of which is firing an action potential, and the other of which is near threshold. Current will pass through the gap junction/resistor into the other cell which, near threshold, will behave electrically like a combination of its membrane capacitance and leakage resistance. The resulting voltage deflection in the near-threshold cell will thus be an attenuated and slowed (i.e., low-pass filtered) version of an action potential. Of course, details of the shape of the "electrotonic potential" will depend on the electrical parameters of the circuit; if the membrane capacitance is small, and the leakage resistance high, the electrotonic potential may still be rather sharp (i.e. have a rapid rate of rise). Suppose, however, that active membrane conductances, such as g_{Na}, are strongly activated near threshold, in addition to there being a high leakage resistance: these factors would be operative, if the gap junction were to be located in an axon. Under these circumstances (and as in Fig. 9.10), the electrotonic potential quickly turns into an active response, perhaps even a full action potential, and the low-pass filter model no longer applies. If the active axonal response propagates back into the soma either as a full antidromic spike, or as a decrementally conducted and partially blocked spike (i.e., a spikelet), then the intracellular voltage trace of Figure 9.12B could be explained, at least partly: each depolarizing response in the recorded neuron—either full action potential or spikelet—corresponds to one or more action potentials in other cells in the population, conducted into the axon of the recorded neuron, and differentially back-propagated to the soma. This model does not explain the population activity as a whole, but does (in principle) account for the shape of the recorded potentials.

We shall return in Chapter 10 to explain how VFO in its entirety could arise; but first we must consider further evidence (this time based on electrophysiology) for the existence of electrical coupling between axons of principal neurons.

Electrophysiological Evidence for Gap Junctions Between Pyramidal Cell Axons

Dietmar Schmitz et al. (2001) adapted the classical technique of evoking short-latency depolarizations (Baker & Llinás, 1971; Gutnick & Prince, 1981; Korn et al., 1973; Logan et al., 1996; MacVicar & Dudek, 1982; Nuñez et al., 1990) to develop an argument for the existence of axonal gap junctions in principal hippocampal neurons (Figs. 9.13–9.16). Schmitz et al. (2001) introduced significant technical refinements in the method, made possible by patch recording of visualized neurons, and by performing most of the experiments in low-Ca^{2+} media, that blocked synaptic transmission. Schmitz et al. also used pharmacological methods not available to earlier investigators. Here, we summarize the findings:

1. Spikelets (short-latency depolarizations), and antidromic spikes with notches, can be evoked in pyramidal neurons by stimulating in stratum oriens, where the plexus of pyramidal cell axons lies. The spikelets and notched spikes have a similar appearance to the potentials occurring during spontaneous

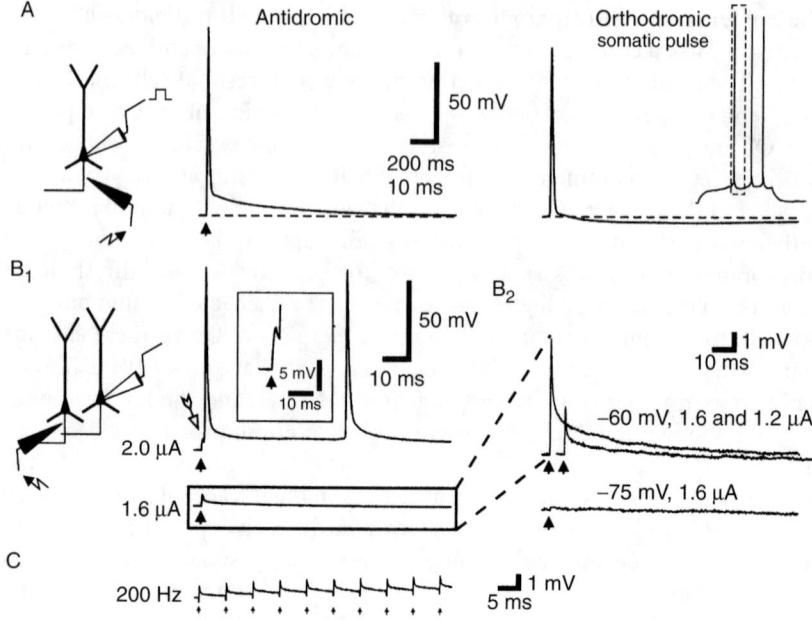

Figure 9.13 Spikelets evoked in a CA1 pyramidal cell by stimulating nearby axons in stratum oriens, suggestive of electrical coupling of the recorded cell to another neuron. Low Ca²⁺ media, whole-cell recordings. The technique of evoking "short latency depolarizations" by antidromic stimulation, in order to show electrotonic coupling, has a long history (e.g., Baker & Llinás, 1971; Korn et al., 1973; Gutnick & Prince, 1981; MacVicar & Dudek, 1982). **A:** Antidromic spike evoked by stratum oriens stimulation has an abrupt onset and depolarizing afterpotential (current-evoked spikes are shown on the right). B_1: Moving the extracellular stimulating electrode laterally (within s. oriens) allows stimuli to elicit antidromic spikes with a notch (compare the last (spontaneous) spike in Fig. 9.12B). Adjusting the stimulus strength leads to evoked spikelets—short latency depolarizations—that could have different amplitudes, and that were blocked by sufficient membrane hyperpolarization (B_2). **C:** Spikelets could follow stimulation at 200 Hz, 1:1 without failures, making it unlikely that dendritic action potentials contribute to spikelet generation. (From Schmitz et al., 2001 with permission.)

ripples (Fig. 9.10). Spikelets follow stimuli at 200 Hz and above, without failures, frequencies characteristic of axonal firing, but not somatic or dendritic firing (Fig. 9.13).

2. Pharmacology and pH manipulations (Fig. 9.14) suggest that a gap junction lies in the path between the extracellular stimulus, and the spikelet recorded in the soma of a particular pyramidal cell.

3. Generation of the spikelet can not be explained with a passive (resistance + capacitance) model, as active membrane conductances are required to produce the spikelet (Fig. 9.15).

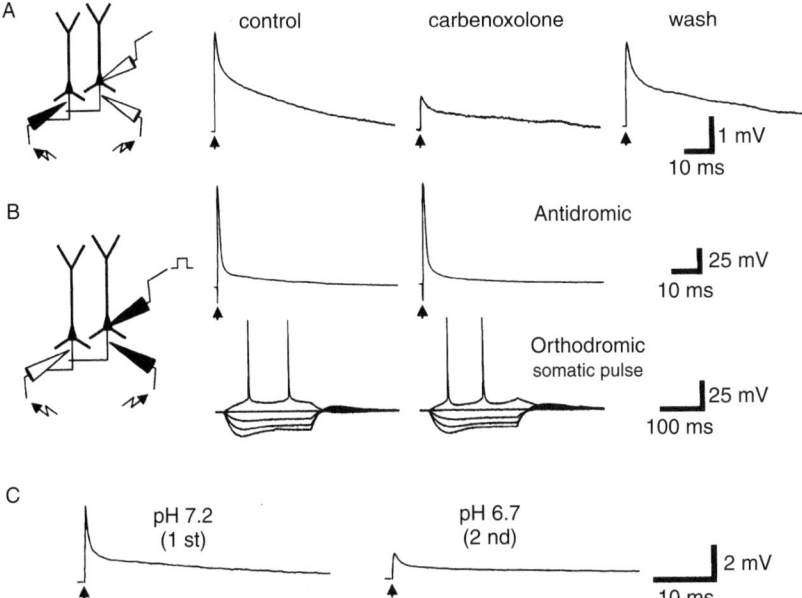

Figure 9.14 The gap junction blocker carbenoxolone reduces the amplitude, and the rate of rise, of s. oriens-evoked spikelets in CA1 pyramidal neurons (low Ca^{2+} media). Experimental arrangements are shown at left, with whole cell recording and current injection from the soma of a pyramidal neuron, and extracellular stimulation in stratum oriens, either near the recorded cell, or more distal. **A:** Spikelet evoked by distal stimulation, recorded under control conditions, and after carbenoxolone (smaller and slower rate of rise). The changes are mostly reversible. **B:** Antidromic spikes (evoked by nearby stimulation), and somatic current-evoked spikes, are not affected by carbenoxolone. **C:** Spikelets were recorded from the same neuron (evoked by distal stimulation), first with a pipette filled with control solution (pH 7.2), then with a pipette filled with acidic solution (pH 6.7, expected to close gap junctions (Spray et al., 1981). Spikelets are smaller and have slower rates of rise in acidic conditions. As a control, a neuron was recorded twice with pipettes having the same pH; spikelets recorded with the second pipette had the same shape as with the first pipette.
(From Schmitz et al., 2001 with permission.)

4. Spikelets—at least those generated under the experimental conditions of Figure 9.13—are conducted antidromically, appearing first in the axon initial segment, next in the soma, and finally in the apical dendrite (Fig. 9.16).

The most economical explanation of the data in Figures 9.13 to 9.16 is this: that the extracellular stratum oriens stimulus evokes action potentials in one or more pyramidal cell axons, but—if the electrode is properly situated and the stimulus not too large—without evoking an action potential in the axon of the cell whose soma is patched. The evoked axonal spikes then

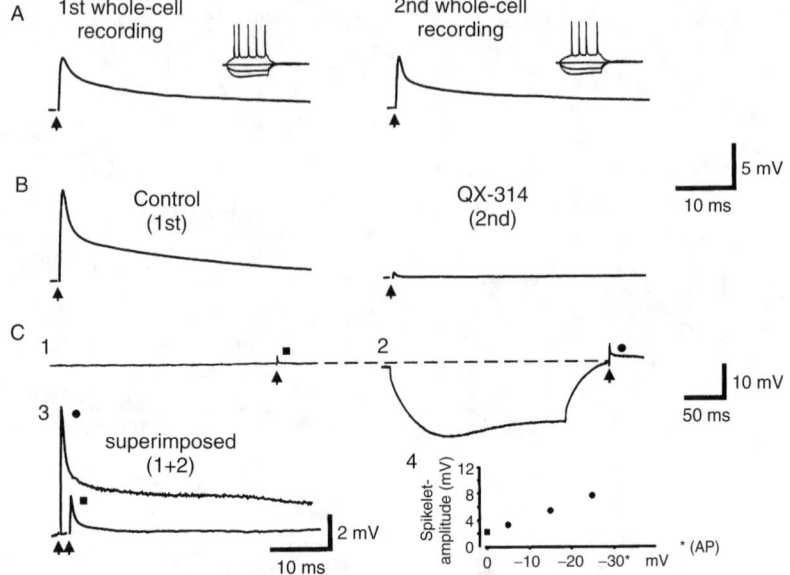

Figure 9.15 Spikelet generation requires active Na$^+$ conductances: gap junction coupling is not always simply "diffusive." Low Ca^{2+} media. **A:** As a control, the same cell was recorded twice in succession, with two different pipettes; this manipulation did not alter spikelet amplitude, or current-evoked action potentials. **B:** The same cell was recorded in succession with 2 pipettes, but the second pipette contained 5 mM QX-314, a Na$^+$ channel blocker. Spikelets were immediately suppressed. [QX-314 has molecular weight ~299, and so conceivably passed through a gap junction to suppress Na$^+$ currents in a different cell. Hence the manipulations used in *c*.] **C:** Spikelets were evoked either without (1) or with (2) a hyperpolarizing prepulse, which would tend to remove inactivation of Na$^+$ channels. Spikelet amplitude is enhanced by the prepulse (3—note the time shift of the smaller spikelet, for clarity); this indicates that gap junction communication, under these experimental conditions, is not purely passive. The increase in spikelet amplitude, as a function of hyperpolarizing prepulse amplitude, is quantitated in (4); a full antidromic spike was evoked when the prepulse was 30 mV.

(From Schmitz et al., 2001 with permission.)

propagate—whether orthodromically or antidromically is hard to determine—and one or more them cross (via a gap junction) into the axon of the recorded neuron, evoking an active response. This response then propagates antidromically into the soma of the recorded cell, evoking there either a spikelet or an antidromic spike, depending on membrane parameters. This explanation explains the frequency-following, the antidromic appearance of the spikes, the gap junction pharmacology, the need for active conductances, and the first appearance of the spikelet in the axon. We are unaware of an alternative model that accounts for the data. Axonal recording techniques, such as described in Chapter 8, may provide further tests of the model.

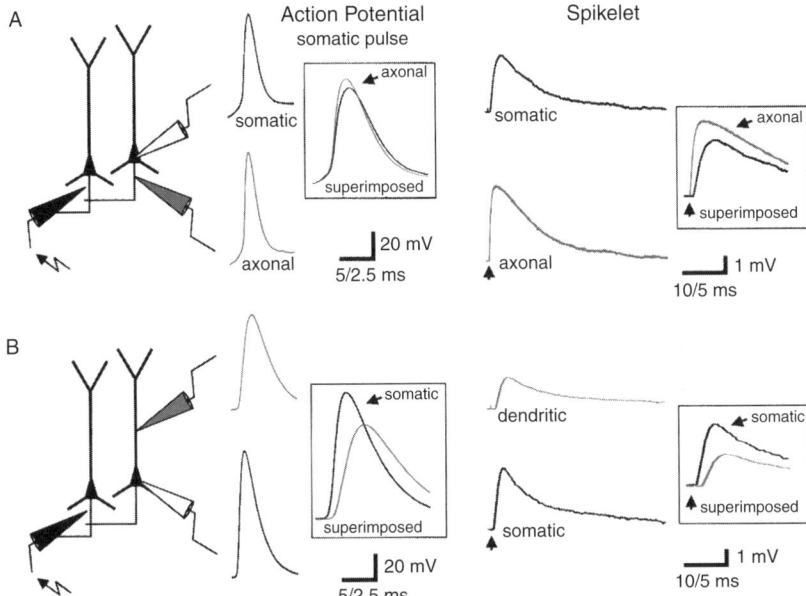

Figure 9.16 Spikelets evoked in CA1 pyramidal cells, by distal s. oriens stimulation, are conducted antidromically along the axon, hence originate in axons. Low Ca²⁺ media. Experimental arrangements shown in the schematics at left: extracellular stimulation in distal s. oriens, with simultaneous soma/axon patch recording (above), or soma/apical dendritic recording (below). The axonal site was 15–30 μm from the soma; experiments were performed at 18°C, to slow conduction. Stimulus artifacts were reduced by subtracting traces obtained in TTX (0.5 μM). Controls included somatic current-evoked spikes (left insets). **A:** The axonal spikelet appears before the spikelet, and the axonal spikelet is larger and has steeper onset. **B:** The somatic spikelet, in turn, appears before the apical dendritic spikelet, and the somatic spikelet is larger and steeper.
(From Schmitz et al., 2001 with permission.)

In the meantime, we consider the existence of axonal gap junctions as our most reasonable hypothesis.

The ability of an action potential in one CA3 pyramidal cell, to trigger a spikelet in a second CA3 pyramidal cell, was reported by MacVicar and Dudek in 1981. DC coupling was shown by these authors as well. These data are important, especially as Schmitz et al. did not record from electrically coupled pairs of pyramidal cells (although they did show dye coupling, see Fig. 9.8B). A more extensive study of simultaneously recorded cell pairs, involving CA1 and neocortical pyramidal cells, has been reported by Mercer et al. (2006). Figure 9.17 shows some examples of their data. Two details are of interest: first, there is a distinct latency between the spike in one cell and the spikelet in the other (Fig. 9.17c), about 0.5 ms in the example shown. This suggests

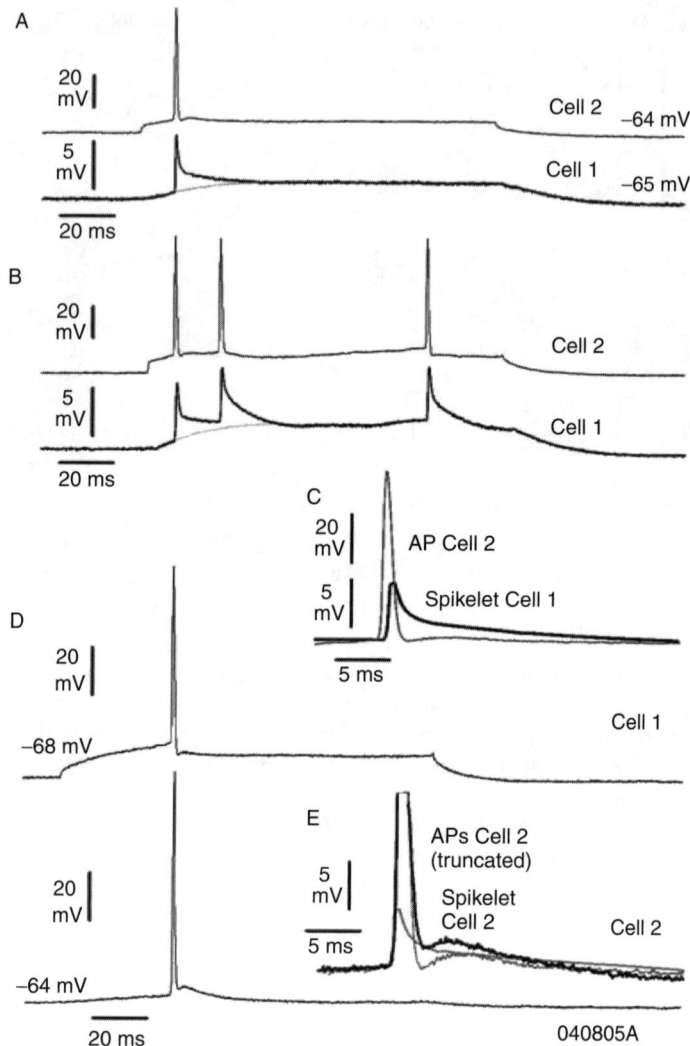

Figure 9.17 Action potentials in one CA1 pyramidal neuron can evoke spikelets in another CA1 pyramidal neuron, with latency ~0.5 ms. If this coupling is between axons, then the delay in crossing from axon to axon is <0.5 ms. Two pyramidal neurons were somatically recorded simultaneously, with sharp electrodes, in standard bathing solution (i.e., no low Ca^{2+}). **A,B:** Current-induced spikes in cell 2 lead to spikelets in cell 1. **C:** The latency from spike in cell 2 to spikelet in cell 1. **D:** A current-induced spike in cell 1 (now shown above) leads to a spike in cell 2. [A correspondence between spikes in one CA3 hippocampal pyramidal cell with spikelets in another had previously been shown by MacVicar and Dudek (1981, 1982).] (From Mercer et al., 2006 with permission.)

conduction of an active response into the cell producing the spikelet. Second, full action potentials, as well as spikelets, can appear in the second cell (Fig. 9.17d). The ability of spikes to propagate from cell to cell is critical for the emergence of collective population oscillations, when the individual cells are not themselves oscillators. Unfortunately, the Mercer et al. (2006) study did not localize the physical site of the electrotonic coupling. The study did, however, demonstrate a relationship between spikes in one pyramidal cell, and spikelets and spikes in another pyramidal cell, in the neocortex.

Gap Junctions and Neuropsychiatric Disease

Gap junctions can be linked to neuropsychiatric pathology in at least four different ways:

1. There are inherited diseases of myelin that result from alterations in connexin genes, expressed in myelin-producing cells (Schwann cells in the peripheral nervous system, oligodendroglial cells in the central nervous system). Examples include a Pelizaeus–Merzbacher disease-like syndrome, due to mutations in the *connexin47* gene, expressed in oligodendroglia (Kleopa et al., 2004; Li et al., 2004; Orthmann-Murphy et al., 2007); and an X-linked form of Charcot–Marie–Tooth disease, a peripheral neuropathy, due to mutations in the Schwann cell connexin-32 (Bergoffen et al., 1993). These disorders do not (so far as known) involve oscillations, and we shall not discuss them further.

2. There is a behavioral phenotype in a connexin-36 knockout mouse, involving learning and memory (Frisch et al., 2005); there are also disruptions in retinal physiology which would be expected to impair vision (Güldenagel et al., 2001).

3. There have been a large number of studies reporting that carbenoxolone and some other gap junction-blocking compounds, suppress seizure discharges, in a variety of epilepsy models, both in vivo and in vitro. In the in vivo cases, the drugs have been administered locally into the brain, intraventricularly, and systemically. A partial list of these studies would include: Gajda et al. (2005), Gareri et al. (2004), Gigout et al. (2006), He et al. (2009), Jahromi et al. (2002), Köhling et al. (2001), Nilsen et al. (2006), Perez Velazquez et al. (1994), Ross et al. (2000), Szente et al. (2002). Seizure activity induced by 4AP is also reduced in a connexin-36 knockout (Maier et al., 2002), although this result is somewhat difficult to interpret: 4AP-induced seizures involve, in part, excessive GABA release; the knockout phenotype may therefore reflect alterations in interneuronal activity. In most of these studies, gap junction blockade reduced (rather than completely abolished) epileptiform activity; however, Grenier et al. (2001, 2003), working on spontaneous seizures in the cat in vivo, showed that halothane virtually completely suppressed seizure activity (Chapter 2, Fig. 9.7).

4. Axotomy (sectioning axons) induces the formation of gap junctions between mammalian spinal motor neurons. Thus, gap junctions are—so far as known—found in immature spinal motor neurons, but not mature ones, in the mammal (Chang et al., 2000a). [Interestingly, gap junctions are easily found between the spinal motor neurons of lower vertebrates, such as frogs (Brenowitz et al., 1983; Collins, 1983; Sotelo & Taxi, 1970).] Nevertheless, axotomy in adult cats (by peripheral nerve section) induced the appearance of dye coupling between spinal motor neurons (Chang et al., 2000b). [Enhanced dye coupling—and presumably electrotonic coupling—after nerve injury may be an ancient evolutionary response: it has been reported in snails (Murphy et al., 1983).] These observations suggest many important possibilities: if something similar happens in the cortex, say after white matter injury, it would imply that certain brain lesions could lead to excessive numbers of gap junctions between principal neurons. Such (hypothetical) gap junctions could in turn contribute to epileptogenesis in the experimentally undercut cortex (Jin et al., 2006; Topolnik et al., 2003), and perhaps also in humans with brain injuries, tumors, and other neuropathologies. We shall return to this hypothesis in Chapter 18.

In Chapter 10, we shall consider different means by which gap junctions can induce the appearance of very fast network oscillations. The emergent VFO behaviors which are seen experimentally require knowledge of individual gap junctions to understand—but they require something else as well.

PART III

Some In Vitro Oscillations

10

Very Fast Oscillations

In this chapter, we consider the contrasting mechanisms of very fast oscillations (VFO, faster than approximately 70 Hz) in telencephalic cortical structures—the hippocampus, neocortex, and entorhinal cortex, for example—and the cerebellar cortex. Both types of VFO require gap junctions, apparently connecting the axons of principal neurons; both types of VFO can be generated without chemical synapses (in particular, without $GABA_A$ receptors), and even seem to be *enhanced* by blockade of chemical synapses; but, with chemical synapses intact, both types of VFO happily coexist with either lower frequency oscillations, or with transient synchronized discharges such as sharp waves. Neither type of VFO can be understood as arising from a system of coupled oscillators. There is, however, a crucial difference in the mathematical properties of the models that account for each respective VFO: telencephalic cortical VFO can be accounted for with a model in which electrical coupling is functionally *strong*, so that a spike in one axon may be able to evoke a spike in a coupled axon; cerebellar VFO can be accounted for with a model in which the coupling is relatively *weak*, so that only when multiple near-simultaneous spikes occur, in axons coupled to a selected axon, will that selected axon itself fire. The difference in models is not merely of interest to physics *cognoscenti*; the difference in models determines what types of experimental tests are suitable for each proposed mechanism and, perhaps most importantly, what type of neuronal population output each sort of model network can generate.

In Chapters 3 and 4, we have already met with some examples of VFO. Here, let us present a catalog of "telencephalic" VFO, divided into four categories:

1. *VFO in reasonably physiological conditions in vivo* can occur spontaneously, superimposed on physiological sharp waves in the hippocampus and

deep entorhinal cortex (Buzsáki et al., 1992; Chrobak & Buzsáki, 1996; Ylinen et al., 1995a), and during the depolarizing phase of the slow oscillation (Grenier et al., 2001); or it can be superimposed on cortical evoked responses that follow brief sensory inputs, in somatosensory (Baker et al., 2003a; Curio et al., 1994; Jones & Barth, 1999; Jones et al., 2000; Okada et al., 2005) and auditory modalities (Lakatos et al., 2005). In all of these cases, populations of cortical neurons become transiently depolarized.

2. *VFO in "nonepileptic" conditions in vitro* can occur (as is true in vivo) superimposed on spontaneous (or also evoked) sharp waves (Maier et al., 2002, 2003; Nimmrich et al., 2005); a difference between in vitro sharp-wave ripples and in vivo ones is that, in the former, many of the pyramidal cells become hyperpolarized (Behrens et al., 2007a; Maier et al., 2002). VFO can also occur in chemically or ionically modified media (Chapter 9, and see also later in this chapter). VFO can likewise occur nested with gamma and beta-2 oscillations in vitro, as described later. Experimentally, it is possible to produce VFO without gamma or beta-2, but it is not clear if the reverse is true, at least for persistent rhythms.

3. *Associated with epileptogenesis in vivo.* Here, VFO can occur leading into an interictal burst or seizure, or superimposed upon burst complexes within a seizure (Chapter 4; Akiyama et al., 2005, 2006; Asano et al., 2005; Fisher et al., 1992; Grenier et al., 2003; Jirsch et al., 2006; Kobayashi et al., 2004; Traub et al., 2005c; Urrestarazu et al., 2006; Worrell et al., 2004). VFO can also occur in isolation in epileptogenic tissue, especially at frequencies above 250 or 300 Hz (so-called "fast ripples"). Fast ripples can be generated in very small volumes of tissue, about 1 mm^3, and there is evidence that they have pathological significance (Bragin et al., 1995, 1999a,b, 2002a,b, 2003, 2005; Staba et al., 2002, 2004b). The gap junction blocking anesthetic, halothane, suppresses the VFO that occurs in association with seizures (Grenier et al., 2003); halothane also suppresses in vivo ripples that occur on sharp waves (Ylinen et al., 1995a).

4. *Associated with epileptogensis in vitro.* As is the case in vivo, VFO can occur leading into an interictal burst or seizure (Khosravani et al., 2005; Pais et al., 2003); superimposed on synchronized bursts (Schwartzkroin & Prince, 1977; Wong & Traub, 1983); or between burst complexes (Traub et al., 2005c).

As discussed in Chapters 5 and 7, VFO also occurs in the cerebellar cortex and subthalamic nucleus, in vivo.

The Generation of VFO Does Not Require GABAergic Interneurons

In this chapter, we try to develop a deeper understanding of how VFO can come about. First, however, it is necessary to discuss how it does *not* come about. The reason for this is that, owing to the way the field has developed historically, certain misleading ideas have become embedded in Neuroscience's collective consciousness.

When "ripple" oscillations were first discovered in the hippocampus (Buzsáki et al., 1992), it was immediately realized that interneurons could fire at the frequency of the ripple (~200 Hz), but that pyramidal cells—at least the somata—did not; and indeed, many pyramidal cell somata seemed not to fire at all. An identified basket cell was then shown to fire at ripple frequency (Ylinen et al., 1995a). Klausberger et al. (2003a, 2003b, 2005) extended the observations on identified interneurons, finding that basket cells and bistratified cells fired in phase with ripples; axoaxonic interneurons fired just before and at the start of a ripple, but then were silent; and cholecystokinin (CCK) basket cells and CCK dendrite-targeting cells hardly fired at all. In addition, Ylinen et al. (1995a) showed that pyramidal cells exhibited phasic inhibitory postsynaptic potentials (IPSPs), several millivolts in amplitude, at ripple frequency; manipulations of membrane potential and intracellular Cl⁻ were consistent with GABA$_A$ receptor-mediated synaptic potentials.

In view of this type of data, it seemed natural to postulate that networks of fast-spiking interneurons actually generated the ripple oscillations, and this idea has become somewhat ingrained in the community. There are several reasons for rejecting the idea, however:

1. In every experimental preparation where it has been possible to examine ripples/VFO with GABA$_A$ receptors blocked, the very fast oscillations have persisted, at similar frequency and possibly longer duration than baseline conditions (Jones et al., 2002; Maier et al., 2003). There are also many examples of VFO occurring with chemical synaptic transmission completely blocked (Chapter 9; see also later).

2. Somatic spiking, in this instance, is not the proper measure of pyramidal cell activity, as there is evidence for an antidromic (i.e., axonal) origin of spikes during VFO(Draguhn et al., 1998; Papatheodoropoulos, 2008). One cannot logically conclude that pyramidal cells are not critical for ripples, simply because their *somata* fire infrequently. An example of the firing ability of a pyramidal cell *axon* is shown in Chapter 8, Fig. 8.11.

3. The observations of Buzsáki et al. (1992), Ylinen et al. (1995a), and Klausberger et al. (2003a,b) can be readily explained with a model in which the pyramidal cell axons generate VFO (Traub & Bibbig, 2000).

4. At least to our knowledge, a pharmacologically isolated network of interneurons (in particular, without phasic synaptic excitation) has never been shown experimentally to generate VFO.

Let us illustrate these ideas further.

Figure 10.1 shows an example of sharp wave/ripple complexes occurring in the hippocampal slice, with GABA$_A$ receptors blocked. The complex has a similar appearance to baseline conditions, but not only are fast phasic inhibitory receptors blocked, but there are no phasic IPSPs in pyramidal cell intracellular recordings. The same study (Nimmrich et al., 2005) showed that field potential ripples could persist with ionotropic glutamate receptors blocked,

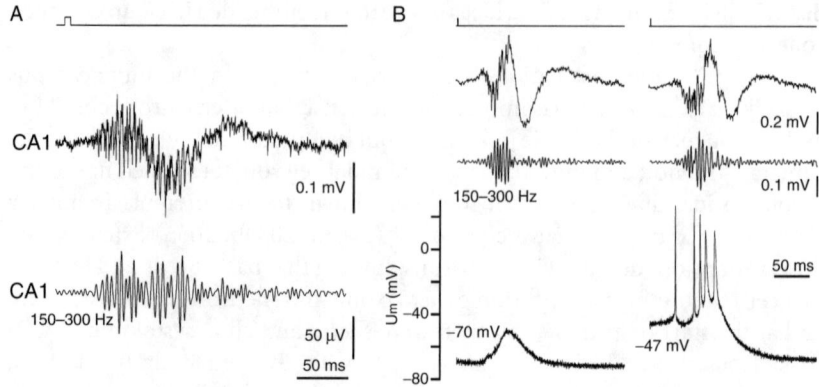

Figure 10.1 Sharp wave ripple complex in vitro (mouse hippocampal CA1 minislice, with CA3 dissected away) do not require GABA$_A$ receptors. When GABA$_A$ receptors are blocked, the complexes do not occur spontaneously, but can be evoked by pressure ejection of KCl (1 M) into stratum radiatum (upper traces). **A:** Stratum radiatum field potential of KCl-evoked sharp wave ripple, in 10 μM gabazine to block GABA$_A$ receptors. 150–300 Hz filtered trace shown below, to highlight the ripple. **B:** Combined extracellular (above) and intracellular (putative pyramidal cell, below) recordings, during KCl-induced sharp wave ripple, while GABA$_A$ receptors are blocked. The cell was held at –70 mV (left), to show the compound glutamatergic EPSP during the sharp wave, without phasic IPSPs. [Phasic IPSPs do occur in baseline conditions, in vivo (Ylinen et al., 1995) and in vitro (Maier et al., 2003).] When the cell was held depolarized at –47 mV (right), it generated a burst during the sharp wave ripple, but again without evidence of phasic IPSPs.

(From Nimmrich et al., 2005 with permission.)

but would not occur in the presence of the gap junction blocker octanol. These data are consistent with earlier data of Draguhn et al. (1998). Jones et al. (2002) also continued to record VFO, superimposed on somatosensory evoked potentials in vivo, after local application of bicuculline, a GABA$_A$ receptor blocker.

How, then, can one explain the ripple-frequency firing of fast-spiking interneurons, and the ripple-frequency IPSPs in pyramidal cells, in vivo? There are two basic principles at work here. First, as we shall describe in the text that follows, a network of pyramidal cell axons, that are electrically coupled, can generate population VFO (Traub et al., 1999b); second, because α-amino-3-hydroxyl-5-methyl-4-isoxazole-propionate (AMPA) receptor-mediated excitatory postsynaptic conductances (EPSCs) in fast-spiking (FS) interneurons are so extremely rapid [decay time of order 1 ms (Geiger et al., 1997)], the output of an oscillating pyramidal cell axonal plexus can *synaptically impose a coherent VFO in a population of FS cells*—and the FS cells, in turn, can impose coherent compound IPSPs on the pyramidal cells (Fig. 10.2, and Traub & Bibbig, 2000).

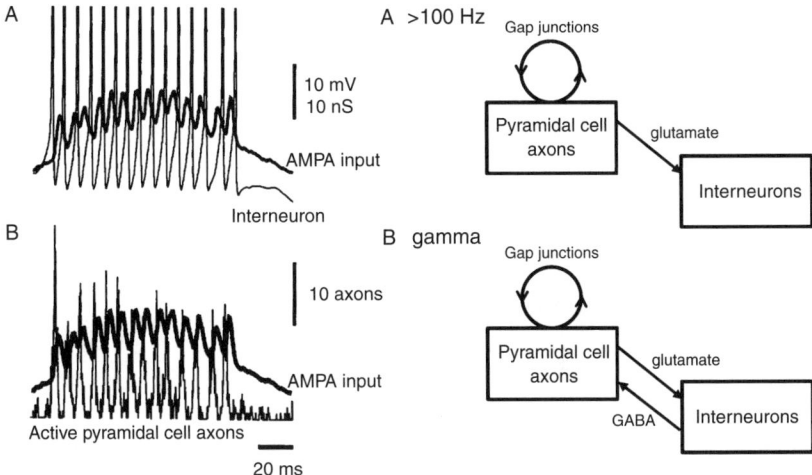

Figure 10.2 VFO generated in the plexus of pyramidal cell axons could plausibly generate coherent VFO in a population of fast-spiking interneurons, and thereby produce coherent phasic IPSPs in the pyramidal cells. This figure shows simulation data, from a model with 3072 pyramidal cells, coupled by gap junctions on their axons (each axon had 1.6 gap junctions, on average; depolarizing pulses, Poisson-distributed in time, occurred in the axons at average frequency 1 Hz per axon; coupling conductance 4.2 nS). The axons synaptically excited FS cells (basket and axoaxonic), with unitary EPSC proportional to $t \times e^{(-t)}$ (t in ms); that is, the decay time constant was 1 ms. The axonal network oscillated at ~140 Hz (B in left of figure), producing a compound phasically oscillating EPSC in each interneuron (A, B in left). Each model FS cell would fire at ~140 Hz, time-locked to this compound EPSC signal (A in left), and so coherent IPSPs could be produced in the pyramidal cells. Right: For this scheme to be stable, recurrent inhibition must be small or absent. A diagrammatically shows the case with no recurrent inhibition at all, and VFO is stable. When recurrent inhibition is too large (B), it interrupts the VFO to produce a gamma oscillation, as explained further on. The mechanism of VFO generation in the axonal plexus will also be explained further on. (From Traub and Bibbig, 2000 with permission.)

When, however, either phasic synaptic excitation, or phasic synaptic inhibition, or both, are blocked, the pyramidal cell axonal plexus can still generate population VFO—provided certain other conditions are met: electrical coupling is present, the axons are sufficiently excitable, and there is at least some degree of spontaneous activity.

Nonsynaptic VFO occurs in other brain areas besides hippocampal pyramidal cell regions, at least in vitro. Figure 10.3 shows VFO in the dentate gyrus, with an intracellular recording from a dentate granule cell (a type of excitatory neuron). The data in Figure 10.3 are extremely relevant to our argument, as they document that the somatic firing rate can be much lower

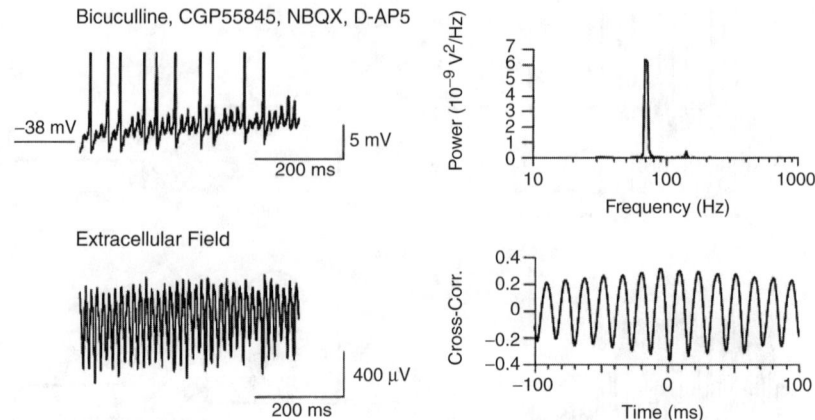

Figure 10.3 Nonsynaptic very fast oscillations in the rat hippocampal dentate gyrus, in vitro. Oscillations were evoked by pressure ejection of 1.5 M potassium methylsulfate into distal stratum moleculare. In this example, phasic synaptic currents (GABA$_A$, GABA$_B$, AMPA/kainate, NMDA) were pharmacologically blocked (bicuculline 10–20 μM, CGP55845 15 μM, NBQX 20 μM, D-AP5 50 μM). On the left are shown simultaneous intracellular (dentate granule cell) and extracellular recordings. (Note how depolarized the granule cell is by the K$^+$ puff, as well as the spikelets). The power spectrum of the field reveals a peak at ~70 Hz, and the intracellular/extracellular cross-correlation shows precise phase-locking. VFO could also be evoked in this preparation in low-calcium media, with frequency ~95 Hz.
(From Towers et al., 2002 with permission.)

than the population oscillation frequency, but that each population wave is accompanied either by a full spike, or by a spikelet, in a given neuron—as originally observed by Draguhn et al. (1998), and illustrated in the previous chapter. VFO in the dentate gyrus, evoked by pressure ejection of hypertonic K$^+$ solution, also occurs in Ca^{2+}-free media, at approximately 95 Hz (Towers et al., 2002). Figure 10.4 illustrates a burst of VFO occurring in layer 5 of a neocortical slice, in conditions where phasic synaptic transmission has been suppressed.

How Might Cortical VFO Be Generated?

A model of VFO must be able to account for these cardinal observations:

1. VFO occurs without chemical synapses, but requires gap junctions.
2. Full somatic action potentials, in any given neuron, do not follow the population frequency.

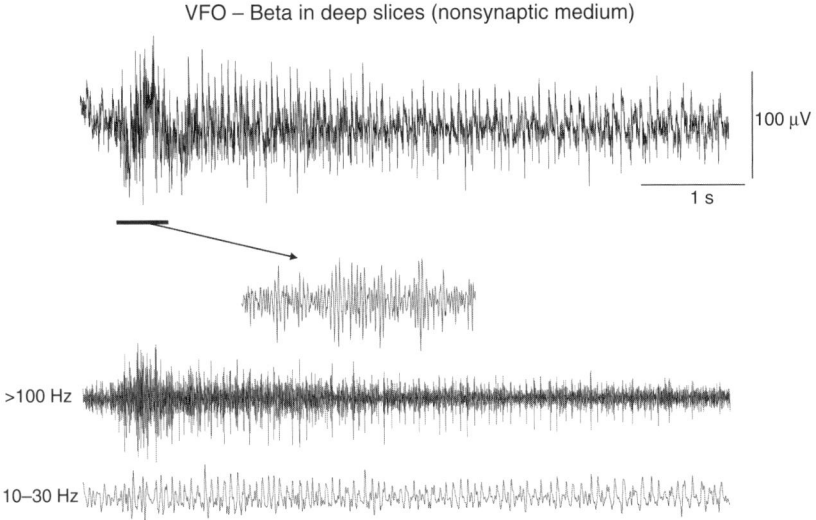

VFO – Beta in deep slices (nonsynaptic medium)

100 μV

1 s

>100 Hz

10–30 Hz

Figure 10.4 VFO can occur spontaneously in a rat neocortical minislice (containing deep layers only), with chemical synapses blocked. The figure shows field potential recordings, raw data above (a segment expanded in the inset), and filtered data below. The background activity is predominantly at beta frequencies (see Chapter 11), but there is a burst of ~1 s of VFO.

(M.A. Whittington and A. Roopun, unpublished data with permission.)

3. The gap junctional coupling between pyramidal cells, as judged by dye coupling, is extremely sparse (Chapter 9), with each cell (at least in hippocampal CA1) coupled to approximately 2 others—in alkaline conditions (Church & Baimbridge, 1991).

4. VFO can be coherent over distances greater than 300 μm (Fig. 10.10) in CA1, that is, it is coherent across populations of hundreds, or thousands, of neurons, despite the extremely low gap junctional connectivity.

The most economical hypothesis for tying these observations together is to suppose that the coupled axons generate VFO. But how do they do it? By analogy with gap junction–mediated approximately 5- to 10-Hz oscillations in the inferior olive, one might naturally first suppose that each axon is an intrinsic oscillator, and that the gap junctions produce phase coupling, so that the whole plexus acts like a series of coupled oscillators. One mathematically tractable version of this type of network was analyzed by Kuramoto (in which, for example, phase coupling between two oscillators is proportional to the sine of their phase differences), and the general category of models are referred to as the "Kuramoto model" (Acebrón et al., 2005). In the original Kuramoto model, coupling between the oscillators was all:all, far from the biological reality; analysis of sparsely coupled systems appears to be difficult indeed.

We would argue that, in any case, one would not expect electrically coupled axons to behave as phase-coupled oscillators, because of the extreme nonlinearity of the axonal membrane (due to the presence of either a high density, or low threshold, or both for the Na^+ channels). We therefore approached this issue by direct simulation of a large network of axonally coupled model neurons, at low coupling densities, simply to see what would happen (Traub et al., 1999b). The results were surprising. Axonally coupled networks could in fact produce VFO with sparse coupling under quite general conditions:

Condition (1) was present when the density of connections (i.e., gap junctions) was higher than one gap junction per axon, on average, the so-called percolation limit; but the density was not so high that all the axons started to fire at maximal frequency. The latter density will, of course, be parameter-sensitive. In our 1999 study, we needed to keep the average density below about three or four gap junctions per axon, and this was consistent with the dye-coupling data.

Condition (2) was present when the conductance of the gap junction (or junctions) between a coupled pair of axons was large enough that an action potential in one axon could induce an action potential in the other (provided the other is not absolutely refractory). The minimum value of this critical conductance will again be parameter-sensitive (e.g., it depends on intrinsic membrane properties, and on how far the gap junction is from the soma), but we have empirically found that conductances of a few nanosiemens will usually work. Again, this range of conductances is consistent with later estimates based on morphology and single gap junction channel properties (Chapter 9).

Condition (3) was present when there was a source of background spontaneous action potentials, so-called ectopic spikes. The frequency of ectopic spikes could be very low (e.g., 0.05 Hz per axon), but not zero. Of course, ectopic spikes are known to occur in the presence of 4AP (Avoli et al., 1998; Keros & Hablitz, 2005; Traub et al., 1995, 2001a) or after tetanic stimulation (Stasheff et al., 1993a,b; Stasheff & Wilson, 1990); and axonal excitability can be increased by transmitters, such as GABA or glutamate, when the appropriate receptors are not blocked (Felts et al., 1995; Kapoor et al., 1997; Liske & Morris, 1989; Sakatani et al., 1991a,b, 1992, 1993, 1994; Semyanov & Kullmann, 2001). In general, however, the source of the proposed ectopic spiking is not known precisely.

The model based on the above postulates—which we refer to as a "*percolation*" model—not only reproduces a number of experimental data, but it also makes some precise experimental predictions. Before proceeding to the predictions, however, it is necessary to give the reader a feel for sparse networks.

The reason for this necessity arises because of the entirely nonintuitive way in which oscillations arise in percolation models, and the way in which the period is determined. By way of contrast: familiar sorts of oscillations have their periods determined by the kinetics of defined membrane channels, channels corresponding to intrinsic voltage- and Ca^{2+}-gated processes, or to ligand-gated currents. For example, some oscillations have their period determined by the relaxation of a K^+ current, and others to relaxation of IPSPs. When, however, spikes can propagate from axon to axon, and intrinsic membrane refractoriness is short (a few milliseconds)—yielding what we shall call a "*one-to-one percolation*" model—then the oscillation period is no longer determined by intrinsic membrane kinetics. Nor is the period, in general, determined by the time it takes for a spike to propagate around a reentrant loop. Instead, the period is determined by more abstract properties of the global network, rather than by the kinetic properties of the individual cells and synaptic currents. In order for the reader to appreciate how this can be so, it is necessary to develop some intuitions about the structure of large networks. These intuitions will prove helpful in understanding detailed, so-called "realistic" simulations of electrically coupled networks of cells, and also in understanding more abstract and idealized sorts of simulations.

Some Properties of Random "Graphs"

The most common meaning of "graph" in mathematics is a method of pictorially representing a function, say $y = f(x)$. The word "graph" has another meaning in mathematics, however: a collection of vertices, or nodes, along with a collection of edges: each edge connects a pair of vertices. A graph of this latter type is an appropriate means to model a network where the network elements are all equivalent, and where interactions between elements occur pairwise, and each interaction has the same properties as any other interaction. For our purposes, a graph would represent a collection of identical axons (ignoring the somata, and pretending that the relevant part of the axon can be condensed to one small bit)—so that the *vertices* are rather idealized axons; and the *edges* correspond to gap junctions, whose properties are assumed (for the moment) to be identical. Graph theory has developed into a rich branch of mathematics. Examples of straightforward questions that one can ask about a graph would include these: is a given graph "connected," that is, does a path exist between any pair of vertices? (A path would consist of a sequence of vertices, starting with the first member of the given pair, and ending with the second member; and the sequence must have the property that consecutive vertices are connected by an edge.) Another question is this: given a graph, can one draw it on a piece of paper, in such a way that none of the edges cross?

If we represent a plexus of electrically coupled axons as a graph, which properties of this graph are relevant to oscillations? Two properties stand out, both related to the connectivity. First, what is the largest ensemble of vertices with the property that any one of these vertices has a path to any other—the "largest connected subgraph"? (Why is this important?—because collective behavior is possible only when axons can interact with each other, either directly or indirectly. If oscillations are to be observed in the population, then this connected ensemble should form a large fraction of the total system.) Second, within this largest ensemble, described above, how long are the paths between different vertices? [Why is this important?—because waves of activity will begin with spontaneous spikes that spread along the edges, or gap junctions, and the period will be determined by how many edges must be crossed (on average) to reach a majority of the population.]

For arbitrary graphs, it is extremely difficult to get a precise handle on these two connectivity properties. There is, however, a theory in which these properties can be described analytically, at least in the limit of ever increasing graph size: the theory of *random graphs*, pioneered by Erdös and Rényi (1960) (see also Bollobás, 2000, 2001). Even though it is unlikely that actual biological axonal networks best correspond to "random graphs," we have found this theory to be extremely instructive, at least as a starting point.

One note of caution, however. The language used to describe "random graphs," at least for the most part, is inherently ambiguous. Strictly speaking, a "random graph" does not even exist. The theory actually has to do with *ensembles* of graphs (all having the same number of vertices, n), with estimates of probabilities that some property attaches to members of the ensembles; and with these probability estimates converging to precise values as $n \to \infty$. Usually, $n \to \infty$ in a particular way: let the number of edges be N and define c by $N = cn$. Then, we let $n \to \infty$ keeping c constant. In doing so, the graph becomes sparser and sparser, in the sense that it is less and less probable that a randomly chosen pair of vertices is joined by an edge; on the other hand, the mean number of edges emanating from a vertex remains constant. (This latter quantity is called the *mean index*, and is equal to $2c$.)

It is common, however, in statistical theories to pretend that a representative of an ensemble has—within itself—average properties that really are average properties of the ensemble as a whole. For example, suppose we construct a large graph with a pseudorandom number generator. Then we expect the mean index of our sample "random graph" to be close to the mean index of the ensemble of graphs having the same value of c. This is a necessary approximation, in order to deal with particular networks (one can not do experiments on "all possible cortical slices with given connectivity"), but it is still an approximation. Figure 10.5 is an example, then, of a "random graph," with mean index 1.6, or $c = 0.8$. (To be precise, Fig. 10.5 only shows part of the graph, that part which forms the "large cluster," defined below. The graph also contains isolated vertices, and small connected subgraphs. These have not been drawn.) Figure 10.5 contains stringy bits—chains of vertices without branches—as well as branching trees, and various cycles.

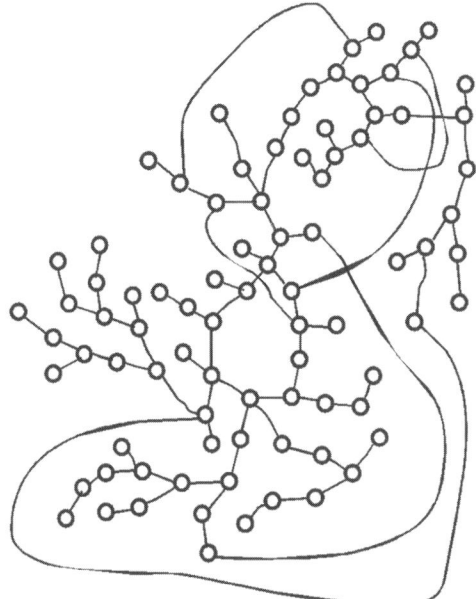

Figure 10.5 Example of the large cluster of a sparse random graph (above the percolation limit). Nodes are sketched as circles, joined by the edges, or lines in this drawing. [For purposes of this chapter, one can think of the nodes as axons, and the edges as gap junctions, although, of course, the layout of the graph on the page (with some of the nodes widely separated, to make drawing the structure easier), does not correspond to such an image.] There are 150 nodes in the original graph, with an average of 1.6 edges attached to each node ($c = 0.8$, in the Erdös-Rényi terminology). Only the large cluster (86 nodes) is shown, and the isolated nodes and other small subgraphs omitted. The size of the large cluster in this example, 57% of the system, is somewhat smaller than would be expected in a very large random graph of this mean index (1.6). Note that the path from one node to another can be quite long.

One of the most profound results in Erdös and Rényi (1960) concerns the expected size of the *largest connected subgraph*, relative to n (the total number of vertices), as a function of the connectivity parameter c. (Note that c is the *only* parameter that characterizes the ensemble of random graphs that we are considering.) They denote this function by $G(c)$ and provide an explicit formula for it, exact in the limit as $n{\rightarrow}\infty$. The function $G(c)$ is plotted in the upper part of Figure 10.6. Note that $G(c) = 0$ when $c < 0.5$, the so-called percolation limit. What does this mean? If $c < 0.5$, then each vertex connects to less than one other vertex, on average. Therefore, one expects the size of clusters to be small. Even if any one of these clusters can get to be large, as n becomes large, the cluster size *relative to the size of the whole graph* becomes arbitrarily small. When, on the other hand, $c > 0.5$, then there is one, and only one, connected subgraph whose size has the "same order" as n, that is, whose

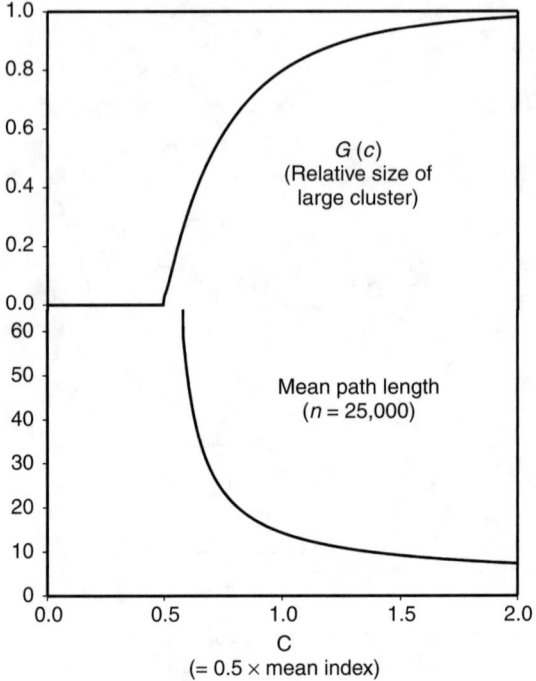

Figure 10.6 Structural properties of large random graphs. In these plots, the parameter c (as defined in Erdös and Rényi, 1961) is the ratio of the number of edges to the number of nodes (or vertices); c is equal to one-half the mean index = the average number of edges emanating from each node. The function $G(c)$ is the asymptotic size of the large cluster, relative to entire graph, as the graph becomes unboundedly large. Erdös and Rényi proved that $G(c) = 1 - x(c) / 2c$, where

$$x(c) = \sum_{k=1}^{\infty} \{[k^{k-1}]/k!\} \times (2c\, e^{-2c})^k$$

The upper plot shows an approximation to $G(c)$, where the series above was summed to 600 terms. Note how surprisingly big the large cluster is, for sparse connectivity; for example, when the mean index is only 2, the large cluster is about 80% of the total.

The lower plot shows the estimated mean path length (on the large cluster, which by definition is connected) in a random graph with 25,000 nodes, using the formula in Newman et al. (2001), where n is the number of nodes and $G(c)$ is as above:

$$< \text{path length} > \sim 1 + [\log(n \times G(c)/2c)]/\log(2c)$$

What is of note here is how surprisingly long the average path lengths are, when the graph is very sparsely connected.

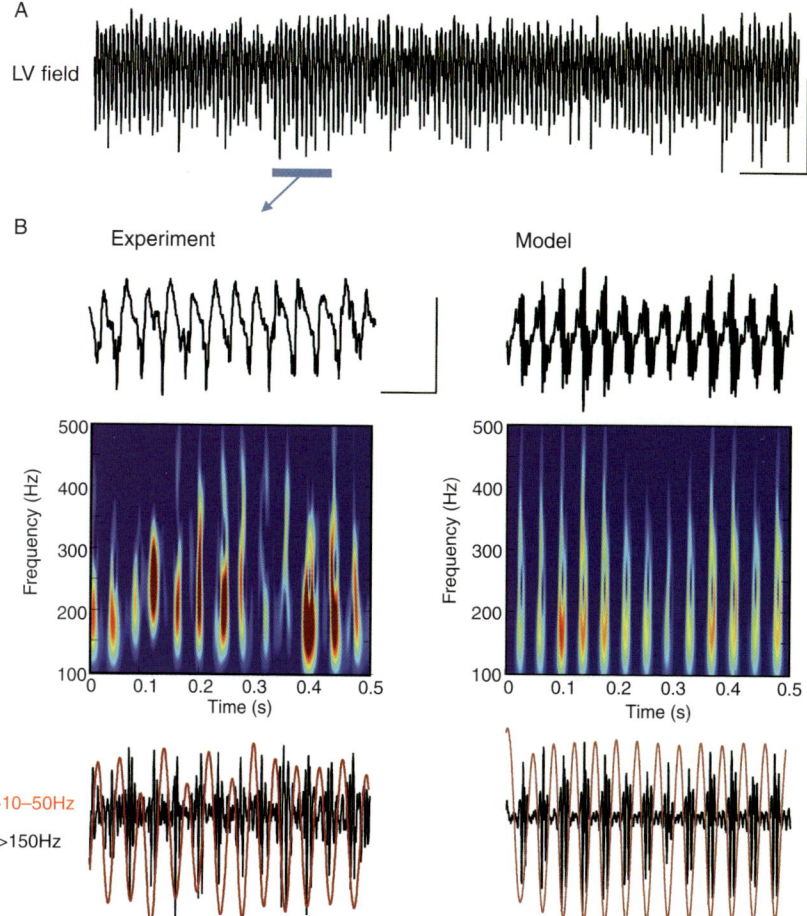

Figure 10.8 VFO can occur nested with a nonsynaptic cortical beta2 oscillation (see Chapter 11). **A:** Field potential recording of a kainate-induced beta-2 oscillation (~30 Hz) recorded in rat somatosensory cortex (layer 5) in vitro. A 500-ms epoch is expanded in **B**, which also shows a time-frequency plot, and two superimposed filtered signals (10–50 Hz, and >150 Hz), to emphasize the nesting of VFO on the beta-2. Corresponding simulation data are shown on the right. The simulation was as in Fig. 10.7, but gap junction conductance was held constant at 4.0 nS, and the density of g_{KM}, the M type of K$^+$ conductance, was ~½ the value used in Fig. 10.7. (Miles A. Whittington and Roger D. Traub, unpublished data.)

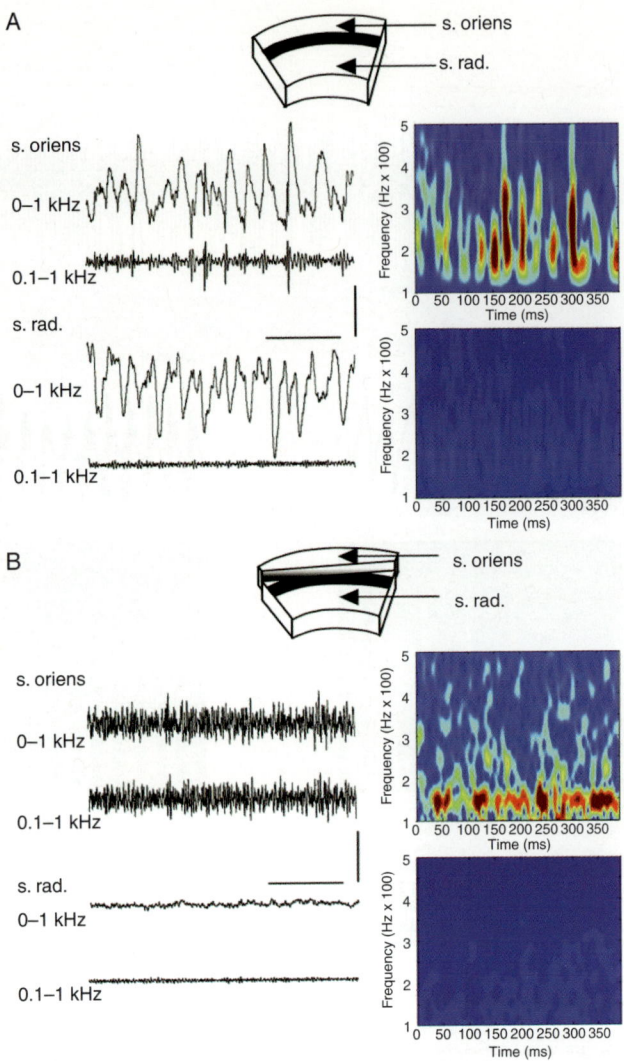

Figure 10.9 VFO occurs nested on persistent gamma oscillations. The preparation was a rat CA1 minislice (with CA3 dissected away, as in the diagram above), bathed in kainate (1 μM). **A:** Field potential recordings from stratum oriens (in which the pyramidal cell axon collaterals lie) show a gamma oscillation (~38 Hz) with superimposed VFO: this can be seen in the 100 Hz–1 kHz signal, and in the time-frequency plot on the right. In contrast, field recordings from stratum radiatum (in which few pyramidal cell axonal collaterals lie) show only the gamma oscillation. (Scale bars = 100 ms, 0.2 mV). **B:** After a cut was made, isolating stratum oriens (as in the diagram), stratum oriens generated a near-continuous VFO, without gamma. In stratum radiatum, there was neither gamma nor VFO. (Scale bars as in **A**.) The VFO is s. oriens is attributed to VFO generated by the pyramidal cell axonal plexus; in intact tissue, this VFO is interrupted by phasic IPSPs to give a gamma oscillation (see Chapter 12). (From Traub, Cunningham et al., 2003 with permission.)

ratio to *n* does not vanish as *n*→∞. This connected subgraph is called the "large cluster" It is truly remarkable how large (relative to the whole ensemble) the large cluster gets, with connectivities that are still quite small, as Figure 10.6 demonstrates. The physical and biological consequence of this mathematical fact (assuming that random graphs are a plausible representation of axonal plexi) is this: that collective phenomena, such as oscillations, are possible in ensembles that are very sparsely connected, as long as they are above the percolation limit, and as long as coupling is strong.

We have also mentioned the importance for oscillations of the *mean path length*, the number of edges that must be traversed (on average) to travel from one vertex to another. (The mean path length is only defined for the large cluster. If one picks two vertices, each from a different cluster, then—by definition—no path exists between them.) The lower part of Figure 10.6 plots an estimate of mean path length for ensembles of randomly constructed graphs having 25,000 vertices, as a function of *c*. It is striking how long the paths are, although—in view of the "stringiness" of Figure 10.5—not utterly surprising. We shall now attempt to explain the physical significance of the path length.

How, then, can VFO arise in a gap-junctionally connected plexus of axons, assuming Conditions 1 to 3 above (Traub et al., 1999b, 2002)? Let us consider first the consequences of a spontaneous ectopic spike in the plexus, arising in the large cluster, and assuming the rest of the plexus is at rest; and let us consider two basic parameters: the mean path length *P* for the plexus (which we can estimate if we suppose the plexus is randomly connected, and we know the mean index), and *T*, the time it takes for a spike in one axon to induce a spike in a connected axon. Our spontaneous ectopic spike will propagate throughout the large cluster, reaching a "typical" axon in time approximately $P \times T$. Then the activity will die away. If this wave of activity is to be part of an oscillation, we would estimate $P \times T$ as about ½ the period. What happens if there are other ectopic spikes? Each of these will initiate its own waves and, as Lewis and Rinzel (2000, 2001) emphasized, the waves can not pass through each other (because of absolute refractoriness), but instead will coalesce, and continue to spread. It is remarkable, however, that over a wide range of ectopic spike rates, the population does indeed oscillate at roughly the period predicted, in random and locally random networks (i.e., where there is a distance constraint on whether connections can form), and in tree structures (*which lack cycles, so that reentry is impossible*) (Traub et al., 1999b), as well as in 2-dimensional lattice arrays (Lewis & Rinzel, 2000). Further mathematical analysis of these types of percolation processes on graphs is, however, needed. The preceding arguments, for example, do not predict mean firing rates, the amplitude of population firing fluctuations, or the shape of the power spectrum of population activity. For the moment at least, these properties are investigated through simulations.

Before illustrating VFO simulation results, we need to consider a biological detail, not implicit in the graph theory itself, but related to it: *T*, the propagation time for spikes to cross from axon to axon, is expected to depend on the gap junction conductance, at least over a certain range. T should decrease as the gap junction conductance increases, at least up to a point. (Of course, with too small a conductance, no spike propagation will be possible.) [The physical reason why T is expected to decrease with increasing gap junction conductance is simple: a spike in one axon will inject more and more current into a coupled axon, as the coupling increases, provided we can ignore shunting effects. Thus, the second axon can be brought to threshold sooner.] Therefore, if the above qualitative arguments are correct, there should be two consequences of increasing gap junction conductance in an axonal network: (1) the population VFO frequency should increase, because propagation from axon-to-axon occurs faster; (2) with more axonal spikes per unit time, at larger gap junction conductances, there should then be more somatic spikelets and/or action potentials per unit time.

The intracellular experimental data in Figure 10.7 (layer 5 of rat neocortex in vitro, with chemical synapses blocked) confirm that somatic spikelet/action potential activity, per unit time, does indeed increase as gap junctions open. This is likewise observed in the simulation data of Figure 10.7. The simulation also predicts that the population oscillation frequency could increase as much as fivefold (in the illustrated case, from ~30 Hz to ~150 Hz), as gap junction conductance is increased—a type of behavior not expected for a system of coupled oscillators. As of this writing, the frequency prediction is under investigation. The resemblance of the simulated somatic potentials to experimental potentials is, however, striking. An interesting detail in Figure 10.7 (and also Fig. 10.4) is that the spikelet amplitudes are relatively uniform—something not seen in cerebellar VFO (see later).

Deviations in Detail from the Random Graph Theory Encountered in Constructing and Running VFO Simulations

We briefly discuss here some of the specific issues that arise in simulating VFO, either with detailed multicompartment models (as in Fig. 10.7), or with more abstract "cellular automaton" models (see the Appendix to Traub et al., 1999b) and Traub et al., (in press).

1. Localization of gap junction connectivity. Dye coupling between pyramidal cells only occurs between pairs of neurons whose somata are up to a few hundreds of micrometers apart (Gutnick et al., 1985; Schmitz et al., 2001). If dye coupling is an accurate marker of gap junctional coupling—and this is not known definitively—then it is possible that gap junctions do not connect cells too spatially separated. That, in turn, would imply that the overall connectivity *is not random*. (It is possible that the connectivity is, at least approximately, locally random—that is, random subject to a distance constraint.) Of course, one can simulate networks with any desired connection topology, but the formulae for $G(c)$ and for path lengths will not apply. To see why the latter is true, suppose cells are only allowed to connect if their somata are less

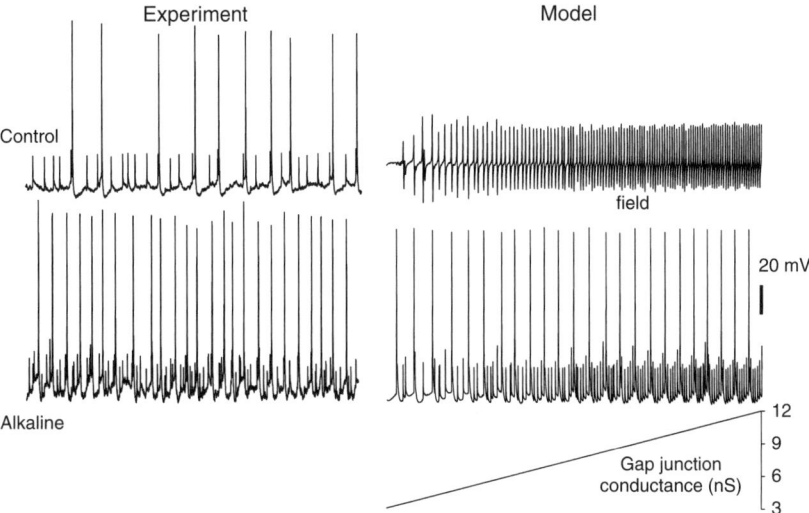

Figure 10.7   Oscillation model based on strong axonal coupling predicts an increase in spikelet frequency as coupling increases, as well as an increase in population frequency. The experiment shows 1-s recordings from a layer 5 intrinsically bursting (IB) cell in vitro, in a medium blocking phasic synaptic excitation and inhibition. Without alkalinization, spikelets occur at ~25 Hz. In more alkaline conditions (opening gap junctions further), spikelets occur at ~50 Hz. The model shows a simulation of 15,000 multicompartment IB cells, with 25,000 gap junctions randomly distributed between pairs of axons ($c = 5/3$, average 3.333 gap junctions lying on each axon); the gap junction conductance was ramped (over 1 s) form 3 nS to 12 nS, while randomly occurring stimuli to the axons were delivered at a mean of 4 Hz per axon, and all other parameters were held fixed. The field frequency increases from ~30 Hz to ~150 Hz, while spikelet frequency also increases (until it saturates). In the model, each somatic spikelet is induced by a full action potential in the axon (not shown).

(Miles A. Whittington and Roger D. Traub, unpublished data.)

than d μm apart. As the network becomes larger and larger, the path length will come to be dominated by the distance constraint; but in a random graph, of any size, the path length depends on c and not on any distance constraint—the latter not even being defined in a random graph. The significance of this consideration will become clearer as data accrue on where, exactly, gap junctions are located on axons.

2. Reflected spikes. In network simulations with detailed models, it can happen that axonal spikes occasionally "reflect" from the soma: due to the broader somatic spike (relative to axonal spikes), an antidromic spike can, on occasion, elicit a second spike that propagates orthodromically. [This phenomenon has been observed experimentally in CA3 pyramidal cells (Traub et al., 1994b)].

Reflected spikes have the effect of amplifying net axonal activity. Whether they alter the fundamental principles of VFO remains to be seen, however.

3. Shunting by gap junctions. In a random graph, the index distribution (i.e., the statistical distribution of the number of edges on a vertex) is Poisson. In a large random graph, there will always be a few vertices with a large number of edges. In physical terms, modeling an axonal plexus with random topology will lead to the existence of a few axons which are gap junctionally connected to many other axons. As Lewis and Rinzel (2001) pointed out, a large number of gap junctions will act as a shunt, so that if one axon fires, its spike may not be communicated. [One can, of course, model networks in which there is a constraint on the maximum number of gap junctions on an axon.] This consideration probably does not have major effects on VFO, however.

4. Reentry. Random graphs, that are sufficiently large, contain cycles of any given order (length). If there is a means for breaking symmetry (i.e, a means for getting a spike to propagate in one direction around a cycle and not the other), then—in principle—reentrant rhythms could occur [also pointed out by Lewis and Rinzel (2001)]. Such rhythms would have a period determined by cycle length, and not by global path lengths—one expects that the re-entrant rhythms would be much faster than the percolation rhythms, because average path lengths are so long in large random graphs, whereas cycles supporting reenty can (in principle) be short. We have never observed stable reentry in detailed multicompartment models, but have in cellular automaton models when propagation is made stochastic (a means of breaking symmetry). It is possible that reentry is involved in the extremely fast oscillations (600 Hz) described by Curio and colleagues (Curio, 2000).

VFO Can Occur Nested with a Slower Oscillation, Beta-2 (20–30 Hz)

In our discussion of VFO basic mechanisms, we have idealized the situation in several ways: in the connection topology, as discussed, but also in ignoring intrinsic membrane currents that are slower than action potentials (i.e., that involve membrane time constants longer than a few milliseconds), and in ignoring synaptic currents. It is central to the themes of this book that VFO continues to exist in several more general, and more physiological, conditions than the idealized state we have described earlier—but with a twist. Very roughly, slower membrane currents and synaptic currents break VFO up into packets. Introducing this idea now not only helps to understand VFO better, but sets the stage for more detailed analyses of certain forms of beta and gamma oscillations in Chapters 11 and 12.

As described in Chapter 8, axons contain a number of K^+ conductances, including the M conductance, g_{KM} . This latter type of conductance helps to shape a beta-2 oscillation found in layer 5 of somatosensory cortex in vitro (and probably also in vivo). As expanded on in Chapter 11, the beta-2 oscillation exists with intact chemical synaptic transmission, but does not depend it. The beta-2 oscillation does, however, depend on gap junctions; and it further is modified by alterations of g_{KM}. In simulations of networks of layer 5 pyramidal cells, without chemical synapses, and varying both gap junction conductance and also g_{KM}, it was found that—in some region of the parameter space—VFO and beta-2 could coexist, one nested on the other (Fig. 10.8). This situation can also be seen experimentally (Fig. 10.8), although in conditions of intact synaptic transmission. The data suggest that an axonal, and

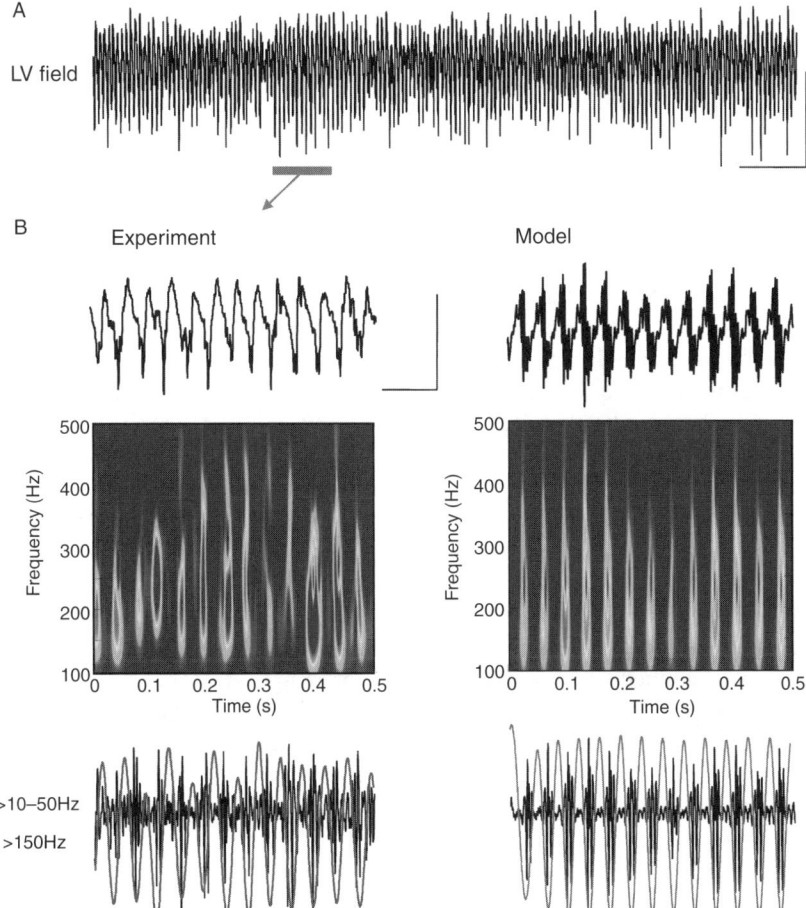

Figure 10.8 VFO can occur nested with a nonsynaptic cortical beta2 oscillation (see Chapter 11). **A:** Field potential recording of a kainate-induced beta-2 oscillation (~30 Hz) recorded in rat somatosensory cortex (layer 5) in vitro. A 500-ms epoch is expanded in **B**, which also shows a time-frequency plot, and two superimposed filtered signals (10–50 Hz, and >150 Hz), to emphasize the nesting of VFO on the beta-2. Corresponding simulation data are shown on the right. The simulation was as in Fig. 10.7, but gap junction conductance was held constant at 4.0 nS, and the density of g_{KM}, the M type of K^+ conductance, was ~½ the value used in Fig. 10.7. Please see color insert.

(Miles A. Whittington and Roger D. Traub, unpublished data.)

perisomatic, phasic hyperpolarizing current can serve to break VFO into short epochs, separated by longer epochs. The duration of the longer epochs is determined by the kinetics of the hyperpolarizing process.

VFO Also Occurs Nested with Persistent Hippocampal (and Medial Entorhinal) Gamma Oscillations

Persistent gamma oscillations (Chapter 12) resemble the somatosensory cortical beta-2 oscillation in also requiring gap junctions; but persistent gamma oscillations differ from that particular beta-2, in that persistent gamma oscillations also require chemical synapses: specifically, both AMPA receptors and GABA$_A$ receptors. As Figure 10.9 illustrates, persistent gamma (in the hippocampal slice) also exhibits VFO packets—in stratum oriens (where pyramidal cell axonal collaterals mostly lie), but not in stratum radiatum (which contains few such collaterals). [A similar nesting phenomenon occurs in medial entorhinal cortex (Cunningham et al., 2004a). The entorhinal cortex gamma also exhibits spikelets and is carbenoxolone-sensitive—see Chapter 12.] When stratum oriens is physically isolated (Fig. 10.9b)—isolating the pyramidal cell axons from the somata and dendrites—a more continuous VFO occurs. This observation, and additional data to be presented in Chapter 12, suggest that synaptic inhibition, in perisomatic regions, is acting to break VFO into short segments—an effect that will be possible if axonal gap junctions are not located too distant from the soma and initial segment, as indeed is suggested by the dye coupling evidence in Schmitz et al. (2001) (see also Chapter 9).

Additional evidence for the relationship between VFO and gamma oscillations derives from the hippocampal slice data in Figure 10.10. In this case, gamma was produced by pressure ejection of hypertonic K$^+$ solution, rather than bath-applied kainate. The admixture of gamma and VFO is apparent in Figure 10.10A and C, for control conditions. When phasic chemical synapses are blocked (Fig. 10.10B), the gamma oscillation disappears, as expected; the VFO, however, remains. It is interesting here to note that the VFO is coherent over distances of at least 300 μm, so that hundreds of cells (at least) must be participating. It is not yet known whether nesting of VFO occurs in neocortical gamma rhythms; this may reflect technical matters concerning the way extracellular fields are measured in neocortex, and differences in "packing" between neocortical neurons, as compared with hippocampal and medial entorhinal cortex neurons.

Summary of One-to-One Percolation

To summarize our views of VFO in the hippocampus, neocortex, and other telencephalic structures: we propose that VFO arises because of relatively rare spontaneous action potentials, percolating from axon to axon across gap

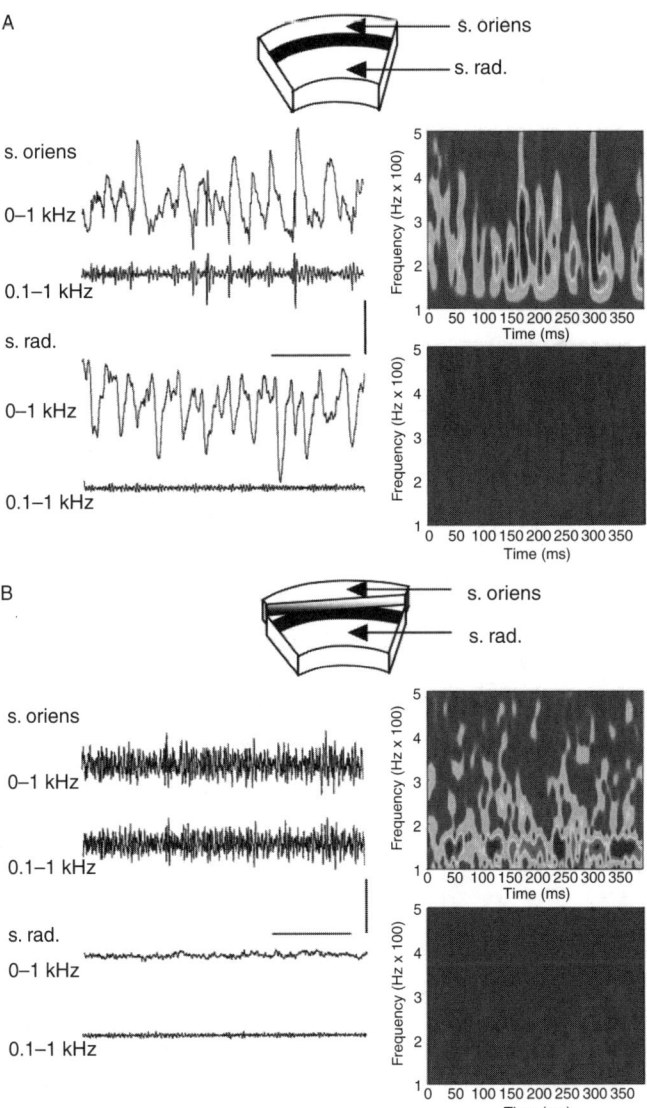

Figure 10.9 VFO occurs nested on persistent gamma oscillations. The preparation was a rat CA1 minislice (with CA3 dissected away, as in the diagram above), bathed in kainate (1 μM). **A:** Field potential recordings from stratum oriens (in which the pyramidal cell axon collaterals lie) show a gamma oscillation (~38 Hz) with superimposed VFO: this can be seen in the 100 Hz–1 kHz signal, and in the time-frequency plot on the right. In contrast, field recordings from stratum radiatum (in which few pyramidal cell axonal collaterals lie) show only the gamma oscillation. (Scale bars = 100 ms, 0.2 mV). **B:** After a cut was made, isolating stratum oriens (as in the diagram), stratum oriens generated a near-continuous VFO, without gamma. In stratum radiatum, there was neither gamma nor VFO. (Scale bars as in **A.**) The VFO is s. oriens is attributed to VFO generated by the pyramidal cell axonal plexus; in intact tissue, this VFO is interrupted by phasic IPSPs to give a gamma oscillation (see Chapter 12). Please see color insert.

(From Traub, Cunningham et al., 2003 with permission.)

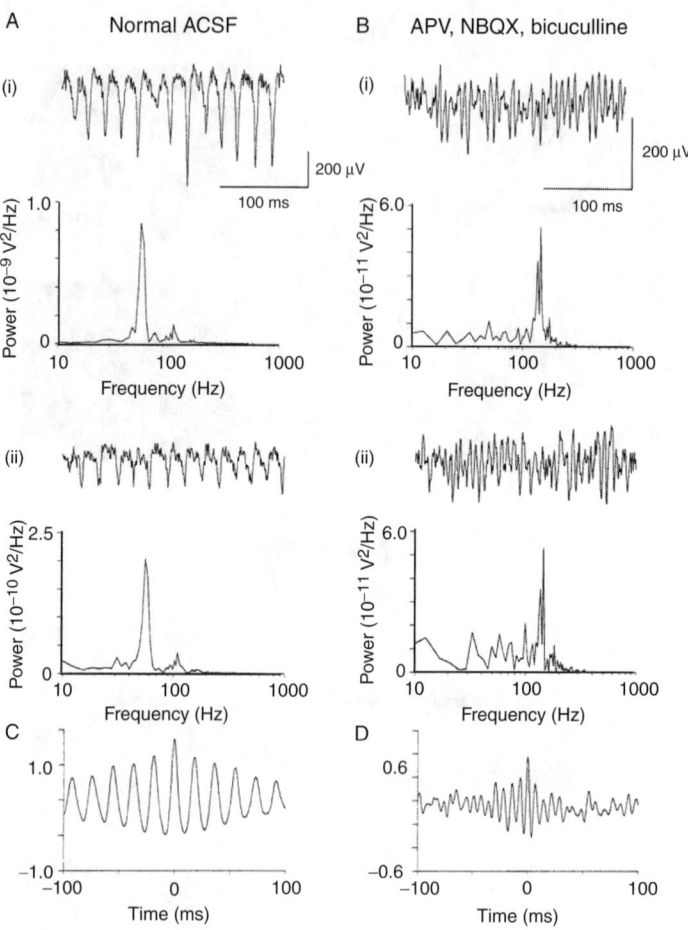

Figure 10.10 Further evidence for nonsynaptically generated VFO co-existing with synaptic gamma oscillations. In this case, oscillations were evoked in rat hippocampal slices (CA1), by pressure ejection into stratum radiatum of 1.5 M potassium methylsulfate. **A, C:** Portions of a ~3 s gamma oscillation, recorded simultaneously with two extracellular electrodes, 300–400 μm apart, just below stratum pyramidale (i and ii). Power spectra show a gamma peak at 57 Hz and a VFO peak at ~110 Hz. The cross-correlation in C shows the two signals to be tightly synchronized. **B, D:** A similar pressure ejection, with phasic synaptic transmission chemically blocked (APV 100 μM, NBQX 20 μM, bicuculline 12 μM). In this case, VFO (~140 Hz), without gamma, lasts about 1 s. The VFO has smaller amplitude than the gamma in **A**. The cross-correlation in **D** shows the VFO to be tightly synchronized between sites.

(Data of F.E.N. LeBeau and E.H. Buhl, from Traub et al., 2001. Reproduced with permission.)

junctions, with the period determined by the global topological network structure, rather than by intrinsic membrane or synaptic conductances. Simulations based on this idea do indeed produce VFO at appropriate frequencies, in conditions where chemical synapses are present or not (as in experiments). The simulations generate patterns of spikelets that are close to the biological data. Finally, a model based on percolation ideas predicts (as shown in a preliminary way in this chapter, and to be developed later) the observed relationships between VFO and other sorts of neuronal oscillations, at gamma and beta frequencies. A prediction of our model is that VFO frequency can be controlled, at least in part, by gap junction conductance. If experimentally verified, that would constitute even stronger evidence that VFO cannot be understood as a system of coupled axonal oscillators, but rather is an emergent property of electrically coupled axonal networks.

We have referred to the model of hippocampal and neocortical VFO as "one-to-one percolation," because the model takes as a postulate that a spike in one axon can induce—across a gap junction—a spike in a coupled axon. In the next part of this chapter, we consider VFO in the cerebellar cortex in vitro. This type of VFO can be replicated with a "many-to-one" percolation model, rather than "one-to-one."

Some Properties of Cerebellar Purkinje Neurons

These cells are remarkable for the large extent of their dendrites, on the order of 160,000 μm^2; the dendrites in addition have a low density of Na^+ channels (Llinás & Sugimori, 1980b; Roth & Häusser, 2001). As a result, the dendrites impose an enormous impedance load on the soma. Action potentials are initiated in the axon, which has an initial segment just tens of μm long (Clark et al., 2005; Kato & Hirano, 1985). In attempting to model these cells, one finds that a high density of axonal Na^+ channels is required (De Schutter & Bower, 1994a; Miyasho et al., 2001; Traub et al., 2008c), reflecting the large dendritic impedance. Correspondingly, in simulating pairs of Purkinje cells, electrically coupled via their axons, we were not able to obtain spike propagation from axon to axon, with coupling conductances up to 6 ns: the current injected by a spike in one axon, into a second axon, was not enough to bring the second axon to threshold in the presence of the dendritic impedance.

Nonsynaptic VFO Occurs in the Cerebellar Cortex In Vitro, Associated with Spikelets in Some of the Purkinje Cells

As we have shown earlier (Chapter 7; Fig. 7.8), bath-applied nicotine induces gamma oscillations in mouse (and also human) cerebellar cortex. The relevant biophysical actions of nicotine in this instance are not known, but may be related to increases in axonal excitability (Kawai et al., 2007). The gamma is admixed with VFO at about 100 to 200 Hz (Fig. 10.11, compare Fig. 10.10). The gamma oscillations do not depend on glutamatergic receptors (Middleton et al., 2008) but, as Chapter 7, Fig. 7.8 shows, they do require $GABA_A$ receptors: blockade of the latter with gabazine eliminates the gamma, but enhances the VFO. As for persistent gamma and VFO in hippocampus, entorhinal cortex and neocortex, in the cerebellar cortex, both gamma and VFO require gap junctions (Middleton et al., 2008), and the VFO also exists without chemical synaptic transmission (e.g., in low Ca^{2+} media).

Ultrastructural evidence for gap junctions between Purkinje cells is not available, but they are dye coupled (Middleton et al., 2008). Whether dye-coupling is mediated by direct gap junctional coupling between Purkinje cells, or through an indirect pathway (as discussed in the previous chapter) is not known. We asked under what conditions of electrical coupling it might be possible to replicate cerebellar VFO, using a network model with 1000 multi-compartment Purkinje cells (Traub et al., 2008c). As one-to-one percolation

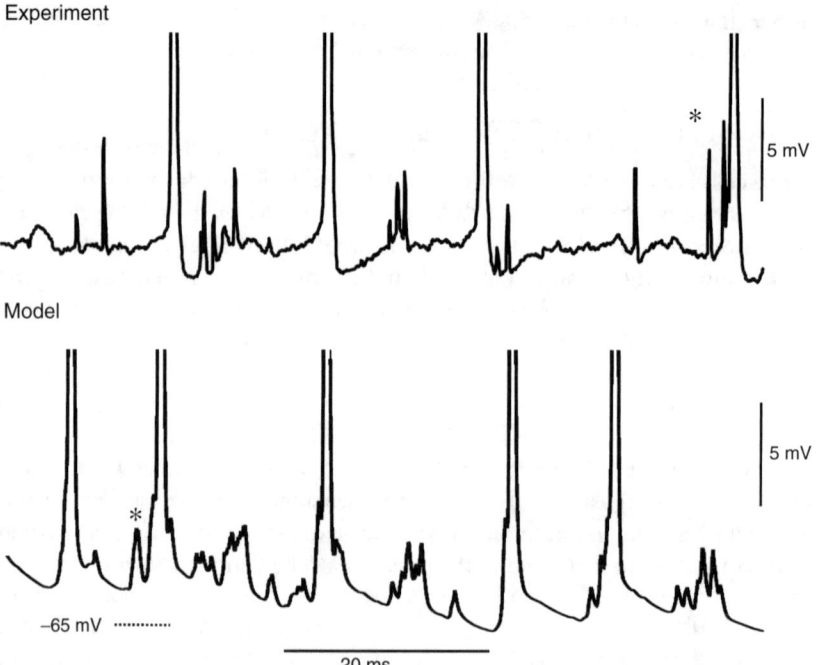

Figure 10.11 Nonsynaptic VFO in the cerebellar cortex in vitro. Intracellular recordings in some Purkinje cells, during VFO (induced by nicotine + gabazine, as in Chapter 7, Fig. 7.8), show high frequencies of spikelets, at many different amplitudes. The simulation data below are from a network model with 1,000 multicompartment Purkinje cells, electrically coupled in their axons.

(From Traub et al., 2008 with permission.)

did not appear to be possible, we wondered whether many-to-one percolation would work: that is, whether oscillations could arise when two or more near-simultaneous spikes were needed to occur in order to trigger a spike in a coupled axon. (Preliminary simulations indicated that electrical coupling between somata or dendrites would not work, as there was no way for spikes to propagate; the low density of dendritic g_{Na} was the main problem). It turned out that a many-to-one model did resemble the data, when connectivity was well above the percolation limit (using a mean index of 5, for example), and when ectopic spike rates were high enough (~15 Hz per axon). A comparison between model and biological data, at the cellular level, is shown Figure 10.11.

In one-to-one models of percolation VFO, which we have investigated, full axonal action potentials propagate through the axonal plexus, and anti-dromically invade the somata to generate either spikelets or full spikes. Somatic spikelets temporally corresponded to full action potentials in the proximal axon. This was not the case for simulated Purkinje VFO. As shown in the left of Figure 10.12 (A1 and A2), somatic spikelets corresponded to somewhat larger axonal spikelets, not to full axonal action potentials. Nevertheless, axonal activity is highly organized, as seen in Figure 10.12B (left) and the raster plot in Figure 10.12 (right). To understand this better (not shown), we picked individual cells, and repeated simulations over and over, each time blocking an additional gap junction lying on these selected cells (Traub et al., 2008c). In this way, it was possible to show that a selected cell's axon would fire *either* ectopically (i.e., spontaneously), *or* when multiple connected cells fired near simultaneously. Thus the many-to-one percolation. It could also happen that only *one* connected cell would fire, thereby leading to a spikelet in the axon, and an attenuated spikelet in the soma.

For one-to-one percolation models, we have argued that the period is determined by topological network properties, and not by intrinsic membrane properties. This is not the case, however, for many-to-one percolation models, at least not the Purkinje VFO model. The latter model predicted that the kinetics of Na^+ channel inactivation would have a major influence on the period; this prediction was verified experimentally (Traub et al., 2008c). The inter-relationship between cerebellar VFO and cerebellar gamma remains to be investigated further, but initial observations suggest fundamental differences. For example, the overt nesting of VFO and gamma rhythms, seen in hippocampus and entorhinal cortex, is not apparent in the admixture of these two rhythms during population gamma in cerebellar cortex.

In summary, there are at least two classes of VFO. They differ primarily in the functional strength of the electrical coupling between principal neurons. [We say "functional" because it is not the absolute value of the gap junction conductance per se which is critical, but rather what matters is whether single spikes can propagate from one axon to another—a property which depends on gap junction conductance (to be sure), but also depends on membrane conductances and the design of the cell—for example, where the gap

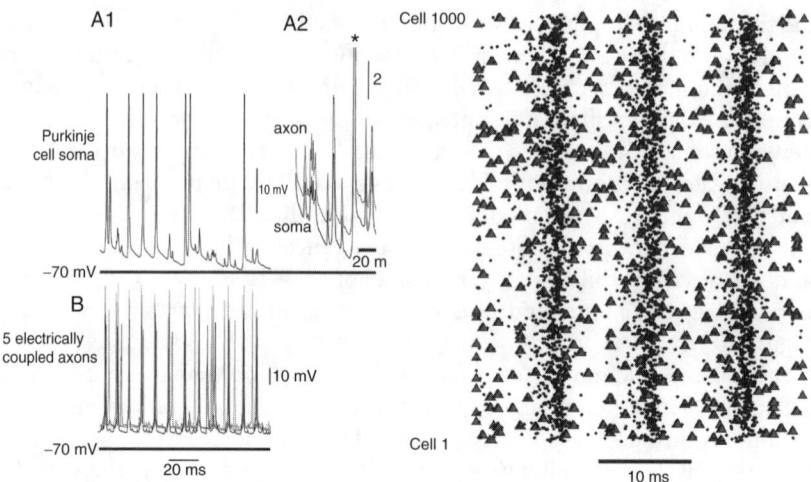

Figure 10.12 Simulated cerebellar VFO is an emergent behavior, dependent on axonal coupling between neurons that are not individually oscillatory, but the coupling is functionally weaker than in the neocortical/hippocampal VFO model. Left: **A1** shows a mixture of somatic action potentials and spikelets of diverse amplitudes. **A2** shows that each somatic spikelet corresponds to a somewhat larger spikelet in the axon. **B** shows that axonal action potentials are, on average, coherent with one another. Right: Raster plots of Purkinje cell action potentials from two simulations: with gap junctions closed (red triangles), and open (black dots). With gap junctions closed, the cells fire as independent Poisson processes, i.e., randomly. With the gap junctions open, firing rates increase significantly and the population organizes. Nevertheless, coupling is not strong enough to allow a spike in one axon to induce a spike in a coupled axon (unlike the neocortical and hippocampal VFO models); instead, it takes two or more near simultaneous spikes. The period of the cerebellar oscillation is therefore determined by intrinsic membrane properties (in this case, by Na+ conductances), rather than as a straightforward consequence of network topology.

(From Traub et al., 2008 with permission.)

junction is located, and the impedance load imposed by the soma and dendrites.] When the functional coupling is *strong enough*, one-to-one percolation can occur. This will allow VFO at low ectopic rates, with (we predict) frequency that is modulated by gap junction conductance, and stereotypic spikelets; gap junction connectivity can also be low, although still it must be above the percolation limit. In contrast, when gap junction coupling is *not strong enough*, many-to-one percolation might still occur, if gap junctional connectivity is sufficiently high, and spontaneous action potentials occur frequently enough. In this case, intrinsic membrane kinetics will make a major contribution to determining oscillation frequency. Spikelet shapes and amplitudes will be heterogeneous.

11

Beta-2 Oscillations

Beta oscillations (10–30 Hz) have been mentioned repeatedly in this book. In Chapter 3, we showed how beta occurred during the slow oscillation of sleep, during the memory phase of a cognitive task (typically associated with behavioral immobility), and in the course of auditory evoked potentials. Beta oscillations can occur during status epilepticus (Chapter 4). Enhanced beta activity occurs in Parkinsonian syndromes, in patients and experimental models, especially in somatosensory, motor, and supplementary motor cortex (Chapter 5), and this beta is, in turn, attenuated with effective treatment. Decreased beta frequency phase synchrony, of cortical oscillations, has been demonstrated in schizophrenia (Chapter 6). Coherent beta frequency oscillations occur in cerebellum and somatosensory cortex, and are likely important for motor coordination (Chapter 7). There are many other observations of beta oscillations occurring in the cortex during sensory stimulation (Haenschel et al., 2000) and cognitive tasks (Tallon-Baudry et al., 1999b). Finally, we showed in Chapter 10 how—at least in vitro—beta-2 oscillations (20–30 Hz) could be nested with very fast oscillations (VFO).

In this chapter, we concentrate on beta-2 oscillations in vitro, in somatosensory and motor cortex. There are several reasons for this focus: this type of oscillation has been modeled and is, at least relatively, understood; the *nonsynaptic* nature of somatosensory beta-2 is truly remarkable, especially as similar types of oscillation may occur in vivo; beta-2 introduces an important physical principle, concerning the interruption of VFO by membrane currents—a principle also relevant to persistent gamma oscillations (Chapter 12); and, finally, the experimental data introduce an instructive physiological principle that must be considered in pharmacological studies of GABA$_A$ receptor manipulation, in vitro and in vivo: specifically, that drugs that act on GABA$_A$ receptors may be producing network effects by effects on distal axons as well as, or instead of, the more traditional effects on perisomatic and dendritic receptors.

We shall begin by summarizing the data in a study by Roopun et al. (2006), of in vitro beta-2 oscillations in secondary somatosensory cortex induced by kainate.

In Secondary Somatosensory Cortex, Gamma Oscillations are Generated in Superficial Layers, and Beta-2 in Deep Layers

Figure 11.1 illustrates oscillations evoked by kainate in 2° somatosensory cortex in vitro, recording field potentials either in deep layers, layer 4, or superficial layers, and analyzing the power spectra of the various signals; the latter are shown as color-coded time frequency plots. The layer 4 recordings pick up oscillations simultaneously from both deep and superficial layers. It is remarkable that in this part of the cortex, at least, superficial and deep layers generate oscillations at different frequencies—gamma superficially, beta-2 deep—and, further, the generating proceeds independently: after a cut that physically separates the superficial and deep layers (Fig. 11.1B), each structure continues to oscillate at its preferred frequency. That Nature has done this simplifies the analysis of mechanisms: it is possible to concentrate on cells in one structure or the other. It is possible to go even further, as intracellular recordings have shown that it is intrinsically bursting (IB) pyramidal cells which robustly participate in the beta$_2$ oscillation (Roopun et al., 2006).

Beta-2 Oscillations are Associated with Spikelets, and are Regulated by the M Type of K$^+$ Conductance, g_{KM}

We have seen in Chapter 10 on VFO that cells may have voltage fluctuations that are time-locked to a population oscillation, and yet the somata fire at much lower rates than the population frequency. (Of course, this phenomenon is quite general, occurring as well in the inferior olive and in other contexts.) As Figure 11.2 demonstrates, layer 5 IB neurons exhibit subthreshold behavior (in this case, spikelets) at beta$_2$ frequency, with intermittent full action potentials. Similar behavior is observed in a network model that includes IB cells, whose axons are both (1) electrically coupled and (2) contain g_{KM}. The model predicted that g_{KM} would partially regulate the oscillation period by modulating how "bursty" the neurons are, and experiments are consistent with this prediction (Fig. 11.2C).

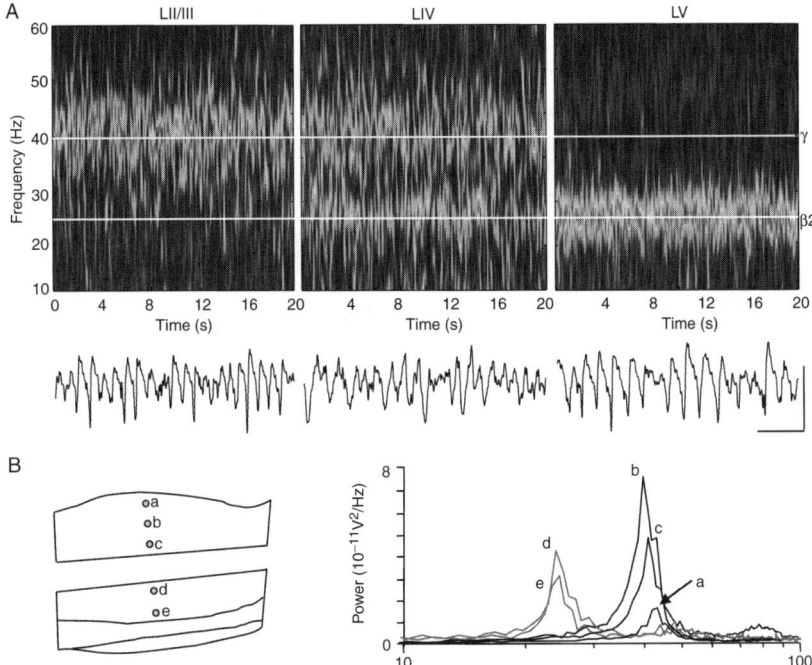

Figure 11.1 Beta-2 oscillations are generated in vitro in deep cortical layers, independent of the superficial layers, in 2° somatosensory cortex. Adult rat neocortical slice, bathed in 400 nM kainate. **A:** Raw data (below) and time-frequency plots (above) for the induced oscillations in superficial layers (layers 2 and 3), layer 4, and layer 5. Superficial layers generate gamma at ~40 Hz. Layer 5 (one of the deep layers) generates beta-2, ~25 Hz. Layer 4 exhibits both oscillations together. (Scale bars = 100 ms, 0.2 mV.) **B:** The deep and superficial layers were separated by a cut at the layer 4/5 border. Field potentials recorded (not concurrently) at the sites indicated in the diagram at left. As the power spectra show, the isolated superficial layers continue to generate gamma, and the isolated deep layers continue to generate beta-2. [Compare this with Chapter 3, Fig. 3.10B, which shows auditory cortex oscillations induced by kainate: In that case, there is superficial gamma without deep beta]. Please see color insert.

(From Roopun et al., 2006 with permission.)

The Somatosensory Cortex Beta-2 Oscillation Requires Gap Junctions

Beta-2 oscillations are eliminated by the gap junction blocking compound carbenoxolone (Fig. 11.3). One especially interesting feature of this phenomenon is that the electrically uncoupled neurons are able to fire action potentials (a control for nonspecific actions of carbenoxolone), but *do not oscillate*. Thus, the beta-2 oscillation, as a population phenomenon, can also not be understood as a system of coupled cellular oscillators, just as is the case for VFO in telencephalic cortex and cerebellar cortex (Chapter 10).

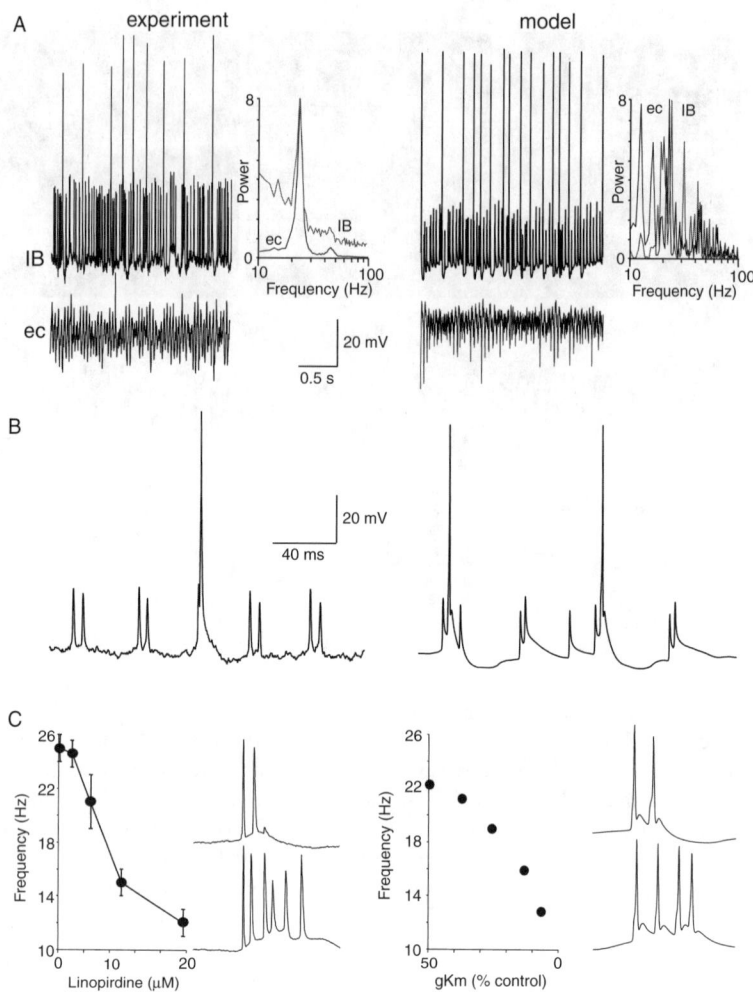

Figure 11.2 The deep beta-2 oscillation (2° somatosensory cortex) is generated by layer 5 tufted IB (intrinsic bursting) neurons, is associated with spikelets, and is modulated by the M current. **A:** In the experiment, beta-2 oscillations were induced with kainate, as in Fig. 11.1A. Beta-2 oscillations appear in an identified intrinsically bursting (IB) layer 5 neuron, and in the field (*ec*, extracellular); power spectra of the intracellular and extracellular signals are shown in the inset. The model traces are from a network model with 2000 layer 5 IB cells (as well as layer 5 RS pyramids, layer 6 pyramids, and several types of interneurons). The IB cells contained axonal gap junctions (conductance 4.0–4.5 nS), 150 μm from the soma, at an average of two lying on each axon. Ectopic spikes occurred at a mean of 4 Hz per axon. Note the abundant spikelets in experiment and model. **B:** Expanded intracellular traces from A, showing that spikelets can occur in doublets. **C:** g_{KM} regulates the frequency of the beta oscillation, in part by modulating the extent to which the IB cells burst (see also Golomb et al., 2006). In the experiment, linopirdine was used to reduce g_{KM}; insets show example intracellular records at 2 μM and 20 μM. In the model, g_{KM} could be regulated directly. The insets show traces at values of 50% and 5% of control conductance.

(From Roopun et al., 2006 with permission.)

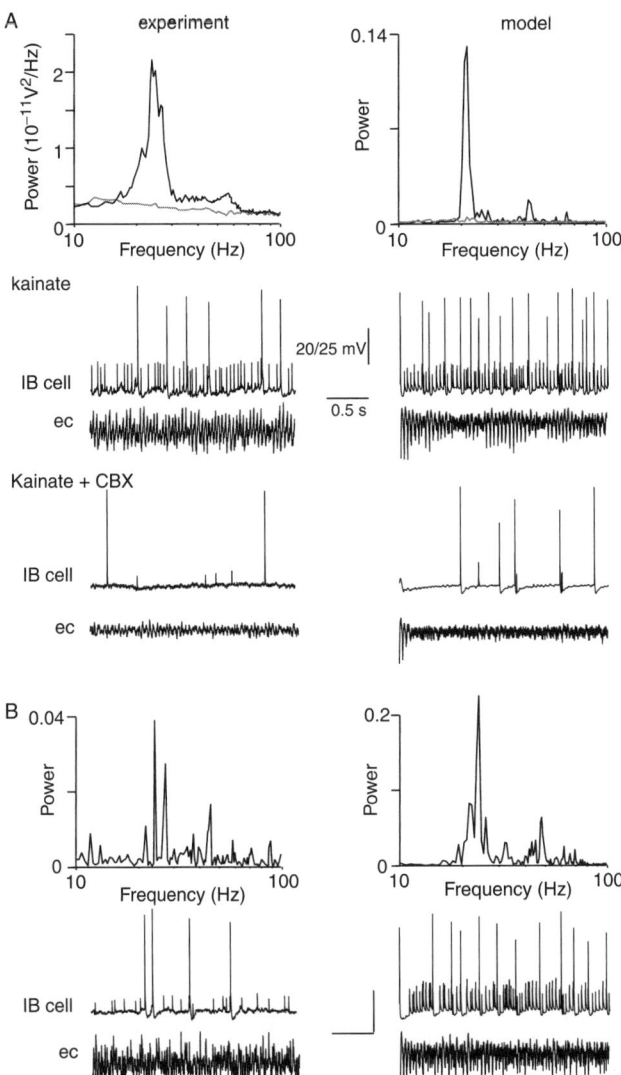

Figure 11.3 The deep beta-2 oscillation (2° somatosensory cortex) is dependent on gap junctions; the individual pyramidal cells do not act as intrinsic oscillators. In the experiments, beta-2 oscillations were evoked with kainate, as in Fig. 11.1, and carbenoxolone (cbx, 200 μM, a gap junction blocker) was added. Simulations were run as in Fig. 11.2, with and without gap junctions. Power spectra are shown in red, for the gap junction blockade conditions. **A:** Gap junction blockade abolishes the beta-2 oscillations. Individual neurons are still able to fire action potentials (suggesting that carbenoxolone is not simply acting nonspecifically), but do so at much lower rates than during the collective oscillation. If the system were simply a collection of coupled oscillators, the neurons would be expected to oscillate at beta-2 frequency after the coupling is removed. Scale bars: 20 mV (experiment), 25 mV (model), 0.5 s. **B:** Beta-2 oscillations can occur with chemical synapses blocked, provided axonal excitability is enhanced. In the experiment, phasic glutamate and GABA receptors were blocked with NBQX (20 μM), d-AP5 (50 μM), bicuculline (20 μM), and CGP55485 (10 μM); and 4AP (40 μM) was added to the bathing medium. In the model, appropriate synaptic conductances were set to zero, and membrane resistivity in the axons was doubled. Scale bars as in A.

(From Roopun et al., 2006 with permission.)

Not only are gap junctions required for beta-2, but chemical synapses appear *not* to be required, at least not beyond providing tonic excitation to the network. Figure 11.3B shows a robust beta-2 oscillation, with intracellular spikelets, occurring in medium that blocks chemical synapses (but with 4-aminopyridine (4AP) added to increase cellular—probably axonal—excitability). We hypothesize that 4AP compensates for the loss of tonic depolarization of axons, presumed to arise from blockade of $GABA_A$ receptors (see Chapter 10, as well as Fig. 11.4, and further data in Chapter 12); this hypothesis remains to be definitively proven, however.

Limited Effects of Specific Chemical Synaptic Blockers on the Beta-2 Oscillation

Figure 11.4 shows the effects of blockade of specific types of chemical synaptic receptors on beta-2 oscillations. The fact that gamma oscillations appear in the very same slices provides both a control to show that the blocking drugs are indeed active, and it provides data on gamma oscillations themselves. We can summarize the results (for the region of cortex at issue, secondary somatosensory, and for oscillation-induction with kainate), along with pointers to the next chapter, as follows:

1. Gamma oscillations in superficial cortical laminae require α-amino-3-hydroxyl-5-methyl-4-isoxazole-propionate (AMPA) receptors, but beta-2 oscillations, occurring concurrently in deep laminae, do not.

2. Gamma oscillations require N-methyl-D-aspartate (NMDA) receptors (at least in this brain region, in the kainate model of persistent cortical rhythms (see Roopun et al., 2008a), but beta-2 oscillations are actually *enhanced* by blockade of NMDA receptors. Our hypothesis is that NMDA receptors tonically depolarize some interneurons (Middleton et al., 2008), and this is why they are needed for gamma, which has an absolute dependence on interneurons. The enhancement of beta-2 with NMDA blockade is hypothetically attributed to one of two mechanisms: First is a proposed destructive interference between the oscillations, so that suppression of one can enhance the other. Second is a reduction in dendritic burst discharges, which have slower time constants than the axonally generated bursts, and which involve calcium entry through voltage-gated calcium channels. Elevating intracellular calcium ion concentration has detrimental effects on gap junction conductances (Chapter 9). Additional attenuation of dendritic electrogenesis by blockade of I_h causes massive increases in beta-2 power (Kramer et al., 2008). However, further studies on how gamma and beta-2 rhythms interact, to influence columnar processing, is difficult, as there is still no method to eliminate beta-2 without also eliminating gamma.

3. Neither neocortical gamma nor beta-2 rhythms depend on $GABA_B$ receptors.

4. It is possible to suppress gamma with partial blockade of $GABA_A$ receptors, while actually enhancing beta-2 (again suggesting destructive interference of gamma on beta2). However, more complete blockade of $GABA_A$

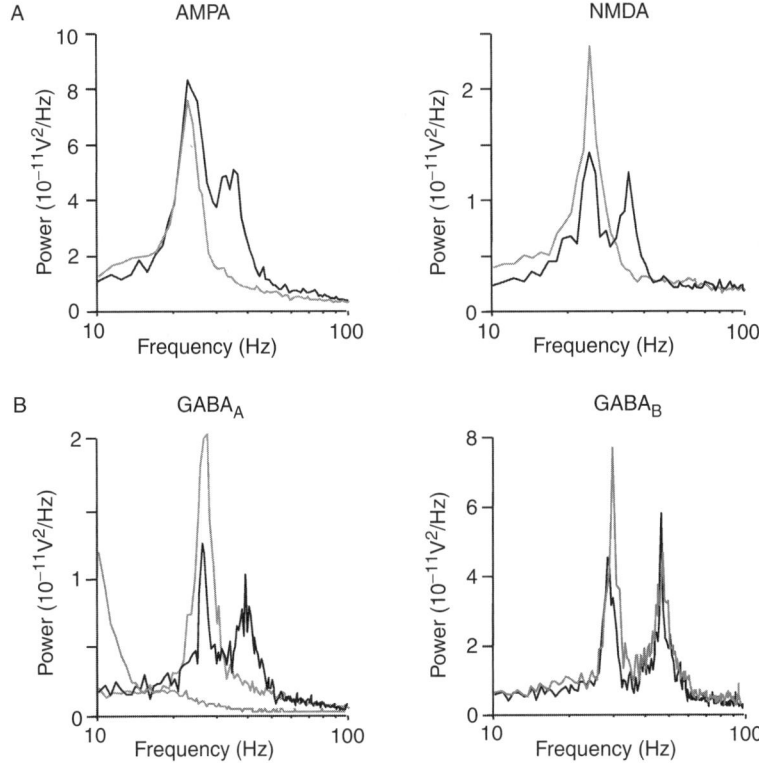

Figure 11.4 The deep beta-2 oscillation (2° somatosensory cortex) does not depend on AMPA, NMDA or GABA_B receptors; there is a weak dependence on GABA_A receptors, whose interpretation is unclear. Oscillations were evoked with kainate as in Fig. 11.1, and field potentials recorded in layer 4, so as to contain both beta-2 and gamma. Power spectra of 60-s data epochs are shown in black for control conditions, and red for drug conditions. **A:** Glutamate receptors. AMPA receptors were blocked with SYM2206 (20 μM), suppressing gamma oscillations, but leaving beta-2 intact. NMDA receptors were blocked with d-AP5 (50 μM), again suppressing gamma oscillations (probably by removing a tonic depolarization of interneurons), but actually enhancing beta-2 activity. **B:** GABA receptors. When GABA_A receptors were partially blocked with gabazine (250 nM), gamma was suppressed, but beta2 enhanced. Further blockade of GABA_A receptors (gabazine 2 μM, blue trace) led to replacement of beta-2 with epileptiform discharges. [These data, along with Fig. 11.3B, indicate that GABA_A receptors cannot be required for beta-2.] GABA_B receptors were blocked with CGP55845 (10 μM). There was little effect on gamma, and beta-2 was enhanced.

(From Roopun et al., 2006 with permission.)

receptors does eliminate persistent beta2, replacing it with epileptiform bursts. As the data in Figure 11.3B make it unlikely that GABA$_A$ receptors are inherently necessary for beta-2, we suspect that *phasic* synaptic inhibition is not the factor at issue during isolated blockade of these receptors; instead, we propose that a *tonic* effect of GABA$_A$ receptors on axons is at play (see Chapter 12). These observations raise an interesting issue regarding the control of local circuit rhythms by interneurons: the GABA$_A$ receptor blocker gabazine does not block beta-2 rhythms in somatosensory cortex at concentrations that decimate synaptic inhibition, and with which hardly any inhibitory postsynaptic potentials (IPSPs) can be seen in layer 5 IB pyramidal cells. Despite this, layer 5 fast-spiking interneurons fire reliably at beta-2 frequencies, and other deep layer pyramids (for example in layer 6) have overt beta-2-frequency trains of IPSPS during the population beta-2 rhythm (Fig. 11.5). It appears, therefore, that the non-synaptically coupled network of layer 5 IB pyramids can influence most other local neuronal populations via synaptic inhibition, but not themselves. Clues as to how this situation may arise come from observations of a peculiar form of inhibitory synaptic plasticity onto layer 5 pyramids. Tonic depolarization of these cells, as seen in kainate-induced beta-2 rhythms, rather paradoxically leads to long-term depression of inhibitory inputs.

Thus, in summary, somatosensory cortex beta-2 appears to be an exclusively gap junction mediated type of oscillation. It is therefore not completely surprising that beta-2 can occur nested with VFO (Chapter 10, Fig. 10.9). Intuitively, we picture the electrically coupled network of layer 5 IB cells as a latent VFO generator. That is, continuous VFO can occur under certain conditions, while under other conditions (as in beta-2), VFO is continually being interrupted by hyperpolarizing membrane currents. In still other conditions—as, for example, when gap junctions are not sufficiently conducting, or synaptic activities are dominating—VFO is effectively shut off. We have conducted extensive parametric explorations, varying gap junction and g_{KM} conductances, that are consistent with this picture.

Experimental in vitro beta-2 oscillations have also been described in primary motor cortex, M1 (Yamawaki et al., 2008). These oscillations have several properties in common with somatosensory cortex beta-2.

Primary Motor Cortex Beta-2, In Vitro, Also Originates in the Deep Layers

Eberhard Buhl et al. (1998) used a combination of kainate and carbachol as activating agents, in order to produce gamma oscillations in somatosensory cortex in vitro. This combination of drugs (but neither alone) also can generate beta-2 oscillations in primary motor cortex in vitro (Fig. 11.6). Analysis of phase delays, and local application of tetrodotoxin in superficial or deep layers, indicate that the M1 oscillation originates in deep layers; unlike 2° somatosensory cortex, however, M1 cortex does not exhibit gamma in the superficial layers, at least

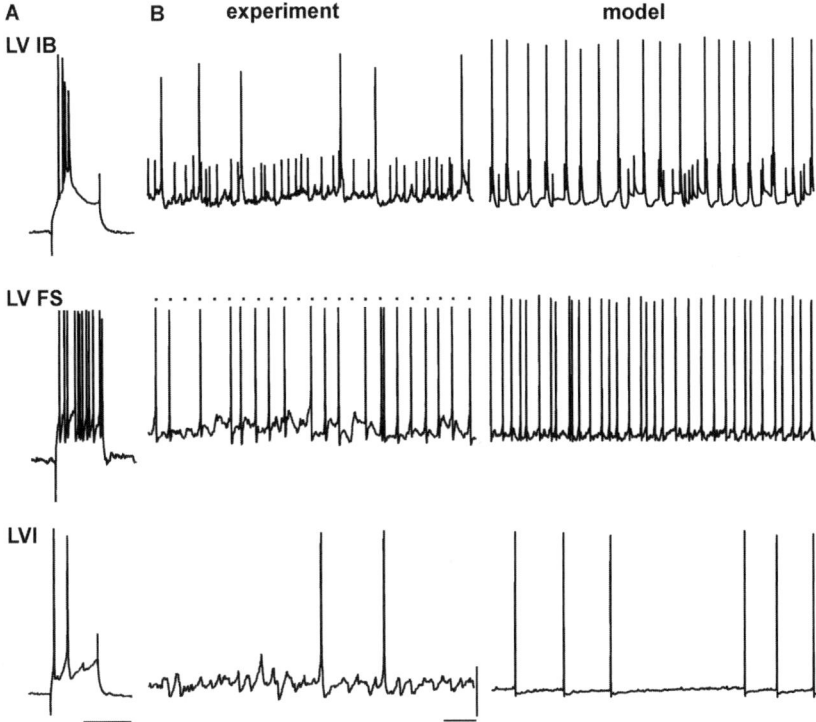

Figure 11.5 Firing patterns of deep cortical neurons during beta-2 oscillations.
A: Response to depolarizing current injection in the absence of kainate in 3 layer
V/VI neurons: layer 5 intrinsically bursting neuron (LV IB, resting membrane
potential, rmp = –70 mV), layer 5 fast spiking neuron (LV FS, rmp = –62 mV) and
layer 6 regular spiking neuron (LVI, rmp = –75 mV). **B:** Spontaneous activity
patterns of the neurons during kainate-induced beta-2 population oscillation
(experiment and simulation). Peak positivity of each period of the concurrent field
recording for FS cell is indicated by dots. Note that occasional periods are not
accompanied by spiking in this neuron, and the single period where a spike doublet
was generated. LV IB average membrane potential, amp = –58 mV, LV FS amp = –55
mV, LVI amp = –70 mV. Scale bars: **A** = 200 ms, **B** = 100 ms, 20 mV.
(From Roopun et al., 2006 with permission.)

under these experimental conditions. Instead, beta-2 is recorded in the superficial
layers, presumably relayed there from deep neurons by synaptic connections.

Pharmacology of M1 Beta-2 Oscillations

The pharmacology of M1 beta-2 resembles that of somatosensory beta-2:
specific blockade of AMPA receptors has little effect, while blocking gap
junctions eliminates it (Fig. 11.7); and blocking GABA$_B$ receptors increases

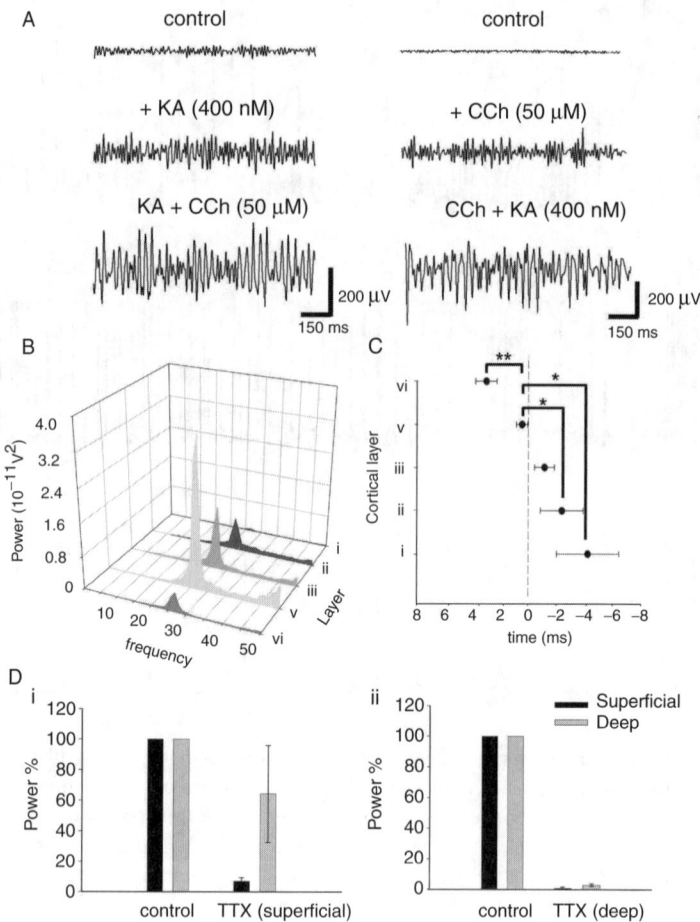

Figure 11.6 A beta-2 oscillation is also generated in vitro in the deep layers of rat primary motor cortex. **A:** Field potential recording show that neither kainate (KA) alone, nor carbachol alone (CCh) induce oscillations, but the combination does. **B:** Power spectra of the field oscillations in different layers show maximum beta-2 amplitude in layer 5. Note the absence of gamma in superficial layers (unlike 2° somatosensory cortex, see Fig. 11.1). **C:** Phase relations of beta-2 in different layers, relative to layer 5. The oscillation appears to originate in deep layers and project to superficial layers. **D:** 10 μM TTX (tetrodotoxin) was applied locally to superficial layers, where it blocked oscillations, but did not block oscillations in the deeper layers (i); however, local application of TTX to the deep layers suppressed oscillations everywhere (ii).

(From Yamawaki et al., 2008 with permission.)

the power. Antagonism of NMDA receptors only affects the frequency, and that slightly. Complete blockade of GABA$_A$ receptors with 50 μM picrotoxin likewise eliminates M1 beta-2 (Yamawaki et al., 2008); partial blockade of these receptors was not reported. The problem is one of interpretation of the picrotoxin data. If, as we have proposed above, it is tonic depolarizing GABA$_A$ receptors which are the relevant ones, then the picrotoxin results are readily explained. If, however, phasic receptors really are important, how might that come about? Recurrent excitation of interneurons, with reciprocal inhibition of pyramidal cells, seems unlikely (but not absolutely impossible), given the lack of effects of AMPA receptor blockade. Yamawaki et al. (2008) consider the possibility of an autonomous generation of the beta-2 rhythm in interneuron networks. At least some interneuron networks (of multipolar bursting cells) can oscillate at frequencies as low as the theta range (Blatow et al., 2003), although the published data concern networks in superficial cortical layers. This possibility does, however, deserve consideration. An experimental test

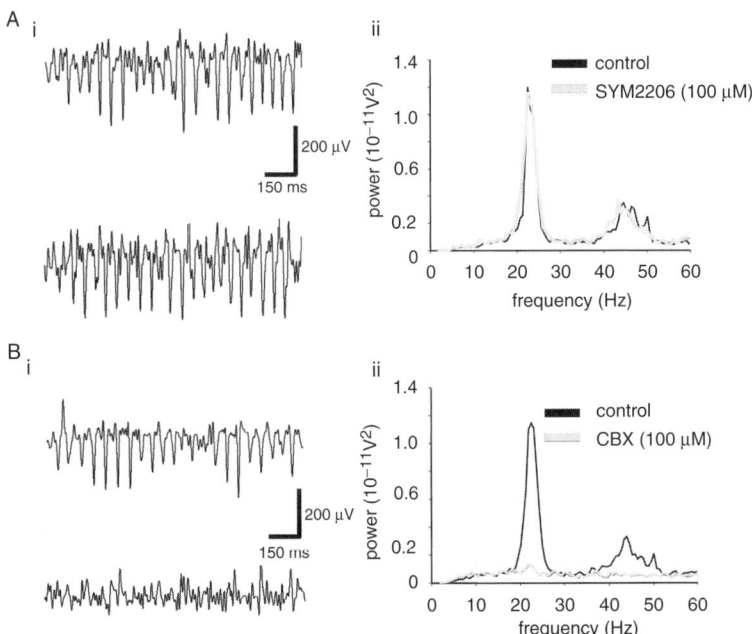

Figure 11.7   The deep beta-2 oscillation (primary motor cortex) does not depend on AMPA receptors, but does require gap junctions. Beta-2 oscillations were evoked as in Fig. 11.5. **A:** Field potential traces (i), and power spectra (ii), showing that the specific AMPA receptor blocker, SYM2206, has no effect on the oscillations (in contrast to persistent gamma oscillations, see Chapter 12). **B:** In contrast, the gap junction blocker carbenoxolone suppresses the beta-2 oscillation in primary motor cortex. [See the text for a discussion of the effects of GABA$_A$ receptor blockade.]

(From Yamawaki et al., 2008 with permission.)

might consist of a protocol similar to Fig. 11.3B: to block GABA$_A$ receptors, and then "rescue" beta-2 by enhancing excitability with 4AP or by pressure ejection of hypertonic K$^+$.

Other Types of Beta Oscillations

At least three other types of in vitro beta oscillations have been described, as follows:

1. A cholinergic cortical beta-2 oscillation in primary auditory cortex, generated by networks of IB pyramidal cells exciting low threshold spiking (LTS) interneurons, thereby providing inhibitory synaptic currents back to IB pyramids. This type of beta-2 rhythm is resistant to M-current blockade, but is sensitive to small concentrations of gabazine. It appears, therefore, that bursting in IB axonal networks may be recruited into beta-2 rhythms by inhibitory synaptic events as well as by intrinsic axonal conductances, in a manner depending on the degree of cortical neuromodulation and the specific region of cortex (M.A. Whittington & A. Roopun, unpublished data). LTS cells, in contrast to basket and other fast spiking interneurons, appear to be able to provide pyramidal cells with inhibition at frequencies lower than gamma (Gibson et al., 1999; Roopun et al., 2008b; Wang et al., 2004).

2. A beta oscillation in the CA1 region of the hippocampus in vitro, in the presence of serotonin (Bibbig et al., 2007). This oscillation is generated roughly this way: in the presence of the metabotropic glutamate receptor agonist DHPG, the CA3 region generates gamma oscillations that are "projected" to CA1 via the Schaffer collaterals; the major synaptic link in CA1 consists of synaptic feed-forward excitation of interneurons, that then impose gamma frequency IPSPs on CA1 neurons. In the presence of serotonin, some CA1 interneurons are recurrently excited by CA1 pyramidal cells, so that they fire extra spikes—spikes that do not result directly from the Schaffer input. The extra interneuronal spikes lead to an overall slowing of the CA1 oscillation. The resulting change in frequency is, in essence, a nonreciprocal analog of the inter-areal interactions leading to gamma–beta transitions discussed in our first monograph (Traub et al., 1999a).

3. A neocortical beta-1 oscillation that results from a rather intricate interaction between gamma and beta-2 network oscillators, by a process called concatenation (Kramer et al., 2008; Roopun et al., 2008b).

Thus, it appears that there are multiple forms of beta rhythm in the cortex, and that most have close associations, in one form or other, with prior or ongoing gamma rhythms. As more and more frequency bands within the EEG range are discovered to have clearly defined and distinct mechanisms (Roopun et al., 2009) one has to ask the question: why? Robust coexistence of gamma and beta-2 rhythms provides an ideal starting point to address this issue of multiple coexistent frequency bands. Figure 11.8 shows the pattern of temporal interactions between deep and superficial layers—generating beta-2 and gamma frequency oscillations respectively. With strong glutamatergic excitation of somatosensory cortex, the two rhythms have no stable phase

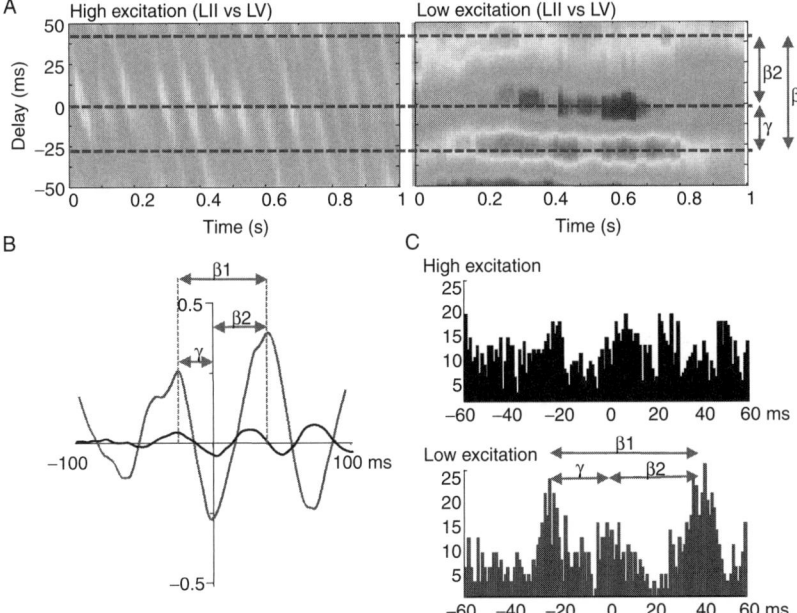

Figure 11.8   Asymmetric local field potential (LFP) and spike cross-correlations suggest period concatenation across laminae. **A:** Cross correlograms from LFP data. During coexpressed gamma and beta-2 rhythms (high excitation condition), the phase relationship between LII and LV varied rhythmically at beta-1 frequency. When beta-1 rhythms were expressed in the local field potential from LII and LV, a phase relationship was seen with LV leading and lagging LII by one gamma and one beta-2 period, respectively. **B:** Mean cross-correlation function between LII and LV (60 epochs of data, $n = 5$ slices) for coexpressed gamma and beta2 field potentials (black) and beta-1 rhythms (red) illustrating the generation of a stable, asymmetric temporal relationship between laminae on reduction in glutamatergic drive. **C:** Graphs show distribution of LV units with respect to adjacent LII units during coexpressed gamma/beta-2 (black) and beta-1 (red) field potential rhythms. Please see color insert. (From Roopun et al., 2008b with permission.)

relationship, suggesting that the beta-2 rhythm is operating to disconnect the main input areas in a cortical column (layer 4 and above) from the primary descending output area (layer 5). However, this situation changes once excitation is reduced. In this case, both the beta-2 and gamma periods are present in individual neurons, but they are organized such that one gamma period is followed by one beta-2 period, repeatedly. The population rhythm in this case is manifest as the sum of these two original periods—resulting in a slower beta1 frequency rhythm. Theoretically at least, such interactions between co-expressed frequencies may occur throughout the dynamic range of the EEG, placing gap junction-mediated network phenomena at the very heart of cortical dynamics. This intriguing idea remains to be shown experimentally.

12

Persistent Gamma Oscillations

Persistent gamma oscillations in vitro were first reported by Fisahn et al. in 1998, who applied carbachol to the solution bathing a hippocampal slice, and who found the maximum amplitude of the oscillations in the CA3 region. The phenomenon was dubbed "persistent oscillations" (alternatively, "pharmacological oscillations") because gamma activity would continue as long as carbachol was in the bath, and the slice remained viable: in other words, for many hours. Carbachol had previously been applied to hippocampal slices, with induction of theta frequency activity that was weakly dependent on (or not at all dependent on) $GABA_A$ receptors (Konopacki et al., 1987; MacVicar & Tse, 1989; Traub et al., 1992; Williams et al., 1997), without observation of gamma oscillations. Eberhard Buhl maintained that the generation of persistent gamma depended upon optimal preservation of the inhibitory circuitry, and for that reason perfused the experimental animals with sucrose, before removal of the brain and slice preparation; perhaps technical considerations, then, explain why gamma had not been previously observed.

Since the original description of persistent, cholinergically induced, gamma in the hippocampus, the experimental model has been expanded in two ways: in finding a whole range of compounds whose bath application induces the oscillation, compounds all of which have metabotropic actions (e.g., kainate, the metabotropic glutamate receptor agonist (S)-3,5-dihydroxyphenylglycine (DHPG), and domoate); and in finding many other brain regions that support the oscillation. Thus, Figure 6.8 of Chapter 6 illustrates a number of neocortical regions that exhibit persistent gamma in vitro, but so also does medial entorhinal cortex (Cunningham et al., 2003), cerebellum (Middleton et al., 2008), hippocampal CA1 (Gillies et al., 2002), and amygdala (Randal, Whittington & Cunningham, unpublished data). "Somewhat" persistent (lasting seconds) gamma oscillations can as well be evoked by pressure ejection of hypertonic K^+ solution (LeBeau et al., 2002; Towers et al., 2002), or of kainate (Gloveli et al., 2005a). Persistent oscillations, produced by bath application of a drug, work best in an interface chamber, which limits the ability to perform patch recordings; oscillations evoked by pressure ejection, however, work in submerged slices, so that patch recordings can be obtained.

The discovery of persistent gamma transformed the discipline of neuronal oscillations. First, the oscillations are extremely robust, making possible a great variety of experiments aimed at uncovering how they work. Second, the oscillations possess a number of features in their phenomenology which—as we shall review in the text that follows—suggest that their mechanisms may be similar to those of several sorts of gamma that occur in vivo, most particularly hippocampal gamma that occurs during the theta state (including active exploration in rats), and probably also the gamma oscillations that occur during the upstates of slow-wave sleep.

In addition, as experiments and modeling of persistent gamma oscillations proceeded, it became clear that in vitro gamma oscillations have—or at least are predicted to have—a property that is truly remarkable: many of the action potentials in principal cells are predicted to be *antidromic*, a property that derives from the intimate association between persistent gamma and VFO, and the role of gap junctions and axons in the generation of the oscillations. If this prediction proves to be true in vivo, or even partly true, it will have revolutionary consequences for our understanding of brain function, at least during certain behavioral states.

What does it all come down to, mechanistically? Amid all the detail that we must shortly delve into, the reader must keep one fact conscious: persistent gamma oscillations occur in conditions where pyramidal cells are not strongly depolarized, or are not even depolarized at all; but at the same time, persistent gamma requires that interneurons be driven to fire by phasic excitation of interneurons. The latter requirement implies that interneurons cannot be too hyperpolarized, nor can they be so depolarized that they fire without phasic synaptic excitatory input. At the same time, persistent gamma is extraordinarily stable, hence can not be sensitive to small parameter changes. Nature accomplishes this feat by having persistent activity, in the pyramidal cell axonal plexus, phasically drive the interneurons; while the interneurons also are not strongly depolarized tonically. Gap junctions keep the axonal plexus active, and very fast oscillations (VFO) appear possibly as an epiphenomenon. Our task is to explain how these pieces of network activity fit together.

First, however, let us review some of the history of gamma, already alluded to in the introduction to this book. Experimental in vitro models of gamma oscillations that were available in 1998, at the time the Fisahn et al. (1998) paper appeared, were these:

1. Interneuron network gamma (ING), in which drugs were used to block phasic excitatory postsynaptic potentials (EPSPs) in interneurons— "pharmacologically isolated interneurons", so-called—while at the same time, providing a source of *tonic* depolarization to the interneurons. [Tonic depolarization was typically achieved by activation of metabotropic glutamate receptors (Whittington et al., 1995; Traub et al., 1996b).] Interneurons continued to interact with each other phasically via mutual synaptic inhibition, and via gap junctions, with the latter playing a role that—experimentally— cannot be neglected (Traub et al., 2001b). Models used to account for ING [including, but not limited to, the citations above, as well as Wang & Buzsáki (1996) and White et al. (1998)] are, at least in part, descendants of earlier models put forth to explain spindle oscillations, putatively arising through mutual synaptic inhibition amongst GABAergic nucleus reticularis thalami neurons (Golomb et al., 1994; Steriade et al., 1987). Variations of the ING models have been proposed recently, using very rapid inhibitory postsynaptic conductance (IPSC) kinetics, and relatively depolarizing inhibitory postsynaptic potential (IPSP) reversal potentials (Bartos et al., 2001, 2002, 2007). With ING, the depolarization of pyramidal cells is not at issue, as these cells do not participate in generating the oscillation, although they "record" the oscillation, in the sense that rhythmical IPSPs in pyramidal cells reflect what the interneuron network is up to. It seems unlikely that ING occurs in vivo, in hippocampus and cortex, under physiological conditions, given the powerful recurrent excitation from pyramidal cells; however, there are regions of the brain that contain extensive networks of GABAergic cells, without recurrent synaptic excitation (at least in the classic sense)—in basal ganglia, cerebellar cortex, nucleus reticularis thalami—and ING might well occur in these regions in vivo, particularly the cerebellum. It should also be noted that all experimental instances of ING, in cortical structures, involve a transient generation of gamma rhythms, lasting no more than a few seconds. Whether or not one considers possible a physiological condition in which fast excitatory synaptic drive to interneurons is absent, evidence indicates that such a situation cannot explain long-lasting epochs of gamma rhythms, which are often seen during complex cognitive tasks in vivo.

2. Tetanic, or stimulus-induced, gamma, which is also transient, but which (unlike ING) does depend on phasic α-amino-3-hydroxyl-5-methyl-4-isoxazole-propionate (AMPA) receptor-mediated excitation of interneurons, and on recurrent synaptic inhibition of pyramidal cells (Traub et al., 1996c; Whittington et al., 1997a, 2001). During this type of oscillation, pyramidal cells are strongly (albeit transiently) depolarized, by about 10 to 20 mV (Whittington et al., 1997). Interneurons are depolarized as well (average approximately17 mV) but, in spite of this, the oscillation is suppressed by the AMPA/kainate receptor blocker NBQX. When a tetanic stimulus is applied in

the absence of neuromodulatory receptors, such as those for serotonin, then pyramidal cell somata fire action potentials at, or near, gamma frequencies. This experimental model captures many of the features of sensory-evoked cortical oscillations in vivo, including the ability to synchronize over long distances (many mm), and to switch to a transient beta oscillation [Traub et al., 1999c; Whittington et al., 1997b—reviewed in our previous monograph (Traub et al., 1999a)].

In contrast, during persistent gamma oscillations, neither pyramidal cells nor interneurons become significantly tonically depolarized: by less than 5 mV for pyramidal cells, and by 0 to 1 mV for interneurons (Traub et al., 2000); yet phasic synaptic excitation is required for the oscillations and—because excitatory postsynaptic potentials (EPSCs) in pyramidal cells are so small during persistent gamma (Fisahn et al., 1998)—it is presumed that it is phasic excitation of *interneurons* (and not of pyramidal cells) that drives the oscillations. (Other data as well, reviewed later, support this latter conclusion.) These are the cardinal facts about persistent gamma for the reader to keep in mind for this chapter.

In the next part of the chapter, we review data on hippocampal persistent gamma (with some examples from neocortex, where the mechanisms are mostly similar), with the primary aim being to understand the most basic mechanisms. Some data has been illustrated already (in Chapter 10) documenting the coexistence of VFO and gamma, but we shall need to expand on the theme. After that, we will consider those features of persistent gamma, in auditory neocortex, which are different from hippocampal gamma.

Persistent Gamma Depends on Phasic Synaptic Inhibition and Synaptic Excitation

The data in Figure 12.1 derive from Fisahn et al. (1998). Parts a and b document not only a requirement for $GABA_A$ receptors (because bicuculline suppresses the oscillation), but also indicate that at least part of this requirement derives from an action on *phasic* inhibitory currents, so that bicuculline is not simply suppressing axonal excitability (see later). The reason for this assertion is that, as shown in part b, pentobarbital slows the frequency of the oscillation. Pentobarbital prolongs the time course of unitary IPSCs in hippocampal neurons (Segal & Barker, 1984). If the major effect of $GABA_A$ receptors were to be tonic rather than phasic, one might expect pentobarbital to enhance gamma power, but without a change in frequency.

As Figure 12.1c demonstrates, carbachol-induced gamma oscillations are also suppressed by the AMPA/kainate receptors: but, are AMPA or kainate receptors the important ones? This question has been addressed by showing that persistent gamma oscillations are suppressed by the specific AMPA receptor blocker SYM2206 (Fig. 12.2). This is the case for gamma evoked by carbachol

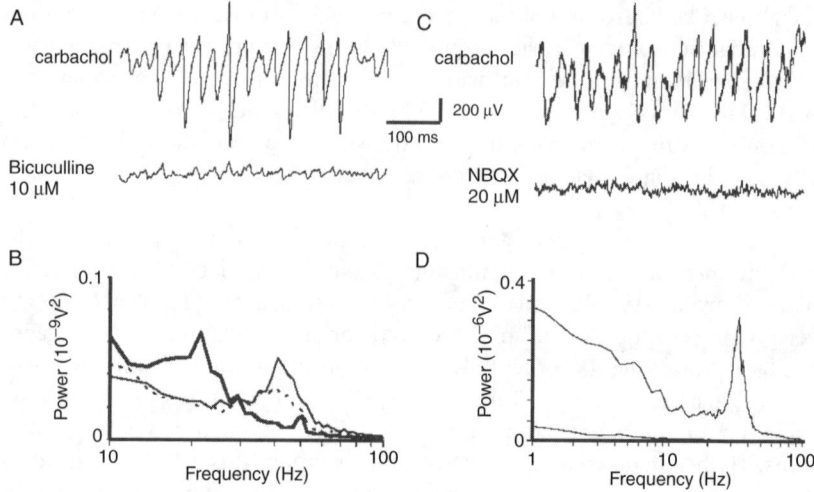

Figure 12.1   Carbachol-induced persistent gamma oscillations in the hippocampal slice, CA3. **A:** Field potential oscillations (in stratum radiatum) induced by carbachol are abolished by bicuculline. The latter GABA$_A$ receptor blocker is acting on phasic synaptic inhibition, at least in part, as shown in *b*, and also because (shown in the original paper, and see Fig. 12.4) gamma frequency IPSCs/IPSPs do exist in pyramidal cells in this experimental model. **B:** Power spectrum of the gamma oscillation (thin line) has a peak at ~40 Hz. The GABA$_A$ agonist pentobarbital (5–20 μM) (thick line) slows the oscillation to ~20 Hz, reversibly (dotted line). **C:** Carbachol gamma is blocked by the AMPA/kainate antagonist NBQX (power spectra below are for the baseline oscillation, and after addition of the antagonist). NBQX is probably acting via block of AMPA receptors on interneurons: The block depends on AMPA receptors, rather than kainate receptors (next figure), and EPSCs in pyramidal cells are small, <50 pA (Fig. 3 of the original paper).

(From Fisahn et al., 1998 with permission.)

or kainate (Fig. 12.2), and also by DHPG in CA1 (Gillies et al., 2002). Another specific AMPA receptor blocker, GYKI 52466 has been used to suppress gamma oscillations evoked by carbachol in CA3 (Fisahn et al., 1998; Mann et al., 2005). Mann et al. (2005) also applied GYKI locally in different hippocampal layers: it was effective near stratum pyramidale, but not in stratum radiatum or molecular/lacunosum, therefore suggesting that phasic excitation of perisomatic-targeting interneurons is the key. Finally, Fuchs et al. (2007), reduced the amplitude of EPSCs in interneurons specifically (and not in pyramidal cells), for example, in a knockout of the glutamate receptor subtype GluR-D (GluR4)—a subtype found almost exclusively in interneurons (see also Chapter 6, where the phenotype of this mouse is discussed). In the knockout mouse, gamma power (elicited by kainate) was reduced in CA3; thus, it appears that phasic excitation of *interneurons* is what

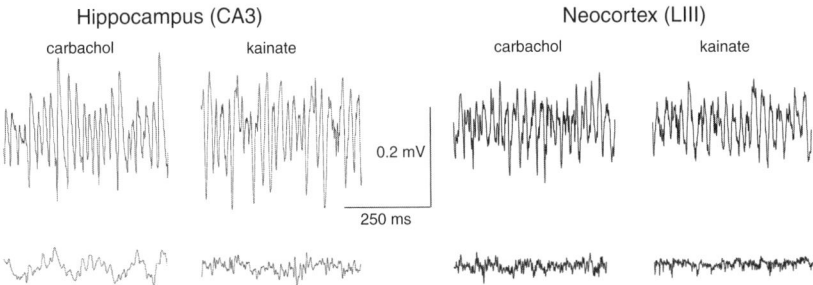

Figure 12.2 Persistent gamma oscillations, induced by carbachol and kainate, in hippocampus (CA3) and superficial layers of neocortex, require AMPA receptors. Traces show field potential oscillations induced by carbachol or kainate (above), and after addition of the specific AMPA receptor blocker SYM2206 (20 μM, below). M.A. Whittington, unpublished data. [Similar findings have been described for persistent gamma oscillations in secondary somatosensory cortex (Roopun et al., 2008); medial entorhinal cortex (Cunningham et al., 2003); and anterior cingulate and agranular insular cortices (Anita Roopun, Ph.D. thesis, University of Leeds, 2005). AMPA receptors are required in each case.]

is essential, not phasic excitation of the pyramidal cells by each other. (See also Morita et al., 2008.)

Carbachol-Induced Persistent Gamma Depends on Activation of M1 Muscarinic Receptors

In vitro gamma oscillations in the cerebellar cortex are induced by nicotine (Middleton et al., 2008), but it is muscarinic receptors (specifically, M1 receptors) that are required in the hippocampal slice (Fig. 12.3; note also the VFO peak in the power spectrum). Fisahn et al. (2002) subsequently showed carbachol would not induce gamma oscillations in hippocampal slices from an M1 receptor knockout mouse. In that study also, it was shown that M1 receptor activation increased both the h-current and a nonspecific cation current, without affecting g_{KM}. It is possible that other actions occur as well.

In the rat hippocampus in vivo, gamma oscillations appear to originate in the CA3 region and project to CA1 (Csicsvari et al., 2003). The same is true with the in vitro model. Not only is gamma power higher in CA3 (Fig. 12.3), but after a cut between the CA3 and CA1 regions, gamma oscillations persist in CA3, but disappear in CA1. DHPG-induced gamma oscillations likewise originate in CA3 and are projected to CA1 (Bibbig et al., 2007). [Gamma oscillations can, however, be induced in the isolated CA1 region with sufficiently high concentrations of kainate (Traub et al., 2003b).]

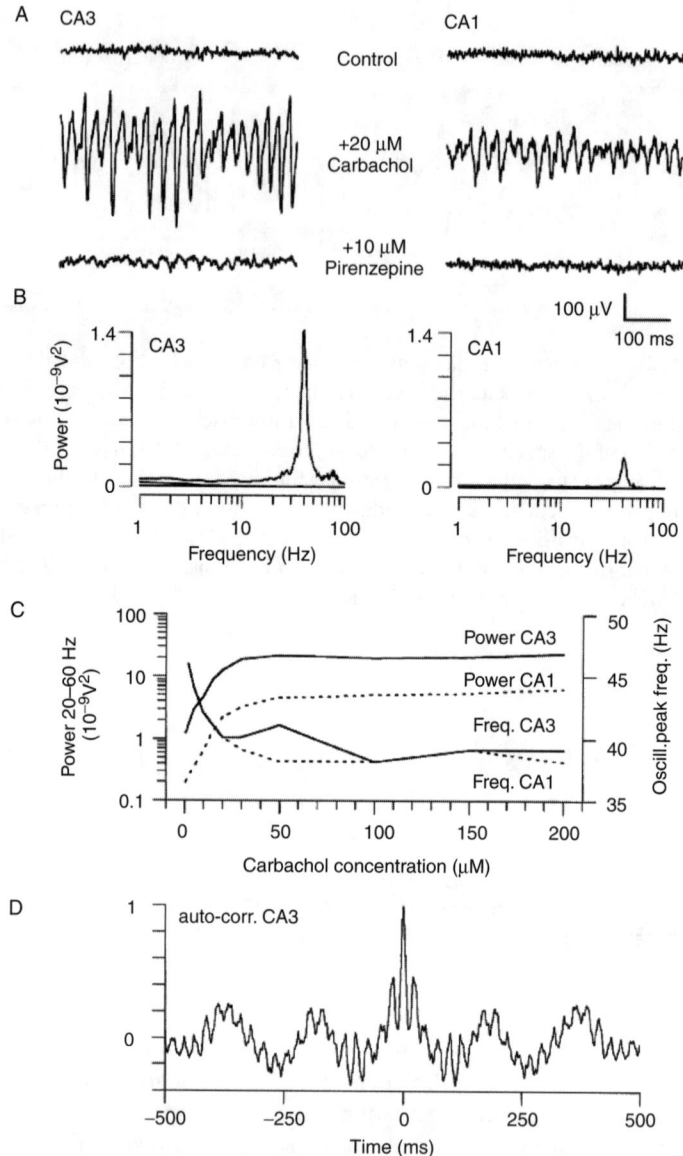

Figure 12.3 Carbachol-induced gamma oscillations depend on muscarinic receptors, and can be nested with theta-frequency oscillations. **A:** Field potentials in hippocampal CA3 (left) and CA1 (right), in control conditions, after addition of carbachol, and after the further addition of the M1 receptor antagonist pirenzapine. **B:** Power spectra (50 seconds of data) quantifying the block of the oscillations by pirenzapine. Note the VFO peak in CA3, also suppressed by pirenzapine. **D:** Nesting of gamma and theta [as also occurs in the hippocampus in vivo (Bragin et al., 1995; Soltesz & Deschênes, 1993)]. The muscarinic nature of hippocampal persistent gamma (after carbachol) is distinct from the nicotinic receptor-induced gamma of cerebellar cortex (Middleton et al., 2008).

(From Fisahn et al., 1998 with permission.)

In the in vivo rat hippocampus, gamma oscillations are especially promi-nent during behavioral states, or anesthetic states, associated with the theta rhythm, wherein gamma waves are superimposed upon, and amplitude-modulated by, theta waves (Bragin et al., 1995; Penttonen et al., 1998; Soltesz & Deschênes, 1993). At least on occasion, gamma waves also appear on theta waves in CA3 carbachol gamma (Fig. 12.3); a similar phenomenon appears in the CA1 region in vitro, in the presence of DHPG (Gillies et al., 2002).

Persistent Gamma Requires Gap Junctions

Perhaps it is not surprising that a form of gamma oscillation should depend on both synaptic excitation and inhibition, given the historical background of oscillation models based on recurrent circuitry between principal cells and interneurons (Freeman, 1979; Traub et al., 1996c; Wilson & Bower, 1992). It was, however, surprising to discover that persistent gamma oscillations require gap junctions (Fig. 12.4), shown first with octanol, but subsequently with other drugs and in other brain regions (see later). The observation of gap junction dependence led naturally to attempts to define better *which* gap junc-tions (whether between interneurons, between principal cells, or both) might be important, and in what particular ways.

The Gap Junctions Required for Persistent Gamma Oscillations Are Those Between Pyramidal Cells

Electrical coupling has been found to be extremely rare between interneurons, in the brains of a connexin36 knockout mouse (Hormuzdi et al., 2001). Such mice, however, nevertheless exhibit persistent gamma oscillations in hip-pocampal slices, in the presence of kainate (Fig. 12.5). [The oscillations are interspersed with synchronized burst discharges, which we have attributed to time-dependent alterations in synaptic potentials, as investigated with DHPG-induced gamma/sharp waves (Traub et al., 2005c). The sharp waves, do not affect our argument.] The data in Figure 12.5 demonstrate that persis-tent gamma oscillations still occur, even with greatly reduced electrical cou-pling between interneurons. The oscillations nevertheless are suppressed by carbenoxolone. The simplest interpretation of these data is that *some other* (*i.e., not between interneurons*) form of electrical coupling is necessary, pre-sumably between principal neurons. This idea seems particularly plausible, as low-calcium approximately 200 Hz ripples (which are produced by electrical coupling between pyramidal cells (see Chapter 10), apparently between their axons) are unchanged in the connexin36 knockout mouse, as compared with wild type mice (Hormuzdi et al., 2001). In other words: *pyramidal cells are electrically coupled in the connexin-36 knockout.*

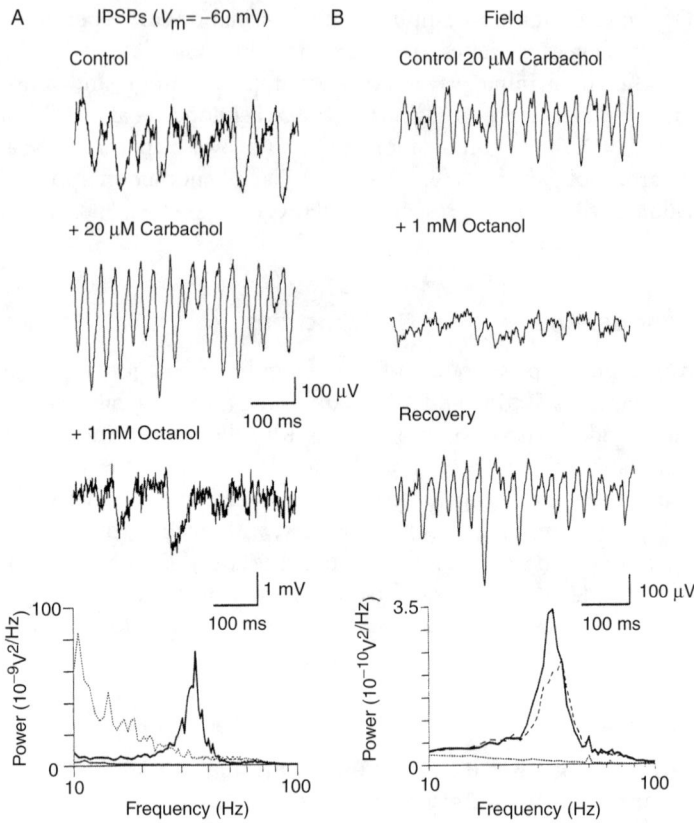

Figure 12.4 Persistent gamma oscillations depend on gap junctions. Oscillations were evoked in the CA3 region in vitro with 20 μM carbachol, and were recorded intracellularly (in a pyramidal cell) to show the gamma-frequency IPSPs (**A**), or with a field electrode (at the stratum radiatum/stratum lacunosum-moleculare border) (**B**). In each case, octanol suppressed the gamma activity. **B** also shows the reversibility of the block by octanol (dashed line in the power spectrum is for the recovery condition).

(From Traub et al., 2000 with permission.)

Here, a difficulty arises, for which we have no ready answer: in Chapter 9, we illustrated a gap junction on a mossy fiber axon, which contained connexin-36. One wonders then, how it is that principal cell axons can be coupled in a connexin-36 knockout? Nevertheless, the data convincingly tell us that the axons are indeed coupled. Presumably—although this remains to be proved experimentally—other gap junction proteins are upregulated in the connexin-36 knockout, between principal neurons, but not between interneurons.

How then can we account for the principal features of persistent gamma: involvement of synaptic excitation and inhibition, along with the need for gap junctions between pyramidal neurons? We shall now illustrate how a network

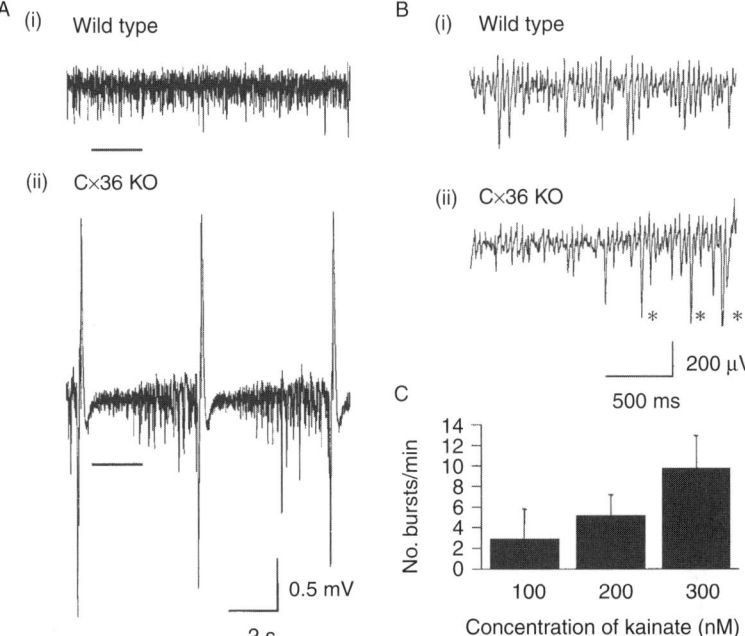

Figure 12.5 Persistent gamma oscillations depend on gap junctions other than those on interneurons, presumably axonal gap junctions. Data is from mouse hippocampal slices derived from a connexin36 knockout mouse, in which electrical coupling between interneurons was virtually absent (Hormuzdi et al., 2001). **A:** Field potential data. Kainate in tissue from this knockout evokes a mixture of gamma oscillations and sharp waves (i). The gap junction blocker carbenoxolone suppresses both the gamma and the sharp waves (**Aii, iii, B, C**). The molecular composition of putative axonal gap junctions in this knockout is not known; in wild-type rats, an axonal gap junction was shown to contain connexin-36 (Hamzei-Sichani et al., 2007). (From Pais et al., 2003 with permission.)

model can—in a straightforward manner—account for these features, and other experimental data as well.

During Persistent Gamma, Pyramidal Cell Firing Is Sparse, but (on average) Leads the Firing of Fast-Spiking Interneurons by a Few Milliseconds

The data in Figure 12.6 derive from a network model, containing pyramidal cells as well as perisomatic-contacting and dendrite-contacting interneurons, along with chemical synaptic excitation and inhibition, as well as gap junctions between the axons of pyramidal cells. The model did not contain gap junctions between interneurons. First, we shall list the important phenomenologies in

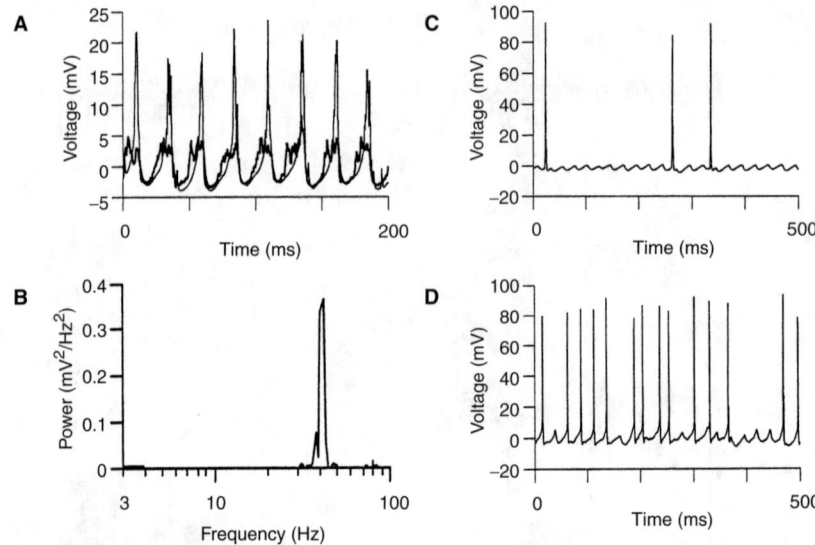

Figure 12.6 During persistent gamma oscillations, pyramidal neurons fire rarely, while fast-spiking interneurons fire >50% of the waves. [Simulation data are shown, but physiological data are similar.] **A:** Local average of pyramidal cell somatic voltages (thick line) and interneurons (thin line). The pyramidal cell activity is temporally more dispersed than for interneurons, and leads the interneurons by a few ms (as in vivo (Csicsvari et al., 2003) and in vitro (Fisahn et al., 1998; Hajos et al., 2004; Mann et al., 2005). **B:** Power spectrum of the mean pyramidal cell signal reveals a peak at 42 Hz. **C:** Somatic voltage of a single pyramidal cell shows rare firing on a background of rhythmic synaptic potentials, mostly IPSPs. (The action potentials are all antidromic, in simulations.) **D:** An interneuron fires on more than half the gamma waves. Synaptic potentials are mostly EPSPs. In this model, the pyramidal cell axonal plexus is the source of EPSPs to interneurons; somatic action potentials in pyramidal cells are not relevant.

(From Traub et al., 2000 with permission.)

the network behavior (both in simulations and experiments), and then provide an account of the basic physical ideas.

1. The timing of pyramidal cell firing is more broadly distributed than for the firing of fast-spiking interneurons (Fig. 12.6A), but (on average), pyramidal cell spikes lead interneuron spikes by a couple of ms (as occurs in vitro and in vivo).
2. Pyramidal cells fire infrequently, on a background of gamma-frequency synaptic potentials (Fig. 12.6C), as occurs experimentally (Fisahn et al., 1998).
3. Fast-spiking interneurons fire at near-gamma frequencies (Fig. 12.6D).

The original source of Figure 12.6 (Traub et al., 2000) further demonstrates that the network model requires synaptic excitation and inhibition, as well as gap junctions, just as the experiments. How does this work? What allows the model to perform so well is a mechanism that—in retrospect, at least—appears utterly simple. The predominant synaptic potentials in pyramidal cells are IPSPs, so it follows that interneurons must, on average, be firing at gamma frequencies. The interneurons are presumably being driven by phasic EPSPs (hence the experimental requirement for AMPA receptors). Where do these EPSPs come from, given the sparse firing of pyramidal cells? One way to produce the EPSPs would be for the pyramidal cell axonal plexus to generate waves of activity, similar to brief ripple oscillations. Such waves would, on the one hand, require gap junctions (explaining the experimental requirement for gap junctions for persistent gamma rhythms). On the other hand, the waves of axonal activity could function to excite the interneurons, without pyramidal cell *somata* needing to fire very much—indeed, in principle, they would not need to fire at all. [This scheme predicts that synaptic excitation to interneurons should contain high-frequency components, which turns to be true (see later); this scheme likewise shows why there is a VFO component to the power spectrum of persistent gamma, at least if the field is recorded in the right place; see later. The somatic averages in Fig. 12.6A are not placed appropriately for the demonstration of VFO.] Nevertheless: why is there a network gamma oscillation, if the pyramidal cell axon plexus "wants" to drive the interneurons in a continuously rippling fashion? The answer appears to be this: dye-coupling data (Schmitz et al., 2001; Chapter 9) indicate that axonal gap junctions in the hippocampus are within about 150 μm of the soma. That is close enough for the gap junction sites to "feel" shunting effects of synaptic inhibition on the soma and axon initial segment. Stated another way: perisomatic IPSPs can block the propagation of spikes from axon to axon, and hence interrupt VFO.

Thus, we visualize gamma as arising from alternating (1) very brief epochs of VFO (generated by the axonal plexus, and phasically exciting interneurons), and (2) population IPSPs, which interrupt the VFO and give rise to the gamma periods (tens of milliseconds). The scheme is somewhat analogous to the beta-2 oscillation, with here a population IPSP playing a role for gamma, a role that was played by intrinsic g_{KM} for the beta-2 oscillation (Chapter 11). In each case, a hyperpolarizing/shunting process repeatedly interrupts VFO, but this process is synaptic in one case, and intrinsic to the principal cell membrane in the other.

Interneuron Gap Junctions Modulate the Power of Persistent Gamma Oscillations

As we have noted, the network model, whose output is illustrated in Figure 12.6, did not contain gap junctions between interneurons. What role might these play? Hormuzdi et al. (2001) reported that hippocampal slices from connexin-36 knockout mice, when bathed in kainate, could still generate persistent

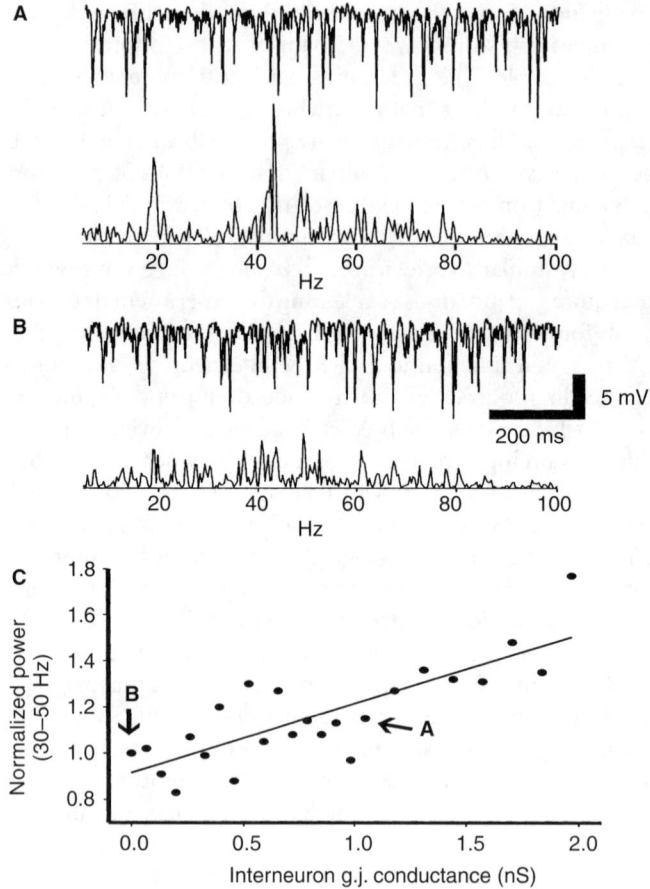

Figure 12.7 A network model of persistent gamma oscillations accounts for the modulatory effect of gap junctions between interneurons [as indicated by connexin-36 knockout data of Hormuzdi et al. (2007)]. The model included excitatory and inhibitory synaptic conductances, gap junctions between the axons of pyramidal cells, and gap junctions between the dendrites of interneurons. Gap junctions between pyramidal cells were an absolute requirement for gamma oscillations, when pyramidal and inhibitory neurons were not strongly depolarized (as is the case with persistent gamma oscillations). If pyramidal cell electrical coupling was present, gamma oscillations were present, with power modulated by interneuron electrical coupling. **A, B:** Examples of "fields" (actually the inverted somatic potential average), and power spectra, for strong and weak interneuron electrical coupling. [Note that this measure of population activity is not suitable for revealing VFO.] **C:** Summary data, showing the correlation between interneuron electrical coupling and power between 30 and 50 Hz. Interneuron electrical coupling "tightened" interneuronal activity and led to additional interneuronal firing, hence to large IPSPs in pyramidal cells.

(From Traub et al., 2003c with permission.)

gamma oscillations, a finding that was confirmed in vivo with recording of hippocampal gamma activity in the same knockout (D.L. Buhl et al., 2003); the power of the gamma oscillations, however, was reduced in the connexin-36 knockout, by roughly 50%. As Figure 12.7 shows, if interneuron gap junctions are incorporated into the network model, they can have a modulatory effect on persistent gamma oscillations, by regulating the extent and synchrony of interneuron firing; but, as is true experimentally, interneuron gap junctions are not essential for gamma to occur at all. It is the electrical coupling between pyramidal cells that is absolutely critical.

GABA$_A$ Receptors Can Excite the Pyramidal Cell Axonal Plexus to Generate VFO

In Chapter 10 (Fig. 10.9), we provided evidence that the CA1 pyramidal cell axonal plexus, excited by kainate, could generate VFO. Interestingly, GABA itself—by a tonic effect—can also generate VFO in this same CA1 stratum oriens minislice preparation (Fig. 12.8). This action of GABA is mediated by GABA$_A$ receptors (Fig. 12.8b,d), and is strongly attenuated (but not eliminated) by the gap junction blocker carbenoxolone (Fig. 12.8c,d). VFO elicited in this way may be related to kainate VFO, because kainate is known to excite the axons of hippocampal interneurons (Semyanov & Kullmann, 2001), and could thereby cause GABA spillover. It would be interesting to know if carbachol and DHPG also excite interneuron axons, perhaps providing a common mechanism for the eliciting of persistent gamma oscillations.

IPSPs Repeatedly Interrupt VFO, to Generate Persistent Gamma Oscillations

Figure 12.9 illustrates further our theme of interacting VFO and phasic synaptic inhibition (tonic effects of GABA are presumed to be acting in the "background," so to speak). Figure 12.9a shows the behavior of a population of pyramidal cells, electrically coupled via their axons, with no other interactions between cells. In the presence of spontaneous axonal spiking, a continuous VFO arises, such as we have described in Chapter 10. When, however, exogenous perisomatic IPSPs are delivered to the pyramidal cells, at 40 Hz, the VFO is continually interrupted, so that only small packets of VFO can occur, separated by intervals of a few tens of ms. The point here is to illustrate a physical principle, the interruption of VFO. During actual persistent gamma, the IPSPs are not exogenous, but are generated by the network itself; Figure 12.6 documents that such a generation indeed can happen in a network model. Figure 12.9b shows the precise correlation of VFO with pyramidal cell IPSPs during experimental persistent gamma oscillations. VFO also correlates with phasic excitation of interneurons, as shown in Figure 12.9c. The EPSPs are

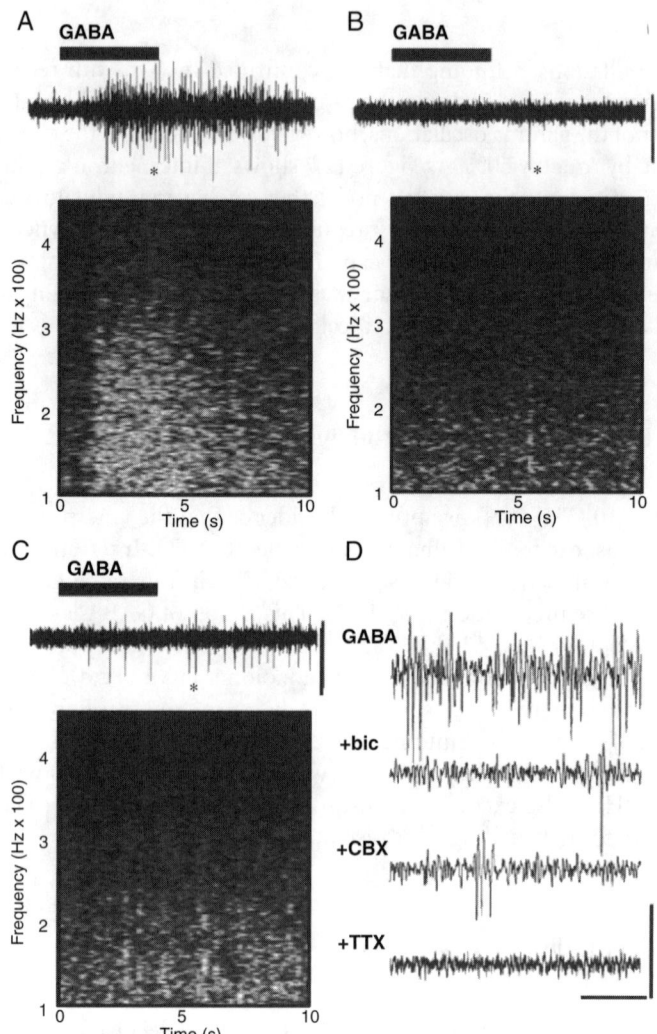

Figure 12.8 A tonic action of GABA, through GABA$_A$ receptors, is to stimulate an axonal plexus to generate VFO. VFO additionally requires that gap junctions be open. The preparation was a CA1 minislice (compare with Chapter 10, Fig. 10.9), but without kainate. The minislice was bathed in blockers of AMPA and NMDA receptors (SYM2206, 20 μM, and d-AP5, 50 μM). Stratum oriens field potentials are plotted along with color-coded time-frequency plots (*a, b, c*). **A:** Application of GABA (0.5 mM) leads to a burst of VFO. **B:** Preperfusion of the minislice with bicuculline (20 μM) prevents GABA application from evoking VFO. Hence, in A, GABA is acting via GABA$_A$ receptors. **C:** Preperfusion of the minislice with the gap junction blocker carbenoxolone (0.2 mM) significantly attenuates the VFO elicited by GABA application. **D:** Expanded-time-scale field potential traces from *A, B, C* (respectively), and from the case where tetrodotoxin (TTX, 2 μM) was preperfused. The Na$^+$ channel blocker TTX blocks spontaneous activity, and also activity elicited by GABA.

(From Traub, Cunningham et al., 2003 with permission.)

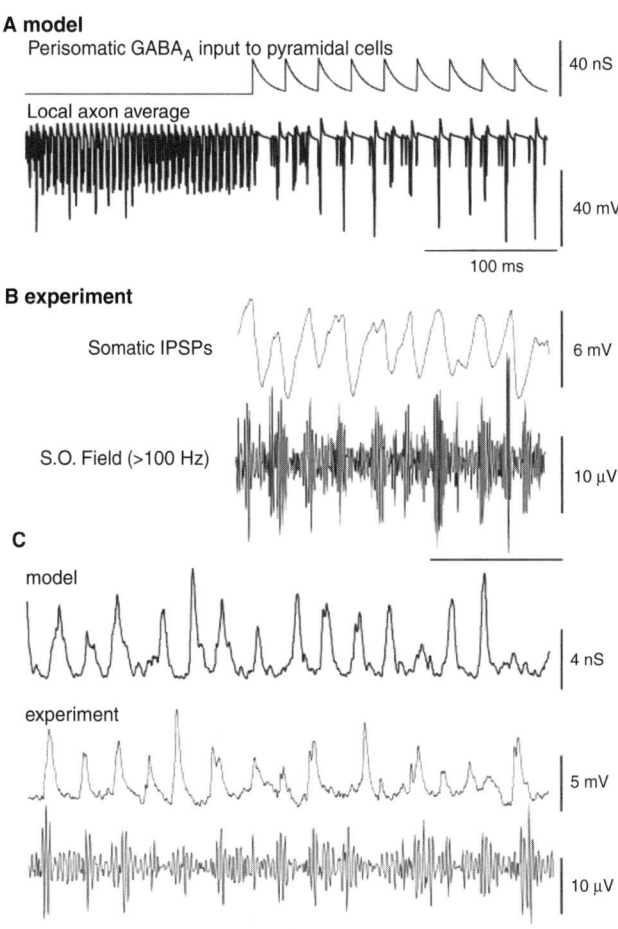

A model
Perisomatic GABA$_A$ input to pyramidal cells 40 nS

Local axon average 40 mV

100 ms

B experiment

Somatic IPSPs 6 mV

S.O. Field (>100 Hz) 10 μV

C

model 4 nS

experiment 5 mV

10 μV

Figure 12.9   Phasic IPSPs periodically interrupt VFO, to account for the nesting of VFO with persistent gamma oscillations: An illustration of the dual role of GABA, favoring VFO (Fig. 12. 8), but also interrupting it. **A:** A "Gedankenexperiment" (thought experiment), in which a network of pyramidal cells is simulated; the cells are electrically coupled in their axons, generating VFO in the first part of the simulation. In the middle of the simulation, large exogenous GABAergic IPSCs are delivered at 40 Hz to perisomatic regions. The IPSCs break the VFO into brief packets (compare Chapter 10, Fig. 10.9). During actual gamma oscillations, the IPSCs are not exogenous, but are generated by the network itself, through synaptic excitation of the interneurons, and recurrent inhibition of the pyramidal cells. **B:** Gamma oscillations were evoked in CA1 with kainate and a pyramidal cell held at ~ –30 mV to reveal oscillating IPSPs (upper trace). The high-passed filtered s. oriens field potential is shown below. Note that packets of VFO occur during the relative intracellular depolarizations, and VFO is attenuated during the hyperpolarizations. **C:** During simulated gamma ("model"), notches and high-frequency components occur in the AMPA receptor-mediated input to an interneuron. Similar notches and high-frequency components occur in experimental EPSPs in an interneuron (middle trace; electrode contains QX314 to block action potentials, and the cell was held at ~ –70 mV to reveal EPSPs). The high-pass filtered (>100 Hz) stratum oriens field is shown below. The notches and high-frequency components are presumed to arise because different unitary EPSPs in the interneuron are locked to different phases of the VFO in the pyramidal cell axonal plexus. Scale bars in A: 40 nS, 40 mV, 100 ms; in B: 6 mV, 10 μV, 100 ms; in C: 4 nS, 5 mV, 10 μV, 100 ms.

(From Traub, Cunningham et al., 2003 with permission.)

inflected and notched (see also Gloveli et al., (2005a). These "notches" correlate well with peak field potential inflections at VFO frequency, and cross-correlation of high-pass filtered EPSPs and concurrent field VFO shows a near-zero phase lag relationship (e.g., Fig. 4 of Cunningham et al., 2004a).

Fast Rhythmic Bursting (FRB) Cells (Chattering Cells) are Necessary for Persistent Gamma in Superficial Layers of Auditory Cortex

As we showed in Chapter 3 (Fig. 3.10), persistent gamma oscillations can be produced in vitro, in neocortical slices, using kainate, particularly in the superficial layers. Figure 12.10 expands on the data of Chapter 3, showing the behavior of interneurons as well as of principal cells. [The firing behavior of the low-threshold-spiking (LTS) interneuron in Fig. 12.10 resembles the firing behavior described for hippocampal oriens/lacunosum-moleculare OLM interneurons during gamma (Gloveli et al., 2005a).] The firing of regular spiking (RS) pyramidal cells during neocortical gamma (Fig. 12.10) strongly resembles the firing of hippocampal pyramidal cells during CA3 gamma; likewise, the firing of fast-spiking (FS) interneurons in neocortical gamma closely resembles that of perisomatic-targeting interneurons in hippocampal gamma (Gloveli et al., 2005a; Hajos et al., 2004; Mann et al., 2005). Neocortex, however, contains fast rhythmic bursting (FRB) cells, which form a minority of the pyramidal neurons [and also interneurons, at least in vivo (Nuñez et al., 1992)]. FRB cells fire singlets and multiplets on about half the gamma waves (Fig. 12.10).

If the axons of the chattering/FRB pyramidal cells are electrically coupled to the axons of RS pyramidal cells (as seems plausible, by virtue of morphological similarity, but nevertheless unproven), forming part of an overall axonal plexus, then the firing patterns of FRB cells lead to profound effects on this axonal plexus: the FRB cells discharge so intensely that even a few of them will "inject" many spikes per unit time into the system. That, in turn means that the RS cells can be relatively hyperpolarized, so as to fire very rare somatic action potentials, while yet there is still enough overall plexus axonal activity to phasically excite the interneurons and keep the gamma oscillation going. (These effects, at least in our network model, depend on the *electrical* outputs of the FRB cells, not on the *chemical synaptic* outputs of the FRB cells.) Indeed, at least up to a point, increasing the number of FRB cells in the network tended to enhance gamma power, up to a saturation point (Fig. 12.11B), while still allowing the network to behave so that RS cells would fire sparse somatic action potentials.

Experiments suggest that FRB cells—even though they form a small fraction of the pyramidal cells—really are critical for persistent gamma oscillations, at least in vitro (Cunningham et al., 2004b): phenytoin [which suppresses persistent g_{Na} and therefore converts FRB into regular spiking behavior

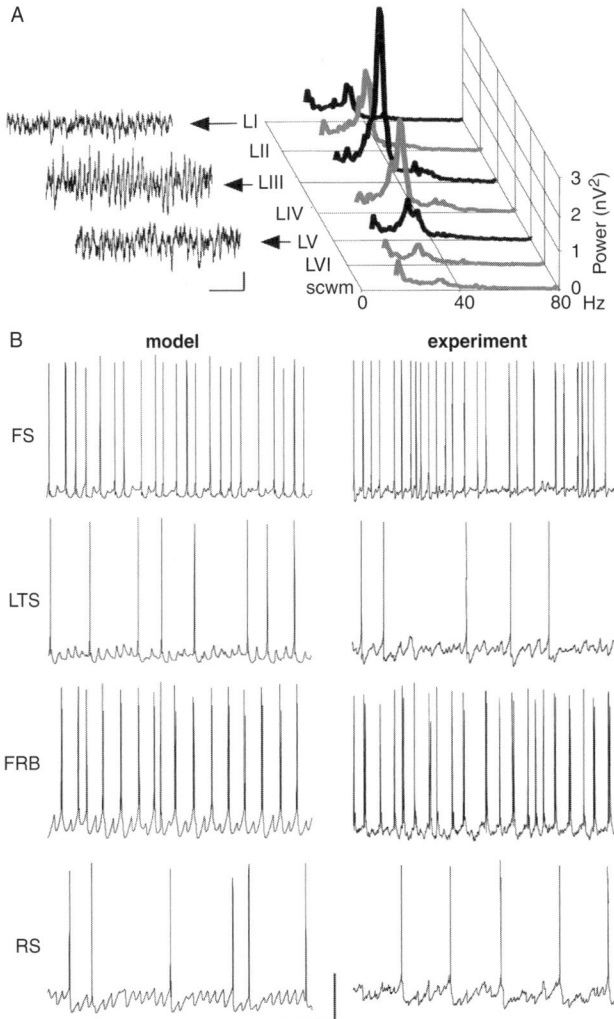

Figure 12.10 Firing patterns of different cell types during persistent gamma oscillations in superficial rat auditory cortex, in vitro, induced by kainate (400 nM): FS interneurons and regular spiking (RS) pyramidal cells behave in similar fashion to hippocampal persistent gamma oscillations, but fast rhythmic bursting (FRB) cells exhibit a different sort of behavior. **A:** Field potential recordings at different cortical depths (scwm = subcortical white matter) show that gamma activity (peak ~33 Hz) has maximal amplitude in layers 2 and 3. Scale bars = 200 ms, 50 μV. **B:** Firing patterns. The network model contained 1,152 pyramidal cells (with a variable proportion of RS, regular spiking, and FRB cells), 96 FS (fast-spiking) basket cells, 96 fast-spiking axo-axonic cells, and 96 LTS (low-threshold spiking) interneurons. In model and experiment, FS cells fired on more than half the gamma waves. LTS cells and RS cells fired intermittently, with prominent synaptic potentials. FRB cells fired 1–2 (model), or 1–4 (experiment) spikes on about half the gamma waves; on the other gamma waves, there were synaptic potentials.

(From Cunningham, Whittington et al., 2004 with permission.)

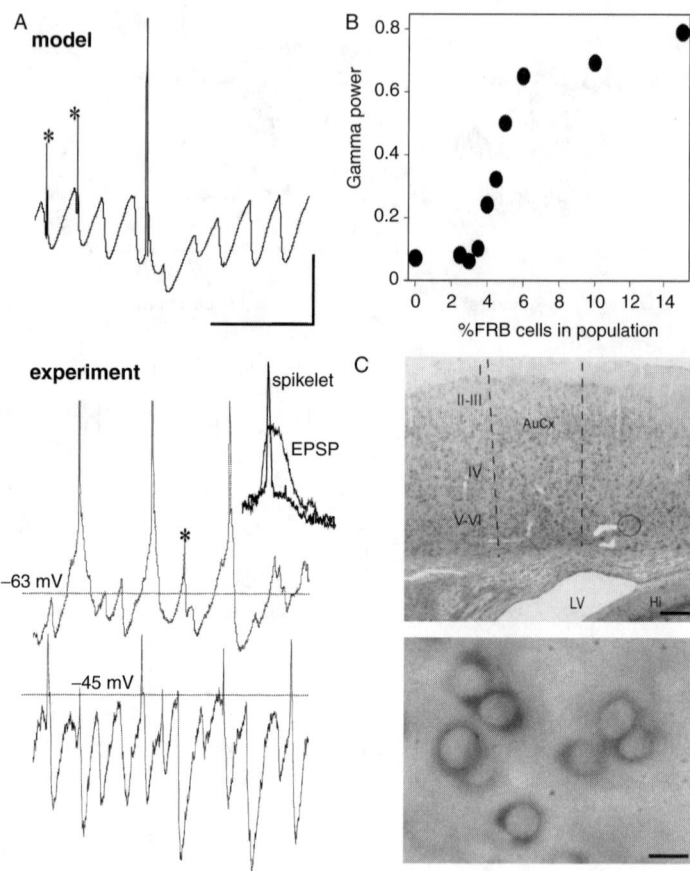

Figure 12.11 Spikelets in regular spiking cells during cortical gamma; predicted necessity for FRB cells; pannexin-2 mRNA in the cortex. **A:** Examples of spikes in model and experimental RS cells. The neuron experiment was recorded at two different membrane potentials: depolarization did not attenuate the spikelets (unlike what would be expected for EPSPs). In the model, spikelets corresponded to decrementally conducted axonal spikes. 10 mV in scale bars is for the model, 5 mV for the experiment. Inset shows average of 10 spikelets and 10 EPSPs (scale bar 40 ms), to emphasize the rapid kinetics of the spikelets. **B:** Summary of simulation data showing how gamma power varies with the fraction of pyramidal cells which are FRB. There is an inflection at about 5%. [The shape of the curve, however, is parameter dependent: if the pyramidal cells are strongly depolarized, gamma occurs without FRB cells, but then the RS somata fire more than observed experimentally.] Experiments (see the original paper) with phenytoin—which suppresses FRB behavior—also suggest that FRB neurons are necessary for gamma in superficial cortex. **C:** Nonradioactive in situ hybridization images showing pannexin-2 mRNA in cortical neurons (scale bar 200 μm above, 10 μm in high-power image below). Pannexin-1 mRNA was also found (not shown here). Pannexin-1/2 heteromers can form function gap junctions in oocytes (Bruzzone et al., 2005). (From Cunningham, Whittington et al., 2004 with permission.)

(Brumberg et al., 2000; Traub et al., 2003a)] acts to suppress persistent gamma in the neocortex, but not in hippocampus (which lacks FRB cells). Unfortunately, it has not proven possible to separate, experimentally, the gap junctional outputs of the FRB cells from their synaptic inputs.

Action Potentials During Persistent Gamma Oscillations are Predicted to be Antidromic

Persistent in vitro gamma oscillations, as also in vitro beta-2 oscillations (Chapter 11) are driven, then, by the activity of the pyramidal cell axonal plexus and, in particular, its tendency to generate VFO. A prediction of this concept is easy to state, but revolutionary in its consequences, if it proves applicable to oscillations in vivo: somatic action potentials in pyramidal cells, during both gamma and beta-2 oscillations, are antidromic. While this is difficult to prove experimentally, there is indirect evidence, in the form of frequent spikelets in pyramidal cells. We have illustrated these for beta-2 oscillations; Fig. 11A shows spikelets occurring during neocortical gamma oscillations. Spikelets have also been described for in vitro gamma oscillations in hippocampus (Fisahn et al., 2004), entorhinal cortex (Cunningham et al., 2003), and cerebellum (Middleton et al., 2008). In the case of entorhinal cortical gamma rhythms, principal cell spikelets are closely phase-locked to the brief epochs of VFO seen on each gamma period (Cunningham et al., 2004a), consistent with the idea of VFO originating in the principal cell axonal plexus.

In the last section of the book, we shall consider what the implications might be for brain function, of the peculiar features of persistent gamma oscillations, most especially the antidromic origin of the pyramidal cell action potentials (see also Chapter 3, Figs. 3.1 and 3.2).

13

Epileptiform Discharges In Vitro

In vitro slice preparations, of hippocampus and other cortical regions, have been of immense value for the study of cellular mechanisms of epilepsy. This field of study has been lively since the 1970s, beginning with the pioneering papers of David A. Prince and his collaborators (including, from those days, Philip A. Schwartzkroin and Robert K.S. Wong) (Schwartzkroin & Prince, 1976, 1977; Wong & Prince, 1981; Wong et al., 1979); and the field is too rich to do full justice to it here (Cohen et al., 2003; Dudek & Sutula, 2007; Jefferys, 2003; Köhling & Avoli, 2006; McCormick & Contreras, 2001; Prince, 1999; Schwartzkroin, 1997; Traub et al., 2005b). We have discussed some of the basic principles of epileptogenesis, derived from in vitro studies, in Chapter 4. In harmony with the *leitmotiven* permeating the present book, we shall concentrate here on the interrelationship between fast oscillations (particularly gamma and VFO) and seizure-like phenomena, keeping in mind these questions:

1. What conditions, in vitro, dually favor either fast oscillations, or synchronized burst discharges (or both together)?
2. To what extent can the in vitro analysis be extrapolated to in vivo experimental epilepsies, and to patients? We defer until the final section of the book, however, discussion of possible therapeutic implications.

The VFO/Electrographic Seizure Transition—as Occurs In Vivo—Can Be Produced In Vitro Under Conditions Favoring Gap Junction Opening

In Chapter 4 (Figs. 4.3–4.5), we illustrated how runs of very fast oscillations (VFO), recordable at the surface of the brain or within brain parenchyma, can precede the onset of electrographic seizure discharges. It is remarkable that similar phenomenology can be elicited in a hippocampal slice (Fig. 13.1),

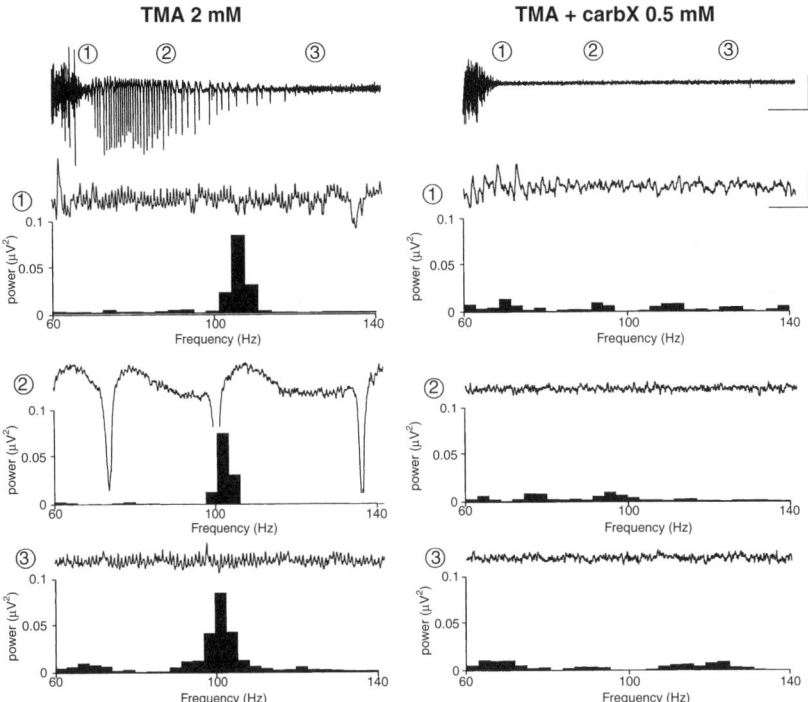

Figure 13.1 VFO before an electrographic seizure discharge, in the rat CA1 hippocampal region in vitro. (Compare Chapter 4, Figs. 4.3, 4.4, 4.5). Data on the left are stratum pyramidale field potentials (and power spectra thereof), in conditions in which trimethylamine (TMA) bathed the slice, alkalinizing the tissue and presumably opening gap junctions; data on the right are from the some conditions, but the gap junction blocker carbenoxolone ("carbX") was also present. In each case, a tetanic electrical stimulus was given (artifacts not shown), which elicited a ~1.5-s gamma oscillation, whether or not carbenoxolone was present. In the TMA condition [but not with carbenoxolone (right column) or in control conditions), the gamma oscillation was followed by a brief epoch of ~110 Hz VFO (1), then an electrographic seizure lasting some seconds and also containing VFO (2), and then further ~100 Hz VFO (3). The phenomenology in the TMA condition resembles that occurring at the onset of seizures in patients and in experimental animals in vivo. Scale bars 1 mV, 2s (top traces); 0.2 mV, 100 ms (expanded traces).

(From Traub et al., 2001 with permission.)

following electrical stimulation of the tissue, provided that the tissue is alkalinized with trimethylamine (TMA), although *not* when the tissue is bathed in
ordinary artificial cerebrospinal fluid. [TMA is presumably acting, at least
for the most part, to open gap junctions (Spray et al., 1981).] The major
effect of the tetanic stimulation, in the experimental conditions illustrated in
Figure 13.1, is to induce a large—but transient—depolarization in pyramidal
cells and interneurons, that is mediated by metabotropic glutamate receptors;
depolarizing $GABA_A$ receptor-mediated responses, and extracellular $[K^+]$
rises, make much smaller contributions to the depolarizations, under the conditions used (Whittington et al., 1997a, 2001). The consequence of these
depolarizations is an induced gamma oscillation (Traub et al., 1996c), which
occurs even when gap junctions are blocked. The gamma oscillation is then
followed by approximately 110 Hz VFO, strikingly like that occurring in vivo.
It is possible that chemical synaptic depression, occurring during the gamma
oscillation, "unmasks" the ability of the tissue to generate VFO (when gap
junctions are open), but this has not been proven for the experimental conditions here illustrated. Remarkably, the VFO is also transient, and is succeeded
by an electrographic seizure (lasting 12–17 seconds, in different experiments).
Both the VFO and the electrographic seizure are prevented by blockade of gap
junctions—presumably of *principal cell* gap junctions, given that other data
(reviewed in Chapters 4 and 10) show that it is principal cells which generate
VFO and burst discharges.

How might the VFO/seizure transition come about? Figure 13.2 illustrates a network simulation of a VFO/seizure transition: one can regard it as
embodying a set of testable hypotheses concerning the physiological situation.
The network model used here is based on that of Traub et al. (2005a); in the
present case, there were 4,750 cortical neurons, including 2000 layer 5 tufted
IB cells, interconnected by axonal gap junctions (2.5 per axon, on average).
(A modification present here, relative to the original study, was in the intrinsic properties of layer 5 pyramidal cell axons, motivated by data from the
D.A. McCormick laboratory that were illustrated in Chapter 8, Figures 8.11 to
8.13. Similar modifications were present for the pyramidal cells whose collective behavior was shown in Chapter 10, Fig. 10.7.) The model conditions
began with gap junctions closed, and all chemical synapses nearly shut off;
as mentioned earlier, the latter would correspond to a hypothetical state of
synaptic depression. In such a parameter regime, the network is mostly quiet.
Over the period of the black bar, axonal gap junctions are opened, and layer
5 pyramids generate a population VFO that appears both in the field, and in
single cells (as spikelets, in this case). Over the period of the red bar, recurrent
synaptic excitation between deep pyramidal cells is allowed to recover, and the
system switches from VFO into a series of synchronized bursts. The latter series
resembles what Steriade et al. (1998a) have called a "fast run." For comparison,
the inset shows an intracellular recording of a fast run in vivo. An interesting
point of comparison between the model and the in vivo data consists of the
occurrence of action potentials of different amplitudes (asterisks): in the

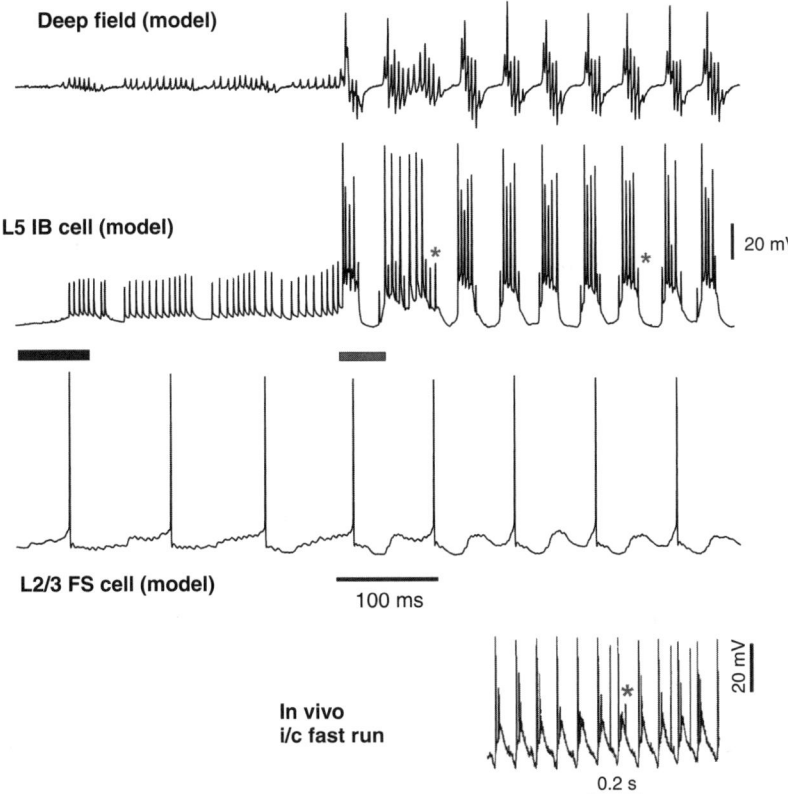

Figure 13.2 Hypothetical mechanism for the transition from VFO to electrographic seizure. Simulation data using a network model similar to that described in Traub et al. (2005). In the present case, there were 4750 model cortical neurons (each multicompartment and with a variety of intrinsic voltage-gated and Ca^{2+}-gated membrane conductances – see Fig. 8.1 of Chapter 8): deep IB pyramidal cells, deep RS pyramidal cells (tufted and nontufted), superficial RS and FRB pyramids, spiny stellates, and four types of interneurons. The critical cells here are the 2000 layer 5 IB pyramidal cells, consistent with experimental observations on the initiation of epileptiform events in, or near, layer 5. The IB cells are interconnected by a total of 2500 axonal gap junctions, so that each axon is coupled, on average, to 2.5 others. Synaptic conductances were set to small values, so that synaptic currents were negligible. IB cell gap junction conductances also began at a negligible value, and then were rapidly increased to 8 nS (dark bar), initiating VFO. Then, recurrent synaptic excitation between deep pyramidal cells increased (red bar): this converted VFO to a rhythmical series of bursts, resembling an in vivo fast run. The inset below shows an intracellular recording from such a fast run. Note that in the simulation, and in the in vivo data, action potentials of multiple amplitudes occur (asterisks): These result in the model from antidromic spikes. Simulated FS cells fire sparsely, as is the case during actual fast runs (Steriade, 2003, his Fig. 5.66).

(R.D. Traub and M.A. Whittington, unpublished data. Inset from Steriade, Amzica, Neckelmann, & Timofeev, 1998. Reproduced with permission.)

model, such differences reflect the tendency of action potentials to originate antidromically. Another point of comparison consists in the sparse firing of fast-spiking interneurons during the fast run, which is also seen in vivo (Steriade, 2003, his Fig. 5.66).

To summarize then, we propose the possibility that VFO can occur in the slice, and in vivo, in conditions of relatively depressed chemical synapses, combined with open axonal gap junctions; and that seizure discharges can arise out of this state, if recurrent synaptic excitation (between principal cells) recovers before inhibitory circuits do.

VFO Also Occurs Interspersed with Synchronized Bursts In Vitro

An association between VFO and hippocampal synchronized bursting, in vitro, has been known for many decades (Schwartzkroin & Prince, 1977; Wong & Traub, 1983). Numerous studies since have indicated a similar association between VFO and interictal spikes, as well as intraseizure spike discharges, in hippocampus and neocortex [reviewed in Chapter 4, and see Chapter 4, Fig. 4.6, as well as Traub et al. (2001c)]. Here, we illustrate two different scenarios by which a VFO/burst discharge association could come about.

We have shown in Chapter 11 how gamma and beta2 oscillations can arise within different layers of the same neocortical region. A logical consequence of this experimental observation is that both superficial pyramidal cells, and deep pyramidal cells, must be electrically connected: superficial cells with superficial cells, and deep cells with deep cells. [It is not possible, at the moment, to draw conclusions as to whether or not electrical coupling exists between deep and superficial cells, although dye coupling data would appear to mitigate against the concept (Gutnick et al., 1985).] Consistent with this idea of electrical coupling in both sets of neocortical layers, we know also that spikelets occur in superficial pyramidal cells (Chapter 12, Fig. 12.11), and that isolated deep layer minislices can generate VFO (Chapter 10, Fig. 10.4). It is therefore possible that the VFO associated with epileptiform events could arise in superficial layers, in deep layers, or both.

Figure 13.3 (left column) shows simulation data for a model cortical column wherein recurrent circuitry in layer 4 leads (in the presence of disinhibition) to sustained depolarization of superficial neurons, VFO in superficial layers, and a pair of deep-neuron bursts. The layer 5 pyramidal cells do not show evident spikelets, while VFO is continuous during the time interval between the two synchronized discharges. A similar pattern of field and cellular activity is seen in disinhibited rat auditory cortex, in vitro, when it is activated by kainate (Fig. 13.4, right column).

In contrast, and as suggested by the simulated fast run in Figure 13.2, Figure 13.4 shows the simulated activity when VFO is generated by the axonal plexus of layer 5 IB neurons. There is now a tight correlation between

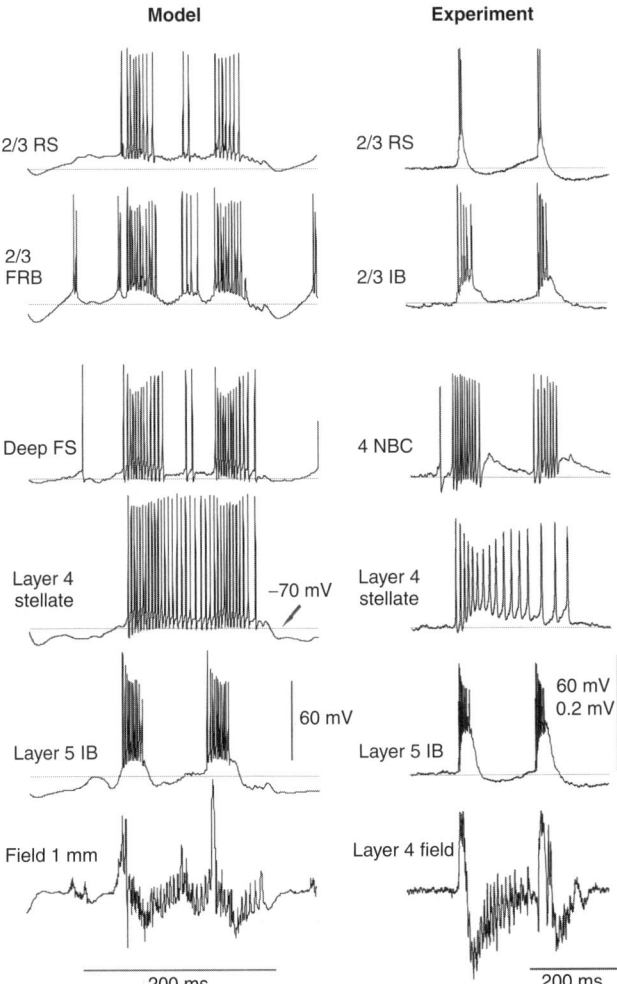

Model   **Experiment**

2/3 RS

2/3 FRB   2/3 RS

Deep FS   2/3 IB

Layer 4 stellate   4 NBC

−70 mV

Layer 5 IB   Layer 4 stellate

60 mV

Field 1 mm   Layer 5 IB

60 mV
0.2 mV

Layer 4 field

200 ms   200 ms

Figure 13.3 Double bursts, an in vitro/simulation model of epileptic polyspikes. Epileptic discharges in patients typically consist of EEG spikes, spike-wave, and polyspike-wave; in the latter case, a short train of EEG spikes occurs in sequence, followed by a slower wave. The simulation data (left column) is from a thalamocortical network model, in which synaptic inhibition was nearly completely blocked, recurrent excitation remained present and contained particularly strong synaptic connections between layer 4 spiny stellate cells. In addition, the spiny stellate cells were electrically coupled via their axons, as were superficial pyramidal cells and layer 6 pyramidal cells, but not layer 5. The result was a double EEG spike notable for a continuous firing of spiny stellate cells; bursting of layer 5 IB cells in phase with the large field transients; and continuous VFO between the large field transients. The field and cellular firing patterns were similar in a neocortical slice experiment (right column). The tissue was rat auditory cortex bathed in kainate (400 nM), and with $GABA_A$ and $GABA_B$ receptors pharmacologically blocked (picrotoxin, 40 μM, and CGP55845A, 10 μM). [The different cells in the experiment were not recorded simultaneously; cellular potentials were aligned to the first large field transient.] It is likely that similar behavior can be driven by electrical coupling between layer 5 pyramidal cells, given that a deep-layer minislice can generate VFO (Chapter 10, Fig. 10.4): see Fig. 13.4 of this chapter. In the present instance one sees no spikelets in layer 5 IB cells; in the model, one can say that this reflects VFO generation in superficial layers, rather than deep layers. NBC cell is a nest basket cell (Wang et al., 2002).

(From Traub et al., 2005 with permission.)

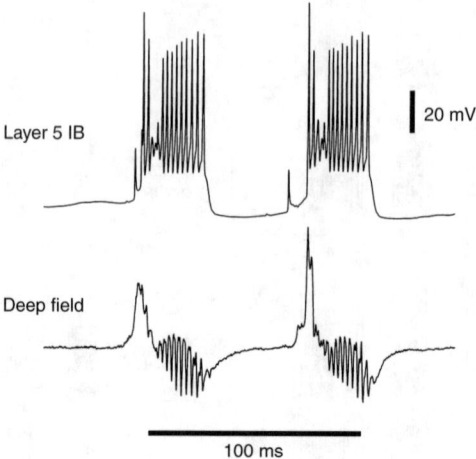

Figure 13.4 Simulation of VFO and synchronized bursts in cortical network model, in which the VFO arises in layer 5. The network was almost completely disinhibited, and there were numerous gap junctions between the axons of layer 5 intrinsically bursting tufted pyramidal cells. In this instance, there is a tight correlation between layer 5 intracellular activity and the field, and spikelets are apparent. (R. D Traub and M.A. Whittington, unpublished data.)

field VFO and layer 5 intracellular activities, either spikelets, or full action potentials.

One conclusion from these data is this: in the in vivo and patient situations, one must keep an open mind as to which type of VFO is operative—or possibly yet unknown types of VFO. Only extremely detailed analyses of field and (where possible) intracellular data will provide the necessary clues.

Gamma Oscillations and Epileptiform Bursts Can Switch Back and Forth

Pais et al. (2003) showed that hippocampal slices from connexin-36 knockout mice exhibited a remarkable type of behavior when bathed in kainate: gamma oscillations alternating with synchronized bursts, the latter associated with VFO, and recurring at intervals of several seconds. A similar type of behavior occurs in hippocampal slices from wild-type mice, when they are bathed in (S)-3,5-dihydroxyphenylglycine (DHPG) (Fig. 13.5). A clue to the mechanism derives from the data in Figure 13.6: when recurrent excitation between pyramidal cells is relatively weak, and recurrent excitation of interneurons strong, gamma oscillations occur, as we discussed in Chapter 11. With the reverse "balance" of excitation of cells—strong to pyramidal cells, weak to interneurons—synchronized discharges with VFO are favored (as also seen in

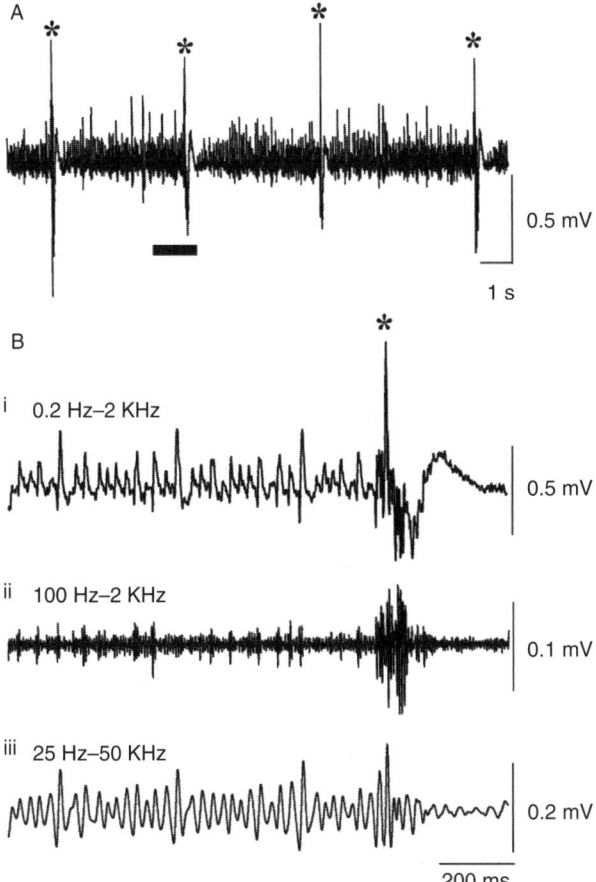

Figure 13.5 Alternating gamma oscillations and epileptiform bursts in mouse hippocampus (CA3) in vitro. This pattern of behavior can be seen when the tissue is bathed with the metabotropic glutamate receptor agonist, DHPG, as in the present case, and is also seen spontaneously in slices from connexin-36 knockout mice (Pais et al., 2003). **A:** Stratum pyramidale field potential recordings, showing synchronized bursts (asterisks) recurring at intervals of several seconds. **B:** The epoch in A marked by a bar is expanded and plotted with three different filter settings. (*i*) Broadband. The burst is marked with an asterisk. (*ii*) Filtered to show brief bursts of VFO associated with many of the gamma waves (see Chapter 11), as well as VFO just before, and during, the burst. (*iii*) Gamma oscillations, speeding up just prior to the burst, and attenuated afterwards (perhaps due to AHP currents).

(From Traub, Pais, et al., 2005 with permission.)

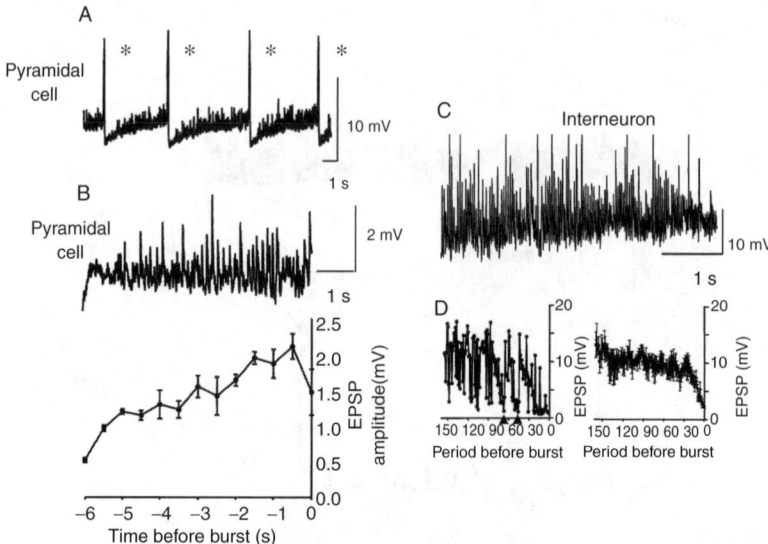

Figure 13.6 (Continuing Figure 13.5.) Alternating gamma and epileptiform bursts are associated with potentiation of EPSPs in pyramidal cells, and depression of EPSPs in interneurons, over the interburst interval. The time-dependent alterations in synaptic efficacies explain, at least partly, the different population behaviors: strong excitation between pyramidal cells, with weak excitation of interneurons, will favor population bursts; while weak excitation of pyramidal cells, along with strong excitation of interneurons, will favor gamma oscillations. **A:** Intracellular recording of a pyramidal cell during the alternating gamma/bursting behavior. Bursts are marked with asterisks; the bursts are associated with large intracellular depolarizations and firing. **B:** A pyramidal cell was held at mean resting potential of –70 mV. The EPSPs, on average, increase in size during the interburst interval. The graph below shows pooled data, where the signals have been high-pass filtered at >0.5 Hz. **C:** A fast-spiking stratum pyramidale interneuron was held at –70 mV, to show the fluctuating attenuation of EPSPs over the interburst interval. Action potentials truncated. **D:** Data from a single interburst interval (left), and pooled data (right), documenting the reduction of interneuron EPSPs over the course of the interburst interval. Bursts themselves are not associated with large depolarizations in interneurons, unlike the case in pyramidal cells.

(From Traub, Pais et al., 2005 with permission.)

Figs. 13.3 and 13.4, in which there is disinhibition). An important, but so far unanswered, question concerns the precise mechanisms by which the synaptic efficacies are modified. Possibilities include repletion of glutamate in presynaptic terminals, for the increasing strength of pyramidal cell EPSPs (Bains et al., 1999); synaptic depression resulting from transmitter depletion, for the declining strength of interneuron EPSPs; and metabotropic regulation of transmitter release at presynaptic terminals.

Thus, the in vitro data suggest the following sort of qualitative "phase portrait" for cortical networks in states when principal cell gap junctions are open: when synaptic excitation and inhibition are both weak, then VFO is favored, although beta-2 oscillations can also occur in certain cortical regions; when synaptic excitation is strong and inhibition weak, then (of course) synchronized bursts are favored, with additional mechanisms coming into play to determine the intervals between bursts (this is the "seizure regime" of the phase portrait); when synaptic excitation of pyramids is weak but of interneurons strong, and IPSPs are present, then gamma oscillations can occur, in hippocampus and superficial neocortex and entorhinal cortex. When all chemical synapses are strong, as may occur in normal physiological states in vivo, then population behavior becomes too complicated to characterize with this sort of simple picture.

PART IV

Implications for Health and Disease

14

Conclusion

Let us summarize—crisply, perhaps a bit polemically—some of the main points of this book. We remind the reader that we have restricted our discussion to selected disorders that involve brain oscillations, and concerning oscillations themselves, we have mostly written about oscillation types that we ourselves have investigated, and that involve gap junctions. Also: it is not easy to add to Mircea Steriade's discourses on spindles and the slow oscillation of sleep (Steriade, 2001, 2003).

1. Epileptic discharges are often preceded by, and intermixed with, very fast oscillations (VFO). VFO require gap junctions between principal cells, and epileptic discharges are facilitated by gap junctions. Whether VFO occurs alone, or is mixed with seizure discharges, is determined, at least in part, by the state of excitatory synapses on pyramidal cells. [4-aminopyridine (4AP) induced epileptogenesis (Avoli, 1996), and certain spontaneous synchronized activities occurring in limbic cortex (Schwartzkroin & Haglund, 1986; Cohen et al., 2002) may be exceptions.]

2. Parkinson's disease is associated with a beta-2 oscillation in motor and somatosensory cortical areas, whose amplitude is positively correlated to the extent of disease symptoms, particularly bradykinesia. It is tempting to speculate that there is a causal relation between excessive beta-2 and bradykinesia, but this is unproved (and see later).

3. A wealth of clinical evidence indicates abnormalities in sensory-induced and cognitive task-related gamma and beta oscillations in schizophrenia. In spite of this, it is hard to identify a specific oscillation question, which can be

investigated in vitro, in the way that one can do for epilepsy and Parkinson's disease (but see Cunningham et al., 2006a).

4. It is likely that the cerebellum interacts with the thalamus and motor cortex through means that involve oscillations; and experimental techniques now exist to study cerebellar cortical oscillations in vivo and in vitro. Nevertheless, this field appears to be in its early stages, and the "right oscillation question"—if there is one—that will open up the clinical issues of cerebellar ataxia, remains to be identified.

Thus, for epilepsy and Parkinson's disease, as compared with other neuropsychiatric conditions, we have a clearer idea what the questions are, even if we lack the answers.

Let us now present an overview of the mechanisms of selected oscillations, as they are understood from in vitro studies, with plausible extrapolations to the in vivo case. (Of course, plausibility, like beauty, is in the eye of the beholder, or perhaps in brain oscillations of the beholder.) Our considerations begin with the presupposition that axonal gap junctions exist. The data in support of this presupposition are, in our opinion, compelling, even if still preliminary. Nor are we aware of models—that can account for the full range of observations on persistent gamma, beta-2, and VFO—that do not presuppose strong electrical coupling between principal cells ("strong" meaning here "allowing action potentials to cross from cell to cell").

1. VFO, in the frequency range up to perhaps 250 Hz (but not necessarily higher) is generated by waves of activity within a plexus of axons that are coupled by gap junctions, sparsely but strongly ("sparsely" here meaning that each axon is coupled to only a few others). A wave of activity is initiated by a spontaneous action potential somewhere in the plexus. The wave of activity grows exponentially in time, at least for a while, because each cell must couple (on average) to more than one other; however, the actual exponent that defines growth is not large, because the connectivity is sparse. That means that propagation must pass over many steps before the bulk of the population fires. The mean period of the oscillation is determined primarily by two factors: a local one, the time it takes for a spike to propagate from one cell to a cell that couples directly to it; and a global one, determined by the overall network structure, and corresponding to the "mean path length." or to some other global property of the graph topology The theory of random graphs is an appropriate mathematical apparatus for beginning the development of intuitions about this type of collective phenomenon; however, biological axonal plexi are undoubtedly not random; and, indeed, nonrandom structural properties of the gap junctional connectivity may be important for biological computation.

2. A collection of pyramidal neurons, electrically coupled in their axons, can generate a beta-2 oscillation when gap junction conductance, and M-current conductance, lie in an appropriate range. Chemical synaptic transmission is not necessary, and indeed, synaptic inputs, from connected networks oscillating at gamma frequencies, act to decrease beta-2 power. VFO can

be seen nested with beta-2, reflecting their common origin in the pyramidal cell axonal plexus. This type of nonsynaptic beta-2 can be viewed as a VFO that is continually interrupted by an intrinsic refractory process, the M-current.

Somatic action potentials, generated by nonsynaptic beta-2, are *antidromic*; spikelets also occur in many of the neurons.

3. Persistent gamma oscillations require gap junctions between pyramidal cells, phasic synaptic excitation of interneurons, and synaptic inhibition of the pyramidal cells. Modeling (Traub et al., 2000) and recent in vivo data (Wulff et al., 2009), suggest that synaptic inhibition *between* interneurons is not required for persistent gamma; and synaptic excitation between pyramidal cells appears to be actively disruptive (Traub et al., 2000, 2005c). One period of a persistent gamma oscillation can be conceptually dissected this way: a few waves of VFO in the pyramidal cell axonal plexus lead to a brief series of coherent, phasic excitatory postsynaptic potentials (EPSPs) in fast-spiking (FS) interneurons; most of the FS interneurons fire, once or twice, near-synchronously; the resulting coherent inhibitory postsynaptic potentials (IPSPs) generate the gamma period. This generation occurs because periso-matic IPSPs produce hyperpolarization, and membrane shunting, and thereby prevent propagation of spikes from axon to axon. That, in turn, halts the VFO, and secondarily stops interneuron firing—until VFO resumes with the fading of the IPSPs. This "deconstruction" of persistent gamma accounts for the suppressive effects of blockers of α-amino-3-hydroxyl-5-methyl-4-isoxazole-propionate (AMPA) and GABA$_A$ receptors, and of gap junctions; the co-occurrence of VFO and persistent gamma, with nesting; the loss of gamma power after molecular manipulations, which either reduce phasic excitation of interneurons, or remove electrical coupling between inter-neurons (Hormuzdi et al., 2001; Traub et al., 2003c); and the occurrence of spikelets during persistent gamma. We are not aware of another model that accounts for this range of data.

The in vitro/computational model of persistent gamma generates neu-ronal firing patterns and phase relations that resemble those occurring in vivo, in the hippocampus during the theta state.

In vitro persistent gamma may (we believe) capture key features of fast oscillations that occur during activated states, such as with a short-term memory task, or superimposed on the depolarizing phases of the slow oscilla-tion of sleep; however, transient gamma oscillations, induced by sensory stim-ulation, probably (we believe) represent a phenomenon that is somewhat different in cellular mechanisms.

During in vitro persistent gamma oscillations, pyramidal cell action potentials are predicted to be *antidromic*, while FS interneuronal action poten-tials are predicted to be synaptically driven.

All of these oscillations—VFO, nonsynaptic beta-2, persistent gamma—require open gap junctions between pyramidal cells. We may ask, though, how a network "decides" to make one oscillation type or another. For this issue, it helps to consider VFO as a "degenerate state," very roughly speaking, into which other oscillation types will convert under certain conditions. The preceding analysis allows us to deduce some conditions favoring conversion to VFO:

1. If the system is in a beta-2 state, with weak chemical synapses, then reduction of M-current is predicted to favor a switch to VFO.

2. If the system is in a gamma state, then reduction of synaptic inhibition should switch the system to VFO. (As noted in Chapter 11, however, this prediction is hard to test with direct block of $GABA_A$ receptors, as the latter receptors appear to contribute to axonal excitability, at a locus somewhere distal to the axon initial segment.)

3. Similarly, if gap junctions could be moved—or re-form—distally along axons, far enough from somata so as to be unaffected by perisomatic inhibition, then the network would be able to generate VFO, but not persistent gamma. A tissue alteration of this sort—if it occurs biologically, say in a disease state—would be expected to disrupt brain function, and to promote epileptogenesis.

4. Increases in axonal excitability are predicted to allow axon–axon spike propagation that is not controlled by local M-currents or synaptic inhibition. Continuous VFO could then occur, but not gamma or beta-2.

5. If, however, axons become excitable *enough*, then the axonal plexus will exhibit extreme degrees of activity, but not organized into VFO. That is, the network activity need not be oscillatory. This will be the case because axonal activity will no longer depend on waves of activity propagating through the plexus. Blockade of gap junctions would then be expected to have little effect on plexus hyperactivity.

What about the possibility of excessive gamma or beta-2 oscillations? We are not aware of excessive gamma oscillations having pathological significance, except as a harbinger of impending seizure, and in a case report of a patient with somatic hallucinations (Baldeweg et al., 1998): although the regulation of gamma oscillations is abnormal in schizophrenia, and it is conceivable that enhanced gamma could be dysfunctional there. (Here, we use "gamma" to mean "30–70 Hz." Some authors use the term "gamma" to include frequencies well above 70 Hz, or else use the term "high gamma" for such fast oscillations. This practice, we believe, causes confusion, because the cellular mechanisms of 30–70 Hz oscillations are, experimentally, quite different than for the faster oscillations.) Excessive *beta* oscillations do, nevertheless, have pathological significance in Parkinson's disease. Therefore, we must ask how excessive beta could come about. Besides the obvious means—increased activation of kainate receptors, proper tuning of gap junction and M-current parameters in layer 5 pyramidal cells—an intriguing possibility is this: that the primary abnormality is actually a *loss of gamma activity* in the superficial layers. [The reader will recall that with in vitro beta-2, suppressing superficial gamma tends to enhance deep beta-2 (Chapter 11, Figs. 11.1, 11.4).] If this possibility proves correct, it may be relevant as well to the cognitive deficits in Parkinson's disease.

In addition, it suggests that dopamine might act to enhance neocortical gamma, rather than to suppress (directly) neocortical beta.

What about the dual issue of "excessive VFO"? We have illustrated and reviewed some of the literature implying that VFO can immediately precede a seizure, and can also exist on its own in brain regions predisposed to epileptogenesis—particularly "fast ripples," in the latter case. A key pathophysiological question is obviously this: what are the causal relations, if any, between the VFO and the seizure? To consider two hypothetical extreme cases:

Case 1. Neural tissue can become diseased in certain ways: interneurons function abnormally, Cl⁻ gradients are not maintained properly, extracellular [K⁺] transients are excessive, pH fluctuates (especially in upward directions, opening gap junctions); or, following axonal injury, excessively many gap junctions form. The reader is free to add possibilities to this list. But in Case 1, we suppose that all of the mechanisms predispose the tissue *both* to VFO *and* to seizures, as independent processes. If Case 1 reflects the reality, one should be able to detect in vivo seizures not preceded by VFO—not as simple, technically, as it sounds, as VFO can be spatially more localized than the electrographic seizure discharges, and thus much harder to detect.

Case 2. Disease mechanisms predispose the tissue to VFO. VFO then tends to (but is not *guaranteed* to) start a seizure. For example, VFO in a pyramidal cell axon plexus is expected to continuously release glutamate from presynaptic terminals—at least until significant synaptic depression sets in—and this could depolarize dendrites and promote bursting. This possibility is addressable experimentally, at least in principle, with intradendritic recording.

Analysis of these dual cases is of practical importance. For example, suppose Case 2 turns out to be closest to the truth. There may be therapies which have no obvious and direct connection to seizures, but which target VFO specifically. Such therapies could prove useful in Case 2, but they would likely be without value in Case 1.

Let us now move back from the details of oscillation mechanisms and consider the cell biology of axonal gap junctions. It is naturally obvious to the reader that far more information about these junctions is required. Some of the essential pieces of information are these:

1. To date, definitive gap junction ultrastructural images have not been produced for pyramidal cell axon–axon couplings. Where are the putative junctions located? How many are there on an axon?

2. Connexin-36 has been localized to a gap junction on a mossy fiber axon (Hamzei-Sichani et al., 2007). Is connexin-36 the *sole* constituent of axonal gap junctions? If that were the case, then how can VFO and persistent gamma oscillations occur in a connexin-36 knockout mouse (Hormuzdi et al., 2001)? If it is not the case, what other gap junction proteins get into the act?

3. How are axonal gap junctions regulated, structurally and functionally, and over what time scales? Structurally, one needs to know if new gap junction

formation could embody a form of learning, in the normal brain; and if new gap junctions form as a dedifferentiation response after tissue injury, in the damaged brain. Functionally, one needs to know if gap junctions are opening and closing dynamically and continuously, and whether their state is correlated with global behavioral states, like the sleep–wake cycle. In the former instance, it would help to understand how oscillations seem to come and go, in vivo. In the latter instance, it might help to explain why certain seizure phenomena are most prone to occur during slow-wave sleep.

Finally, we return to an idea described at the beginning of this book (Chapter 3, Fig. 3.1), that networks of neurons can operate in two modes: a *conventional* one, in which neurons (mostly dendrites) integrate synaptic input signals, and produce (or "compute," as some authors write) axonal outputs, that in turn influence the behavior of other neurons; and an *unconventional* one, in which network activity is primarily driven by an electrically coupled axonal plexus, whereby somatic spikes—when they occur at all—are antidromic, and not the result of synaptic integration. If the reader has made it this far into the book, without a rage attack and without drifting into slow-wave sleep, then perhaps he or she will be convinced by the presented data that the unconventional mode actually exists, at least in vitro. But does this unconventional mode occur in vivo, in pure form, or close to it? Or—as we suspect—are normal brain operations a yet-to-be-understood hybrid of the two modes, sometimes leaning toward one, sometimes the other? If so, then the field of neural networks is not a mature one. The field is still in a fetal stage.

References

Abeles M (1982) Role of the cortical neuron. Integrator or coincidence detector. Israel J. Med. Sci. 18 83–92.

Acebrón JA, Bonilla LL, Pérez Vicente CJ, Ritort F, Spigler R (2005) The Kuramoto model: A simple paradigm for synchronization phenomena. Rev. Modern Physics 77: 137–185.

Achermann P, Borbély AA (1997) Coherence analysis of the human sleep electroencephalogram. Neuroscience 85: 1195–1208.

Achermann P, Borbély AA (1998) Low-frequency (<1 Hz) oscillations in the human sleep electroencephalogram. Neuroscience 81: 213–222.

Achim AM, Lepage M (2005) Episodic memory-related activation in schizophrenia: meta-analysis. Br. J. Psychiatry 187: 500–509.

Adams PR, Brown DA (1982) Synaptic inhibition of the M-current: slow excitatory post-synaptic potential mechanism in bullfrog sympathetic neurones. J. Physiol. 332: 263–272.

Adams PR, Brown DA, Constanti A (1982) Pharmacological inhibition of the M-current, J. Physiol. 332: 223–262.

Adrian ED (1936) The spread of activity in the cerebral cortex. J. Physiol. 88: 127–161.

Adrian ED (1942) Olfactory reactions in the brain of the hedgehog. J. Physiol. 100: 459–473.

Adrian ED (1943) Afferent areas in the cerebellum connected with the limbs. Brain 66: 289–315.

Adrian ED (1950) The electrical activity of the mammalian olfactory bulb. Electroencephalogr. Clin. Neurophysiol. 2: 377–388.

Adrian ED, Matthews BH (1934) The interpretation of potential waves in the cortex. J. Phsyiol. 81: 440–471.

Adrian ED, Cattell M, Hoagland H (1931) Sensory discharges in single cutaneous nerve fibres. J. Physiol. 72: 377–391.

Agmon A, Connors BW (1989) Repetitive burst-firing neurons in the deep layers of mouse somatosensory cortex. Neurosci. Lett. 99: 137–141.

Agmon A, Connors BW (1992) Correlation between intrinsic firing patterns and thalamocortical synaptic responses of neurons in mouse barrel cortex. J. Neurosci. 12: 319–329.

Agrawal R, Daniel EE (1986) Control of gap junction formation in canine trachea by arachidonic acid metabolites. Am. J. Physiol. 250: C4895–505.

Ahlskog JE (2007) Beating a dead horse: dopamine and Parkinson disease. Neurology 69: 1701–1711.

Ajima A, Tanaka S (2006) Spatial patterns of excitation and inhibition evoked by lateral connectivity in layer 2/3 of rat barrel cortex. Cereb. Cortex 16: 1202–1211.

Akbarian S, Kim JJ, Potkin SG, Hagman JO, Tafazzoli A, Bunney WE Jr, Jones EG (1995) Gene expression for glutamic acid decarboxylase is reduced without loss of neurons in prefrontal cortex of schizophrenics. Arch. Gen. Psychiatry 52: 258–266.

Akiyama T, Otsubo H, Ochi A, Ishiguro T, Kadokura G, Nair RR, Weiss SK, Rutka JT, Snead OC III (2005) Focal cortical high-frequency oscillations trigger epileptic spasms: confirmation by digital video subdural EEG. Clin. Neurophysiol. 116: 2819–2825.

Akiyama T, Otsubo H, Ochi A, Galicia ES, Weiss SK, Donner EJ, Rutka JT, Snead OC III (2006) Topographic movie of ictal high-frequency oscillations on the brain surface using subdural EEG in neocortical epilepsy. Epilepsia 47: 1953–1957.

Akkal D, Dum RP, Strick PL (2007) Supplementary motor area and presupplementary motor areas: targets of basal ganglia and cerebellar output. J. Neurosci. 27: 10659–10673.

Alarcon G, Binnie CD, Elwes D, Polkey CE (1995) Power spectrum and intracranial EEG patterns at seizure onset in partial epilepsy. Electroencephalogr. Clin. Neurophysiol. 94: 326–337.

Albus JS (1971) A theory of cerebellar function. Math. Biosciences 10: 25–61.

Aldenkamp A, Vigevano F, Arzimanoglou A, Covanis A (2006) Role of valproate across the ages. Treatment of epilepsy in children. Acta Neurol. Scand. Suppl. 184: 1–13.

Aleksic B, Ishihara R, Takahashi N, Maeno N, Ji X, Saito S, Inada T, Ozaki N (2007) Gap junction coding genes and schizophrenia: a genetic association study. J. Hum. Genet. 52: 498–501.

Aleman A, Hijman R, de Haan EH, Kahn RS (1999) Memory impairment in schizophrenia: a meta-analysis. Am. J. Psychiatry 156: 1358–1366.

Alford S, Christenson J, Grillner S (1991) Presynaptic $GABA_A$ and $GABA_B$ receptor-mediated phasic modulation in axons of spinal motor interneurons. Eur. J. Neurosci. 3: 107–117.

Ali AB (2003) Involvement of post-synaptic kainate receptors during synaptic transmission between unitary connections in rat neocortex. Eur. J. Neurosci. 17: 2344–2350.

Ali AB, Rossier J, Staiger JF, Audinat E (2001) Kainate receptors regulate unitary IPSCs elicited in pyramidal cells by fast-spiking interneurons in the neocortex. J. Neurosci. 21: 2992–2999.

Allen GI, Korn H, Oshima T, Toyama K (1975) The mode of synaptic linkage in the cerebro-ponto-cerebellar pathway of the cat. II. Responses of single cells in the pontine nuclei. Exp. Brain Res. 24: 15–36.

Allen GI, Oshima T, Toyama K (1977) The mode of synaptic linkage in the cerebro-ponto-cerebellar pathway investigated with intracellular recording from pontine nuclei cells of the cat. Exp. Brain Res. 29: 123–136.

Allen PJ, Fish DR, Smith SJM (1992) Very high-frequency rhythmic activity during SEEG suppression in frontal lobe epilepsy. Electroenceph. Clin. Neurophysiol. 82: 155–159.

Allen TJ, Brown DA (2004) Modulation of the excitability of cholinergic basal forebrain neurones by K_{ATP} channels. J. Physiol. 554: 353–370.

Almeida AN, Martinez V, Feindel W (2005) The first case of invasive EEG monitoring for the surgical treatment of epilepsy: historical significance and context. Epilepsia 46: 1082–1085.

Al-Mudallal AS, LaManna JC, Lust WD, Harik SI (1996) Diet-induced ketosis does not cause cerebral acidosis. Epilepsia 37: 258–261.

Alsen C (1983) Biological significance of peptides from *Anemonia sulcata*. Fed. Proc. 42: 101–108.

Alusi SH, Aziz TZ, Glickman S, Jahanshahi M, Stein JF, Bain PG (2001a) Stereotactic lesional surgery for the treatment of tremor in multiple sclerosis: a prospective case-controlled study. Brain 124: 1576–1589.

Alusi SH, Worthington J, Glickman S, Bain PG (2001b) A study of tremor in multiple sclerosis. Brain 124: 720–730.

Alviña K, Walter JT, Kohn A, Ellis-Davies G, Khodakhah K (2008) Questioning the role of rebound firing in the cerebellum. Nat. Neurosci. 11: 1256–1258.

Alzheimer A (1907) Über eine eigenartige Erkrankung der Hirnrinde. Allg. Z. Pzychiat. Psych.-Gerichtl. Med. 64: 146–148.

Alzheimer A (1911) Über eigenartige Krankheitsfälle des späteren Alters. Zbl. Ges. Neurol. Psych. 4: 356–385.

Amit DJ (1998) Simulation in neurobiology: theory or experiment? Trends Neurosci. 21: 231–237.

Amit DJ, Brunel N (1997) Model of global spontaneous activity and local structured activity during delay periods in the cerebral cortex. Cereb. Cortex 7: 237–252.

Amit DJ, Gutfreund H, Sompolinsky H (1985) Spin-glass models of neural networks. Physiol. Rev. A 32: 1007–1018.

Amitai Y, Friedman A, Connors BW, Gutnick MJ (1993) Regenerative activity in apical dendrites of pyramidal cells in neocortex. Cereb. Cortex 3: 26–38.

Amzica F, Steriade M (1995a) Short- and long-range neuronal synchronization of the slow (<1 Hz) cortical oscillation. J. Neurophysiol. 73: 20–38.

Amzica F, Steriade M (1995b) Disconnection of intracortical synaptic linkages disrupts synchronization of a slow oscillation. J. Neurosci. 15: 4658–4677.

Andermann ML, Moore CI (2006) A somatotopic map of vibrissa motion direction within a barrel column. Nat. Neurosci. 9: 543–551.

Andersen P, Andersson SA (1968) Physiological Basis of the Alpha Rhythm. Appleton-Century-Crofts, New York.

Andreasen NC (1999) A unitary model of schizophrenia: Bleuler's "fragmented phrene" as schizencephaly. Arch. Gen. Psychiatry 56: 781–787.

Andreuccetti P, Barone Lumaga MR, Cafiero G, Filosa S, Parisi E (1987) Cell junctions during the early development of the sea urchin embryo (*Paracentrotus lividus*). Cell Differ. 20: 137–146.

Andrew RD, Taylor CP, Snow RW, Dudek FE (1982) Coupling in rat hippocampal slices: dye transfer between CA1 pyramidal cells. Brain Res. Bull. 8: 211–222.

Antic SD (2003) Action potentials in basal and oblique dendrites of rat neocortical pyramidal neurons. J. Physiol. 550: 35–50.

Antkowiak B, Hentschke H (1997) Cellular mechanisms of gamma rhythms in rat neocortical brain slices probed by the volatile anaesthetic isoflurane. Neurosci. Lett. 231: 87–90.

Arbib MA, Amari S (1985) Sensori-motor transformations in the brain (with a critique of the tensor theory of cerebellum). J. Theor. Biol. 112: 123–155.

Arellano RO, Rivera A, Ramon F (1990) Protein phosphorylation and hydrogen ions modulate calcium-induced closure of gap-junction channels. Biophys. J. 57: 363–367.

Ariav G, Polsky A, Schiller J (2003) Submillisecond precision of the input-output transformation function mediated by fast sodium dendritic spikes in basal dendrites of CA1 pyramidal neurons. J. Neurosci. 23: 7750–7758.

Armstrong CM, Hille B (1998) Voltage-gated ion channels and electrical excitability. Neuron 20: 371–380.

Armstrong DM, Rawson JA (1979) Activity patterns of cerebellar cortical neurones and climbing fibre afferents in the awake cat. J. Physiol. 289: 425–448.

Armstrong DM, Saper CB, Levey AI, Wainer BH, Terry RD (1983) Distribution of cholinergic neurons in rat brain: demonstrated by the immunocytochemical localization of choline acetyltransferase. J. Comp. Neurol. 216: 53–68.

Arnold SE (2000) Cellular and molecular neuropathology of the parahippocampal region in schizophrenia. Ann. NY Acad. Sci. 911: 275–292.

Arnsten AF, Li BM (2005) Neurobiology of executive functions: catecholamine influences on prefrontal cortical functions. Biol. Psychiatry 57: 1377–1384.

Aroniadou VA, Keller A (1993) The patterns and synaptic properties of horizontal intracortical connections in the rat motor cortex. J. Neurophysiol. 70: 1553–1569.

Asada H, Kawamura Y, Maruyama K, Kume H, Ding RG, Kanbara N, Kuzume H, Sanbo M, Yagi T, Obata K (1997) Cleft palate and decreased brain gamma-aminobutyric acid in mice lacking the 67–kDa isoform of glutamic acid decarboxylase. Proc. Natl. Acad. Sci. USA 94: 6496–6499.

Asano E, Juhász C, Shah A, Muzik O, Chugani DC, Shah J, Sood S, Chugani HT (2005) Origin and propagation of epileptic spasms delineated on electrocorticography. Epilepsia 46: 1086–1097.

Asanuma H, Hunsperger RW (1975) Functional significance of projection from the cerebellar nuclei to the motor cortex in the cat. Brain Res. 98: 73–92.

Asanuma C, Thach WT, Jones EG (1983) Brain stem and spinal projections of the deep cerebellar nuclei in the monkey, with observations on the brain stem projections of the dorsal column nuclei. Brain Res. Rev. 5: 299–322.

Ascoli GA, Alonso-Nanclares L, Anderson SA, Barrionuevo G, Benavides-Piccione R, Burkhalter A, Buzsáki G, Cauli B, Defelipe J, Fairén A, Feldmeyer D, Fishell G, Fregnac Y, Freund TF, Gardner D, Gardner EP, Goldberg JH, Helmstaedter M, Hestrin S, Karube F, Kisvárday ZF, Lambolez B, Lewis DA, Marin O, Markram H, Muñoz A, Packer A, Petersen CC, Rockland KS, Rossier J, Rudy B, Somogyi P, Staiger JF, Tamas G, Thomson AM, Toledo-Rodriguez M, Wang Y, West DC, Yuste R. (2008) Petilla terminology: nomenclature of features of GABAergic interneurons of the cerebral cortex. Petilla Interneuron Nomenclature Group, Nat Rev Neurosci. 2008 Jul; 9(7): 557-568.

Ashby P, Lang AE, Lozano AM, Dostrovsky JO (1995) Motor effects of stimulating the human cerebellar thalamus. J. Physiol. 489: 287–298.

Astman N, Gutnick MJ, Fleidervish IA (1998) Activation of protein kinase C increases neuronal excitability by regulating persistent Na+ current in mouse neocortical slices. J. Neurophysiol. 80: 1547–1551.

Astman N, Gutnick MJ, Fleidervish IA (2006) Persistent sodium current in layer 5 neocortical neurons is primarily generated in the proximal axon. J. Neurosci. 26: 3465–3473.

Ault B, Hildebrand LM (1994) GABA$_A$ receptor-mediated excitation of nociceptive afferents in the rat isolated spinal cord-tail preparation. Neuropharmacology 33: 109–114.

Aumann TD, Fetz EE (2004) Oscillatory activity in forelimb muscles of behaving monkeys evoked by microstimulation in the cerebellar nuclei. Neurosci. Lett. 361: 106–110.

Avila M, Thaker G, Adami H (2001) Genetic epidemiology and schizophrenia: a study of reproductive fitness. Schizophr. Res. 47: 233–241.

Avoli M (1996) GABA-mediated synchronous potentials and seizure generation. Epilepsia 37: 1035–1042.

Avoli M, Gloor P (1981) The effects of transient functional depression of the thalamus on spindles and on bilateral synchronous epileptic discharges of feline generalized penicillin epilepsy. Epilepsia 22: 443–452.

Avoli M, Gloor P (1982) Role of the thalamus in generalized penicillin epilepsy: observations on decorticated cats. Exp. Neurol. 77: 386–402.

Avoli M, Gloor P, Kostopoulos G, Gotman J (1983) An analysis of penicillin-induced generalized spike and wave discharges using simultaneous recordings of cortical and thalamic single neurons. J. Neurophysiol. 50: 819–837.

Avoli M, Louvel J, Pumain R, Olivier A (1987) Seizure-like discharges induced by lowering [Mg^{2+}]$_o$ in the human epileptogenic neocortex maintained in vitro. Brain Res. 417: 199–203.

Avoli M, Drapeau C, Louvel J, Pumain R, Olivier A, Villemure J-G (1991) Epileptiform activity induced by low extracellular magnesium in the human cortex maintained in vitro. Ann. Neurol. 30: 589–596.

Avoli M, Methot M, Kawasaki H (1998) GABA-dependent generation of ectopic action potentials in the rat hippocampus. Eur. J. Neurosci. 10: 2714–2722.

Axelsson R, Ohman R (1987) Patterns of response to neuroleptic treatment: factors influencing the amelioration of individual symptoms in psychotic patients. Acta Psychiatr. Scand. 76: 707–714.

Ayala GF, Matsumoto H, Gumnit RJ (1970) Excitability changes and inhibitory mechanism in neocortical neurons during seizures. J. Neurophysiol. 33: 73–85.

Azmitia EC, Gannon PJ, Kheck NM, Whitaker-Azmitia PM (1996) Cellular localization of the 5–HT1A receptor in primate brain neurons and glial cells. Neuropsychopharmacology 14: 35–46.

Azouz R, Gray CM (2000) Dynamic spike threshold reveals a mechanism for synaptic coincidence detection in cortical neurons *in vivo*. Proc. Natl. Acad. Sci. USA 97: 8110-8115.

Azouz R, Gray CM, Nowak LG, McCormick DA (1997) Physiological properties of inhibitory interneurons in cat striate cortex. Cereb. Cortex 7: 534–545.

Bacci A, Rudolph U, Huguenard JR, Prince DA (2003) Major differences in inhibitory synaptic transmission onto two neocortical interneuron subclasses. J. Neurosci. 23: 9664–9674.

Bacci A, Huguenard JR, Prince DA (2005) Modulation of neocortical interneurons: extrinsic influences and exercises in self-control. Trends Neurosci. 28: 602–610.

Baer SM, Rinzel J, Carrillo H (1995) Analysis of an autonomous phase model for neuronal parabolic bursting. J. Math. Biol. 33: 309–333.

Bagetta G, Nisticò G (1992) Glutamate transmission is involved in the mechanisms of neuronal degeneration produced by intrahippocampal tetanus toxin in rats. Toxicol. Lett. 64–65: 447–453.

Bains JS, Longacher JM, Staley KJ (1999) Reciprocal interactions between CA3 network activity and strength of recurrent collateral synapses. Nat. Neurosci. 2: 720–726.

Baker M, Bostock H (1992) Ectopic activity in demyelinated spinal root axons of the rat. J. Physiol. 451: 539–552.

Baker R, Llinás R (1971) Electrotonic coupling between neurones in the rat mesencephalic nucleus. J. Physiol. 212: 45–63.

Baker SN (2007) Oscillatory interactions between sensorimotor cortex and the periphery. Curr. Opin. Neurobiol. 17: 649–655.

Baker SN, Olivier E, Lemon RN (1997) Coherent oscillations in monkey motor cortex and hand muscle EMG show task-dependent modulation. J. Physiol. 501: 225–241.

Baker SN, Kilner JM, Pinches EM, Lemon RN (1999) The role of synchrony and oscillations in the motor output. Exp. Brain Res. 128: 109–117.

Baker SN, Curio G, Lemon RN (2003a) EEG oscillations at 600 Hz are macroscopic markers for cortical spike bursts. J. Physiol. 550: 529–534.

Baker SN, Pinches EM, Lemon RN (2003b) Synchronization in monkey motor cortex during a precision grip task. II. Effect of oscillatory activity on corticospinal output. J. Neurophysiol. 89: 1941–1953.

Baker SN, Chiu M, Fetz EE (2006) Afferent encoding of central oscillations in the monkey arm. J. Neurophysiol. 95: 3904–3910.

Baks-Te Bulte L, Wouterlood FG, Vinkenoog M, Witter MP (2005) Entorhinal projections terminate onto principal neurons and interneurons in the subiculum: a quantitative electron microscopical analysis in the rat. Neuroscience 136: 729–739.

Bal T, McCormick DA (1993) Mechanisms of oscillatory activity in guinea-pig nucleus reticularis thalami *in vitro*: a mammalian pacemaker. J. Physiol. 468: 669–691.

Bal T, von Krosigk M, McCormick DA (1995a) Synaptic and membrane mechanisms underlying synchronized oscillations in the ferret lateral geniculate nucleus *in vitro*. J. Physiol. 483: 641–663.

Bal T, von Krosigk M, McCormick DA (1995b) Role of the ferret perigeniculate nucleus in the generation of synchronized oscillations *in vitro*. J. Physiol. 483: 665–685.

Baldeweg T, Spence S, Hirsch SR, Gruzelier J (1998) γ-band electroencephalographic oscillations in a patient with somatic hallucinations. Lancet 352: 620–621.

Baldwin KM, Hakim RS (1981) Freeze-fracture analysis of gap and septate junctions in embryos of a moth. Tissue Cell 13: 691–699.

Ballanyi K, Onimaru H, Homma I (1999) Respiratory network function in the isolated brainstem-spinal cord of newborn rats. Prog. Neurobiol. 59: 583–634.

Balslev T, Cortez MA, Blaser SI, Haslam RH (1997) Recurrent seizures in metachromatic leukodystrophy. Pediatr. Neurol. 17: 150–154.

Bamford NS, Zhang H, Zchmitz Y, Wu NP, Cepeda C, Levine MS, Schmauss C, Zakharenko SS, Zablow L, Sulzer D (2004) Heterosynaptic dopamine neurotransmission selects sets of corticostriatal terminals. Neuron 42: 653–663.

Bandyopadhyay S, Gonzalez-Islas C, Hablitz JJ (2005) Dopamine enhances spatiotemporal spread of activity in rat prefrontal cortex. J. Neurophysiol. 93: 864–872.

Banks MA, Pearce RA (2000) Kinetic differences between synaptic and extrasynaptic GABA$_A$ receptors in CA1 pyramidal cells. J. Neurosci. 20: 937–948.

Banks MI, Li TB, Pearce RA (1998) The synaptic basis of GABA$_{A,slow}$. J. Neurosci. 18: 1305–1317.

Bao L, Locovei S, Dahl G (2004) Pannexin membrane channels are mechanosensitive conduits for ATP. FEBS Lett. 572: 65–68.

Barabási A-L, Albert R (1999) Emergence of scaling in random networks. Science 286: 509–512.

Baranova A, Ivanov D, Skoblov M, Pestova A, Kelmanson I, Shagin D, Usman N, Statoverov D, Lukyanov S, Panchin Y (2002) Mammalian pannexin family homologous to invertebrate gap-junction proteins are differentially expressed in nervous tissue. Soc. Neurosci. Abstr. 836.9.

Barbe MT, Monyer H, Bruzzone R (2006) Cell-cell communication beyond connexins: the pannexin channels. Physiol. (Bethesda) 21: 103–114.

Barch DM (2005) The cognitive neuroscience of schizophrenia. Annu. Rev. Clin. Psychol. 1: 321–353.

Barclay J, Balaguero N, Mione M, Ackerman SL, Letts VA, Brodbeck J, Canti C, Meir A, Page KM, Kusumi K, Perez-Reyes E, Lander ES, Frankel WN, Gardiner RM, Dolphin AC, Rees M (2001) Ducky mouse phenotype of epilepsy and ataxia is associated with mutations in the Cacna2d2 gene and decreased calcium channel current in cerebellar Purkinje cells. J. Neurosci. 212: 6095–6104.

Barth DS (2003) Submillisecond synchronization of fast electrical oscillations in neocortex. J. Neurosci. 23: 2502–2510.

Barth DS, MacDonald KD (1996) Thalamic modulation of high-frequency oscillating potentials in auditory cortex. Nature 383: 78–81.

Barth DS, Di S, Baumgartner C (1989) Laminar cortical interactions during epileptic spikes studied with principal component analysis and physiological modeling. Brain Res. 484: 13–35.

Barth DS, Baumgartner C, Di S (1990) Laminar interactions rat motor cortex during cyclical excitability changes of the penicillin focus. Brain Res. 508: 105–117.

Bartos M, Vida I, Frotscher F, Geiger JRP, Jonas P (2001) Rapid signaling at inhibitory synapses in a dentate gyrus interneuron network. J. Neurosci. 21: 2687–2698.

Bartos M, Vida I, Frotscher M, Meyer A, Monyer A, Geiger JRP, Jonas P (2002) Fast synaptic inhibition promotes synchronized gamma oscillations in hippocampal interneuron networks. Proc. Natl. Acad. Sci. USA 99: 13222–13227.

Bartos M, Vida I, Jonas P (2007) Synaptic mechanisms of synchronized gamma oscillations in inhibitory interneuron networks. Nat. Rev. Neurosci. 8: 45–56.

Bast T, Oezkan O, Rona S, Stippich C, Seitz A, Rupp A, Fauser S, Zentner J, Rating D, Scherg M (2004) EEG and MEG source analysis of single and averaged interictal spikes reveals intrinsic epileptogenicity in focal cortical dysplasia. Epilepsia 4: 621–631.

Baudoux S, Empson RM, Richards CD (2003) Pentobarbitone modulates calcium transients in axons and synaptic boutons of hippocampal CA1 neurons. Br. J. Pharmacol. 140: 971–9.

Bauer M, Oostenveld R, Peeters M, Fries P (2006) Tactile spatial attention enhances gamma-band activity in somatosensory cortex and reduces low-frequency activity in parieto-occipital areas. J. Neurosci. 26: 490–501.

Bazhenov M, Timofeev I, Steriade M, Sejnowski TJ (2000) Spiking-bursting activity in the thalamic reticular nucleus initiates sequences of spindle oscillations in thalamic networks. J. Neurophysiol. 84: 1076–1087.

Bazhenov M, Stopfer M, Rabinovich M, Huerta R, Abarbanel HD, Sejnowski TJ, Laurent G (2001) Model of transient oscillatory synchronization in the locust antennal lobe. Neuron 30: 553–567.

Bazil CW, Morrell MJ, Pedley TA (2005) Epilepsy. In: Rowland LP (Ed.), Merritt's Neurology, 11th ed. Lippincott Williams & Wikins, Philadelphia, pp. 990–1016.

Beaussart M, Faou R (1976) Paroxysmal rolandic epilepsy pre-study of 293 cases. Lille Med. 21: 414–422.

Behr J, Gloveli T, Schmitz D, Heinemann U (2000) Dopamine depresses excitatory synaptic transmission onto rat subicular neurons via presynaptic D1–like dopamine receptors. J. Neurophysiol. 84: 112–119.

Behrens CJ, van den Boom LP, de Hoz L, Friedman A, Heinemann U (2005) Induction of sharp wave-ripple complexes *in vitro* and reorganization of hippocampal networks. Nat. Neurosci. 8: 1560–1567.

Behrens CJ, van den Boom LP, Heinemann U (2007a) Effects of $GABA_A$ receptor antagonists bicuculline and gabazine on stimulus-induced sharp wave-ripple complexes in adult rat hippocampus in vitro. Eur. J. Neurosci. 25: 2170–2181.

Behrens MM, Ali SS, Dao DN, Lucero J, Shekhtman G, Quick KL, Dugan LL (2007b) Ketamine-induced loss of phenotype of fast-spiking interneurons is mediated by NADPH-oxidase. Science 318: 1645–1647.

Beierlein M, Gibson JR, Connors BW (2000) A network of electrically coupled interneurons drives synchronized inhibition in neocortex. Nat. Neurosci. 3: 904–910.

Bekkers JM (2000a) Properties of voltage-gated potassium currents in nucleated patches from large layer 5 cortical pyramidal neurons of the rat. J. Physiol. 525: 593–609.

Bekkers JM (2000b) Distribution and activation of voltage-gated potassium channels in cell-attached and outside-out patches from large layer 5 cortical pyramidal neurons of the rat. J. Physiol. 525: 611–620.

Bell K, Churchill L, Kalivas PW (1995) GABAergic projection from the ventral pallidum and globus pallidus to the subthalamic nucleus. Synapse 20: 10–18.

Benabid AL, Pollak P, Louveau A, Henry S, de Rougemont J (1987) Combined (thalamotomy and stimulation) stereotactic surgery of the VIM thalamic nucleus for bilateral Parkinson disease. Appl. Neurophysiol. 50: 344–346.

Benabid AL, Koudsie A, Benazzouz A, Vercueil L, Fraix V, Chabardes S, Lebas JF, Pollak P (2001) Deep brain stimulation of the corpus luysi (subthalamic nucleus) and other targets in Parkinson's disease. Extension to new indications such as dystonia and epilepsy. J. Neurol. 248 Suppl. 3: III37–47.

Benedeczky I, Molnar E, Somogyi P (1994) The cisternal organelle as a Ca^{2+}-storing compartment associated with GABAergic synapses in the axon initial segment of hippocampal pyramidal neurones. Exp. Brain Res. 101: 216–230.

Benifla M, Otsubo H, Ochi A, Weiss SK, Donner EJ, Shroff M, Chuang S, Hawkins C, Drake JM, Elliott I, Smith ML, Snead OC 3rd, Rutka JT (2006) Temporal lobe surgery for intractable epilepsy in children: an analysis of outcomes in 126 children. Neurosurgery 59: 1203–1213.

Bennett MVL (1966) Physiology of electrotonic junctions. Ann. NY Acad. Sci. 137: 509–539.

Bennett MVL, Verselis VK (1992) Biophysics of gap junctions. Semin. Cell Biol. 3: 29–47.

Bennett MVL, Aljure E, Nakajima Y, Pappas GD (1963) Electrotonic junctions between teleost spinal neurons: electrophysiology and ultrastructure. Science 141: 262–264.

Berger B, Tassin JP, Blanc G, Moyne MA, Thierry A-M (1974) Histochemical confirmation for dopaminergic innervation of the rat cerebral cortex after destruction of the noradrenergic ascending pathways. Brain Res. 81: 332–337.

Berger B, Thierry A-M, Tassin JP, Moyne MA (1976) Dopaminergic innervation of the rat prefrontal cortex: a fluorescence histochemical study. Brain Res. 106: 133–145.

Berger B, Verney C, Alvarez C, Vigny A, Helle KB (1985) New dopaminergic terminal fields in the motor, visual (area 18b) and retrosplenial cortex in the young and adult rat. Immunocytochemical and catecholamine histochemical analyses. Neuroscience 15: 983–998.

Berger H (1929) Über das Elektroenkephalogramm des Menschen. Arch. Psychiatr. Nervenkrankh. 87: 527–570.

Berger T, Larkum ME, Lüscher H-R (2001) High I_h channel density in the distal apical dendrite of layer V pyramidal cells increases bidirectional attenuation of EPSPs. J. Neurophysiol. 85: 855–868.

Berger T, Senn W, Lüscher HR (2003) Hyperpolarization-activated current I_h disconnects somatic and dendritic spike initiation zones in layer V pyramidal neurons. J. Neurophysiol. 2428–2437.

Bergman H, Wichmann T, DeLong MR (1990) Reversal of experimental parkinsonism by lesions of the subthalamic nucleus. Science 249: 1436–1438.

Bergman H, Wichmann T, Karmon B, DeLong MR (1994) The primate subthalamic nucleus. II. Neuronal activity in the MPTP model of parkinsonism. J. Neurophysiol. 72: 507–520.

Bergoffen J, Scherer SS, Wang S, Scott MO, Bone LJ, Paul DL, Chen K, Lensch MW, Chance PF, Fischbeck KH (1993) Connexin mutations in X-linked Charcot-Marie-Tooth disease. Science 262: 2039–2042.

Berkovic SF, Howell RA, Hay DA, Hopper JL (1998) Epilepsies in twins: genetics of the major epilepsy syndromes. Ann. Neurol. 43: 435–445.

Berkovic SF, Serratosa JM, Phillips HA, Xiong L, Andermann E, Díaz-Otero F, Gómez-Garre P, Martín M, Fernández-Bullido Y, Andermann F, Lopes-Cendes I, Dubeau F, Desbiens R, Scheffer I, Wallace RH, Mulley JC, Pandolfo M (2004) Familial partial epilepsy with variable foci: clinical features and linkage to chromosome 22q12. Epilepsia 45: 1054–1060.

Berkovic SF, Mulley JC, Scheffer IE, Petrou S (2006) Human epilepsies: interaction of genetic and acquired factors. Trends Neurosci. 29: 391–397.

Bernard A, Khrestchatisky M (1994) Assessing the extent of RNA editing in the TMII regions of GluR5 and GluR6 kainate receptors during brain development. J. Neurochem. 62: 2057–2060.

Bertrand O, Tallon-Baudry C (2000) Oscillatory gamma activity in humans: a possible role for object representation. Int. J. Psychophysiol. 38: 211–223.

Bessaïh T, Bourgeais L, Badiu CI, Carter DA, Tóth TI, Ruano D, Lambolez B, Crunelli V, Leresche N (2006) Nucleus-specific abnormalities of GABAergic synaptic transmission in a genetic model of absence seizures. J. Neurophysiol. 96: 3074–3081.

Beurrier C, Congar P, Bioulac B, Hammond C (1999) Subthalamic nucleus neurons switch from single-spike activity to burst-firing mode. J. Neurosci. 19: 599–609.

Beurrier C, Bioulac B, Hammond C (2000) Slowly inactivating sodium current (I_{NaP}) underlies single-spike activity in rat subthalamic neurons. J. Neurophysiol. 83: 1951–1957.

Bevan MD, Magill PJ, Hallworth NE, Bolam JP, Wilson CJ (2002a) Regulation of the timing and pattenr of action potential generation in rat subthalamic neurons in vitro by GABA$_A$ IPSPs. J. Neurophysiol. 87: 1348–1362.

Bevan MD, Magill PJ, Terman D, Bolam JP, Wilson CJ (2002b) Move to the rhythm: oscillations in the subthalamic nucleus-external globus pallidus network. Trends Neurosci. 25: 525–531.

Bevan MD, Hallworth NE, Baufreton J (2007) GABAergic control of the subthalamic nucleus. Prog. Brain Res. 160: 173–188.

Beyer KH (1947) New concept of competitive inhibition of the renal tubular excretion of penicillin. Science 105: 94–95.

Bhanushali MJ, Tuite PJ (2004) The evaluation and management of patients with neuroleptic malignant syndrome. Neurol. Clin. 22: 389–411.

Bhisitkul RB, Villa JE, Kocsis JD (1987) Axonal GABA receptors are selectively present on normal and regenerated sensory fibers in rat peripheral nerve. Exp. Brain Res. 66: 659–663.

Bhugra D (2005) The global prevalence of schizophrenia. PloS Med. 2(5): e151 doi:10.1371/journal.pmed.0020151.

Bi GQ, Poo MM (1998) Synaptic modifications in cultured hippocampal neurons: dependence on spike timing, synaptic strength, and postsynaptic cell type. J. Neurosci. 18: 10464–10472.

Bibbig A, Faulkner HJ, Whittington MA, Traub RD (2001) Self-organized synaptic plasticity contributes to the shaping of γ and β oscillations in vitro. J. Neurosci. 21: 9053–9067.

Bibbig A, Traub RD, Whittington MA (2002) Characteristics of long-range syn-chronization of γ and β oscillations and the plasticity of excitatory and inhibitory synapses: a network model. J. Neurophysiol. 88: 1634–1654.

Bibbig A, Middleton S, Racca C, Gillies MJ, Garner H, LeBeau FEN, Davies CH, Whittington MA (2007) Beta rhythms (15–20 Hz) generated by nonreciprocal communication in hippocampus. J. Neurophysiol. 97: 2812–2823.

Bidzan L (2007) A review of the effects of nicotine on schizophrenia. Psychatr. Pol. 41: 737–744.

Binder DK (2004) A history of Todd and his paralysis. Neurosurgery 54: 480–486.

Binshtok AM, Fleidervish IA, Sprengel R, Gutnick M (2006) NMDA receptors in layer 4 spiny stellate cells of the mouse barrel cortex contain the NR2C subunit. J. Neurosci. 26: 708–715.

Bishop GA, Chen YF, Burry RW, King JS (1993) An analysis of GABAergic afferents to basket cell bodies in the cat's cerebellum. Brain Res. 623: 293–298.

Bittman K, Becker DL, Cicirata F, Parnavelas JG (2002) Connexin expression in homotypic and heterotypic cell coupling in the developing cerebral cortex. J. Comp. Neurol. 443: 201–212.

Black JA, Kocsis JD, Waxman SG (1990) Ion channel organization of the myelinated fiber. Trends Neurosci. 13: 48–54.

Blackwood W, McMenemy WH, Meyer A, Norman RM, Russell DS (1967) Greenfield's Neuropathology. Williams & Wilkins, Baltimore, 1967.

Blatow M, Rozov A, Katona I, Hormuzdi SG, Meyer AH, Whittington MA, Caputi A, Monyer H (2003) A novel network of multipolar bursting interneurons generates theta frequency oscillations in neocortex. Neuron 38: 805–817.

Bleakman D, Lodge D (1998) Neuropharmacology of AMPA and kainate receptors. Neuropharmacology 37: 1187–1204.

Blenkinsop TA, Lang EJ (2006) Block of inferior olive gap junctional coupling decreases Purkinje cell complex spike synchrony and rhythmicity. J. Neurosci. 26: 1739–1748.

Blethyn KL, Hughes SW, Tóth TI, Cope DW, Crunelli V (2006) Neuronal basis of the slow (<1 Hz) oscillation in neurons of the nucleus reticularis thalami *in vitro*. J. Neurosci. 26: 2474–2486.

Bleuler E (1911) Dementia Praecox oder Gruppe der Schizophrenien. Arts & Boeve, Nijmegen.

Blume WT (2001) Current trends in electroencephalography. Curr. Opin. Neurol. 14: 193–197.

Boeve BF, Silber MH, Ferman TJ, Lucas JA, Parisi JE (2001) Association of REM sleep behavior disorder and neurodegenerative disease may reflect an underlying synucleinopathy. Mov. Disord. 16: 622–630.

Bolam JP, Hanley JJ, Booth PAC, Bevan MD (2000) Synaptic organization of the basal ganglia. J. Anat. 196: 527–542.

Bollobás B (2000) Modern Graph Theory. Graduate Texts in Mathematics 184. Springer, New York.

Bollobás B (2001) Random Graphs, 2nd ed. Cambridge Studies in Advanced Mathematics 73. Cambridge University Press, Cambridge, U.K.

Bolstad I, Leergaard TB, Bjaalie JG (2007) Branching of individual somatosensory cerebropontine axons in rat: evidence of divergence. Brain Struct. Funct. 212: 85–93.

Bonati LH, Naegelin Y, Wieser HG, Fuhr P, Ruegg S (2006) Beta activity in status epilepticus. Epilepsia 47: 207–210.

Boor R, Jacobs J, Hinzmann A, Bauermann T, Scherg M, Boor S, Vucurevic G, Pfleiderer C, Kutschke G, Stoeter P (2007) Combined spike-related functional MRI and multiple source analysis in the non-invasive spike localization of benign rolandic epilepsy. Clin. Neurophysiol. 118: 901–909.

Borck C, Jefferys JGR (1999) Seizure-like events in disinhibited ventral slices of adult rat hippocampus. J. Neurophysiol. 82: 2130–2142.

Börgers C, Kopell N (2003) Synchronization in networks of excitatory and inhibitory neurons with sparse, random connectivity. Neural Comput. 15: 509–538.

Borst JG, Sakmann B (1998) Calcium current during a single action potential in a large presynaptic terminal of the rat brainstem. J. Physiol. 506: 143–157.

Bostanci MO, Bagirici F (2007) Anticonvulsive effects of quinine on penicillin-induced epileptiform activity: an in vivo study. Seizure 16: 166–172.

Bou-Flores C, Berger AJ (2001) Gap junctions and inhibitory synapses modulate inspiratory motoneuron synchronization. J. Neurophysiol. 85: 1543–1551.

Bourrat F, Sotelo C (1983) Postnatal development of the inferior olivary complex in the rat. I. An electron microscopic study of the medial accessory olive. Brain Res. 284: 291–310.

Braak H, Bohl JR, Müller CM, Rüb U, de Vos RA, Del Tredici K (2006) Stanley Fahn Lecture 2005: The staging procedure for the inclusion body pathology associated with sporadic Parkinson's disease reconsidered. Mov. Disord. 21: 2042–2051.

Bracci E, Vreugdenhil M, Hack SP, Jefferys JGR (1999) On the synchronizing mechanisms of tetanically induced hippocampal oscillations. J. Neurosci. 19: 8104–8113.

Bragin A, Jandó G, Nádasdy Z, Hctkc J, Wise K, Buzsáki G (1995) Gamma (40–100 Hz) oscillation in the hippocampus of the behaving rat. J. Neurosci. 15: 47–60.

Bragin A, Penttonen M, Buzsáki G (1997) Termination of epileptic afterdischarge in the hippocampus. J. Neurosci. 17: 2567–2579.

Bragin A, Engel J Jr, Wilson CL, Fried I, Buzsáki G (1999a) High-frequency oscillations in the human brain. Hippocampus 9: 137–142.

Bragin A, Engel J Jr, Wilson CL, Fried I, Mathern GW (1999b) Hippocampal and entorhinal cortex high-frequency oscillations (100–500 Hz) in human epileptic brain and in kainic acid-treated rats with chronic seizures. Epilepsia 40: 127–137.

Bragin A, Mody I, Wilson CL, Engel J Jr (2002a) Local generation of fast ripples in epileptic brain. J.Neurosci. 22: 2012–2021.

Bragin A, Wilson CL, Staba RJ, Reddick M, Fried I, Engel J Jr (2002b) Interictal high-frequency oscillations (80–500 Hz) in the human epileptic brain: entorhinal cortex. Ann. Neurol. 52: 407–415.

Bragin A, Wilson CL, Engel J Jr (2003) Spatial stability over time of brain areas generating fast ripples in the epileptic rat. Epilepsia 44: 1233–1237.

Bragin A, Azizyan A, Almajano J, Wilson CL, Engel J Jr (2005) Analysis of chronic seizure onsets after intrahippocampal kainic acid injection in freely moving rats. Epilepsia 46: 1592–1598.

Braitenberg V (1961) Functional interpretation of cerebellar histology. Nature 190: 539–540.

Braitenberg V, Schüz A (1998) Cortex: Statistics and Geometry of Neuronal Connectivity. Springer, Heidelberg.

Breier A, Wolkowitz OM, Doran AR, Roy A, Boronow J, Hommer DW, Pickar D (1987) Neuroleptic responsivity of negative and positive symptoms in schizophrenia. Am. J. Psychiatry 144: 1549–1555.

Brenowitz GL, Collins WF III, Erulkar SD (1983) Dye and electrical coupling between frog motoneurons. Brain Res. 274: 371–375.

Brett B, Barth DS (1997) Subcortical modulation of high-frequency (gamma band) oscillating potentials in auditory cortex. J. Neurophysiol. 78: 573–581.

Bringuier V, Frégnac Y, Baranyi A, Debanne D, Shulz DE (1997) Synaptic origin and stimulus dependency of neuronal oscillatory activity in the primary visual cortex of the cat. J. Physiol. 500: 751–774.

Brock LG, Coombs JS, Eccles JC (1952) The recording of potentials from motoneurones with an intracellular electrode. J. Physiol. 117: 431–460.

Brodmann K (1909) Vergleichende Lokalisationslehre der Großhirnrinde. Barth, Leipzig.

Brosch M, Bauer R, Eckhorn R (1995) Synchronous high-frequency oscillations in cat area 18. Eur. J. Neurosci. 7: 86–95.

Brown DA, Adams PR (1980) Muscarinic suppression of a novel voltage-sensitive K^+ current in a vertebrate neurone. Nature 283: 673–676.

Brown EM (1994) French Psychiatry's initial reception of Bayle's discovery of general paresis of the insane. Bull. Hist. Med. 68: 235–253.

Brown P (2000) Cortical drives to human muscle: the Piper and related rhythms. Prog. Neurobiol. 60: 97–108.

Brown P (2003) Oscillatory nature of human basal ganglia activity: relationship to the pathophysiology of Parkinson's disease. Mov. Disord. 18: 357–363.

Brown P (2006) Bad oscillations in Parkinson's disease. J. Neural Transm. Suppl. (70): 27–30.

Brown P, Williams D (2005) Basal ganglia local field potential activity: character and functional significance in the human. Clin. Neurophysiol. 116: 2510–2519.

Brown P, Salenius S, Rothwell JC, Hari R (1998) Cortical correlate of the Piper rhythm in humans. J. Neurophysiol. 80: 2911–2917.

Brown P, Oliviero A, Mazzone P, Insola A, Tonali P, Di Lazzaro V (2001) Dopamine dependency of oscillations between subthalamic nucleus and pallidum in Parkinson's disease. J. Neurosci. 21: 1033–1038.

Brown P, Kupsch A, Magill PJ, Sharott A, Harnack D, Meissner W (2002) Oscillatory local field potentials recorded from the subthalamic nucleus of the alert rat. Exp. Neurol. 177: 581–585.

Brown-Séquard CE (1856) Recherches expérimentales sur la production d'une affection convulsive épileptiforme, á la suite de lésions de la moelle épinière. Comptes Rendus Acad. Sci. 42: 86–89.

Brückner G, Szeöke S, Pavlica S, Grosche J, Kacza J (2006) Axon initial segment ensheathed by extracellular matrix in perineuronal nets. Neuroscience 138: 365–375.

Brumberg JC, Nowak LG, McCormick DA (2000) Ionic mechanisms underlying repetitive high-frequency burst firing in supragranular cortical neurons. J. Neurosci. 20: 4829–4843.

Bruns A, Eckhorn R (2004) Task-related coupling from high- to low-frequency signals among visual cortical areas in human subdural recordings. Int. J. Psychophysiol. 51: 97–116.

Brusa R, Zimmermann F, Koh D-S, Feldmeyer D, Gass P, Seeburg PH, Sprengel R (1995) Early-onset epilepsy and postnatal lethality associated with an editing-deficient GluR-B allele in mice. Science 270: 1677–1680.

Bruzzone R, Dermietzel R (2006) Structure and function of gap junctions in the developing brain. Cell Tissue Res. 326: 239–248.

Bruzzone R, Hormuzdi SG, Barbe MT, Herb A, Monyer H (2003) Pannexins, a family of gap junction proteins expressed in brain. Proc. Natl. Acad. Sci. 100: 13644–13649.

Bruzzone R, Barbe MT, Jakob NJ, Monyer H (2005) Pharmacological properties of homomeric and heteromeric pannexin hemichannels expressed in Xenopus oocytes. J. Neurochem. 95: 1033–1043.

Bucci P, Mucci A, Merlotti E, Volpe U, Galderisi S (2007) Induced gamma activity and event-related coherence in schizophrenia. Clin. EEG. Neurosci. 38: 96–104.

Buchhalter JR (1993) Animal models of inherited epilepsy. Epilepsia 34 Suppl. 3: S31–41.

Buhl DL, Buzsáki G (2005) Developmental emergence of hippocampal fast-field "ripple" oscillations in the behaving rat pups. Neuroscience 134: 1423–1430.

Buhl DL, Harris KD, Hormuzdi SG, Monyer H, Buzsáki G (2003) Selective impairment of hippocampal gamma oscillations in connexin-36 knock-out mouse in vivo. J. Neurosci. 23: 1013–1018.

Buhl EH, Han Z-S, Lörinczi Z, Stezhka VV, Karnup SV, Somogyi P (1994) Physiological properties of anatomically identified axo-axonic cells in the rat hippocampus. J. Neurophysiol. 71: 1289–1307.

Buhl EH, Tamás G, Szilágyi T, Stricker C, Paulsen O, Somogyi P (1997) Effect, number and location of synapses made by single pyramidal cells onto aspiny interneurones of cat visual cortex. J. Physiol. 500: 689–713.

Buhl EH, Tamás G, Fisahn A (1998) Cholinergic activation and tonic excitation induce persistent gamma oscillations in mouse somatosensory cortex in vitro. J. Physiol. 513: 117–126.

Bulloch AGM, Kater SB, Miller HR (1984) Stability of new electrical connections between adult *Helisoma* neurons is influenced by preexisting neuronal interactions. J. Neurosci. 52: 1094–1105.

Bunney BS, Chiodo LA, Grace AA (1991) Midbrain dopamine system electrophysiological functioning: a review and new hypothesis. Synapse 9: 79–94.

Burchell TR, Faulkner HJ, Whittington MA (1998) Gamma frequency oscillations gate temporally coded afferent inputs in the rat hippocampal slice. Neurosci. Lett. 255: 1–4.

Burke D, Hagbarth KE, Skuse NE (1978) Recruitment order of human spindle endings in isometric voluntary contractions. J. Physiol. 285: 101–112.

Burkhalter A (2008) Many specialists for suppressing cortical excitation. Front. Neurosci. 2: 155–167.

Burmeister M, McInnis MG, Zöllner S (2008) Psychiatric genetics: progress amid controversy. Nat. Rev. Genet. 9: 527–540.

Burnashev N, Monyer H, Seeburg PH, Sakmann B (1992) Divalent ion permeability of AMPA receptor channels is dominated by the edited form of a single subunit. Neuron 8: 189–198.

Burnashev N, Zhou Z, Neher E, Sakmann B (1995) Fractional calcium currents through recombinant GluR channels of the NMDA, AMPA and kainate receptor subtypes. J. Physiol. 485: 403–418.

Burns RS, Chiueh CC, Markey SP, Ebert MH, Jacobowitz DM, Kopin IJ (1983) A primate model of parkinsonism: selective destruction of dopaminergic neurons in the pars compacta of the substantia nigra by N-methyl-4–phenyl-1,2,3,6–tetrahydropyridine. Proc. Natl. Acad. Sci. USA 80: 4546–4550.

Burns RS, LeWitt PA, Ebert MH, Pakkenberg H, Kopin IJ (1985) The clinical syndrome of striatal dopamine deficiency. Parkinsonism induced by 1–methyl-4–phenyl-1,2,3,6–tetrahydropyridine (MPTP). N. Engl. J. Med. 312: 1418–1421.

Buwalda B, de Groote L, Van der Zee EA, Matsuyama T, Luiten PG (1995) Immunocytochemical demonstration of developmental distribution of muscarinic acetylcholine receptors in rat parietal cortex. Brain Res. Dev. Brain Res. 84: 185–191.

Buzsáki G (1986) Hippocampal sharp waves: their origin and significance. Brain Res. 398: 242–252.

Buzsáki G (2002) Theta oscillations in the hippocampus. Neuron 33: 325–340.

Buzsáki G (2006) Rhythms of the Brain. Oxford University Press, Oxford.

Buzsáki G, Horváth Z, Urioste R, Hetke J, Wise K (1992) High-frequency network oscillation in the hippocampus. Science 256:1025–1027.

Cai X, Gu Z, Zhong P, Ren Y, Yan Z (2002) Serotonin 5–HT1A receptors regulate AMPA receptor channels through inhibiting Ca^{2+}/calmodulin-dependent kinase II in prefrontal cortical pyramidal neurons. J. Biol. Chem. 277: 36553–36562.

Camfield P, Camfield C (2002) Epileptic syndromes in childhood: clinical features, outcomes, and treatment. Epilepsia 43 Suppl. 3: 27–32.

Camperi M, Wang X-J (1998) A model of visuospatial working memory in prefrontal cortex: recurrent network and cellular bistability. J. Comput. Neurosci. 5: 383–405.

Canolty RT, Edwards E, Dalal SS, Soltani M, Nagarajan SS, Kirsch HE, Berger MS, Barbaro NM, Knight RT (2006) High gamma power is phase-locked to theta oscillations in human neocortex. Science 313: 1626–1628.

Canteras NS, Shammah-Lagnado SJ, Silva BA, Ricardo JA (1988) Somatosensory inputs to the subthalamic nucleus: a combined retrograde and anterograde horseradish peroxidase study in the rat. Brain Res. 458: 53–64.

Caplan JB, Madsen JR, Schulze-Bonhage A, Aschenbrenner-Scheibe R, Newman EL, Kahana MJ (2003) Human theta oscillations related to sensorimotor integration and spatial learning. J. Neurosci. 23: 4726–4736.

Caporale N, Dan Y (2008) Spike timing-dependent plasticity: a Hebbian learning rule. Annu. Rev. Neurosci. 31: 25–46.

Cardin JA, Palmer LA, Contreras D (2005) Stimulus-dependent γ (30–50 Hz) oscillations in simple and complex fast rhythmic bursting cells in primary visual cortex. J. Neurosci. 25: 5339–5350.

Cassim F, Labyt E, Devos D, Defebvre L, Destée A, Derambure P (2002) Relationship between oscillations in the basal ganglia and synchronization of cortical activity. Epileptic Disord. 4 Suppl. 3: S31–S45.

Castaldo P, del Giudice EM, Coppola G, Pascotto A, Annunziato L, Taglialatela M (2002) Benign familial neonatal convulsions caused by altered gating of KCNQ2/KCNQ3 potassium channels. J. Neurosci. 22: RC199.

Castillo PE, Malenka RC, Nicoll RA (1997) Kainate receptors mediate a slow postsynaptic current in hippocampal CA3 neurons. Nature 388: 182–186.

Castro-Alamancos MA (2000) Origin of synchronized oscillations induced by neocortical disinhibition in vivo. J. Neurosci. 20: 9195–9206.

Castro-Alamancos MA, Rigas P (2002) Synchronized oscillations caused by disinhibition in rodent neocortex are generated by recurrent synaptic activity mediated by AMPA receptors. J. Physiol. 542: 567–581.

Caton R (1875) The electric currents of the brain. Br. Med. J. 2: 278.

Catterall WA (1999) Molecular properties of brain sodium channels: an important target for anticonvulsant drugs. Adv. Neurol. 79: 441–456.

Cauli B, Audinat E, Lambolez B, Angulo MC, Ropert N, Tsuzuki K, Hestrin S, Rossier J (1997) Molecular and physiological diversity of cortical nonpyramidal cells. J. Neurosci. 17: 3894–3906.

Caveney S, Podgorski C (1975) Intercellular communication in a positional field. Ultrastructural correlates and tracer analysis of communication between insect epidermal cells. Tissue Cell 7: 559–574.

Chagnac-Amitai Y, Connors BW (1989a) Horizontal spread of synchronized activity in neocortex, and its control by GABA-mediated inhibition. J. Neurophysiol. 61: 747–758.

Chagnac-Amitai Y, Connors BW (1989b) Synchronized excitation and inhibition driven by intrinsically bursting neurons in neocortex. J. Neurophysiol. 62: 1149–1162.

Chagnac-Amitai Y, Luhmann HJ, Prince DA (1990) Burst generating and regular spiking layer 5 pyramidal neurons of rat neocortex have different morphological features. J. Comp. Neurol. 296: 598–613.

Chambers WW, Sprague JM (1955) Functional localization in the cerebellum. II. Somatotopic organization in cortex and nuclei. Arch. Neurol. Psychiatry 74: 653–680.

Chang Q, Balice-Gordon RJ (2000) Gap junctional communication among developing and injured motor neurons. Brain Res. Brain Res. Rev. 32: 242–249.

Chang Q, Gonzalez M, Pinter MJ, Balice-Gordon RJ (1999) Gap junctional coupling and patterns of connexin expression among neonatal rat lumbar spinal motor neurons. J. Neurosci. 19: 10813–10828.

Chang Q, Pereda A, Pinter MJ, Balice-Gordon RJ (2000) Nerve injury induces gap junctional coupling among axotomized adult motor neurons. J. Neurosci. 20: 674–684.

Chapman JP (1966) The early symptoms of schizophrenia. Br. J. Med. Psychol. 112: 225–251.

Chatt AB, Ebersole JS (1982) The laminar sensitivity of cat striate cortex to penicillin induced epileptogenesis. Brain Res. 241: 382–387.

Chen AC, Herrmann CS (2001) Perception of pain coincides with the spatial expansion of electroencephalographic dynamics in human subjects. Neurosci. Lett. 297: 183–186.

Chen G, Greengard P, Yan Z (2004) Potentiation of NMDA receptor currents by dopamine D1 receptors in prefrontal cortex. Proc. Natl. Acad. Sci. USA 101: 2596–2600.

Chen JC, Chesler M (1991) Extracellular alkalinization evoked by GABA and its relationship to activity-dependent pH shifts in turtle cerebellum. J. Physiol. 442: 431–446.

Chen S, Wang J, Siegelbaum SA (2001) Properties of hyperpolarization-activated pacemaker current defined by coassembly of HCN1 and HCN2 subunits and basal modulation by cyclic nucleotide. J. Gen. Physiol. 117: 491–503.

Cheron G, Gall D, Servais L, Dan B, Maex R, Schiffmann SN (2004) Inactivation of calcium-binding protein genes induces 160 Hz oscillations in the cerebellar cortex of alert mice. J. Neurosci. 24: 434–441.

Cheron G, Servais L, Dan B, Gall D, Roussel C, Schiffmann SN (2005a) Fast oscillation in the cerebellar cortex of calcium binding protein-deficient mice: a new sensorimotor arrest rhythm. Prog. Brain Res. 148: 165–180.

Cheron G, Servais L, Wagstaff J, Dan B (2005b) Fast cerebellar oscillation associated with ataxia in a mouse model of Angelman syndrome. Neuroscience 130: 631–637.

Cheron G, Servais L, Dan B (2008) Cerebellar network plasticity: from genes to fast oscillation. Neuroscience 153: 1–19.

Chervin RD, Pierce PA, Connors BW (1988) Periodicity and directionality in the propagation of epileptiform discharges across neocortex. J. Neurophysiol. 60: 1695–1713.

Chevrie JJ, Aicardi J (1972) Childhood epileptic encephalopathy with slow spike-wave. A statistical study of 80 cases. Epilepsia 13: 259–271.

Chiba M (1980) Patterns of organization of the corticopontine projection in the cat with the pontocerebellar projection. J. Hirnforsch. 21: 89–99.

Cho RY, Konecky RO, Carter CS (2006) Impairments in frontal cortical (syn-chrony and cognitive control in schizophrenia. Proc. Natl. Acad. Sci. USA 103: 19878–19883.

Chow CC, Kopell N (2000) Dynamics of spiking neurons with electrical coupling. Neural Comput. 12: 1643–1678.

Chow CC, White JA, Ritt J, Kopell N (1998) Frequency control in synchronized networks of inhibitory neurons. J. Comput. Neurosci. 5: 407–420.

Chrobak JJ, Buzsáki G (1994) Selective activation of deep layer (V-VI) retrohippocampal cortical neurons during hippocampal sharp waves in the behaving rat. J. Neurosci. 14: 6160–6170.

Chrobak JJ, Buzsáki G (1996) High-frequency oscillations in the output networks of the hippocampal-entorhinal axis of the freely behaving rat. J. Neurosci. 16: 3056–3066.

Chrobak JJ, Buzsáki G (1998) Gamma oscillations in the entorhinal cortex of the freely behaving rat. J. Neurosci. 18: 388–398.

Church J, Baimbridge KG (1991) Exposure to high-pH medium increases the incidence and extent of dye coupling between rat hippocampal CA1 pyramidal neurons in vitro. J. Neurosci. 11: 3289–3295.

Cioni B (2007) Motor cortex stimulation for Parkinson's disease. Acta Neurochir. Suppl. 97: 233–238.

Clark GD, Clifford DB, Zorumski CF (1990) The effect of agonist concentration, membrane voltage and calcium on N-methyl-D-aspartate receptor desensitization. Neuroscience 39: 787-797.

Clark BA, Monsivais P, Branco T, London M, Häusser M (2005) The site of action potential initiaiton in cerebellar Purkinje neurons. Nat. Neurosci. 8: 137–139.

Clementz BA, Keil A, Kissler J (2004) Aberrant brain dynamics in schizophrenia: delayed buildup and prolonged decay of the visual steady-state response. Cogn. Brain Res. 18: 121–129.

Clifford DB, Olney JW, Maniotis A, Collins RC, Zorumski CF (1987) The functional anatomy and pathology of lithium pilocarpine and high-dose pilocarpine seizures. Neuroscience 23: 953–968.

Clough JF, Phillips C, Sheridan JD (1971) The short-latency projection from the baboon's motor cortex to fusimotor neurones of the forearm and hand. J. Physiol. 216: 257–279.

Cobb SR, Buhl EH, Halasy K, Paulsen O, Somogyi P (1995) Synchronization of neuronal activity in hippocampus by individual GABAergic interneurons. Nature 378: 75–78.

Coenen AM, Van Luijtelaar EL (2003) Genetic animal models for absence epilepsy: a review of the WAG/Rij strain of rats. Behav. Genet. 33: 635–655.

Cohen I, Navarro V, Clemenceau S, Baulac M, Miles R (2002) On the origin of interictal activity in human temporal lobe epilepsy in vitro. Science 298: 1418–1421.

Cohen I, Navarro V, Le Duigou C, Miles R (2003) Mesial temporal lobe epilepsy: a pathological replay of developmental mechanisms? Biol. Cell. 95: 329–333.

Cohen I, Huberfeld G, Miles R (2006) Emergence of disinhibition-induced synchrony in the CA3 region of the guinea pig hippocampus in vitro. J. Physiol. 570: 583–594.

Colbert CM, Johnston D (1996) Axonal action-potential initiation and Na^+ channel densities in the soma and axon initial segment of subicular pyramidal neurons. J. Neurosci. 16: 6676–6686.

Colbert CM, Pan E (2002) Ion channel properties underlying axonal action potential initiation in pyramidal neurons. Nat. Neurosci. 5: 533–538.

Colder BW, Wilson CL, Frysinger RC, Chao LC, Harper RM, Engel J Jr (1996) Neuronal synchrony in relation to burst discharge in epileptic human temporal lobes. J. Neurophysiol. 75: 2496–2508.

Cole AE, Nicoll RA (1984) The pharmacology of cholinergic excitatory responses in hippocampal pyramidal cells. Brain Res. 305: 283–290.

Colling SB, Man WD, Draguhn A, Jefferys JGR (1996) Dendritic shrinkage and dye-coupling between rat hippocampal CA1 pyramidal cells in the tetanus toxin model of epilepsy. Brain Res. 741: 38–43.

Collingridge GL (1995) The brain slice preparation: a tribute to the pioneer Henry McIlwain. J. Neurosci. Methods 59: 5–9.

Collins DR, Pelletier JG, Paré D (2001) Slow and fast (gamma) neuronal oscillations in the perirhinal cortex and lateral amygdala. J. Neurophysiol. 85: 1661–1672.

Collins WF III Organization of electrical coupling between frog lumbar motoneurons, J. Neurophysiol. 49 (1983) 730–744.

Combs CM (1954) Electro-anatomical study of cerebellar localization: stimulation of various afferents. J. Neurophysiol. 17: 123–143.

Compte A, Brunel N, Goldman-Rakic PS, Wang X-J (2000) Synaptic mechanisms and network dynamics underlying spatial working memory in a cortical network model. Cereb. Cortex 10: 910–923.

Condorelli DF, Parenti R, Spinella F, Salinaro AT, Belluardo N, Cardile V, Cicirata F (1998) Cloning of a new gap junction gene (Cx36) highly expressed in mammalian brain neurons. Eur. J. Neurosci. 10: 1202–1208.

Condorelli DF, Belluardo N, Trovato-Salinaro A, Mudó G (2000) Expression of Cx36 in mammalian neurons. Brain Res. Brain Res. Rev. 32: 72–85.

Condorelli DF, Trovato-Salinaro A, Mudó G, Mirone MB, Belluardo N (2003) Cellular expression of connexins in the rat brain: neuronal localization, effects of kainate-induced seizures and expression in apoptotic neuronal cells. Eur. J. Neurosci. 18: 1807–1827.

Connors BW (1984) Initiation of synchronized neuronal bursting in neocortex. Nature 310: 685–687.

Connors BW, Gutnick MJ, Prince DA (1982) Electrophysiological properties of neocortical neurons in vitro. J. Neurophysiol. 48: 1302–1320.

Connors BW, Benardo LS, Prince DA (1983) Coupling between neurons of the developing rat neocortex. J. Neurosci. 3: 773–782.

Connors BW, Benardo LS, Prince DA (1984) Carbon dioxide sensitivity of dye coupling among glia and neurons of the neocortex. J. Neurosci. 4: 1324–1330.

Conrad B, Brooks VB (1974) Effects of dentate cooling on rapid alternating arm movements. J. Neurophysiol. 37: 792–804.

Constantinidis C, Steinmetz MA (1996) Neuronal activity in posterior parietal area 7a during the delay periods of a spatial memory task. J. Neurophysiol. 76: 1352–1355.

Contreras D, Steriade M (1995) Cellular basis of EEG slow rhythms: a study of dynamic corticothalamic relationships. J. Neurosci. 15: 604–622.

Contreras D, Steriade M (1996) Spindle oscillation in cats: the role of corticothalamic feedback in a thalamically generated rhythm. J. Physiol. 490: 159–179.

Contreras D, Curró Dossi R, Steriade M (1993) Electrophysiological properties of cat reticular thalamic neurones in vivo. J. Physiol. 470: 273–294.

Contreras D, Destexhe A, Sejnowski TJ, Steriade M (1996a) Control of spatiotemporal coherence of a thalamic oscillation by corticothalamic feedback. Science 274: 771–774.

Contreras D, Timofeev I, Steriade M (1996b) Mechanisms of long-lasting hyper-polarizations underlying slow sleep oscillations in cat corticothalamic networks. J. Physiol. 494: 251–264.

Cooley JW, Tukey JW (1965) An algorithm for the machine calculation of complex Fourier series. Math. Comput. 19: 297–301.

Coombs JS, Eccles JC, Fatt P (1955a) The specific ionic conductances and the ionic movements across the motoneuronal membrane that produce the inhibitory postsynaptic potential. J. Physiol. 130: 326–373.

Coombs JS, Eccles JC, Fatt P (1955b) Excitatory synaptic action in motoneurones. J. Physiol. 130: 374–395.

Coombs JS, Curtis DR, Eccles JC (1957) The generation of impulses in motoneurones. J. Physiol. 139: 232–249.

Cooper EC, Harrington E, Jan YN, Jan LY (2001) M channel KCNQ2 subunits are localized to key sites for control of neuronal network oscillations and synchronization in mouse brain. J. Neurosci. 21: 9529–9540.

Correll CU, Schenk EM (2008) Tardive dyskinesia and new antipsychotics. Curr. Opin. Psychiatry 21: 151–156.

Courtemanche R, Lamarre Y (2005) Local field potential oscillations in primate cerebellar cortex: synchronization with cerebral cortex during active and passive expectancy. J. Neurophysiol. 93: 2039–2052.

Courtemanche R, Fujii N, Graybiel AM (2003) Synchronous, focally modulated beta-band oscillations characterize local field potential activity in the striatum of awake behaving monkeys. J. Neurosci. 23: 11741–11752.

Covanis A (2006) Panayiotopoulos syndrome: a benign childhood autonomic epilepsy frequently imitating encephalitis, syncope, migraine, sleep disorder, or gastroenteritis. Pediatrics 118: e1237–1243.

Cramer NP, Keller A (2006) Cortical control of a whisking central pattern generator. J. Neurophysiol. 96: 209–217.

Crochet S, Fuentealba P, Timofeev I, Steriade M (2004) Selective amplification of neocortical neuronal output by fast prepotentials in vivo. Cereb. Cortex 14: 1110–1121.

Cruikshank SJ, Hopperstad M, Younger M, Connors BW, Spray DC, Srinivas M (2004) Potent block of Cx36 and Cx50 gap junction channels by mefloquine. Proc. Natl. Acad. Sci. USA 101: 12364–12369.

Crunelli V, Leresche N (2002) Childhood absence epilepsy: genes, channels, neurons and networks. Nat. Rev. Neurosci. 3: 371–382.

Crunelli V, Lightowler S, Pollard CE (1989) A T-type Ca^{2+} current underlies low-threshold Ca^{2+} potentials in cells of the cat and rat lateral geniculate nucleus. J. Physiol. 413: 543–561.

Csibra G, Davis G, Spratling MW, Johnson MH (2000) Gamma oscillations and object processing in the infant brain. Science 290: 1582–1585.

Csicsvári J, Hirase H, Czurko A, Buzsáki G (1998) Reliability and state dependence of pyramidal cell-interneuron synapses in the hippocampus: an ensemble approach in the behaving rat. Neuron 21: 179–189.

Csicsvári J, Hirase H, Czurkó A, Mamiya A, Buzsáki G (1999a) Oscillatory coupling of hippocampal pyramidal cells and interneurons in the behaving rat. J. Neurosci. 19: 274–287.

Csicsvari J, Hirase H, Czurkó A, Mamiya A, Buzsáki G (1999b) Fast network oscillations in the hippocampal CA1 region of the behaving rat. J. Neurosci. 19: 1–4.

Csicsvari J, Jamieson B, Wise KD, Buzsáki G (2003) Mechanisms of gamma oscillations in the hippocampus of the behaving rat. Neuron 37: 311–322.

Cunha RA, Malva JO, Ribeiro JA (1999) Kainate receptors coupled to G_i/G_o proteins in the rat hippocampus. Mol. Pharmacol. 56: 429–433.

Cunningham MO, Davies CH, Buhl EH, Kopell N, Whittington MA (2003) Gamma oscillations induced by kainate receptor activation in the entorhinal cortex in vitro. J. Neurosci. 23: 9761–9769.

Cunningham MO, Halliday DM, Davies CH, Traub RD, Buhl EH, Whittington MA (2004a) Coexistence of gamma and high-frequency oscillations in the medial entorhinal cortex in vitro. J. Physiol. 559: 347–353.

Cunningham MO, Hunt J, Middleton S, LeBeau FEN, Gillies MG, Davies CH, Maycox PR, Whittington MA, Racca C (2006a) Region-specific reduction in entorhinal gamma oscillations and parvalbumin-immunoreactive neurons in animal models of psychiatric illness. J. Neurosci. 26: 2767–2776.

Cunningham MO, Pervouchine D, Racca C, Kopell NJ, Davies CH, Jones RSG, Traub RD, Whittington MA (2006b) Neuronal metabolism governs cortical response state. Proc. Natl. Acad. Sci. USA 103: 5597–5601.

Cunningham MO, Whittington MA, Bibbig A, Roopun A, LeBeau FEN, Vogt A, Monyer H, Buhl EH, Traub RD (2004b) A role for fast rhythmic bursting neurons in cortical gamma oscillations in vitro. Proc. Natl. Acad. Sci. USA 101: 7152–7157.

Curio G (2000) Linking 600–Hz "spikelike" EEG/MEG wavelets ("sigma-bursts") to cellular substrates: concepts and caveats. J. Clin. Neurophysiol. 17: 377–396.

Curio G, Mackert BM, Burghoff M, Koetitz R, Abraham-Fuchs K, Harer W (1994) Localization of evoked neuromagnetic 600 Hz activity in the cerebral somatosensory system. Electroencephalogr. Clin. Neurophysiol. 91: 483–487.

Curró Dossi R, Nuñez A, Steriade M (1992) Electrophysiology of a slow (0.5–4 Hz) intrinsic oscillation of cat thalamocortical neurones in vivo. J. Physiol. 447: 215–234.

Curtis DR (1973) Bicuculline, GABA and central inhibition. Proc. Aust. Assoc. Neurol. 9: 145–153.

Dan Y, Poo MM (2004) Spike timing-dependent plasticity of neural circuits. Neuron 44: 23–30.

Davidson JS, Baumgarten IM (1988) Glycyrrhetinic acid derivatives: a novel class of inhibitors of gap-junctional intercellular communication. Structure-activity relationships. J. Pharamacol. Exp. Ther. 246: 1104–1107.

Davies J, Evans RH, Francis AA, Watkins JC (1979) Excitatory amino acid receptors and synaptic excitation in the mammalian central nervous system. J. Physiol. (Paris) 75: 641–654.

Deans MR, Gibson JR, Sellitto C, Connors BW, Paul DL (2001) Synchronous activity of inhibitory networks in neocortex requires electrical synapses containing connexin36. Neuron 31: 477–485.

Debanne D, Guérineau NC, Gähwiler BH, Thompson SM (1997) Action-potential propagation gated by an axonal I_A-like K^+ conductance in hippocampus. Nature 389: 286–289.

Delgado-Escueta AV, Enrile-Bascal F (1984) Juvenile myoclonic epilepsy of Janz. Neurology 34: 285–294.

DeLorey TM, Handforth A, Anagnostaras SG, Homanics GE, Minassian BA, Asatourian A, Fanselow MS, Delgado-Escueta A, Ellison GD, Olsen RW (1998) Mice lacking the β_3 subunit of the $GABA_A$ receptor have the epilepsy phenotype and many of the behavioral characteristics of Angelman syndrome. J. Neurosci. 18: 8505–8514.

Demirbilek V, Dervent A (2004) Panayiotopoulos syndrome: video-EEG illustration of a typical seizure. Epileptic Disord. 6: 121–124.

Deniau JM, Mailly P, Maurice N, Charpier S (2007) The pars reticulata of the substantia nigra: a window to basal ganglia output. Prog. Brain Res. 160: 151–172.

de Ruyter van Steveninck RR, Lewen GD, Strong SP, Koberle R, Bialek W (1997) Reproducibility and variability in neural spike trains. Science 275: 1805–1808.

Deschênes M (1981) Dendritic spikes induced in fast pyramidal tract neurons by thalamic stimulation. Exp. Brain Res. 43: 304–308.

Deschênes M, Labelle A, Landry P (1979) A comparative study of ventrolateral and recurrent excitatory postsynaptic potentials in large pyramidal tract cells in the cat. Brain Research 160: 37–46.

Deschênes M, Paradis M, Roy JP, Steriade M (1984) Electrophysiology of neurons of lateral thalamic nuclei in cat: resting properties and burst discharges. J. Neurophysiol. 51: 1196–1219.

Deschênes M, Bourassa J, Pinault D (1994) Corticothalamic projections from layer V cells in rat are collaterals of long-range corticofugal axons. Brain Res. 664: 215–219.

De Schutter E, Bower JM (1994a) An active membrane model of the cerebellar Purkinje cell I. Simulation of current clamps in slice. J. Neurophysiol. 71: 375–400.

De Schutter E, Bower JM (1994b) An active membrane model of the cerebellar Purkinje cell II. Simulation of synaptic responses. J. Neurophysiol. 71: 401–419.

de Solages C, Szapiro G, Brunel N, Hakim V, Isope P, Buisseret P, Rousseau C, Barbour B, Léna C (2008) High-frequency organization and synchrony of activity in the Purkinje cell layer of the cerebellum. Neuron 58: 775–788.

Destexhe A, McCormick DA, Sejnowski TJ (1993) A model of 8–10 Hz spindling in interconnected thalamic relay and reticular neurons. Biophys. J. 65: 2473–2477.

Destexhe A, Contreras D, Sejnowski TJ, Steriade M (1994) A model of spindle rhythmicity in the isolated thalamic reticular nucleus. J. Neurophysiol. 72: 803–818.

Destexhe A, Bal T, McCormick DA, Sejnowski TJ (1996a) Ionic mechanisms underlying synchronized oscillations and propagating waves in a model of ferret thalamic slices. J. Neurophysiol. 76: 2049–2070.

Destexhe A, Contreras D, Steriade M, Sejnowski TJ, Huguenard JR (1996b) In vivo, in vitro, and computational analysis of dendritic calcium currents in thalamic reticular neurons. J. Neurosci. 16: 169–185.

Destexhe A, Contreras D, Steriade M (1998a) Mechanisms underlying the synchronizing action of corticothalamic feedback through inhibition of thalamic relay cells. J. Neurophysiol. 79: 999–1016.

Destexhe A, Neubig M, Ulrich D, Huguenard J (1998b) Dendritic low-threshold calcium currents in thalamic relay cells. J. Neurosci. 18: 3574–3588.

Destexhe A, Contreras D, Steriade M (1999a) Spatiotemporal analysis of local field potentials and unit discharges in cat cerebral cortex during natural wake and sleep states. J. Neurosci. 19: 4595–4608.

Destexhe A, Contreras D, Steriade M (1999b) Cortically-induced coherence of a thalamic-generated oscillation. Neuroscience 92: 427–443.

Destexhe A, Hughes SW, Rudolph M, Crunelli V (2007) Are corticothalamic 'up' states fragments of wakefulness? Trends Neurosci. 30: 334–342.

Deuchars J, Thomson AM (1995a) Single axon fast inhibitory postsynaptic potentials elicited by a sparsely spiny interneuron in rat neocortex. Neuroscience 65: 935–942.

Deuchars J, Thomson AM (1995b) Innervation of burst firing spiny interneurons by pyramidal cells in deep layers of rat somatomotor cortex: paired intracellular recordings with biocytin filling. Neuroscience 69: 739–755.

Deuchars J, Thomson AM (1996) CA1 pyramid-pyramid connections in rat hippocampus in vitro: dual intracellular recordings with biocytin filling. Neuroscience 74: 1009–1018.

Deuchars J, West DC, Thomson AM (1994) Relationships between morphology and physiology of pyramid-pyramid single axon connections in rat neocortex *in vitro*. J. Physiol. 478: 423–435.

Devaux JJ, Kleopa KA, Cooper EC, Scherer SS (2004) KCNQ2 is a nodal K^+ channel. J. Neurosci. 24: 1236–1244.

Devinsky O (1999) Patients with refractory seizures. N. Engl. J. Med. 340: 1565–1570.

Devos D, Labyt E, Derambure P, Bourriez JL, Cassim F, Reyns N, Blond S, Guieu JD, Destée A, Defebvre L (2004) Subthalamic nucleus stimulation modulates motor cortex oscillatory activity in Parkinson's disease. Brain 127: 408–419.

Dezawa M, Mutoh T, Dezawa A, Adachi-Usami E (1998) Putative gap junctional communication between axon and regenerating Schwann cells during mammalian peripheral nerve regeneration. Neuroscience 85: 663–667.

De Zeeuw CI, Hertzberg EL, Mugnaini E (1995) The dendritic lamellar body: a new neuronal organelle putatively associated with dendrodendritic gap junctions. J. Neurosci. 15: 1587–1604.

Dichter MA (2006) Models of epileptogenesis in adult animals available for antiepileptogenesis drug screening. Epilepsy Res. 68: 31–35.

Dichter M, Spencer WA (1969a) Penicillin-induced interictal discharges from the cat hippocampus. I. Characteristics and topographical features, J. Neurophysiol. 32: 649–662.

Dichter M, Spencer WA (1969b Penicillin-induced interictal discharges from the cat hippocampus. II. Mechanisms underlying origin and restriction. J. Neurophysiol. 32: 663–687.

Dickinson R, Awaiz S, Whittington MA, Lieb WR, Franks NP (2003) The effects of general anaesthetics on carbachol-evoked gamma oscillations in the rat hippocampus in vitro. Neuropharmacology 44: 864–872.

Dickson CT, Biella G, de Curtis M (2000) Evidence for spatial modules mediated by temporal synchronization of carbachol-induced gamma rhythm in medial entorhinal cortex. J. Neurosci. 20: 7846–7854.

Dietrichs E (2008) Clinical manifestation of focal cerebellar disease as related to the organization of neural pathways. Acta Neurol. Scand. Suppl. 188: 6–11.

Diño MR, Mugnaini E (2008) Distribution and phenotypes of unipolar brush cells in relation to the granule cell systems of the rat cochlear nucleus. Neuroscience 154: 29–50.

Di Pasquale E, Keegan KD, Noebels JL (1997) Increased excitability and inward rectification in layer V cortical pyramidal neurons in the epileptic mutant mouse Stargazer. J. Neurophysiol. 77: 621–631.

Dobbins KR, Saul RF (2000) Transient visual loss after licorice ingestion. J. Neuroophthalmol. 20: 38–41.

Docherty JR, Starke K (1981) Postsynaptic alpha-adrenceptor subtypes in rabbit blood vessels and rat anococcygeus muscle studied in vitro. J. Cardiovasc. Pharmacol. 3: 854–866.

Dodge FA Jr, Cooley JW (1973) Action potential of the motoneuron. IBM J. Res. Dev. 17: 219–229.

Dodge FA Jr, Frankenhaeuser B (1958) Membrane currents in isolated frog nerve fibre under voltage clamp conditions. J. Physiol. 143: 76–90.

Doesburg SM, Kitajo K, Ward LM (2005) Increased gamma-band synchrony precedes switching of conscious preceptual objects in binocular rivalry. NeuroReport 16: 1139–1142.

Doheny HC, Faulkner HJ, Gruzelier JH, Baldeweg T, Whittington MA (2000) Pathway-specific habituation of induced gamma oscillations in the hippocampal slice. NeuroReport 11: 2629–2633.

Domann R, Uhlig S, Dorn T, Witte OW (1991) Participation of interneurons in penicillin-induced epileptic discharges. Exp. Brain Res. 83: 683–686.

Donoghue JP, Ebner FF (1981) The laminar distribution and ultrastructure of fibers projecting from three thalamic nuclei to the somatic sensorz-motor cortex of the opossum. J. Comp. Neurol. 198: 389–420.

Donoghue JP, Sanes JN, Hatsopoulos NG, Gaál G (1998) Neural discharge and local field potential oscillations in primate motor cortex during voluntary movements. J. Neurophysiol. 79: 159–173.

Doose H, Gerken H, Leonhardt R, Völzke E, Völz C (1970) Centrencephalic myoclonic-astatic petit mal. Neuropaediatrie 2: 59–78.

Dow RS (1939) Cerebellar action potentials in response to stimulation of various afferent connections. J. Neurophysiol. 2: 543–555.

Dow RS, Fernández-Guardiola A, Manni E (1962) The production of cobalt experimental epilepsy in the rat. Electroenceph. Clin. Neurophysiol. 14: 399–407.

Draguhn A, Traub RD, Schmitz D, Jefferys JGR (1998) Electrical coupling underlies high-frequency oscillations in the hippocampus in vitro. Nature 394: 189–192.

Dudek FE, Sutula TP (2007) Epileptogenesis in the dentate gyrus: a critical perspective. Prog. Brain Res. 163: 755–773.

Duncan JS (2007) Epilepsy surgery. Clin. Med. 7: 137–142.

Dupont E, Hanganu IL, Kilb W, Hirsch S, Luhmann HJ (2006) Rapid developmental switch in the mechanisms driving early cortical columnar networks. Nature 439: 79–83.

Durán-Ferreras E, Redondo L, Izquierdo G (2006) Acute generalised chorea following bilateral pallidal infarction due to cerebral anoxia. Rev. Neurol. 42: 767–768.

Dye J (1991) Ionic and synaptic mechanisms underlying a brainstem oscillator: An in vitro study of the pacemaker nucleus of Apteronotus. J. Comp. Physiol. A 168: 521–532.

Dye J, Heiligenberg W (1987) Intracellular recording in the medullary pacemaker nucleus of the weakly electric fish, Apteronotus, during modulatory behaviors. J. Comp. Physiol. A 161: 187–200.

Eadie MJ, Bladin PF (2001) A Disease Once Sacred. A History of the Medical Understanding of Epilepsy. John Libbey, Eastleigh, U.K.

Ebersole JS (1977) Initial abnormalities of neuronal responses during epileptogenesis in visual cortex. J. Neurophysiol. 40: 514–526.

Ebersole JS, Chatt AB (1984) Laminar interactions during neocortical epileptogenesis. Brain Res. 298: 253–272.

Ebersole JS, Pedley TA (Eds.) (2003) Current Practice of Clinical Electroencephalography. Lippincott Williams & Wilkins, Philadelphia, pp 639–680.

Eccles JC, Llinás R, Sasaki K (1964) Excitation of cerebellar Purkinje cells by the climbing fibres. Nature 203: 245–246.

Eccles JC, Ito M, Szentágothai J (1967) The Cerebellum as a Neuronal Machine. Springer, Heidelberg.

Eccles JC, Llinás R, Sasaki K (1966) Parallel fibre stimulation and the responses induced thereby in the Purkinje cells of the cerebellum. Exp. Brain Res. 1: 17–39.

Eckhorn R (1994) Oscillatory and non-oscillatory synchronization in the visual cortex and their possible roles in associations of visual features. Prog. Brain Res. 102: 405–426.

Eckhorn R, Bauer R, Jordan W, Brosch M, Kruse W, Munk M, Reithboeck HG (1988) Coherent oscillations: a mechanism of feature linking in the visual cortex? Biol. Cybern. 60: 121–130.

Edwards E, Soltani M, Deouell LY, Berger MS, Knight RT (2005) High gamma activity in response to deviant auditory stimuli recorded directly from human cortex. J. Neurophysiol. 94: 4269–4280.

Eeckman FH, Freeman WJ (1990) Correlations between unit firing and EEG in the rat olfactory system. Brain Res. 528: 238–244.

Egger V, Feldmeyer D, Sakmann B (1999) Coincidence detection and changes of synaptic efficacy in spiny stellate neurons in rat barrel cortex. Nat. Neurosci. 2: 1098–1105.

Elekes K, Szábo T (1985) Synaptology of the medullary command (pacemaker) nucleus of the weakly electric fish (Apteronotus leptorhynchus) with particular reference to comparative aspects. Exp. Brain Res. 60: 509–520.

Elias LA, Kriegstein AR (2008) Gap junctions: multifaceted regulators of embryonic cortical development. Trends Neurosci. 31: 243–250.

Empson RM, Amitai Y, Jefferys JGR, Gutnick MJ (1993) Injection of tetanus toxin into the neocortex elicits persistent epileptiform activity but only transient impairment of GABA release. Neuroscience 57: 235–239.

Engel A, Gaub HE (2008) Structure and mechanics of membrane proteins. Annu. Rev. Biochem. 77: 127–148.

Engel AK, Singer W (2001) Temporal binding and the neural correlates of sensory awareness. Trends Cogn. Sci. 5: 16–25.

Engel AK, König P, Kreiter AK, Singer W (1991a) Interhemispheric synchronization of oscillatory neuronal responses in cat visual cortex. Science 252: 1177–1179.

Engel AK, Kreiter AK, König P, Singer W (1991b) Synchronization of oscillatory neuronal responses between striate and extrastriate visual cortical areas of the cat. Proc. Natl. Acad. Sci. USA 88: 6048–6052.

Engel AK, Roelfsema PR, Fries P, Brecht M, Singer W (1997) Role of the temporal domain for response selection and perceptual binding. Cereb. Cortex 7: 571–582.

Engel J Jr, Pedley TA (1998) Epilepsy. A Comprehensive Textbook. 3 Vols. Lippincott-Raven, Philadelphia.

Erdös P, Rényi A (1960) On the evolution of random graphs. Publ. Math. Instit. Hungar. Acad. Sci. 5: 17–61.

Ermentrout GB, Kopell N (1998) Fine structure of neural spiking and synchronization in the presence of conduction delays. Proc. Natl. Acad. Sci. USA 95: 1259–1264.

European Chromosome 16 Tuberous Sclerosis Consortium (1993) Identification and characterization of the tuberous sclerosis gene on chromosome 16. Cell 75: 1305–1315.

Euston DR, Tatsuno M, McNaughton BL (2007) Fast-forward playback of recent memory sequences in prefrontal cortex during sleep. Science 318: 1147–1150.

Ewert TAS, Vahle-Hinz C, Engel AK (2008) High-frequency whisker vibration is encoded by phase-locked responses of neurons in the rat's barrel cortex. J. Neurosci. 28: 5359–5368.

Falconer MA, Taylor DC (1968) Surgical treatment of drug-resistant epilepsy due to mesial temporal sclerosis. Etiology and significance. Arch. Neurol. 19: 353–361.

Faltus F, Hynek K, Dolezalovà, V, Kumnickovà Z, Zemek P (1973) Experience in the treatment of schizophrenia with clozapine. Act. Nerv. Super. (Praha) 15: 95.

Fanselow EE, Richardson KA, Connors BW (2008) Selective, state-dependent activation of somatostatin-expressing inhibitory interneurons in mouse neocortex. J. Neurophysiol. 100: 2640–2652.

Featherstone RE, Kapur S, Fletcher PJ (2007) The amphetamine-induced sensitized state as a model of schizophrenia. Prog. Neuropsychopharmacol. Biol. Psychiatry 31: 1556–1571.

Feldmeyer D, Sakmann B (2000) Synaptic efficacy and reliability of excitatory connections between the principal neurones of the input (layer 4) and output layer (layer 5) of the neocortex. J. Physiol. 525: 31–39.

Feldmeyer D, Egger V, Lübke J, Sakmann B (1999) Reliable synaptic connections between pairs of excitatory layer 4 neurones within a single 'barrel' of developing somatosensory cortex. J. Physiol. 521: 169–190.

Feldmeyer D, Lübke J, Silver RA, Sakmann B (2002) Synaptic connections between layer 4 spiny neurone-layer 2/3 pyramidal cell pairs in juvenile rat barrel cortex: physiology and anatomy of interlaminar signalling within a cortical column. J. Physiol. 538: 803–822.

Feldmeyer D, Roth A, Sakmann B (2005) Monosynaptic connections between pairs of spiny stellate cells in layer 4 and pyramidal cells in layer 5A indicate that lemniscal and paralemniscal afferent pathways converge in the infragranular somatosensory cortex. J. Neurosci. 25: 3423–3431.

Feldmeyer D, Lübke J, Sakmann B (2006) Efficacy and connectivity of intracolumnar pairs of layer 2/3 pyramidal cells in the barrel cortex of juvenile rats. J. Physiol. 575: 583–602.

Fell J, Klaver P, Lehnertz K, Grünwald T, Schaller C, Elger CE, Fernandez G (2001) Human memory formation is accompanied by rhinal-hippocampal coupling and decoupling. Nat. Neurosci. 4: 1259–1264.

Fell J, Klaver P, Elger CE, Fernandez G (2002) Suppression of EEG gamma activity may cause the attentional blink. Conscious Cogn. 11: 114–122.

Fell J, Fernandez G, Klaver P, Elger CE, Fries P (2003) Is synchronized neuronal gamma activity relevant for selective attention? Brain Res. Brain Res. Rev. 42: 265–272.

Felts PA, Kapoor R, Smith KJ (1995) A mechanism for ectopic firing in central demyelinated axons. Brain 118: 1225–1231.

Ferri R, Elia M, Musumeci SA, Pettinato S (2000) The time course of high-frequency bands (15–45 Hz) in all-night spectral analysis of sleep EEG. Clin. Neurophysiol. 111: 1258–1265.

Few WP, Scheuer T, Catterall WA (2007) Dopamine modulation of neuronal Na^+ channels requires binding of A kinase-anchoring protein 15 and PKA by a modified leucine zipper motif. Proc. Natl. Acad. Sci. USA 104: 5187–5192.

Findley LJ, Gresty MA, Halmagyi GM (1981) Tremor, the cogwheel phenomenon, and clonus in Parkinson's disease. J. Neurol. Neurosurg. Psychiatry 44: 534–546.

Finnerty GT, Jefferys JGR (2000) 9–16 Hz oscillation precedes secondary generalization of seizures in the rat tetanus toxin model of epilepsy. J. Neurophysiol. 83: 2217–2226.

Fisahn A, Pike FG, Buhl EH, Paulsen O (1998) Cholinergic induction of network oscillations at 40 Hz in the hippocampus in vitro. Nature 394: 186–189.

Fisahn A, Yamada M, Duttaroy A, Gan J-W, Deng C-X, McBain CJ, Wess J (2002) Muscarinic induction of hippocampal gamma oscillations requires coupling of the M1 receptor to two mixed cation currents. Neuron 33: 615–624.

Fisahn A, Contractor A, Traub RD, Buhl EH, Heinemann SF, McBain CJ (2004) Distinct roles for the kainate receptor subunits GluR5 and GluR6 in kainate-induced hippocampal gamma oscillations. J. Neurosci. 24: 9658–9668.

Fisahn A, Heinemann SF, McBain CJ (2005) The kainate receptor subunit GluR6 mediates metabotropic regulation of the slow and medium AHP currents in mouse hippocampal neurones. J. Physiol. 562: 199–203.

Fischer MH, Löwenbach H (1934) Aktionsströme des Zentralnervensystems unter der Einwirkung von Krampfgiften. I. Mitteilung Strychnin und Pikrotoxin. Arch. Exp. Pathol. Pharmakol. 174: 357–382.

Fisher RS (1989) Animal models of the epilepsies. Brain Res. Rev. 14: 245–278.

Fisher RS, Prince DA (1977a) Spike-wave rhythms in cat cortex induced by parenteral penicillin. I. Electroencephalographic features. Electroenceph. Clin. Neurophysiol. 42: 608–624.

Fisher RS, Prince DA (1977b) Spike-wave rhythms in cat cortex induced by parenteral penicillin. II. Cellular features. Electroenceph. Clin. Neurophysiol. 42: 625–639.

Fisher RS, Webber WRS, Lesser RP, Arroyo S, Uematsu S (1992) High-frequency EEG activity at the start of seizures. J. Clin. Neurophysiol. 9: 441–448.

Fleidervish IA, Binshtok AM, Gutnick MJ (1998) Functionally distinct NMDA receptors mediate horizontal connectivity within layer 4 of mouse barrel cortex. Neuron 21: 1055–1065.

Flint AC, Connors BW (1996) Two types of network oscillations in neocortex mediated by distinct glutamate receptor subtypes and neuronal populations. J. Neurophysiol. 75: 951–956.

Flint AC, Maisch US, Weishaupt JH, Kriegstein AR, Monyer H (1997) NR2A subunit expression shortens NMDA receptor synaptic currents in developing neocortex. J. Neurosci. 17: 2469–2476.

Foffani G, Priori A, Egidi M, Rampini P, Tamma F, Caputo E, Moxon KA, Cerutti S, Barbieri S (2003) 300–Hz subthalamic oscillations in Parkinson's disease. Brain 126: 2153–2163.

Foix C, Nicolesco J (1925) Les Noyaux Gris Centraux e la Région Mésencephalo-sous-optique. Masson, Paris.

Ford JM, Roach BJ, Faustman WO, Mathalon DH (2007) Synch before you speak: auditory hallucinations in schizophrenia. Am. J. Psychiatry 164: 458–466.

Forsythe ID (1994) Direct patch recording from identified presynaptic terminals mediating glutamatergic EPSCs in the rat CNS, *in vitro*. J. Physiol. 479: 381–387.

Forsythe ID, Westbrook GL (1988) Slow excitatory postsynaptic currents mediated by N-methyl-D aspartate receptors on cultured mouse central neurones. J. Physiol. 396: 515–533.

Frankenhaeuser B, Hodgkin A L (1957) The action of calcium on the electrical properties of squid axons. J. Physiol. 137: 218–244.

Franowicz MN, Barth DS (1995) Comparison of evoked potentials and high-frequency (gamma-band) oscillating potentials in rat auditory cortex. J. Neurophysiol. 74: 96–112.

Fredette BJ, Mugnaini E (1991) The GABAergic cerebello-olivary projection in the rat. Anat. Embryol. (Berl.) 184: 225–243.

Freeman JA, Nicholson CN (1970) Space-time transformation in the frog cerebellum through an intrinsic tapped delay-line. Nature 226: 640–642.

Freeman WJ (1979) Nonlinear dynamics of paleocortex manifested in the olfactory EEG. Biol. Cybern. 35: 21–34.

Freeman WJ, Schneider W (1982) Changes in spatial patterns of rabbit olfactory EEG with conditioning to odors. Psychophysiology 19: 44–56.

Freeman WJ, Rogers LJ, Holmes MD, Silbergeld DL (2000) Spatial spectral analysis of human electrocorticograms including the alpha and gamma bands. J. Neurosci. Methods 95: 111–121.

French CR, Sah P, Buckett KJ, Gage PW (1990) A voltage-dependent persistent sodium current in mammalian hippocampal neurons. J. Gen. Physiol. 95: 1139–1157.

Frerking M (2004) When astrocytes signal, kainate receptors respond. Proc. Natl. Acad. Sci. USA. 101: 2649-2650.

Frerking M, Malenka RC, Nicoll RA (1998) Synaptic activation of kainate receptors on hippocampal interneurons. Nature Neurosci. 1: 479–486.

Freund TF, Martin KAC, Smith AD, Somogyi P (1983) Glutamate decarboxylase-immunoreactive terminals of Golgi-impregnated axoaxonic cells and of presumed basket cells in synaptic contact with pyramidal neurons of the cat's visual cortex. J. Comp. Neurol. 221: 263–278.

Freund TF, Martin KA, Soltesz I, Somogyi P, Whitteridge D (1989) Arborisation pattern and postsynaptic targets of physiologically identified thalamocortical afferents in striate cortex of the macaque monkey. J. Comp. Neurol. 289: 315–336.

Frick A, Feldmeyer D, Helmstaedter M, Sakmann B (2007) Monosynaptic connections between pairs of L5A pyramidal neurons in columns of juvenile rat somatosensory cortex. Cereb. Cortex. 18: 397-406.

Friedlander WJ (1986a) Who was 'the father of bromide treatment of epilepsy'? Arch. Neurol. 43: 505–507.

Friedlander WJ (1986b) Putnam, Merritt and the discovery of Dilantin. Epilepsia 27 Suppl. 3: S1–S20.

Friedlander WJ (2001) The History of Modern Epilepsy: The Beginning, 1865–1914. Greenwood Press, Westport, CT.

Friedman JI, Tang C, Carpenter D, Buchsbaum M, Schmeidler J, Flanagan L, Golembo S, Kanellopoulou I, Ng J, Hof PR, Harvey PD, Tsopelas ND, Stewart D, Davis KL (2008) Diffusion tensor imaging findings in first-episode and chronic schizophrenia patients. Am. J. Psychiatry 165: 1024–1032.

Friedman-Hill S, Maldonado PE, Gray CM (2000) Dynamics of striate cortical activity in the alert macaque: I. Incidence and stimulus-dependence of gamma-band neuronal oscillations. Cereb. Cortex 10: 1105–1116.

Friedrich RW, Laurent G (2001) Dynamic optimization of odor representations by slow temporal patterning of mitral cell activity. Science 291: 889–894.

Frien A, Eckhorn R (2000) Functional coupling shows stronger stimulus dependency for fast oscillations than for low-frequency components in striate cortex of awake monkey. Eur. J. Neurosci. 12: 1466–1478.

Frien A, Eckhorn R, Bauer R, Woelbern T, Kehr H (1994) Stimulus-specific fast oscillations at zero phase between visual areas V1 and V2 of awake monkey. NeuroReport 5: 2273–2277.

Frien A, Eckhorn R, Bauer R, Woelbern T, Gabriel A (2000) Fast oscillations display sharper orientation tuning than slower components of the same recordings in striate cortex of the awake monkey. Eur. J. Neurosci. 12: 1453–1465.

Fries P, Roelfsema PR, Engel AK, König P, Singer W (1997) Synchronization of oscillatory responses in visual cortex correlates with perception in interocular rivalry. Proc. Natl. Acad. Sci. USA 94: 12699–12704.

Fries P, Reynolds JH, Rorie AE, Desimone R (2001) Modulation of oscillatory neuronal synchronization by selective visual attention. Science 291: 1560–1563.

Fries P, Schröder J-H, Roelfsema PR, Singer W, Engel AK (2002) Oscillatory neuronal synchronization in primary visual cortex as a correlate of stimulus selection. J. Neurosci. 22: 3739–3754.

Frisch C, De Souza-Silva MA, Söhl G, Güldenagel M, Willecke K, Huston JP, Dere E (2005) Stimulus complexity dependent memory impairment and changes in motor performance after deletion of the neuronal gap junction protein connexin36 in mice. Behav. Brain Res. 157: 177–185.

Frith C, Dolan R (1996) The role of the prefrontal cortex in higher cognitive functions. Cogn. Brain Res. 5: 175–181.

Fróes MM, Correia AHP, Garcia-Abreu J, Spray DC, De Carvalho ACC, Neto VM (1999) Gap-junctional coupling between neurons and astrocytes in primary central nervous system cultures. Proc. Natl. Acad. Sci. USA 96: 7541–7546.

Fuchs E, Doheny HC, Faulkner HJ, Caputi A, Traub RD, Bibbig A, Kopell N, Whittington MA, Monyer H (2001) Genetically altered AMPA-type glutamate receptor kinetics in interneurons disrupt long-range synchrony of gamma oscillation. Proc. Natl. Acad. Sci. USA 98: 3571–3576.

Fuchs EC, Zivkovic AR, Cunningham MO, Middleton S, LeBeau FEN, Bannerman DM, Rozov A, Whittington MA, Traub RD, Rawlins JNP, Monyer H (2007) Recruitment pattern of parvalbumin-positive interneurons determines hippocampal function and associated behaviour. Neuron 53: 591–604.

Fuchs SA, De Barse MM, Scheepers FE, Cahn W, Dorland L, d Sain-van der Velden MG, Klomp LW, Berger R, Kahn RS, de Koning TJ (2008) Cerebrospinal fluid D-serine and glycine concentrations are unaltered and unaffected by olanzapine therapy in male schizophrenic patients. Eur. Neuropsychopharmacol. 18: 333–338.

Fuentealba P, Crochet S, Timofeev I, Bazhenov M, Sejnowski TJ, Steriade M (2004a) Experimental evidence and modeling studies support a synchronizing role for electrical coupling in the cat thalamic reticular neurons in vivo. Eur. J. Neurosci. 20: 111–119.

Fuentealba P, Crochet S, Timofeev I, Steriade M (2004b) Synaptic interactions between thalamic and cortical inputs onto cortical neurons in vivo. J. Neurophysiol. 91: 1990–1998.

Fuentealba P, Timofeev I, Steriade M (2004c) Prolonged hyperpolarizing potentials precede spindle oscillations in the thalamic reticular nucleus. Proc. Natl. Acad. Sci. USA 101: 9816–9821.

Fuentealba P, Bequm R, Capogna M, Jinno S, Márton LF, Csicsvari J, Thomson A, Somogyi P, Klausberger T (2008) Ivy cells: a population of nitric-oxide-producing, slow-spiking GABAergic neurons and their involvement in hippocampal network activity. Neuron 57: 917–929.

Fujita M (1982) Adaptive filter model of the cerebellum. Biol. Cybern. 45: 195–206.

Fukuda T (2009) Network architecture of gap junction-coupled neuronal linkage in the striatum. J. Neurosci. 29: 1235–1243.

Fukuda T, Kosaka T (2000) Gap junctions linking the dendritic network of GABAergic interneurons in the hippocampus. J. Neurosci. 20: 1519–1528.

Fukuda T, Kosaka T (2003) Ultrastructural study of gap junctions between dendrites of parvalbumin-containing GABAergic neurons in various neocortical areas of the adult rat. Neuroscience 120: 5–20.

Fukuda T, Kosaka T, Singer W, Galuske RAW (2006) Gap junctions among dendrites of cortical GABAergic neurons establish a dense and widespread intercolumnar network. J. Neurosci. 26: 3434–3443.

Furshpan EJ, Potter DD (1959) Transmission at the giant motor synapses of the crayfish. J. Physiol. 145: 289–325.

Fuster JM, Alexander GE (1971) Neuron activity related to short-term memory. Science 173: 652–654.

Gähwiler BH, Brown DA (1987) Muscarine affects calcium-currents in rat hippocampal pyramidal cells in vitro. Neurosci. Lett. 76: 301–306.

Gajda Z, Szupera Z, Blazsó G, Szente M (2005) Quinine, a blocker of neuronal Cx36 channels, suppresses seizure activity in rat neocortex in vivo. Epilepsia 46: 1581–1591.

Gajdusek DC, Zigas V (1957) Degenerative disease of the central nervous system in New Guinea; the endemic occurrence of kuru in the native population. N. Engl. J. Med. 257: 974–978.

Gajdusek DC, Gibbs CJ Jr., Alpers M (1967) Transmission and passage of experimental "kuru" to chimpanzees. Science 155: 212–214.

Galambos R, Makeig S, Talmachoff P (1981) A 40 Hz auditory potential recorded from the human scalp. Proc. Natl. Acad. Sci. 78: 2643–2647.

Galarreta M, Hestrin S (1999) A network of fast-spiking cells in the neocortex connected by electrical synapses. Nature 402: 72–75.

Galarreta M, Hestrin S (2002) Electrical and chemical synapses among parvalbumin fast-spiking GABAergic interneurons in adult mouse neocortex. Proc. Natl. Acad. Sci. USA 99: 12438–12443.

Gale JT, Amirnovin R, Williams ZM, Flaherty AW, Eskandar EN (2007) From symphony to cacophony: pathophysiology of the human basal ganglia in Parkinson disease. Neurosci. Biobehav. Rev. doi:10.1016/j.neubiorev.2006.11.005.

Garcia-Cairasco N (2002) A critical review on the participation of inferior colliculus in acoustic-motor and acoustic-limbic networks involved in the expression of acute and kindled audiogenic seizures. Hear. Res. 168: 208–222.

Gardiner M (2005) Genetics of idopathic generalized epilepsies. Epilepsia 46 Suppl. 9: 15–20.

Gareri P, Condorelli D, Belluardo N, Russo E, Loiacono A, Barrresi V, Trovato-Salinato A, Mirone MB, Ferreri Ibbadu G, De Sarro G (2004) Anticonvulsant effects of carbenoxolone in genetically epilepsy prone rats (GEPRs). Neuropharmacology 47: 1205–1216.

Garris SS, Oles KS (2005) Impact of topiramate on serum bicarbonate concentrations in adults. Ann. Pharmacother. 39: 424–426.

Garwicz M, Ekerot C-F, Jörntell H (1998) Orgnaizational principles of cerebellar neuronal circuitry. News Physiol. Sci. 13: 26–32.

Gaspar P, Berger B, Febvreet A, Vigny A, Henry JP (1989) Catecholamine innervation of the human cerebral cortex as revealed by comparative immunohistochemistry of tyrosine hydroxylase and dopamine-β-hydroxylase. J. Comp. Neurol. 279: 249–271.

Gasparini S, Migliore M, Magee JC (2004) On the initiation and propagation of dendritic spikes in CA1 pyamidal neurons. J. Neurosci. 24: 11046–11056.

Gasser HS, Grundfest H (1939) Axon diameters in relation to the spike dimensions and the conduction velocity in mammalian fibers. Am. J. Physiol. 127: 393–414.

Gatev P, Darbin O, Wichmann T (2006) Oscillations in the basal ganglia under normal conditions and in movement disorders. Mov. Disord. 21: 1566–1577.

Geiger JRP, Jonas P (2000) Dynamic control of presynaptic Ca^{2+} inflow by fast-inactivating K^+ channels in hippocampal mossy fiber boutons. Neuron 28: 927–939.

Geiger JRP, Melcher T, Koh D-S, Sakmann B, Seeburg PH, Jonas P, Monyer H (1995) Relative abundance of subunit mRNAs determines gating and Ca^{2+} permeability of AMPA receptors in principal neurons and interneurons in rat CNS. Neuron 15: 193–204.

Geiger JRP, Lübke J, Roth A, Frotscher M, Jonas P (1997) Submillisecond AMPA receptor-mediated signaling at a principal neuron-interneuron synapse. Neuron 18: 1009–1023.

Geijo-Barrientos E (2000) Subthreshold inward membrane currents in guinea-pig frontal cortex neurons. Neuroscience 95: 965–972.

Geijo-Barrientos E, Pastore C (1995) The effects of dopamine on the subthreshold electrophysiological responses of rat prefrontal cortex neurons in vitro. Eur. J. Neurosci. 7: 358–366.

George AL (2004) Inherited channelopathies associated with epilepsy. Epil. Curr. 4: 65–70.

Géraud M (2007) Emil Kraepelin: a pioneer of modern psychiatry. On the occasion of the hundred and fiftieth anniversary of his birth. Encephale 33: 561–567.

Giaretta D, Avoli M, Gloor P (1987) Intracellular recordings in pericruciate neurons during spike and wave discharges of feline generalized penicillin epilepsy. Brain Res. 405: 68–79.

Gibbs FA, Davis H, Lennox WG (1935) The electroencephalogram in epilepsy and in conditions of impaired consciousness. Arch. Neurol. Psychiatry 34: 1133–1148.

Gibbs FA, Lennox WG, Gibbs EL (1936) The electroencephalogram in diagnosis and in localization of epileptic seizures. Arch. Neurol. Psychiatry 36: 1225–1235.

Gibson JR, Beierlein M, Connors BW (1999) Two networks of electrically coupled inhibitory neurons in neocortex. Nature 402: 75–79.

Gibson JR, Beierlein M, Connors BW (2005) Functional properties of electrical synapses between inhibitory interneurons of neocortical layer 4. J. Neurophysiol. 93: 467–480.

Gieselmann V (2008) Metachromatic leukodystrophy: genetics, pathogenesis and therapeutic options. Acta Paediatr. Suppl. 97: 15–21.

Gigout S, Louvel J, Pumain R (2006) Effects in vitro and in vivo of a gap junction blocker on epileptiform activities in a genetic model of absence epilepsy. Epilep. Res. 69: 15–29.

Gillies M, Traub RD, LeBeau FEN, Davies CH, Gloveli T, Buhl EH, Whittington MA (2002) A model of atropine-resistant theta oscillations in hippocampal area CA1. J. Physiol. 543: 779–793.

Gilliland MA, Bergmann BM, Rechtschaffen A (1989) Sleep deprivation in the rat: VIII. High EEG amplitude sleep deprivation. Sleep 12: 53–59.

Gilman S (1985) The cerebellum: its role in posture and movement. In: Swash M, Kennard C (Eds), Scientific Basis of Clinical Neurology. Churchill-Livingstone, Edinburgh, pp 36–55.

Gilula NB, Branton D, Satir P (1970) The septate junction: a structural basis for intercellular coupling. Proc. Natl. Acad. Sci. USA 67: 213–220.

Giuffrida R, Licata F, Li Volsi G, Perciavalle V (1982) Motor responses evoked by microstimulation of cerebellar interpositus nucleus in cats submitted to dorsal rhizotomy. Neurosci. Lett. 30: 241–244.

Gladwell SJ, Jefferys JG (2001) Second messenger modulation of electrotonic coupling between region CA3 pyramidal cell axons in the rat hippocampus. Neurosci. Lett. 300: 1–4.

Glasscock E, Qian J, Yoo JW, Noebels JL (2007) Masking epilepsy by combining two epilepsy genes. Nat. Neurosci. 10: 1554–1558.

Gloor P (1969) Epileptogenic action of penicillin. Ann. NY Acad. Sci. 166: 350–360.

Gloor P, Testa G, Guberman A (1973) Brain-stem and cortical mechanisms in an animal model of generalized corticoreticular epilepsy. Trans. Am. Neurol. Assoc. 98: 203–205.

Gloveli T, Dugladze T, Saha S, Monyer H, Heinemann U, Traub RD, Whittington MA, Buhl EH (2005a) Differential involvement of oriens/pyramidale interneurons in hippocampal network oscillations *in vitro*. J. Physiol. 562: 131–147.

Gloveli T, Dugladze T, Rotstein HG, Traub RD, Monyer H, Heinemann U, Whittington MA, Kopell NJ (2005b) Orthogonal arrangement of rhythm-generating microcircuits in the hippocampus. Proc. Natl. Acad. Sci. USA 102: 13295–13300.

Gnanalingham KK, Smith LA, Hunter AJ, Jenner P, Marsden CD (1993) Alterations in striatal and extrastriatal D-1 and D-2 dopamine receptors in the MPTP-treated common marmoset: an autoradiographic study. Synapse 14: 184–194.

Gobbelé R, Buchner H, Curio G (1998) High-frequency (600 Hz) SEP activities originating in the subcortical and cortical human somatosensory system. Electroencephalogr. Clin. Neurophysiol. 108: 182–189.

Gobbi G, Bruno L, Pini A, Giovanardi Rossi P, Tassinari CA (1987) Periodic spasms: an unclassified type of epileptic seizure in childhood. Dev. Med. Child Neurol. 29: 766–775.

Gobel S (1971) Axo-axonic septate junctions in the basket formations of the cat cerebellar cortex. J. Cell Biol. 51: 328–333.

Goddard GV (1969) A permanent change in brain function resulting from daily electrical stimulation of the rat. Exp. Neurol. 25: 295–303.

Goetz C (2007) Textbook of Clinical Neurology, 3rd ed., Saunders, Oxford.

Goldberg JA, Boraud T, Maraton S, Haber SN, Vaadia E, Bergman H (2002) Enhanced synchrony among primary motor cortex neurons in the 1–methyl-4–phenyl-1,2,3,6–tetrahydropyridine primate model of Parkinson's disease. J. Neurosci. 22: 4639–4653.

Goldberg JH, Yuste R, Tamás G (2003) Ca^{2+} imaging of mouse neocortical interneurone dendrites: contribution of Ca^{2+}-permeable AMPA and NMDA receptors to subthreshold Ca^{2+} dynamics. J. Physiol. 551: 67–78.

Goldensohn ES (2001) Cellular electrical phenomena in focal epilepsy. In: Lüders H, Comair YG (Eds), Epilepsy Surgery. Lippincott Williams Wilkins, Philadelphia, pp. 1–18.

Goldensohn ES, Purpura DP (1963) Intracellular potentials of cortical neurons during focal epileptogenic discharges. Science 139: 840–842.

Goldman-Rakic PS (1994) Working memory dysfunction in schizophrenia. J. Neuropsychiatry Clin. Neurosci. 6: 348–357.

Goldman-Rakic PS (1998) The cortical dopamine system: role in memory and cognition. Adv. Pharmacol. 42: 707–711.

Goldman-Rakic PS (1999) The physiological approach: functional architecture of working memory and disordered cognition in schizophrenia. Biol. Psychiatry 46: 650–661.

Golomb D (1998) Models of neuronal transient synchrony during propagation of activity through neocortical circuitry. J. Neurophysiol. 79: 1–12.

Golomb D, Amitai Y (1997) Propagating neuronal discharges in neocortical slices: computational and experimental study. J. Neurophysiol. 78: 1199–1211.

Golomb D, Wang X-J, Rinzel J (1994) Synchronization properties of spindle oscillations in a thalamic reticular nucleus model. J. Neurophysiol. 72: 1109–1126.

Golomb D, Yue C, Yaari Y (2006) Contribution of persistent Na$^+$ current and M-type K$^+$ current to somatic bursting in CA1 pyramidal cells: combined experimental and modeling study. J. Neurophysiol. 96: 1912–1926.

Gómez-Lira G, Trillo E, Ramírez M, Asai M, Sitges M, Gutiérrez R (2002) The expression of GABA in mossy fiber synaptosomes coincides with the seizure-induced expression of GABAergic transmission in the mossy fiber synapse. Exp. Neurol. 177: 276–283.

González-Nieto D, Gómez-Hernández JM, Larrosa B, Gutiérrez C, Muñoz MD, Fasciani I, O'Brien J, Zappalà A, Cicirata F, Barrio LC. (2008) Regulation of neuronal connexin-36 channels by pH. Proc Natl Acad Sci U S A. 105: 17169-74.

Gonzalez-Burgos G, Lewis DA (2008) GABA neurons and the mechanisms of network oscillations: implications for understanding cortical dysfunction in schizophrenia. Schizophr. Bull. 34: 944–961.

Goodenough DA, Paul DL (2003) Beyond the gap: functions of unpaired connexon channels. Nat. Rev. Mol. Cell Biol. 4: 285–294.

Gorelova N, Seamans JK, Yang CR (2002) Mechanisms of dopamine activation of fast-spiking interneurons that exert inhibition in rat prefrontal cortex. J. Neurophysiol. 88: 3150–3166.

Gothelf D (2007) Velocardiofacial syndrome. Child Adolesc. Psychiatr. Clin. N. Am. 16: 677–693.

Grace AA (1991) Phasic versus tonic dopamine release and the modulation of dopamine system responsivity: a hypothesis for the etiology of schizophrenia. Neuroscience 41: 1–24.

Graeber MB, Kösel S, Egensperger R, Banati RB, Müller U, Bise K, Hoff P, Möller HJ, Fujisawa K, Mehraein P (1997) Rediscovery of the case described by Alois Alzheimer in 1911: historical, histological and molecular genetic analysis. Neurogenetics 1: 73–80.

Gray CM (1994) Synchronous oscillations in neuronal systems: mechanisms and functions. J. Comput. Neurosci. 1: 11–38.

Gray CM, McCormick DA (1996) Chattering cells: superficial pyramidal neurons contributing to the generation of synchronous oscillations in the visual cortex. Science 274: 109–113.

Gray CM, Singer W (1989) Stimulus-specific neuronal oscillations in orientation columns of cat visual cortex. Proc. Natl. Acad. Sci. USA 86: 1698–1702.

Gray CM, König P, Engel AK, Singer W (1989) Oscillatory responses in cat visual cortex exhibit inter-columnar synchronization which reflects global stimulus properties. Nature 338: 334–337.

Gray CM, Engel AK, König P, Singer W (1990) Stimulus-dependent neuronal oscillations in cat visual cortex: receptive field properties and feature dependence. Eur. J. Neurosci. 2: 607–619.

Gray CM, Engel AK, König P, Singer W (1992) Synchronization of oscillatory neuronal responses in cat striate cortex: temporal properties. Vis. Neurosci. 8: 337–347.

Gray CM, Viana Di Prisco G (1997) Stimulus-dependent neuronal oscillations and local synchronization in striate cortex of the alert cat. J. Neurosci. 17: 3239–3253.

Gray J (2004) Consciousness. Creeping Up on the Hard Problem. Oxford University Press, Oxford.

Green JD, Maxwell DS, Schindler WH, Stumpf C (1960) Rabbit EEG "theta" rhythm: its anatomical source and relation to activity in single neurons. J. Neurophysiol. 23: 403–420.

Green MF (1996) What are the functional consequences of neurocognitive deficits in schizophrenia? Am. J. Psychiatry 153: 321–330.

Greenberg DA, Pal DK (2007) The state of the art in the genetic analysis of the epilepsies. Curr. Neurol. Neurosci. Rep. 7: 320–328.

Greenberg RM, Kellner CH (2005) Electroconvulsive therapy: a selected review. Am. J. Geriatr. Psychiatry 13: 268–281.

Greene R (2001) Circuit analysis of NMDAR hypofunction in the hippocampus, in vitro, and psychosis of schizophrenia. Hippocampus 11: 569–577.

Grenier F, Timofeev I, Steriade M (2001) Focal synchronization of ripples (80–200 Hz) in neocortex and their neuronal correlates. J. Neurophysiol. 86: 1884–1898.

Grenier F, Timofeev I, Steriade M (2003) Neocortical very fast oscillations (ripples, 80–200 Hz) during seizures: intracellular correlates. J. Neurophysiol. 89: 841–852.

Grillner S (1985) Neurobiological bases of rhythmic motor acts in vertebrates. Science 228: 143–149.

Grimby L, Hannerz J, Borg J, Hedman B (1981) Firing properties of single human motor units on maintained maximal voluntary effort. Ciba Found. Symp. 82: 157–177.

Griniasty M, Tsodyks MV, Amit DJ (1993) Conversion of temporal correlations between stimuli to spatial correlations between attractors. Neural Comput. 5, 1–17.

Gritchenko II, Chesler M (1996) Calcium and barium-dependent extracellular alkaline shifts evoked by electrical activity in rat hippocampal slices. Neuroscience 75: 1117–1126.

Gross DW, Gotman J (1999) Correlation of high-frequency oscillations with the sleep-wake cycle and cognitive activity in humans. Neuroscience 94: 1005–1018.

Gross RA (1992) A brief history of epilepsy and its therapy in the Western Hemisphere. Epilep. Res. 12: 65–74.

Grüsser-Cornehis U, Bäurle J (2001) Mutant mice as a model for cerebellar ataxia. Prog. Neurobiol. 63: 489–540.

Gu QA, Bear MF, Singer W (1989) Blockade of NMDA-receptors prevents ocularity changes in kitten visual cortex after reversed monocular deprivation. Brain Res. Dev. Brain Res. 47: 281–288.

Guan D, Lee JC, Tkatch T, Surmeier DJ, Armstrong WE, Foehring RC (2006) Expression and biophysical properties of Kv1 channels in supragranular neocortical pyramidal neurones. J. Physiol. 571: 371–389.

Guckenheimer J, Holmes P (1983) Nonlinear Oscillations, Dynamical Systems, and Bifurcations of Vector Fields. Springer-Verlag, New York.

Guidotti A, Auta J, Davis JM, Di-Giorgi-Gerevini V, Dwivedi Y, Grayson DR, Impagnatiello F, Pandey G, Pesold C, Sharma R, Uzunov D, Costa E (2000) Decrease in reelin and glutamic acid decarboxylase67 (GAD67) expression in schizophrenia and bipolar disorder: a postmortem brain study. Arch. Gen. Psychiatry 57: 1061–1069.

Güldenagel M, Ammermüller J, Feigenspan A, Teubner B, Degen J, Söhl G, Willlecke K, Weiler R (2001) Visual transmission deficits in mice with targeted disruption of the gap junction gene connexin36. J. Neurosci. 21: 6036–6044.

Gulledge AT, Stuart GJ (2003) Action potential initiation and propagation in layer 5 pyramidal neurons of the rat prefrontal cortex: absence of dopamine modulation. J. Neurosci. 23: 11363–11372.

Gulledge AT, Stuart GJ (2005) Cholinergic inhibition of neocortical neurons. J. Neurosci. 25: 10308–10320.

Gulyás AI, Miles R, Sik A, Tóth K, Tamamaki N, Freund TF (1993) Hippocampal pyramidal cells excite inhibitory neurons through a single release site. Nature 366: 683–687.

Gumbiner B (1987) Structure, biochemistry, and assembly of epithelial tight junctions. Am. J. Physiol. 253: C749–C758.

Güngör S, Yalnizoğlu D, Turanli G, Saatçi I, Ergoğan –Bakar E, Topçu M (2007) Malformations of cortical development and epilepsy: evaluation of 101 cases (part II). Turk. J. Pediatr. 49: 131–140.

Gupta A, Wang Y, Markram H (2000) Organizing principles for a diversity of GABAergic interneurons and synapses in the neocortex. Science 287: 273–278.

Gur RE, Keshavan MS, Lawrie SM (2007) Deconstructing psychosis with human brain imaging. Schizophr. Bull. 33: 921–931.

Gutiérrez R (2002) Activity-dependent expression of simultaneous glutamatergic and GABAergic neurotransmission from the mossy fibers in vitro. J. Neurophysiol. 87: 2562–2570.

Gutnick MJ, Lobel-Yaakov R (1983) Carbon dioxide uncouples dye-coupled neuronal aggregates in neocortical slices. Neurosci. Lett. 42: 197–200.

Gutnick MJ, Prince DA (1972) Thalamocortical relay neurons: antidromic invasion of spikes from a cortical epileptogenic focus. Science 176: 424–426.

Gutnick MJ, Prince DA (1981) Dye coupling and possible electrotonic coupling in the guinea pig neocortical slice. Science 211: 67–70.

Gutnick MJ, Connors BW, Prince DA (1982) Mechanisms of neocortical epilepto-genesis in vitro. J. Neurophysiol. 48:1321–1335.

Gutnick MJ, Lobel-Yaakov R, Rimon G (1985) Incidence of neuronal dye-coupling in neocortical slices depends on the plane of section. Neuroscience 15: 659–666.

Gutnick MJ, Amitai Y, Barkai E (1992) Chronic models of cortical epilepsy: exper-imental manipulations leading to long-term reorganization of local neocortical circuitry. In: Engel J Jr, C Wasterlain C, Cavalheiro EA, Heinemann U, Avanzini, G (Eds), Molecular Neurobiology of Epilepsy (Epilepsy Res. Suppl. 9), Elsevier, Philadelphia, pp 221–229.

Haas HL, Jefferys JGR (1984) Low-calcium field burst discharges of CA1 pyramidal neurones in rat hippocampal slices. J. Physiol. 354: 185–201.

Haas LF (2003) Hans Berger (1873–1941), Richard Caton (1842–1926), and electroencephalography. J. Neurol. Neurosurg. Psychiatry 74: 9.

Haddad PM, Dursun SM (2008) Neurological complications of psychiatric drugs: clinical features and management. Hum. Psychopharmacol. 23 Suppl. 1: 15–26.

Haddad PM, Sharma SG (2007) Adverse effects of atypical antipsychotics: differential risk and clinical implications. CNS Drugs 21: 911–936.

Hadley RD, Kater SB, Cohan CS (1983) Electrical synapse formation depends on interaction of mutually growing neurites. Science 221: 466–468.

Haenschel C, Baldeweg T, Croft RJ, Whittington M, Gruzelier J (2000) Gamma and beta frequency oscillations in response to novel auditory stimuli: A comparison of human electroencephalogram (EEG) data with *in vitro* models. Proc. Natl. Acad. Sci. USA 97: 7645–7650.

Haenschel C, Bittner RA, Haertling F, Rotarska-Jagiela A, Maurer K, Singer W, Linden DE (2007) Contribution of impaired early-stage visual processing to working memory dysfunction in adolescents with schizophrenia: a study with event-related potentials and functional magnetic resonance imaging. Arch. Gen. Psychiatry 64: 1229–1240.

Hafting T, Fyhn M, Molden S, Moser MB, Moser EI (2005) Microstructure of a spatial map in the entorhinal cortex. Nature 436: 801–806.

Hagiwara S, Morita H (1962) Electrotonic transmission between two nerve cells in leech ganglion. J. Neurophysiol. 25: 721–731.

Haider B, Duque A, Hasenstaub AR, McCormick DA (2006) Neocortical network activity in vivo is generated through a dynamic balance of excitation and inhibition. J. Neurosci. 26: 435–445.

Haig AR, Gordon E, De Pascalis V, Meares RA, Bahramali H, Harris A (2000) Gamma activity in schizophrenia: evidence of impaired network binding. Clin. Neurophysiol. 111: 1461–1468.

Hajos N, Palhalmi J, Mann EO, Nemeth B, Paulsen O, Freund TF (2004) Spike timing of distinct types of GABAergic interneuron during hippocampal gamma oscillations in vitro. J. Neurosci. 24: 9127–9137.

Håkansson K, Galdi S, Hendrick J, Snyder G, Greengard P, Fisone G (2006) Regulation of phosphorylation of the GluR1 AMPA receptor by dopamine D2 receptors. J. Neurochem. 96: 482–488.

Halliday AM (1967) Changes in the form of cerebral evoked responses in man associated with various lesions of the nervous system. Electroenceph. Clin. Neurophysiol. Suppl. 25: 178

Halliwell JV, Adams PR (1982) Voltage-clamp analysis of muscarinic excitation in hippocampal neurons. Brain Res. 250: 71–92.

Hamada Y, Miyashita E, Tanaka H (1999) Gamma-band oscillations in the "barrel cortex" rat's exploratory whisking. Neuroscience 88: 667–671.

Hammond C, Feger J, Bioulac B, Souteyrand JP (1979) Experimental hemiballism in the monkey produced by unilateral kainic acid lesion in corpus Luysii. Brain Res. 171: 577–580.

Hampson EC, Vaney DI, Weiler R (1992) Dopaminergic modulation of gap junction permeability between amacrine cells in mammalian retina. J. Neurosci. 12: 4911–4922.

Hamzei-Sichani F, Kamasawa N, Janssen WGM, Yasamura T, Davidson KGV Hof PR, Wearne SL, Stewart MG, Young SR, Whittington MA, Rash JE, Traub RD (2007) Gap junctions on hippocampal mossy fiber axons demonstrated by thin-section electron microscopy and freeze-fracture replica immunogold labeling. *Proc. Natl. Acad. Sci. USA* 104: 12548–12553.

Hamzei-Sichani F, Vivar C, Nagy JI, Rash JE, Yasumura T, Davidson KGV, Janssen WGM, Wearne SL, Hof PR, Traub RD, Gutiérrez R. (in prep.) Mixed Chemical-Electrical Synapses between Hippocampal Mossy Fibers and Pyramidal Cells.

Hanna RB, Keeter JS, Pappas GD (1978) The fine structure of a rectifying electrotonic synapses. J. Cell Biol. 79: 764–773.

Haring JH, Yan W, Faber KM (1997) Neuronal dye coupling in the developing rat fascia dentata. Brain Res. Dev. Brain Res. 103: 205–208.

Harris EW, Ganong AH, Cotman CW (1984) Long-term potentiation in the hippocampus involves activation of N-methyl-D-aspartate receptors. Brain Res. 323: 132–137.

Harris-Warrick RM, Marder E, Selverston AI, Moulins M (1992) Dynamic Biological Networks.

The Stomatogastric Nervous System. MIT Press, Cambridge, MA.

Hartline HK (1938) The responses of single optic nerve fibers of the vertebrate eye to illumination of the retina. Am. J. Physiol. 121: 400–415.

Hartline HK,Ratliff F (1958) Spatial summation of inhibitory influences in the eye of Limulus, and the mutual interactions of receptor units. J. Gen. Physiol. 41: 1049–1066.

Hartmann MJ, Bower JM (1998) Oscillatory activity in the cerebellar hemispheres of unrestrained rats. J. Neurophysiol. 80: 1598–1604.

Hartwell CE (1996) The schizophrenogenic mother concept in American psychiatry. Psychiatry 59: 274–297.

Harvey AL (1997) Recent studies on dendrotoxins and potassium ion channels. Gen. Pharmacol. 28: 7–12.

Harvey AL, Robertson B (2004) Dendrotoxins: structure-activity relationships and effects on potassium ion channels. Curr. Med. Chem. 11: 3065–3072.

Harvey RJ, Napper RM (1988) Quantitative study of granule and Purkinje cells in the cerebellar cortex of the rat. J. Comp. Biol. 274: 151–157.

Hasenstaub A, Shu Y, Haider B, Kraushaar U, Duque A, McCormick DA (2005) Inhibitory postsynaptic potentials carry synchronized frequency information in active cortical networks. Neuron 47: 423–435.

Hashimoto T, Morita H, Tada T, Maruyama T, Yamada Y, Ikeda S (2001) Neuronal activity in the globus pallidus in chorea caused by striatal lacunar infarction. Ann. Neurol. 50: 528–531.

Hashimoto T, Volk DW, Eggan SM, Mirnics K, Pierri JN, Sun Z, Sampson AR, Lewis DA (2003) Gene expression deficits in a subclass of GABA neurons in the prefrontal cortex of subjects with schizophrenia. J. Neurosci. 23: 6315–6326.

Hashimoto T, Bergen SE, Nguyen QL, Xu B, Monteggia LM, Pierri JN, Sun Z, Sampson AR, Lewis DA (2005) Relationship of brain-derived neurotrophic factor and its receptor TrkB to altered inhibitory prefrontal circuitry in schizophrenia. J. Neurosci. 25: 372–383.

Hashimoto T, Bazmi HH, Mirnics K, Wu Q, Sampson AR, Lewis DA (2008) Conserved regional patterns of GABA-related transcript expression in the neocortex of subjects with schizophrenia. Am. J. Psychiatry 165: 479–489.

Hassler R (1938) Zur Pathologie der Paralysis agitans und des postenzephalitischen Parkinsonismus. J. Psychol. Neurol. 48: 387–476.

Haueisen J,l Heuer T, Nowak H, Liepert J, Weiller C, Okada Y, Curio G (2000) The influence of lorazepam on somatosensory-evoked fast frequency (600 Hz) activity in MEG. Brain Res. 874: 10–14.

Hauser WA, Annegers JF, Rocca WA (1996) Descriptive epidemiology of epilepsy: contributions of population based studies from Rochester, Minnesota. Mayo Clin. Proc. 71: 576–586.

He J, Hsiang H-L, Wu C, Mylvagnanam S, Carlen PL, Zhang L (2009) Cellular mechanisms of cobalt-induced hippocampal epileptiform discharges. Epilepsia 50: 99–115.

Hebb DO (1949) The Organization of Behavior: A Neuropsychological Theory. Lawrence Erlbaum, Mahwah, NJ

Hecht-Nielsen R (1987) Kolmogorov's mapping neural network existence theorem. IEEE International Conference on Neural Networks, San Diego, SOS Printing, 2: 11–14.

Heck DH, Thach WT, Keating JG (2007) On-beam synchrony in the cerebellum as the mechanism for the timing and coordination of movement. Proc. Natl. Acad. Sci. USA 104: 7658–7663.

Heckers S, Stone D, Walsh J, Shick J, Koul P, Benes FM (2002) Differential hippocampal expression of glutamic acid decarboxylase 65 and 67 messenger RNA in bipolar disorder and schizophrenia. Arch. Gen. Psychiatry 59: 521–529.

Heinrichs RW, Zakzanis KK (1998) Neurocognitive deficit in schizophrenia: a quantitative review of the evidence. Neuropsychology 12: 426–445.

Heise CE, Mitrofanis J (2005) Reduction in parvalbumin expression in the zona incerta after 6OHDA lesion in rats. J. Neurocytol. 34: 421–434.

Heisenberg W (1977) Die Bedeutung des Schönen in der exakten Naturwissenschaft. Reprinted in Quantentheorie und Philosophie (1983) Reclam, Stuttgart.

Hempelmann A, Heils A, Sander T (2006) Confirmatory evidence for an association of the connexin-36 with juvenile myoclonic epilepsy. Epilepsy Res. 71: 223–228. (single nucleotide polymorphism. Statistical evidence, not so easy to interpret.)

Henson RA, Urich H (1982) Cancer and the Nervous System: The Neurological Manifestations of Systemic Malignant Disease. Oxford University Press, Oxford.

Henze DA, González-Burgos GR, Urban NN, Lewis DA, Barrionuevo G (2000) Dopamine increases excitability of pyramidal neurons in primate prefrontal cortex. J. Neurophysiol. 84: 2799–2809.

Heresco-Levy U, Javitt DC, Ebstein R, Vass A, Lichtenberg P, Bar G, Catinari S, Ermilov M (2005) D-serine efficacy as add-on pharmacotherapy to resperidone and olanzapine for treatment-refractory schizophrenia. Biol. Psychiatry 57: 577–585.

Herculano-Houzel S, Munk MH, Neuenschwander S, Singer W (1999) Precisely synchronized oscillatory firing patterns require electroencephalographic activation. J. Neurosci. 19: 3992–4010.

Herning RI, Jones RT, Hooker WD, Mendelson J, Blackwell L (1985) Cocaine increases EEG beta: a replication and extension of Hans Berger's historic experiments. Electroenceph. Clin. Neurophysiol. 60: 470–477.

Herrmann CS, Demiralp T (2005) Human EEG gamma oscillations in neuropsychiatric disorders. Clin. Neurophysiol. 116: 2719–2733.

Hillman D, Chen S, Aung TT, Cherksey B, Sugimori M, Llinás RR (1991) Localization of P-type calcium channels in the central nervous system. Proc. Natl. Acad. Sci. USA 88: 7076–7080.

Hippius H, Müller N (2008) The work of Emil Kraepelin and his research group in München. Eur. Arch. Psychiatry Clin. Neurosci. 258 Suppl. 2: 3–11.

Hirose S, Zenri F, Akiyoshi H, Fukuma G, Iwata H, Inoue T, Yonetani M, Tsutsumi M, Muranaka H, Kurokawa T, Hanai T, Wada K, Kaneko S, Mitsudome A (2000) A novel mutation of KCNQ3 (c.925T→C) in a Japanese family with benign familial neonatal convulsions. Ann. Neurol. 47: 822–826.

Hirtz D, Thurman DJ, Gwinn-Hardy K, Mohamed M, Chaudhuri AR, Zalutsky R (2007) How common are the "common" neurologic disorders? Neurology 68: 326–337.

Hochner B, Spira ME, Werman R (1976) Penicillin decreases chloride conductance in crustacean muscle: a model for the epileptic neuron. Brain Res. 107: 85–103.

Hodgkin AL, Huxley AF (1952) A quantitative description of membrane current and its application to conduction and excitation in nerve. J. Physiol. 117: 500–544.

Hoffman DA, Magee JC, Colbert CM, Johnston D (1997) K^+ channel regulation of signal propagation in dendrites of hippocampal neurons. Nature 387: 869–875.

Hoffman SN, Salin PA, Prince DA (1994) Chronic neocortical epileptogenesis in vitro. J. Neurophysiol. 71: 1762–1773.

Hollmann M, Heinemann S (1994) Cloned glutamate receptors. Annu. Rev. Neurosci. 17: 31–108.

Hollmann M, Hartley M, Heinemann S (1991) Ca^{2+} permeability of KA-AMPA-gated glutamate receptor channels depends on subunit composition. Science 252: 851–853.

Holmes G (1917) The symptoms of acute cerebellar injuries due to gunshot injuries. Brain 40: 461–535.

Hong LE, Summerfelt A, McMahon RP, Thaker GK, Buchanan RW (2004) Gamma/beta oscillation and sensory gating deficit in schizophrenia. NeuroReport 15: 155–159.

Hoogenboom N, Schoffelen JM, Oostenveld R, Parkes LM, Fries P (2006) Localizing human visual gamma-band activity in frequency, time and space. Neuroimage 29: 764–773.

Hopfield, J.J., Neural networks and physical systems with emergent collective computational abilities. Proc. Natl Acad. Sci. 79 (1982) 2554–2558.

Hopfield JJ, Tank DW (1986) Computing with neural circutis: a model. Science 233: 625–633.

Horikawa J, Tanahashi A, Suga N (1994) After-discharges in the auditory cortex of the mustached bat: no oscillatory discharges for binding auditory information. Hearing Res. 76: 45–52.

Hormuzdi SG, Pais I, LeBeau FEN, Towers SK, Rozov A, Buhl EH, Whittington MA, Monyer H (2001) Impaired electrical signaling disrupts gamma frequency oscillations in connexin 36–deficient mice. Neuron 31: 487–495.

Hormuzdi SG, Filippov MA, Mitropoulou G, Monyer H, Bruzzone R (2004) Electrical synapses: a dynamic signaling system that shapes the activity of neuronal networks. Biochim. Biophys. Acta 1662: 113–137.

Horne MK, Butler EG (1995) The role of the cerebello-thalamo-cortical pathway in skilled movement. Prog. Neurobiol. 46: 199–213.

Hornykiewicz O (2006) The discovery of dopamine deficiency in the parkinsonian brain. J. Neural. Transm. Suppl. 70: 9–15.

Hosseinzadeh H, Asl MN, Parvardeh S, Tagi Mansouri SM (2005) The effects of carbenoxolone on spatial learning in the Moris water maze task in rats. Med. Sci. Monit. 11: BR88–94.

Howard MW, Rizzuto DS, Caplan JP, Madsen JR, Lisman J, Aschenbrenner-Scheibe R, Schulze-Bonhage A, Kahana MJ (2003) Gamma oscillations correlate with working memory load in humans. Cereb. Cortex 13: 1369–1374.

Hrachovy RA, Frost JD Jr. (2006) The EEG in selected generalized seizures. J. Clin. Neurophysiol. 23: 312–332.

Hu EH, Bloomfield SA (2003) Gap junctional coupling underlies the short-latency spike synchrony of retinal α ganglion cells. J. Neurosci. 23: 6768–6777.

Hu H, Vervaeke K, Storm JF (2007) M-channels (Kv7/KCNQ channels) that regulate synaptic integration, excitability, and spike pattern of CA1 pyramidal cells are located in the perisomatic region. J. Neurosci. 27: 1853–1867.

Hubel DH, Wiesel TN (1959) Receptive fields of single neurones in the cat's striate cortex. J. Physiol. 148: 574–591.

Hubel DH, Wiesel TN (1962) Receptive fields, binocular interaction and functional architecture in the cat's visual cortex. J. Physiol. 160: 106–154.

Hubel DH, Wiesel TN (1963) Shape and arrangement of columns in cat's striate cortex. J. Physiol. Lond. 165: 559–568.

Huffman J, Kossoff EH (2006) State of the ketogenic diet(s) in epilepsy. Curr. Neurol. Neurosci. Rep. 6: 332–340.

Hughes SW, Blethyn KL, Cope DW, Crunelli V (2002a) Properties and origin of spikelets in thalamocortical neurones in vitro. Neuroscience 110: 395–401.

Hughes SW, Cope DW, Blethyn KL, Crunelli V (2002b) Cellular mechanisms of the slow (<1 Hz) oscillation in thalamocortical neurons in vitro. Neuron 33: 947–958.

Hughes SW, Lörincz M, Cope DW, Blethyn KL, Kékesi KA, Parri HR, Juhász G, Crunelli V (2004) Synchronized oscillations at α and θ frequencies in the lateral geniculate nucleus. Neuron 42: 1–20.

Humphries SV (1963) Convulsant effect of penicillin on the cerebral cortex. Lancet 1: 115–116.

Hutchison WD, Dostrovsky JO, Walters JR, Courtemanche R, Boraud T, Goldberg J, Brown P (2004) Neuronal oscillations in the basal ganglia and movement disorders: evidence from whole animal and human recordings. J. Neurosci. 24: 9240–9243.

Hwa GGC, Avoli M, Olivier A, Villemure JG (1991) Bicuculline-induced epileptogenesis in the human neocortex maintained in vitro. Exp. Brain Res. 83: 329–339.

Hyde TM, Ziegler JC, Weinberger DR (1992) Psychiatric disturbances in metachromatic leukodystrophy. Insights into the neurobiology of psychosis. Arch Neurol. 49: 401–406.

Ikeda H, Leyba L, Bartolo A, Wang Y, Okada YC (2002) Synchronized spikes of thalamocortical axonal terminals and cortical neurons are detectable outside the pig brain with MEG. J. Neurophysiol. 87: 626–630.

Ikeda H, Wang Y, Okada YC (2005) Origins of the somatic N20 and high-frequency oscillations evoked by trigeminal stimulation in the piglets. Clin. Neurophysiol. 116: 827–841.

Impagnatiello F, Guidotti AR, Pesold C, Dwivedi Y, Caruncho H, Pisu MG, Uzunov DP, Smalheiser NR, Davis JM, Pandey GN, Pappas GD, Tueting P, Sharma RP, Costa E (1998) A decrease of reelin expression as a putative vulnerability factor in schizophrenia. Proc. Natl. Acad. Sci. USA 95: 15718–15723.

Inoue M, Peeters BW, van Luijtelaar EL, Vossen JM, Coenen AM (1990) Spontaneous occurrence of spike-wave discharges in five inbred strains of rats. Physiol. Behav. 48: 199–201.

Insausti R, Herrero MT, Witter MP (1997) Entorhinal cortex of the rat: cytoarchitectonic subdivision and the origin and distribution of cortical efferents. Hippocampus 7: 146–183.

Ioannides AA, Fenwick PB (2005) Imaging cerebellum activity in real time with magnetoencephalographic data. Prog. Brain Res. 148: 139–150.

Isope P, Barbour B (2002) Properties of unitary granule cell-Purkinje cell synapses in adult rat cerebellar slices. J. Neurosci. 22: 9668–9678.

Isope P, Dieudonné S, Barbour B (2002) Temporal organization of activity in the cerebellar cortex: a manifesto for synchrony. Ann. NY Acad. Sci. 978: 164–174.

Itazawa S-I, Isa T, Ozawa S (1997) Inwardly rectifying and Ca^{2+}-permeable AMPA-type glutamate receptor channels in rat neocortical neurons. J. Neurophysiol. 78: 2592–2605.

Ito M (1982) Experimental verification of Marr-Albus' plasticity assumption for the cerebellum. Acta Biol. Acad. Sci. Hung. 33: 189–199.

Ito M (1985) Processing of vibrissa sensory information within the rat neocortex. J. Neurophysiol. 54: 479–490.

Ito M (2006) Cerebellar circuitry as a neuronal machine. Prog. Neurobiol. 78: 272–303.

Iversen L (2006) Neurotransmitter transporters and their impact on the development of psychopharmacology. Br. J. Pharmacol. 147 Suppl. 1: S82–S88.

Jablensky A (1987) Multicultural studies and the nature of schizophrenia: a review. J. R. Soc. Med. 80: 162–167.

Jabs R, Kirchhoff F, Kettenmann H, Steinhauser C (1994) Kainate activates Ca^{2+}-permeable glutamate receptors and blocks voltage-gated K^+ currents in glial cells of mouse hippocampal slices. Pflügers Arch. 426: 310–319.

Jackson A, Spinks RL, Freeman TC, Wolpert DM, Lemon RN (2002) Rhythm generation in monkey motor cortex explored using pyramidal tract stmulation. J. Physiol. 541: 685–699.

Jackson ME, Homayoun H, Moghaddam B (2004) NMDA receptor hypofunction produces concomitant firing rate potentiation and burst activity reduction in the prefrontal cortex. Proc. Natl. Acad. Sci. USA 101: 8467–8472.

Jackson-Lewis V, Jakowec M, Burke RE, Przedborski S (1995) Time course and morphology of dopaminergic neuronal death caused by the neurotoxin 1–methyl-4–phenyl-1,2,3,6–tetrahydropyridine. Neurodegeneration 4: 257–269.

Jacobs J, LeVan P, Chander R, Hall J, Dubeau F, Gotman J (2008) Interictal high-frequency oscillations (80–500 Hz) are an indicator of seizure onset areas independent of spikes in the human epileptic brain. Epilepsia 49: 1893–1907.

Jagadeesh B, Gray CM, Ferster D (1992) Visually evoked oscillations of membrane potential in cells of cat visual cortex. Science 257: 552–554.

Jaggy A, Faissler D, Gaillard C, Srenk P, Graber H (1998) Genetic aspects of idiopathic epilepsy in Labrador retrievers. J. Small Anim. Pract. 39 275–280.

Jahnsen H, Llinás R (1984a) Electrophysiological properties of guinea-pig thalamic neurones: an in vitro study. J. Physiol. 349: 205–226.

Jahnsen H, Llinás R (1984b) Ionic basis for the electroresponsiveness and oscillatory properties of guinea-pig thalamic neurones in vitro. J. Physiol. 349: 227–247.

Jahr CE, Stevens CF (1990a) A quantitative description of NMDA receptor-channel kinetic behavior. J. Neurosci. 10: 1830–1837.

Jahr CE, Stevens CF (1990b) Voltage dependence of NMDA-activated macroscopic conductances predicted by single-channel kinetics. J. Neurosci. 10: 3178–3182.

Jahr CE, Stevens CF (1993) Calcium permeability of the N-methyl-D-aspartate receptor channel in hippocampal neurons in culture. Proc. Natl. Acad. Sci. USA 90: 11573–11577.

Jahromi SS, Wentlandt K, Piran S, Carlen PL (2002) Anticonvulsant actions of gap junctional blockers in an in vitro seizure model. J. Neurophysiol. 88: 1893–1902.

Janz D (1985) Epilepsy with impulsive petit mal (juvenile myoclonic epilepsy). Acta Neurol. Scand. 72: 449–459.

Jasper HH (1936) Cortical excitatory state and variability in human brain rhythms. Science 83: 259-260.

Javoy-Aqid F, Aqid Y (1980) Is the mesocortical dopaminergic system involved in Parkinson disease? Neurology 30: 1326–1330.

Jayalakshmi SS, Mohandas S, Sailaja S, Borgohain R (2006) Clinical and electroencephalographic study of first-degree relatives and probands with juvenile myoclonic epilepsy. Seizure 15: 177–183.

Jefferys JGR (2003) Models and mechanisms of experimental epilepsies. Epilepsia 44 Suppl. 12: 44–50.

Jellinger KA (2003) Neuropathological spectrum of synucleinopathies. Mov. Disord. 18 Suppl. 6: S2–S12.

Jen JC, Graves D, Hess EJ, Hanna MG, Griggs RC, Baloh RW, CINCH Investigators (2007) Primary episodic ataxias: diagnosis, pathogenesis and treatment. Brain 130: 2484

Jenny AB (1979) Commissural projections of the cortical hand motor area in monkeys. J. Comp. Neurol. 188: 137–145.

Jensen MS, Azouz R, Yaari Y (1996) Spike after-depolarization and burst generation in adult rat hippocampal CA1 pyramidal cells. J. Physiol. 492: 199–210.

Jeschke M, Lenz D, Budinger E, Herrmann CS, Ohl F (2008) Gamma oscillations in gerbil auditory cortex during a target-discrimination task reflect matches with short-term memory. Brain Res. 1220: 70-80.

Ji D, Wilson MA (2007) Coordinated memory replay in the visual cortex and hippocampus during sleep. Nat. Neurosci. 10: 100–107.

Jiang C, Haddad GG (1997) Modulation of K^+ channels by intracellular ATP in human neocortical neurons. J. Neurophysiol. 77: 93–102.

Jiao Y, Nadler JV (2007) Stereological analysis of GluR2–immunoreactive hilar neurons in the pilocarpine model of temporal lobe epilepsy: correlation of cell loss with mossy fiber sprouting. Exp. Neurol. 205: 569–582.

Jin X, Prince DA, Huguenard JR (2006) Enhanced excitatory synaptic connectivity in layer V pyramidal neurons of chronically injured epileptogenic neocortex in rats. J. Neurosci. 26: 4891–4900.

Jirsch JD, Urrestarazu E, LeVan P, Olivier A, Dubeau F, Gotman J (2006) High-frequency oscillations during human focal seizures. Brain 129: 1593–1608.

Johnson JW, Ascher P (1987) Glycine potentiates the NMDA response in cultured mouse brain neurons. Nature 325: 529–531.

Johnston D, Brown TH (1981) Giant synaptic potential hypothesis for epileptiform activity. Science 211: 294–297.

Johnston D, Magee JC, Colbert CM, Christie BR (1996) Active properties of neuronal dendrites. Annu. Rev. Neurosci. 19: 165–186.

Jokeit H, Makeig S (1994) Different event-related patterns of γ-band power in brain waves of fast- and slow-reacting subjects. Proc. Natl. Acad. Sci. USA 91: 6339–6343.

Joliot M, Ribary U, Llinás R (1994) Human oscillatory brain activity near 40 Hz coexists with cognitive temporal binding. Proc. Natl. Acad. Sci. USA 91: 11748–11751.

Jonas P, Monyer H (Eds.) (1999) Ionotropic Glutamate Receptors in the CNS (Handbook of Experimental Pharmacology. Springer, Berlin.

Jonas P, Koh DS, Kampe K, Hermsteiner M, Vogel W (1991) ATP-sensitive and Ca-activated K channels in vertebrate axons: novel links between metabolism and excitability. Pflügers Arch. Eur. J. Physiol. 418: 68–73.

Jonas P, Racca C, Sakmann B, Seeburg PH, Monyer H (1994) Differences in Ca^{2+} permeability of AMPA-type glutamate receptor channels in neocortical neurons caused by differential GluR-B subunit expression. Neuron 12: 1281–1289.

Jones EG (1995) Cortical development and neuropathology in schizophrenia. Ciba Found. Symp. 193: 277–295.

Jones EG, Powell TPS (1969) Synapses on the axon hillocks and initial segments of pyramidal cell axons in the cerebral cortex. J. Cell Sci. 5: 495–507.

Jones MS, Barth DS (1997) Sensory-evoked high-frequency (gamma-band) oscillating potentials in somatosensory cortex of the unanesthetized rat. Brain Res. 768: 167–176.

Jones MS, Barth DS (1999) Spatiotemporal organization of fast (>200 Hz) electrical oscillations in rat vibrissa/barrel cortex. J. Neurophysiol. 82: 1599–1609.

Jones MS, Barth DS (2002) Effects of bicuculline methiodide on fast (>200 Hz) electrical oscillations in rat somatosensory cortex. J. Neurophysiol. 88: 1016–1025.

Jones MS, MacDonald KD, Choi B, Dudek FE, Barth DS (2000) Intracellular correlates of fast (>200 Hz) electrical oscillations in rat somatosensory cortex. J. Neurophysiol. 84: 1505–1518.

Jones RSG, Buhl EH (1993) Basket-like interneurones in layer II of the entorhinal cortex exhibit a powerful NMDA-mediated synaptic excitation. Neurosci. Lett. 149: 35–39.

Jörntell H, Ekerot C-F (2006) Properties of somatosensory synaptic integration in cerebellar granule cells in vivo. J. Neurosci. 26: 11786–11797.

Jörntell H, Hansel C (2006) Synaptic memories upside down: bidirectional plasticity at cerebellar parallel fiber-Purkinje cell synapses. Neuron 52: 227–238.

Jung R, Berger W (1979) Fiftieth anniversary of Hans Berger's publication of the electroencephalogram. His first records in 1924–1931. Arch. Psychiatr. Nervenkr. 227: 279–300.

Kahana MJ, Sekuler R, Caplan JB, Kirschen M, Madsen JR (1999) Human theta oscillations exhibit task dependence during virtual maze navigation. Nature 399: 781–784.

Kalinichenko SG, Okhotin VE (2005) Unipolar brush cells – a new type of excitatory interneuron in the cerebellar cortex and cochlear nuclei of the brainstem. Neurosci. Behav. Physiol. 35: 21–36.

Kamasawa N, Furman CS, Davidson KG, Sampson JA, Magnie AR, Gebhardt BR, Kamasawa M, Yasumura T, Zumbrunnen JR, Pickard GE, Nagy JI, Rash JE (2006) Abundance and ultrastructural diversity of neuronal gap junctions in the OFF and ON sublaminae of the inner plexiform layer of rat and mouse retina. Neuroscience 142: 1093–1117.

Kamasawa N, Hamzei-Sichani F, Yasumura T, Janssen WGM, Davidson KGV, Wearne SL, Hof PR, Traub RD, Rash JE (2007) Ultrastructural evidence for mixed synapses in hippocampal principal neurons using thin-section and freeze-fracture replica immunogold labeling (FRIL) electron microscopy. Soc. Neurosci. Abstr. 581.12.

Kamiya H, Ozawa S (1998) Kainate receptor-mediated inhibition of presynaptic Ca^{2+} influx and EPSP in area CA1 of the rat hippocampus. J. Physiol. 509: 833–845.

Kampa BM, Stuart GJ (2006) Calcium spikes in basal dendrites of layer 5 pyramidal neurons during action potential bursts. J. Neurosci. 26: 7424–7432.

Kampa BM, Letzkus JJ, Stuart GJ (2006) Requirement of dendritic calcium spikes for induction of spike-timing-dependent synaptic plasticity. J. Physiol. 574: 283–290.

Kandel E, Schwartz JH, Jessell TM (2000) Principles of Neural Science. McGraw-Hill, New York.

Kanemoto K, Tsuji T, Kawasaki J (2001) Reexamination of interictal psychoses based on DSM IV psychosis classification and international epilepsy classification. Epilepsia 42: 98–103.

Kanno Y, Loewenstein WR (1964) Intercellular diffusion. Science 143: 959–960.

Kapoor R, Li YG, Smith KJ (1997) Slow sodium-dependent potential oscillations contribute to ectopic firing in mammalian demyelinated axons. Brain 120: 647–652.

Kass JI, Mintz IM (2006) Silent plateau potentials, rhythmic bursts, and pacemaker firing: Three patterns of activity that coexist in quadristable subthalamic neurons. Proc. Natl. Acad. Sci. USA 103: 183–188.

Kato T, Hirano A (1985) A Golgi study of the proximal portion of the human Purkinje cell axon. Acta Neuropathol. 68: 191–195.

Katsumaru H, Kosaka T, Heizmann CW, Hama K (1988) Gap junctions on GABAergic neurons containing the calcium-binding protein parvalbumin in the rat hippocampus (CA1 region). Exp Brain Res 72: 363–370.

Kaufmann PY (1912) Electrical phenomena in cerebral cortex [Russian]. Obz. Psikiatr. Nev. Exsper. Psikl. 7–8: 403.

Kawaguchi S, Samejima A, Yamamoto T (1983) Post-natal development of the cerebello-cerebral projection in kittens. J. Physiol. 343: 215–232.

Kawaguchi Y (1993) Groupings of nonpyramidal and pyramidal cells with specific physiological and morphological characteristics in rat frontal cortex. J. Neurophysiol. 69: 416–431.

Kawaguchi Y (1995) Physiological subgroups of nonpyramidal cells with specific morphological characteristics in layer II/III of rat frontal cortex. J. Neurosci. 15: 2638–2655.

Kawaguchi Y (1997) Selective cholinergic modulation of cortical GABAergic cell subtypes. J. Neurophysiol. 78: 1743–1747.

Kawaguchi Y (2001) Distinct firing patterns of neuronal subtypes in cortical synchronized activities. J. Neurosci. 21: 7261–7272.

Kawaguchi Y, Kubota Y (1993) Correlation of physiological subgroupings of nonpyramidal cells with parvalbumin- and Calbindin$_{D28k}$-immunoreactive neurons in layer V of rat frontal cortex. J. Neurophysiol. 70: 387–396.

Kawai H, Lazar R, Metherate R (2007) Nicotinic control of axon excitability regulates thalamocortical transmission. Nat. Neurosci. 10: 1168–1175.

Kay AR, Sugimori M, Llinás R (1998) Kinetic and stochastic properties of a persistent sodium current in mature guinea pig cerebellar Purkinje cells. J. Neurophysiol. 80: 1167–1179.

Kebir O, Tabbane K (2008) Working memory in schizophrenia: A review. Encephale 34: 289–298.

Keil A, Müller MM, Ray WJ, Gruber T, Elbert T(1999) Human gamma band activity and perception of a gestalt. J. Neurosci. 19: 7152–7161.

Keinänen K, Wisden W, Sommer B, Werner P, Herb A, Verdoorn TA, Sakmann B, Seeburg PH (1990) A family of AMPA selective glutamate receptors. Science 249: 556–560.

Kelly RM, Strick PL (2003) Cerebellar loops with motor cortex and prefrontal cortex of a nonhuman primate. J. Neurosci. 23: 8432–8444.

Keros S, Hablitz JJ (2005) Ectopic action potential generation in cortical interneurons during synchronized GABA responses. Neuroscience 131: 833–842.

Kha HT, Finkelstein DI, Pow DV, Lawrence AJ, Horne MK (2000) Study of projections from the entopeduncular nucleus to the thalamus of the rat. J. Comp. Neurol. 426: 366–377.

Khaliq ZM, Raman IM (2005) Axonal propagation of simple and complex spikes in cerebellar Purkinje neurons. J. Neurosci. 25: 454–463.

Khosravani H, Pinnegar CR, Mitchell JR, Bardakjian BL, Federico P, Carlen PL (2005) Increased high-frequency oscillations precede in vitro low-Mg seizures. Epilepsia 46: 1188–1197.

Khosravani H, Mehrotra N, Rigby M, Hader WJ, Pinnegar CR, Pillay N, Wiebe S, Federico P (2009) Spatial localization and time-dependent changes of electrographic high frequency oscillations in human temporal lobe epilepsy. Epilepsia 50: 605-616.

Kiesmann M, Marescaux C, Vergnes M, Micheletti G, Depaulis A, Warter JM (1988) Audiogenic seizures in Wistar rats before and after repeated auditory stimuli: clinical, pharmacological, and electroencephalographic studies. J. Neural Transm. 72: 235–244.

Kilner JM, Salenius S, Baker SN, Jackson A, Hari R, Lemon RN (2003) Task-dependent modulations of cortical oscillatory activity in human subjects during a bimanual precision grip task. Neuroimage 18: 67–73.

King JS, Martin GF, Bowman MH (1975) The direct spinal area of the inferior olivary nucleus: an electron microscopic study. Exp. Brain Res. 22: 13–24.

Kinney JW, Davis CN, Tabarean I, Conti B, Bartfai T, Behrens MM (2006) A specific role for NR2A-containing NMDA receptors in the maintenance of parvalbumin and GAD67 immunoreactivity in cultured interneurons. J. Neurosci. 26: 1604–1615.

Kistler WM, van Hemmen JL, De Zeeuw CI (2000) Time window control: a model for cerebellar function based on synchronization, reverberation, and time slicing. Prog. Brain Res. 124: 275–297.

Klausberger T, Magill PJ, Márton LF, Roberts JDB, Cobden PM, Buzsáki G, Somogyi P (2003a) Brain-state- and cell-type-specific firing of hippocampal interneurons in vivo. Nature 421: 844–848.

Klausberger T, Márton LF, Baude A, Roberts JDB, Magill PJ, Somogyi P (2003b) Spike timing of dendrite-targeting bistratified cells during hippocampal network oscillations in vivo. Nat. Neurosci. 7: 41–47.

Klausberger T, Márton LF, O'Neill J, Huck JH, Dalezios Y, Fuentealba P, Suen WY, Papp E, Kaneko T, Watanabe M, Csicsvari J, Somogyi P (2005) Complementary roles of cholecystokinin- and parvalbumin-expressing GABAergic neurons in hippocampal network oscillations. J. Neurosci. 25: 9782–9793.

Kleckner NW, Dingledine R (1988) Requirement for glycine in activation of NMDA-receptors in Xenopus oocytes. Science 241: 835–837.

Kleopa KA, Orthmann JL, Enriquez A, Paul DL, Scherer SS (2004) Unique distributions of the gap junction proteins connexin29, connexin32, and connexin47 in oligodendrocytes. Glia 47: 346–357.

Kobayashi K, Suzuki H (2007) Dopamine selectively potentiates hippocampal mossy fiber to CA3 synaptic transmission. Neuropharmacology 52: 552–561.

Kobayashi K, Oka M, Akiyama T, Inoue T, Abiru K, Ogino T, Yoshinaga H, Ohtsuka Y, Oka E (2004) Very fast rhythmic activity on scalp EEG associated with epileptic spasms. Epilepsia 45: 488–496.

Kocsis JD, Waxman SG (1982) Intra-axonal recordings in rat dorsal column axons: membrane hyperpolarization and decreased excitability precede the primary afferent depolarization. Brain Res. 238: 222–227.

Koehler PJ (1994) Brown-Séquard's spinal epilepsy. Med. Hist. 38: 189–203.

Koeze TH (1973) Thresholds of cortical activation of baboon alpha and gamma-motoneurones during halothane anaesthesia. J. Physiol. 229: 319–337.

Koeze TH, Phillips CG, Sheridan JD (1968) Thresholds of cortical activation of muscle spindles and alpha motoneurones of the baboon's hand. J. Physiol. 195: 419–449.

Koh DS, Burnashev N, Jonas P (1995a) Block of native Ca^{2+}-permeable AMPA receptors in rat brain by intracellular polyamines generates double rectification. J. Physiol. 486: 305–312.

Koh DS, Geiger JR, Jonas P, Sakmann B (1995b) Ca^{2+}-permeable AMPA and NMDA receptor channels in basket cells of rat hippocampal dentate gyrus.

Kohl F (1999) The beginning of Emil Kraepelin's classification of psychoses. A historica-methodological reflection on the occasion of the 100th anniversary of his "Heidelberg Address" 27 November 1898 on "nosologic dichotomy" of endogenous psychoses. Psychiatr. Prax. 26: 105–111.

Köhling R, Avoli M (2006) Methodological approaches to exploring epileptic disorders in the human brain in vitro. J. Neurosci. Methods 155: 1–19.

Köhling R, Lücke A, Straub H, Speckmann EJ, Tuxhorn I, Wolf P, Pannek H, Oppel F (1998) Spontaneous sharp waves in human neocortical slices excised from epileptic patients. Brain 121: 1073–1087.

Köhling R, Gladwell SJ, Bracci E, Vreugdenhil M, Jefferys JGR (2001) Prolonged epileptiform bursting induced by $0–Mg^{2+}$ in rat hippocampal slices depends on gap junctional coupling. Neuroscience 105: 579–587.

Kole MH, Hallermann S, Stuart GJ (2006) Single I_h channels in pyramidal neuron dendrites: properties, distribution, and impact on action potential output. J. Neurosci. 26: 1677–1687.

Kole MH, Letzkus JJ, Stuart GJ (2007) Axon initial segment Kv1 channels control axonal action potential waveform and synaptic efficacy. Neuron 55: 633–647.

Kole MHP, Ilschner SU, Kampa BM, Williams SR, Ruben PC, Stuart GJ (2008) Action potential generation requires a high sodium channel density in the axon initial segment. Nat. Neurosci. 11: 178–186.

Kolmogorov AN (1957) On the representation of continuous functions of several variables by superposition of continuous functions of one variable and addition. Dokl. Akad. Nauk SSSR 114: 369–373.

Kolomiets BP, Deniau JM, Mailly P, Ménétrey A, Glowinski J, Thierry AM (2001) Segregation and convergence of information flow through the cortico-subthalamic pathways. J. Neurosci. 21: 5764–5772.

König P, Engel AK (1995) Correlated firing in sensory-motor systems. Curr. Opin. Neurobiol. 5: 511–519.

König P, Schillen TB (1991) Stimulus-dependent assembly formation of oscillatory responses: I. Synchronization. Neural Comput. 3: 155–166.

König P, Engel AK, Roelfsema PR, Singer W (1995a) How precise is neuronal synchronization? Neural Comput. 7: 469–485.

König P, Engel AK, Singer W (1995b) Relation between oscillatory activity and long-range synchronization in cat visual cortex. Proc. Natl. Acad. Sci. 92: 290–294.

König P, Engel AK, Singer W (1996) Integrator or coincidence detector? The role of the cortical neuron revisited. Trends Neurosci. 19: 130–137.

Konnerth A, Heinemann U, Yaari Y (1984) Slow transmission of neural activity in hippocampal area CA1 in absence of active chemical synapses. Nature 307: 69–71.

Konnerth A, Heinemann U, Yaari Y (1986) Nonsynaptic epileptogenesis in the mammalian hippocampus in vitro. I. Development of seizurelike activity in low extracellular calcium. J. Neurophysiol. 56: 409–423.

Konnerth A, Dreessen J, Augustine GJ (1992) Brief dendritic calcium signals initiate long-lasting synaptic depression in cerebellar Purkinje cells. Proc. Natl. Acad. Sci. USA 89: 7051–7055.

Konopacki J, MacIver MB, Bland BH, Roth SH (1987) Carbachol-induced EEG 'theta' activity in hippocampal brain slices. Brain Res. 405: 196–198.

Konopacki J, Kowalczyk T, Golebiewski H (2004) Electrical coupling underlies theta oscillations recorded in hippocampal formation slices. Brain Res. 1019: 270–274.

Koós T, Tepper JM (1999) Inhibitory control of neostriatal projection neurons by GABAergic interneurons. Nat. Neurosci. 2: 467–472.

Kopell N, Ermentrout G (1998) Fine structure of neural spiking and synchronization in the presence of conduction delays. Proc. Natl. Acad. Sci. USA 95: 1259–1264.

Kopell N, Ermentrout G (2004) Chemical and electrical synapses perform complementary roles in the synchronization of interneuronal networks. Proc. Natl. Acad. Sci. USA 101: 15482–15487.

Kopell N, Ermentrout GB, Whittington MA, Traub RD (2000) Gamma rhythms and beta rhythms have different synchronization properties. Proc. Natl. Acad. Sci. USA 97: 1867–1872.

Kopniczky Z, Dochnal R, Mácsai M, Pál A, Kiss G, Mihály A, Szabó G (2006) Alterations of behaviour and spatial learning after unilateral entorhinal ablation of rats. Life Sci. 78: 2683–2688.

Koppel B, Samkoff L, Daras M (1996) Relation of cocaine use to seizures and epilepsy. Epilepsia 37: 875–878.

Korff CM, Nordli DR Jr. (2006) Epilepsy syndromes in infancy. Pediatr. Neurol. 34: 253–263.

Korn H, Axelrad H (1980) Electrical inhibition of Purkinje cells in the cerebellum of the rat. Proc. Natl. Acad. Sci. USA 77: 6244–6247.

Korn H, Sotelo C, Crépel F (1973) Electrotonic coupling between neurons in the rat lateral vestibular nucleus. Exp. Brain Res. 16: 255–275.

Kosaka T (1980) The axon initial segment as a synaptic site: ultrastructure and synaptology of the initial segment of the pyramidal cell in the rat hippocampus (CA3 region). J. Neurocytol. 9: 861–882.

Kosaka T (1983a) Gap junctions between non-pyramidal cell dendrites in the rat hippocampus (CA1 and CA3 regions). Brain Res. 271:157–161.

Kosaka T (1983b) Axon initial segments of the granule cell in the rat dentate gyrus: synaptic contacts on bundles of axon initial segments. Brain Res. 274: 129–134.

Kosaka T (1983c) Neuronal gap junctions in the polymorph layer of the rat dentate gyrus. Brain Res. 277: 347–351.

Kosaka T, Kosaka K (2005) Intraglomerular dendritic link connected by gap junctions and chemical synapses in the mouse main olfactory bulb: electron microscopic serial section analysis. Neuroscience 131: 611–625.

Koutroumanidis M (2007) Panayiotopoulos syndrome: an important electroclinical example of benign childhood system epilepsy. Epilepsia 48: 1044–1053.

Krack P, Batir A, Van Blercom N, Chabardes S, Fraix V, Ardouin C, Koudsie A, Limousin PD, Benazzouz A, LeBas JF, Benabid AL, Pollak P (2003) Five-year follow-up of bilateral stimulation of the subthalamic nucleus in advanced Parkinson's disease. N. Engl. J. Med. 349: 1925–1934.

Kraepelin E (1913a) Psychiatrie. Ein Lehrbuch für Studierende und Ärtze. Achte, vollständig umgearbeitete Auflage. III Band. Barth Verlag, Leipzig.

Kraepelin E (1913b) Dementia praecox and paraphrenia. (Translator R.M. Barclay, 1919). Livingstone, Edinburgh.

Kramer MA, Roopun AK, Carracedo LM, Traub RD, Whittington MA, Kopell NJ (2008) Rhythm generation through period concatenation in rat somatosensory cortex. PloS Comput. Biol. 4(9): e1000169.

Kreiter AK, Singer W (1992) Oscillatory neuronal responses in the visual cortex of the awake macaque monkey. Eur. J. Neurosci. 4: 369–375.

Kriebel ME, Bennett MVL, Waxman SG, Pappas GD (1969) Oculomotor neurons in fish: electrotonic coupling and multiple sites of impulse initiation. Science 166: 520–524.

Kriegstein AR, Connors BW (1986) Cellular physiology of the turtle visual cortex: synaptic properties and intrinsic circuitry. J. Neurosci. 6: 178–191.

Kristensen BW, Noraberg J, Zimmer J (2001) Comparison of excitotoxic profiles of ATPA, AMPA, KA and NMDA in organotypic hippocampal slice cultures. Brain Res. 917: 21–44.

Kröner S, Krimer LS, Lewis DA, Barrionuevo G (2007) Dopamine increases inhibition in the monkey dorsolateral prefrontal cortex through cell type-specific modulation of interneurons. Cereb. Cortex 17: 1020–1032.

Krystal JH, Karper LP, Seibyl JP, Freeman GK, Delaney R, Bremner JD, Heninger GR, Bowers MB Jr., Charney DS (1994) Subanesthetic effects of the noncompetitive NMDA antagonist, ketamine, in humans. Psychotomimetic, perceptual, cognitive, and neuroendocrine responses. Arch. Gen. Psychiatry 51: 199–214.

Kühn AM, Kempf F, Brücke C, Gaynor-Doyle L, Martinez-Torres, Pogosyan A, Trottenberg T, Kupsch A, Schneider G-H, Hariz MI, Vandenberghe W, Nuttin B, Brown P (2008) High-frequency stimulation of the subthalamic nucleus suppresses oscillatory β activity in patients with Parkinson's disease in parallel with improvement in motor performance. J. Neurosci. 28: 6165–6173.

Kullmann DM, Semyanov A (2002) Glutamatergic modulation of GABAergic signaling among hippocampal interneurons: novel mechanisms regulating hippocampal excitability. Epilepsia 43 Suppl. 5: 174–178.

Kumar R, Lang AE, Rodriguez-Oroz MC, Lozano AM, Limousin P, Pollak P, Benabid AL, Guridi J, Ramos E, van der Linden C, Vandewalle A, Caemaert J, Lannoo E, van den Abbeele D, Vingerhoets G, Wolters M, Obeso JA (2000) Deep brain stimulation of the globus pallidus pars interna in advanced Parkinson's disease. Neurology 55 Suppl. 6: S34–39,

Kumar SS, Huguenard JR (2003) Pathway-specific differences in subunit composition of synaptic NMDA receptors on pyramidal neurons in neocortex. J. Neurosci. 23: 10074–10083.

Kumar SS, Bacci A, Kharazia V, Huguenard JR (2002) A developmental switch of AMPA receptor subunits in neocortical pyramidal neurons. J. Neurosci. 22: 3005–3015.

Kumbier E, Haack K (2002) Alfred Hauptmann – the fate of a German neurologist of Jewish origin. Fortschr. Neurol. Psychiatr. 70: 204–209.

Kuner T, Schoepfer R (1996) Multiple structural elements determine subunit specificity of Mg^{2+} block in NMDA receptor channels. J. Neurosci. 16: 3549–3558.

Kwon JS, O'Donnell BF, Wallenstein GV, Greene RW, Hirayasu Y, Nestor PG, Hasselmo ME, Potts GF, Shenton ME, McCarley RW (1999) Gamma frequency-range abnormalities to auditory stimulation in schizophrenia. Arch. Gen. Psychiatry 56: 1001–1005.

Kyriakopoulos M, Bargiotas T, Barker GJ, Frangou S (2008) Diffusion tensor imaging in schizophrenia. Eur. Psychiatry 23: 255–273.

Kyuhou S-I, Okada YC (1993) Detection of magnetic evoked fields associated with synchronous population activities in the transverse CA1 slice of the guinea pig. J. Neurophysiol. 70: 2665–2668.

Labhardt F (1954) Largactil therapy in schizophrenia and other psychotic conditions. Schweiz. Arch. Neurol. Psychiatr. 73: 309–338.

Lachaux JP, Rodriguez E, Martinerie J, Adam C, Hasboun D, Varela FJ (2000) A quantitative study of gamma-band activity in human intracranial recordings triggered by visual stimuli. Eur. J. Neurosci. 12: 2608–2622.

Lagier S, Carleton A, Lledo P-M (2004) Interplay between local GABAergic interneurons and relay neurons generates γ oscillations in the rat olfactory bulb. J. Neurosci. 24: 4382–4392.

Lahti AC, Weiler MA, Tamara Michaelidis BA, Parwani A, Tamminga CA (2001) Effects of ketamine in normal and schizophrenic volunteers. Neuropsychopharmacology 25: 455–467.

Lahtinen H, Palva JM, Sumanen S, Voipio J, Kaila K, Taira T (2002) Postnatal development of rat hippocampal gamma rhythm in vivo. J. Neurophysiol. 88: 1469–1474.

Lakatos P, Chen C-M, O'Connell MN, Mills A, Schroeder CE (2007) Neuronal oscillations and multisensory interaction in primary auditory cortex. Neuron 53: 279–292.

Lakatos P, Shah AS, Knuth KH, Ulbert I, Karmos G, Schroeder CE (2005) An oscillatory hierarchy controlling neuronal excitability and stimulus processing in the auditory cortex. J. Neurophysiol. 94: 1904–1911.

Lakatos P, Karmos G, Mehta AD, Ulbert I, Schroeder CE (2008) Entrainment of neuronal oscillations as a mechanism of attentional selection. Science 320: 110–113.

Lalande M, Minassian BA, DeLorey TM, Olsen RW (1999) Parental imprinting and Angelman syndrome. In: Delgado-Escueta AV, Wilson WA, Olsen RW, Porter RJ (Eds), Jasper's Basic Mechanisms of the Epilepsies, Third Edition: Advances in Neurology, Vol. 79. Lippincott Williams & Wilkins, Philadelphia, pp. 421–429.

Lamarre Y, de Montigny C, Dumont M, Weiss M (1971) Harmaline-induced rhythmic activity of cerebellar and lower brain stem neurons. Brain Res. 32: 246–250.

Lamme VAF, Spekreijse H (1998) Neuronal synchrony does not represent texture segregation. Nature 396: 362–366.

Lampl I, Schwindt P, Crill W (1998) Reduction of cortical pyramidal neuron excitability by the action of phenytoin on persistent Na$^+$ current. J. Pharmacol. Exp. Ther. 284: 228–237.

Landisman CE, Long MA, Beierlein M, Deans MR, Paul DL, Connors BW (2002) Electrical synapses in the thalamic reticular nucleus. J. Neurosci. 22: 1002–1009.

Lang EJ (2003) Excitatory afferent modulation of complex spike synchrony. Cerebellum 2: 165–170.

Lang PM, Burgstahler R, Sippel W, Irnich D, Schlotter-Weiger B, Grafe P (2003) Characterization of neuronal nicotinic acetylcholine receptors in the membrane of unmyelinated C-fiber axons by in vitro studies. J. Neurophysiol. 90: 3295–3303.

Larkman A, Mason (1990) Correlations between morphology and electrophysiology of pyramidal neurons in slices of rat visual cortex. I. Establishment of cell classes. J. Neurosci. 10: 1407–1414.

Larkman AU, Major G, Stratford KJ, Jack JJB (1992) Dendritic morphology of pyramidal neurones of the visual cortex of the rat. IV: Electrical geometry. J. Comp. Neurol. 323: 137–152.

Larkum ME, Zhu JJ (2002) Signaling of layer 1 and whisker-evoked Ca^{2+} and Na^+ action potentials in distal and terminal dendrites of rat neocortical pyramidal neurons in vitro and in vivo. J. Neurosci. 22: 6691–7005.

Larkum ME, Kaiser KMM, Sakmann B (1999a) Calcium electrogenesis in distal apical dendrites of layer 5 pyramidal cells at a critical frequency of back-propagating action potentials. Proc. Natl. Acad. Sci. USA 96: 14600–14604.

Larkum ME, Zhu JJ, Sakmann B (1999b) A new cellular mechanism for coupling inputs arriving at different cortical layers. Nature 398: 338–341.

Larkum ME, Zhu JJ, Sakmann B (2001) Dendritic mechanisms underlying the coupling of the dendritic with the axonal action potential initiation zone of adult rat layer 5 pyramidal neurons. J. Physiol. 533: 447–466.

Larkum ME, Waters J, Sakmann B, Helmchen F (2007) Dendritic spikes in apical dendrites of neocortical layer 2/3 pyramidal neurons. J. Neurosci. 27: 8999–9008.

Larkum ME, Watanabe S, Lasser-Ross N, Rhodes P, Ross WN (2008) Dendritic properties of turtle pyramidal neurons. J. Neurophysiol. 99: 683–694.

Larsen WJ (1977) Structural diversity of gap junctions. A review. Tissue Cell 9: 373–394.

Lau D, Vega-Saenz de Miera EC, Contreras D, Ozaita A, Harvey M, Chow A, Noebels JL, Paylor R, Morgan JI, Leonard CS, Rudy B (2000) Impaired fast-spiking, suppressed cortical inhibition, and increased susceptibility to seizures in mice lacking Kv3.2 K^+ channel proteins. J. Neurosci. 20: 9071–9085.

Laurent G (1996) Dynamical representation of odors by oscillating and evolving neural assemblies. Trends Neurosci. 19: 489–496.

Laurent G (2002) Olfactory network dynamics and the coding of multidimensional signals. Nat. Rev. Neurosci. 3: 884–895.

Laurent G, Davidowitz H (1994) Encoding of olfactory information with oscillating neural assemblies. Science 265: 1872–1875.

Laurent G, Wehr M, Davidowitz H (1996) Temporal representations of odors in an olfactory network. J. Neurosci. 16: 3837–3847.

Laurie SE, Delany CJ, Clarke VR, Bortolotto ZA, Ornstein PL, Isaac J, Collingridge GL (2001) Synaptic activation of a presynaptic kainate receptor facilitates AMPA receptor-mediated synaptic transmission at hippocampal mossy fibre synapses. Neuropharmacology 41: 907–915.

Lawrence JJ, Saraga F, Churchill JF, Statland JM, Travis KE, Skinner FK, McBain CJ (2006) Somatodendritic Kv7/KCNQ/M channels control interspike intervals in hippocampal interneurons. J. Neurosci. 26: 12325–12338.

Lawrie SM, Abukmeil SS (1998) Brain abnormality in schizophrenia. A systematic and quantitative review of volumetric magnetic resonance imaging studies. Br. J. Psychiatry 172: 110–120.

LeBeau FEN, Towers SK, Traub RD, Whittington MA, Buhl EH (2002) Fast network oscillations induced by potassium transients in the rat hippocampus *in vitro*. J. Physiol. 542: 167–179.

Lebedev MA, Wise SP (2000) Oscillations in the premotor cortex: single-unit activity from awake, behaving monkeys. Exp. Brain Res. 130: 195–215.

Lee D (2003) Coherent oscillations in neuronal activity of the supplementary motor area during a visuomotor task. J. Neurosci. 23: 6798–6809.

Lee H, Simpson GV, Logothetis NK, Rainer G (2005) Phase locking of single neuron activity to theta oscillations during working memory in monkey extrastriate visual cortex. Neuron 45: 147–16.

Lee J, Park S (2005) Working memory impairments in schizophrenia: a meta-analysis. Abnorm. Psychol. 114: 599–611.

Lee KH, McCormick DA (1995) Acetylcholine excites GABAergic neurons of the ferret perigeniculate nucleus through nicotinic receptors. J. Neurophysiol. 73: 2123–2128.

Lee KH, Williams LM, Breakspear M, Gordon E (2003) Synchronous gamma activity: a review and contribution to an integrative neuroscience model of schizophrenia. Brain Res. Brain Res. Rev. 41: 57–78.

Lee SA, Spencer DD, Spencer SS (2000) Intracranial EEG seizure-onset patterns in neocortical epilepsy. Epilepsia 41: 297–307.

Lee S-H, Blake R (1999) Visual form created solely from temporal structure. Science 284: 1165–1168.

Lee SK, Kim JY, Hong KS, Nam HW, Park SH, Chung CK (2000) The clinical usefulness of ictal surface EEG in neocortical epilepsy. Epilepsia 41: 1450–1455.

Lee SK, Lee SY, Kim KK, Hong KS, Lee DS, Chung CK (2005) Surgical outcome and prognostic factors of cryptogenic neocortical epilepsy. Ann. Neurol. 58: 525–532.

Leite JP, Garcia-Cairasco N, Cavalheiro EA (2002) New insights from the use of pilocarpine and kainate models. Epilepsy Res. 50: 93–103.

Leniger T, Thone J, Wiemann M (2004) Topiramate modulates pH of hippocampal CA3 neurons by combined effects on carbonic anhydrase and Cl^-/HCO_3^- exchange. *Br. J. Pharmacol.* 142: 831–842.

Lenz FA, Kwan HC, Martin RL, Tasker RR, Dostrovsky JO, Lenz YE (1994) Single unit analysis of the human ventral thalamic nuclear group. Tremor-related activity in functionally identified cells. Brain 117: 531–543.

Lerche H, Weber YG, Jurkat-Rott K, Lehmann-Horn F (2005) Ion channel defects in idiopathic epilepsies. Curr. Pharm. Des. 11: 2737–2752.

Leresche N, Lightowler S, Soltesz I, Jassik-Gerschenfeld D, Crunelli V (1991) Low-frequency oscillatory activities intrinsic to rat and cat thalamocortical cells. J. Physiol. 441: 155–174.

Lerma J (2006) Kainate receptor physiology. Curr. Opin. Pharmacol. 6: 89–97.

Lettvin JY, Maturana HR, McCulloch WS, Pitts WH (1959) What the frog's eye tells the frog's brain. Proc. IRE 47: 1940–1959.

Lewis DA, Hashimoto T, Volk DW (2005) Cortical inhibitory neurons and schizophrenia. Nat. Rev. Neurosci. 6: 312–324.

Lewis TJ, Rinzel J (2000) Self-organized synchronous oscillations in a network of excitable cells coupled by gap junctions. Network: Comput. Neural Syst. 11: 299–320.

Lewis TJ, Rinzel J (2001) Topological target patterns and population oscillations in a network with random gap junctional coupling. Neurocomputing 38–40: 763–768.

Leznik E, Llinás R (2005) Role of gap junctions in synchronized neuronal oscillations in the inferior olive. J. Neurophysiol. 94: 2447–2456.

Li C-L, McIlwain (1957) Maintenance of resting membrane potentials in slices of mammalian cerebral cortex and other tissues in vitro. J. Physiol. 139: 178–190.

Li L, Bischofberger J, Jonas P (2007) Differential gating and recruitment of P/Q-, N-, and R-type Ca^{2+} channels in hippocampal mossy fiber boutons. J. Neurosci. 27: 13420–13429.

Li S, Arbuthnott GW, Jutras MJ, Goldberg JA, Jaeger D (2007) Resonant antidromic cortical circuit activation as a consequence of high-frequency subthalamic deep-brain stimulation. J. Neurophysiol. 98: 3525–3537.

Li X, Pearce RA (2000) Effects of halothane on $GABA_A$ receptor kinetics: evidence for slowed agonist unbinding. J. Neurosci. 20: 899–907.

Li X, Ionescu AV, Lynn BD, Lu S, Kamasawa N, Morita M, Davidson KG, Yasumura T, Rash JE, Nagy JI (2004) Connexin47, connexin29 and connexin32 co-expression in oligodendrocytes and Cx47 association with zonula occludens-1 (ZO-1) in mouse brain. Neuroscience 126: 611–630.

Li X, Kamasawa N, Ciolofan C, Olson CO, Lu S, Davidson KG, Yasumura T, Shigemoto R, Rash JE, Nagy JI (2008) Connexin45–containing neuronal gap junctions in rodent retina also contain connexin36 in both apposing hemiplaques, forming bihomotypic gap junctions, with scaffolding contributed by zonula occludens-1. J. Neurosci. 28: 9769–9789.

Li Y, Gamper N, Hilgemann DW, Shapiro MS (2005) Regulation of Kv7 (KCNQ) K^+ channel open probability by phosphatidylinositol 4,5–bisphosphate. J. Neurosci. 25: 9825–9835.

Lian G, Sheen V (2006) Cerebral developmental disorders. Curr. Opin. Pediatr. 18: 614–620.

Liddle PF (1987) The symptoms of chronic schizophrenia. A re-examination of the positive-negative dichotomy. Br. J. Psychiatry 151: 145–151.

Light GA, Hsu JL, Hsieh MH, Meyer-Gomes K, Sprock J, Swerdlow NR, Braff DL (2006) Gamma band oscillations reveal neural network cortical coherence dysfunction in schizophrenia patients. Biol. Psychiatry 60: 1231–1240.

Limousin P, Pollak P, Benazzouz A, Hoffmann D, Le Bas JF, Broussole E, Perret JE, Benabid AL (1995) Effect of parkinsonian signs and symptoms of bilateral subthalamic nucleus stimulation. Lancet 345: 91–95.

Limousin P, Greene J, Pollak P, Rothwell J, Benabid AL, Frackowiak R (1997) Changes in cerebral activity pattern due to subthalamic nucleus or internal pallidum stimulation in Parkinson's disease. Ann. Neurol. 42: 283–291.

Limousin P, Krack P, Pollak P, Benazzouz A, Ardouin C, Hoffmann D, Benabid AL (1998) Electrical stimulation of the subthalamic nucleus in advanced Parkinson's disease. N. Engl. J. Med. 339: 1105–1111.

Lindquist CE, Dalziel JE, Cromer BA, Birnir B (2004) Penicillin blocks human $\alpha 1 \beta 1$ and $\alpha 1 \beta 1 \gamma 2S$ $GABA_A$ channels that open spontaneously. Eur. J. Pharmacol. 496: 23–32.

Liske S, Morris ME (1989) Effects of GABA, THIP, and potassium on excitability of myelinated axons of isolated amphibian spinal roots. Can. J. Physiol. Pharmacol. 67: 682–685.

Lisman JE, Coyle JT, Green RW, Javitt DC, Benes FM, Heckers S, Grace AA (2008) Circuit-based framework for understanding neurotransmitter and risk gene interactions in schizophrenia. Trends Neurosci. 31: 234–242.

Liss B, Roeper J (2001) Molecular physiology of neuronal K-ATP channels. Mol. Membr. Biol. 18: 117–127.

Litt B, Esteller R, Echauz J, D'Alessandro M, Shor R, Henry T, Pennell P, Epstein C, Bakay R, Dichter, Vachtsevanos G (2001) Epileptic seizures may begin hours in advance of clinical onset: a report of five patients. Neuron 30: 1–64.

Litvin O, Tiunova A, Connell-Alberts Y, Panchin Y, Baranova A (2006) What is hidden in the pannexin treasure trove: the sneak peek and the guesswork. J. Cell. Mol. Med. 10: 613–634.

Liu X-B, Coble J, van Luijtelaar G, Jones EG (2007) Reticular nucleus-specific changes in α3 subunit protein at GABA synapses in genetically epilepsy-prone rats. Proc. Natl. Acad. Sci. USA 104: 12512–12517.

Liu Y, Zhang LI, Tao HW (2007) Heterosynaptic scaling of developing GABAergic synapses: dependence on glutamatergic input and developmental stage. J. Neurosci. 27: 5301–5312.

Livingstone MS (1996) Oscillatory firing and interneuronal correlations in squirrel monkey striate cortex. J. Neurophysiol. 75: 2467–2485.

Llinás R (1978) The role of calcium in neuronal function. In: *The Neurosciences: Fourth Study Program*, edited by F. O. Schmitt and F. G. Worden. M.I.T. Press, Cambridge, Mass.

Llinás R (1988) The intrinsic electrophysiological properties of mammalian neurons: insights into central nervous system function. Science 242: 1654–1664.

Llinás R, Hess R (1976) Tetrodotoxin-resistant dendritic spikes in avian Purkinje cells. Proc. Natl. Acad. Sci. USA 73: 2520–2523.

Llinás R, Jahnsen H (1982) Electrophysiology of mammalian thalamic neurones *in vitro*. Nature 297: 406–408.

Llinás R, Mühlethaler (1988) Electrophysiology of guinea-pig cerebellar nuclear cells in the *in vitro* brain stem-cerebellar preparation. J. Physiol. 404: 241–258.

Llinás R, Nicholson C (1971) Electrophysiological properties of dendrites and somata in alligator Purkinje cells. J. Neurophysiol. 34: 532–551.

Llinás RR, Paré D (1991) Of dreaming and wakefulness. Neuroscience 44: 521–535.

Llinás R, Ribary U (1993) Coherent 40–Hz oscillation characterizes dream state in humans. Proc. Natl. Acad. Sci. USA 90: 2078–2081.

Llinás R, Sugimori M (1980a) Electrophysiological properties of *in vitro* Purkinje cell somata in mammalian cerebellar slices. J. Physiol. 305: 171–195.

Llinás R, Sugimori M (1980b) Electrophysiological properties of *in vitro* Purkinje cell dendrites in mammalian cerebellar slices. J. Physiol. 305: 197–213.

Llinás R, Yarom Y (1981a) Electrophysiology of mammalian inferior olivary neurones *in vitro*. Different types of voltage-dependent ionic conductances. J. Physiol. 315: 549–567.

Llinás R, Yarom Y (1981b) Properties and distribution of ionic conductances generating electroresponsiveness of mammalian inferior olivary neurones in vitro. J. Physiol. 315: 569–584.

Llinás R, Yarom Y (1986) Oscillatory properties of guinea-pig inferior olivary neurones and their pharmacological modulation: an in vitro study. J. Physiol. 376: 163–182.

Llinás R, Baker R, Sotelo C (1974) Electrotonic coupling between neurons in cat inferior olive. J. Neurophysiol. 37: 560–571.

Llinás R, Yarom Y, Sugimori M (1981) Isolated mammalian brain in vitro: new technique for analysis of electrical activity of neuronal circuit function. Fed. Proc. 40: 2240–2245.

Llinás RR, Grace AA, Yarom Y (1991) *In vitro* neurons in mammalian cortical layer 4 exhibit intrinsic oscillatory activity in the 10- to 50-Hz range. Proc. Natl. Acad. Sci. USA 88: 897–901.

Llinás R, Ribary U, Contreras D, Pedroarena C (1998) The neuronal basis for consciousness. Philos. Trans. R. Soc. Lond. B Biol. Sci. 353: 1841–1849.

Llinás RR, Leznik E, Urbano FJ (2002) Temporal binding via cortical coincidence detection of specific and nonspecific thalamocortical inputs: a voltage-dependent dye-imaging study in mouse brain slices. Proc. Natl. Acad. Sci. USA 99: 449–454.

Loewenstein WR (1981) Junctional intercellular communication: the cell-to-cell membrane channel. Physiol. Rev. 61: 829–913.

Loewenstein WR, Kanno Y, Socolar SJ (1978) Quantum jumps of conductance during formation of membrane channels at cell-cell junction. Nature 274: 133–136.

Logan SD, Pickering AE, Gibson IC, Nolan MF, Spanswick D (1996) Electrotonic coupling between rat sympathetic preganglionic neurones *in vitro*. J. Physiol. 495: 491–502.

Loiseau P, Pestre M, Dartigues JF, Commenges D, Barberger-Gateau C, Cohadon S (1983) Long-term prognosis in two forms of childhood epilepsy: typical absence seizures and epilepsy with rolandic (centrotemporal) EEG foci. Ann. Neurol. 13: 642–648.

London M, Häusser M (2005) Dendritic computation. Annu. Rev. Neurosci. 28: 503–532.

Long MA, Cruikshank SCJ, Jutras MJ, Connors BW (2005) Abrupt maturation of a spike-synchronizing mechanism in neocortex. J. Neurosci. 25: 7309–7316.

Long MA, Deans MR, Paul DL, Connors BW (2002) Rhythmicity without synchrony in the electrically uncoupled inferior olive. J. Neurosci. 22: 10898–10905.

Long MA, Landisman CE, Connors BW (2004) Small clusters of electrically coupled neurons generate synchronous rhythms in the thalamic reticular nucleus. J. Neurosci. 24: 341–349.

Loomis AL, Harvey EN, Hobart G (1935) Potential rhythms of the cerebral cortex during sleep. Science 81: 597–598.

López-Bayghen E, Rosas S, Castelán F, Ortega A (2007) Cerebellar Bergmann glia: an important model to study neuron-glia interactions. Neuron Glia Biol. 3: 155–167.

López-Muñoz F, Ucha-Udabe R, Alamo-González C (2004) A century of barbiturates in neurology. Rev. Neurol. 39: 767–775.

López-Muñoz F, Boya J, Alamo C (2006) Neuron theory, the cornerstone of neuroscience, on the centenary of the Nobel Prize award to Santiago Ramón y Cajal. Brain Res. Bull. 70: 391–405.

Lorente de Nó R. Analysis of the activity of the chains of internuncial neurons. J. Neurophysiol. 1 (1938) 207–244.

Lörincz ML, Crunelli V, Hughes S (2008) Cellular dynamics of cholinergically induced alpha (8–13 Hz) in sensory thalamic nuclei in vitro. J. Neurosci. 28: 660–671.

Louis ED, Williamson PD, Darcey TM (1990) Chronic focal epilepsy induced by microinjection of tetanus toxin into the cat motor cortex. Electroenceph. Clin. Neurophysiol. 75: 548–557.

Loup F, Wieser H-G, Yonekawa Y, Aguzzi A, Fritschy J-M (2000) Selective alterations in GABA$_A$ receptor subtypes in human temporal lobe epilepsy. J. Neurosci. 20: 5401–5419.

Loussouarn G, Park KH, Bellocq C, Baro I, Charpentier F, Escande D (2003) Phosphatidylinositol-4,5-bisphosphate, PIP2, controls KCNQ1/KCNE1 voltage-gated potassium channels: a functional homology between voltage-gated and inward rectifier K$^+$ channels. EMBO J. 22: 5412–5421.

Löwel S, Singer W (1992) Selection of intrinsic horizontal connections in the visual cortex by correlated neuronal activity. Science 255: 209–212.

Lüders HO (2008) Textbook of Epilepsy Surgery. Informa Healthcare, New York.

Luhmann HJ, Mittmann T, van Luijtelaar G, Heinemann U (1995) Impairment of intracortical GABAergic inhibition in a rat model of absence epilepsy. Epilepsy Res. 22: 43–51.

Lutzenberger W, Ripper B, Busse L, Birbaumer N, Kaiser J (2002) Dynamics of gamma-band activity during an audiospatial working memory task in humans. J. Neurosci. 22: 5630–5638.

Ma J, Leung LS (2002) Metabotropic glutamate receptors in the hippocampus and nucleus accumbens are involved in generating seizure-induced hippocampal gamma waves and behavioral hyperactivity. Behav. Brain Res. 133: 45–56.

Ma JY, Catterall WA, Scheuer T (1997) Persistent sodium currents through brain sodium channels induced by G protein βγ subunits. Neuron 19: 443–452.

MacDonald KD, Brett B, Barth DS (1996) Inter- and intra-hemispheric spatiotemporal organization of spontaneous electrocortical oscillations. J. Neurophysiol. 76: 423–437.

MacDonald KD, Fifkova E, Jones MS, Barth DS (1998) Focal stimulation of the thalamic reticular nucleus induces focal gamma waves in cortex. J. Neurophysiol. 79: 474–477.

MacDonald MC, Almor A, Henderson VW, Kempler D, Andersen ES (2001) Assessing working memory and language comprehension in Alzheimer's disease. Brain Lang. 78: 17–42.

Macdonald RL, Barker JL (1978) Specific antagonism of GABA-mediated postsynaptic inhibition in cultured mammalian spinal cord neurons: a common mode of convulsant action. Neurology 28: 325–330.

MacKay WA, Mendonça AJ (1995) Field potential oscillatory bursts in parietal cortex before and during reach. Brain Res. 704: 167–174.

MacLeod K, Laurent G (1996) Distinct mechanisms for synchronization and temporal patterning of odor-encoding neural assemblies. Science 274: 976–980.

MacLeod K, Bäcker A, Laurent G (1998) Who reads temporal information contained across synchronized and oscillatory spike trains? Nature 395: 693–698.

MacVicar BA, Dudek FE (1980) Dye-coupling between CA3 pyramidal cells in slices of rat hippocampus. Brain Res. 196: 494–497.

MacVicar BA, Dudek FE (1981) Electrotonic coupling between pyramidal cells: a direct demonstration in rat hippocampal slices. Science 213: 782–785.

MacVicar BA, Dudek FE (1982) Electrotonic coupling between granule cells of the rat dentate gyrus: physiological and anatomical evidence. J. Neurophysiol. 47: 579–592.

MacVicar BA, Jahnsen H (1985) Uncoupling of CA3 pyramidal neurons by propionate. Brain Res. 330: 141–145.

MacVicar BA, Tse FWY (1989) Local neuronal circuitry underlying cholinergic rhythmical slow activity in CA3 area of rat hippocampal slices. J. Physiol. 417: 197–212.

MacVicar BA, Ropert N, Krnjevic K (1982) Dye-coupling between pyramidal cells of rat hippocampus in vivo. Brain Res. 238: 239–244.

Magariño-Ascone C, Payo JH, Macadar O, Buño W (2002) High-frequency stimulation of the subthalamic nucleus silences subthalamic neurones: a possible cellular mechanism in Parkinson's disease. Neuroscience 115: 1109–1117.

Magill PJ, Sharott A, Bevan MD, Brown P, Bolam JP (2004) Synchronous unit activity and local field potentials evoked in the subthalamic nucleus by cortical stimulation. J. Neurophysiol. 92: 700–714.

Maier N, Güldenagel M, Söhl G, Siegmund H, Willecke K, Draguhn A (2002) Reduction of high-frequency network oscillations (ripples) and pathological network discharges in hippocampal slices from connexin 36–deficient mice. J. Physiol. 541: 521–528.

Maier N, Nimmrich V, Draguhn A (2003) Cellular and network mechanisms underlying spontaneous sharp wave-ripple complexes in mouse hippocampal slices. J. Physiol. 550: 873–887.

Maingay M, Romero-Ramos M, Carta M, Kirik D (2006) Ventral tegmental area dopamine neurons are resistant to human mutant alpha-synuclein overexpression. Neurobiol. Dis. 23: 522–532.

Makowski L, Caspar DLD, Phillips WC, Goodenough DA (1984) Gap junction structures. VI.

Variation and conservation in connexon conformation and packing. Biophys. J. 45: 208–218.

Maldonado HM, Delgado Escueta AV, Walsh GO, Swartz BE, Rand RW (1988) Complex partial seizures of hippocampal and amygdalar origin. Epilepsia 29: 420–433.

Maldonado P, Friedman-Hill S, Gray CM (2000) Dynamics of striate cortical activity in the alert macaque: II. Fast time scale synchronization. Cereb. Cortex 10: 1117–1131.

Mallet N, Pogosyan A, Sharott A, Csicsvari J, Bolam JP, Brown P, Magill PJ (2008a) Disrupted dopamine transmission and the emergence of exaggerated beta oscillations in subthalamic nucleus and cerebral cortex. J. Neurosci. 28: 4795–4806.

Mallet N, Pogosyan A, Márton LF, Bolam JP, Brown P, Magill PJ (2008b) Parkinsonian beta oscillations in the external globus pallidus and their relationship with subthalamic nucleus activity. J. Neurosci. 28: 14245–14258.

Maloney KJ, Cape EG, Gotman J, Jones BE (1997) High-frequency gamma electroencephalogram activity in association with sleep-wake states and spontaneous behaviors in the rat. Neuroscience 76: 541–555.

Mann EO, Suckling JM, Hajos N, Greenfield SA, Paulsen O (2005) Perisomatic feedback inhibition underlies cholinergically induced fast network oscillations in the rat hippocampus in vitro. Neuron 45: 105–117.

Mann-Metzer P, Yarom Y (1999) Electrotonic coupling interacts with intrinsic properties to generate synchronized activity in cerebellar networks of inhibitory interneurons. J. Neurosci. 19: 3298–3306.

Manni E, Petrosini L (2004) A century of cerebellar somatotopy: a debated rep resentation. Nat. Rev. Neurosci. 5: 241–249.

Mao B-Q, Hamzei-Sichani F, Aronov D, Froemke RC, Yuste R (2001) Dynamics of spontaneous activity in neocortical slices. Neuron 32: 883–898.

Marescaux C, Vergnes M, Depaulis A (1992) Genetic absence epilepsy in rats from Strasbourg - a review. J. Neural Transm. Suppl. 35: 37–69.

Marder E (1984) Roles for electrical coupling in neural circuits as revealed by selective neuronal deletions. J. Exp. Biol. 112: 147–167.

Marder E, Calabrese RL (1996) Principles of rhythmic motor pattern generation. Physiol. Rev. 76: 687–717.

Markram H (1997) A network of tufted layer 5 pyramidal neurons. Cereb Cortex 7: 523–533.

Markram H, Lübke J, Frotscher M, Roth A, Sakmann B (1997a) Physiology and anatomy of synaptic connections between thick tufted pyramidal neurones in the developing rat neocortex. J. Physiol. 500: 409–440.

Markram H, Lübke J, Frotscher M, Sakmann B (1997b) Regulation of synaptic efficacy by coincidence of postsynaptic APs and EPSPs. Science 275: 213–215.

Markram H, Wang Y, Tsodyks M (1998) Differential signaling via the same axon of neocortical pyramidal neurons. Proc. Natl. Acad. Sci. USA 95: 5323–5328.

Markram H, Toledo-Rodriguez M, Wang Y, Gupta A, Silberberg G, Wu C (2004) Interneurons of the neocortical inhibitory system. Nat. Rev. Neurosci. 5: 793–807.

Marr D (1969) A theory of cerebellar cortex. J. Physiol. 202: 437–470.

Marsden JF, Ashby P, Limousin-Dowsey P, Rothwell JC, Brown P (2000) Coherence between cerebellar thalamus, cortex and muscle in man: cerebellar thalamus interactions. Brain 123: 1459–1470.

Marsden JF, Limousin-Dowsey P, Ashby P, Pollak P, Brown P (2001) Subthalamic nucleus, sensorimotor cortex and muscle interrelationships in Parkinson's disease. Brain 124: 378–388.

Marshall SP, Lang EJ (2004) Inferior olive oscillations gate transmission of motor cortical activity to the cerebellum. J. Neurosci. 24: 11356–11367.

Marson AG, Al-Kharusi AM, Alwaidh M, Appleton R, Baker GA, Chadwick DW, Cramp C, Cockerell OC, Cooper PN, Doughty J, Eaton B, Gamble C, Goulding PJ, Howell SJ, Hughes A, Jackson M, Jacoby A, Kellett M, Lawson GR, Leach JP, Nicolaides P, Roberts R, Shackley P, Shen J, Smith DF, Smith PE, Smith CT, Vanoli A, Williamson PR; SANAD Study group (2007) The SANAD study of effectiveness of carbamazepine, gabapentin, lamotrigine, oxcarbazepine, or topiramate for tratment of partial epilepsy: an unblinded randomised controlled trial. Lancet 369: 1000–1015.

Martin AR, Pilar G (1963) Dual mode of synaptic transmission in the avian ciliary ganglion. J. Physiol. 168: 443–463.

Martin FC, Handforth A (2006) Carbenoxolone and mefloquine suppress tremor in the harmaline mouse model of essential tremor. Mov. Disord. 21: 1641–1649.

Martina M, Jonas P (1997) Functional differences in Na^+ channel gating between fast-spiking interneurones and principal neurones of rat hippocampus. J. Physiol. 505: 593–603.

Martina M, Schultz JH, Ehmke H, Monyer H, Jonas P (1998) Functional and molecular differences between voltage-gated K^+ channels of fast-spiking interneurons and pyramidal neurons of rat hippocampus. J. Neurosci. 18: 8111–8125.

Martina M, Vida I, Jonas P (2000) Distal initiation and active propagation of action potentials in interneuron dendrites. Science 287: 295–300.

Martina M, Krasteniakov NV, Bergeron R (2003) D-Serine differently modulates NMDA receptor function in rat CA1 hippocampal pyramidal cells and interneurons. J. Physiol. 548: 411–423.

Mas C, Taske N, Deutsch S, Guipponi M, Thomas P, Covanis A, Friis M, Kjeldsen MJ, Pizzolato GP, Villemure J-G, Buresi C, Rees M, Malafosse A, Gardiner M, Anonarakis SE, Meda P (2004) Association of the connnexin36 gene with juvenile myoclonic epilepsy. J. Med. Genet. 41: e93.

Mason A, Larkman A (1990) Correlations between morphology and electrophysiology of pyramidal neurons in slices of rat visual cortex. II. Electrophysiology. J. Neurosci. 10: 1415–1428.

Mason A, Nicoll A, Stratford K (1991) Synaptic transmission between individual pyramidal neurons of the rat visual cortex *in vitro*. J. Neurosci. 11: 72–84.

Massimini M, Huber R, Ferrarelli F, Hill S, Tononi G (2004) The sleep slow oscillation as a traveling wave. J. Neurosci. 24: 6862–6870.

Massimini M, Ferrarelli F, Huber R, Esser SK, Singh H, Tononi G (2005) Breakdown of cortical effective connectivity during sleep. Science 309: 2228–2232.

Massimini M, Ferrarelli F, Esser SK, Riedner BA, Huber R, Murphy M, Peterson MJ, Tononi G (2007) Triggering sleep slow waves by transcranial magnetic stimulation. Proc. Natl. Acad. Sci. USA 104: 8496–8501.

Matsuki N, Quandt FN, ten Eick RE, Yeh JZ (1984) Characterization of the block of sodium channels by phenytoin in mouse neuroblastoma cells. J. Pharmacol. Exp. Ther. 228: 523–530.

Matsumoto H, Ajmone Marsan C (1964a) Cortical cellular phenomena in experimental epilepsy: interictal manifestations. Exp. Neurol. 9: 286–304.

Matsumoto H, Ajmone Marsan C (1964b) Cortical cellular phenomena in experimental epilepsy: ictal manifestations. Exp. Neurol. 9: 305–326.

Matsumoto H, Ajmone Marsan C (1964c) Cellular mechanisms in experimental epileptic seizures. Science 144: 193–194.

Matussek P (1952) Studies in delusional perception. [translated and condensed]. In: Cutting J, Sheppard M (Eds) (1987) Clinical Roots of the Schizophrenia Concept. Translations of Seminal European Contributions on Schizophrenia. Cambridge University Press, Cambridge, U.K.

Maxeiner S, Krüger O, Schilling K, Traub O, Urschel S, Willecke K (2003) Spatiotemporal transcription of connexin45 during brain development results in neuronal expression in adult mice. Neuroscience 119: 689–700.

McConnell HW, Mitchell SC, Smith RL, Brewster M (1997) Trimethylaminuria associated with seizures and behavioural disturbance: a case report. Seizure 6: 317–321.

McCormick DA, Contreras D (2001) On the cellular and network bases of epileptic seizures. Annu. Rev. Physiol. 63: 815–846.

McCormick DA, Pape H-C (1990a) Properties of a hyperpolarization-activated cation current and its role in rhythmic oscillation in thalamic relay neurones. J. Physiol. 431: 291–318.

McCormick DA, Pape H-C (1990b) Noradrenergic and serotonergic modulation of a hyperpolarization-activated cation current in thalamic relay neurones. J. Physiol. 431: 319–342.

McCormick DA, Prince DA (1986) Acetylcholine induces burst firing in thalamic reticular neurones by activating a potassium conductance. Nature 319: 402–405.

McCormick DA, Prince DA (1987) Actions of acetylcholine in the guinea-pig and cat medial and lateral geniculate nuclei, in vitro. J. Physiol. 392: 147–165.

McCormick DA, von Krosigk M (1992) Corticothalamic activation modulates thalamic firing through glutamate "metabotropic" receptors. Proc. Natl. Acad. Sci. 89: 2774–2778.

McCormick DA, Connors BW, Lighthall JW, Prince DA (1982) Comparative electrophysiology of pyramidal and sparsely spiny stellate neurons of the neocortex. J. Neurophysiol. 54: 782–806.

McCulloch W (1988) Embodiments of Mind. With a foreword by Jerome Y. Lettvin. MIT Press, Cambridge, MA.

McCulloch W, Pitts W (1943) A logical calculus of ideas immanent in nervous activity. Bull. Math. Biophys. 5: 115–133.

McDonald I (2007) Gordon Holmes lecture: Gordon Holmes and the neurological heritage. Brain 130: 288–298.

McGehee DS, Heath MJS, Gelber S, Devay P, Role LW (1995) Nicotine enhancement of fast excitatory synaptic transmission in CNS by presynaptic receptors. Science 269: 1692–1696.

McGurk SR, Carter C, Goldman R, Green MF, Marder SR, Xie H, Schooler NR, Kane JM (2005) The effects of clozapine and risperidone on spatial working memory in schizophrenia. Am. J. Psyhciatry 162: 1013–1016.

McIntyre DC (1979) Effects of focal vs. generalized kindled convulsions from anterior neocortex or amygdala on CER acquisition in rats. Physiol. Behav. 23: 855–859.

McIntyre DC, Goddard GV (1973) Transfer, interference and spontaneous recovery of convulsions kindled from the rat amygdala. Electroenceph. Clin. Neurophysiol. 35: 533–543.

McIntyre DC, Poulter MO, Gilby K (2002) Kindling: some old and some new. Epilepsy Res. 50: 79–92.

McKenna PJ (1997) Schizophrenia and Related Syndromes. Psychology Press, Hove, U.K.

McLaren S, Cookson JC, Silverstone T (1992) Positive and negative symptoms, depression and social disability in chronic schizophrenia: a comparative trial of bromperidol and fluphenazine decanoates. Int. Clin. Psychopharmacol. 7: 67–72.

Melchitzky DS, Gonzalez-Burgos G, Barrionuevo G, Lewis DA (2001) Synaptic targets of the intrinsic axon collaterals of supragranular pyramidal neurons in monkey prefrontal cortex. J. Comp. Neurol. 430: 209–221.

Mellanby J, George G, Robinson A, Thompson P (1977) Epileptiform syndrome in rats produced by injecting tetanus toxin into the hippocampus. J. Neurol. Neurosurg. Psychiatr. 40: 404–414.

Mellers JD, Toone BK, Lishman WA (2000) A neuropsychological comparison of schizophrenia and schizophrenia-like psychosis of epilepsy. Psychol. Med. 30: 325–335.

Melloni L, Molina C, Pena M, Torres D, Singer W, Rodriguez E (2007) Synchronization of neural activity across cortical areas correlates with conscious perception. J. Neurosci. 27: 2858–2865.

Melyan Z, Wheal HV, Lancaster B (2002) Metabotropic-mediated kainate receptor regulation of I_{sAHP} and excitability in pyramidal cells. Neuron 34: 107–114.

Melyan Z, Lancaster B, Wheal HV (2004) Metabotropic regulation of intrinsic excitability by synaptic activation of kainate receptors. J. Neurosci. 24: 4530–4534.

Memmesheimer RM, Timme M (2006) Designing the dynamics of spiking neural networks. Phys. Rev. Lett. 97: 188101

Mendez JS, Finn BW (1975) Use of 6–hydroxydopamine to create lesions in catecholamine neurons in rats. J. Neurosurg. 42: 166–173.

Menon V, Freeman WJ, Cutillo BA, Desmond JE, Ward MF, Bressler SL, Laxer KD, Barbaro N, Gevins AS (1996) Spatio-temporal correlations in human gamma band electrocorticograms. Electroenceph. Clin. Neurophysiol. 98: 89–102.

Menzies L, Ooi C, Kamath S, Suckling J, McKenna P, Fletcher P, Bullmore E, Stephenson C (2007) Effects of gamma-aminobutyric acid-modulating drugs on working memory and brain function in patients with schizophrenia. Arch. Gen. Psychiatry 64: 156–167.

Mercer A, Bannister AP, Thomson AM (2006) Electrical coupling between pyramidal cells in adult cortical regions. Brain Cell Biol. 35: 13–27.

Merritt HH, Putnam TJ (1938) Sodium diphenyl hydantoinate in the treatment of convulsive disorders. JAMA 111: 1068–1073.

Metherate R, Cruikshank SJ (1999) Thalamocortical inputs trigger a propagating envelope of gamma-band activity in auditory cortex in vitro. Exp. Brain Res. 126: 160–174.

Metrakos K, Metrakos JD (1961) Is the centrencephalic EEG inherited as a dominant? Electroenceph. Clin. Neurophysiol. 13: 289.

Meyer AH, Katona I, Blatow M, Rozov A, Monyer H (2002) *In vivo* labeling of parvalbumin-positive interneurons and analysis of electrical coupling in identified neurons. J. Neurosci. 22: 7055–7064.

Meyer J, Mai M, Ortega G, Mössner R, Lesch KP (2002) Mutational analysis of the connexin36 gene (CX36) and exclusion of the coding sequence as a candidate region for catatonic schizophrenia in a large pedigree. Schizophr. Res. 58: 87–91.

Middleton SJ (2005) Oscillatory activity in the mouse cerebellum *in vitro*. Ph.D. Thesis, University of Leeds.

Middleton SJ, Racca C, Cunningham MO, Traub RD, Monyer H, Knöpfel T, Schofield IS, Jenkins A, Whittington MA (2008) High-frequency network oscillations in cerebellar cortex. Neuron 58: 763–774.

Miles R (1990) Synaptic excitation of inhibitory cells by single CA3 hippocampal pyramidal cells of the guinea-pig *in vitro*. J. Physiol. 428: 61–77.

Miles R (2000) Diversity in inhibition. Science 287: 244–246.

Miles R, Poncer J-C (1993) Metabotropic glutamate receptors mediate a post-tetanic excitation of guinea-pig hippocampal inhibitory neurones. J. Physiol. 463: 461–473.

Miles R, Wong RKS (1983) Single neurones can initiate synchronized population discharge in the hippocampus. Nature 306: 371–373.

Miles R, Wong RKS (1986) Excitatory synaptic interactions between CA3 neurones in the guinea-pig hippocampus. J. Physiol. 373: 397–418.

Miles R, Wong RKS (1987a) Inhibitory control of local excitatory circuits in the guinea-pig hippocampus. J. Physiol. 388: 611–629.

Miles R, Wong RKS (1987b) Latent synaptic pathways revealed after tetanic stimulation in the hippocampus. Nature 329: 724–726.

Miles R, Wong RKS, Traub RD (1984) Synchronized afterdischarges in the hippocampus: contribution of local synaptic interaction. Neuroscience 12: 1179–1189.

Miles R, Traub RD, Wong RKS (1988) Spread of synchronous firing in longitudinal slices from the CA3 region of the hippocampus. J. Neurophysiol. 60: 1481–1496.

Miles R, Tóth K, Gulyás AI, Hajos N, Freund TF (1996) Differences between somatic and dendritic inhibition in the hippocampus. Neuron 16: 815–823.

Miller EK, Erickson CA, Desimone R (1996) Neural mechanisms of visual working memory in prefrontal cortex of the macaque. J. Neurosci. 16: 5154–5167.

Miller MW (1987) The origin of corticospinal projection neurons in rat. Exp. Brain Res. 67: 339–351.

Millichap JG, Bickford RG, Klass DW, Backus RE (1962) Infantile spasms, hypsarhythmia, and mental retardation. A study of etiologic factors in 61 patients. Epilepsia 3: 188–197.

Mills RJ, Yap L, Young CA (2008) Treatment for ataxia in multiple sclerosis. Young Cochrane Database of Systematic Reviews, DOI: 10.1002/14651858.CD005029. pub2.

Milojkovic BA, Radojicic SM, Goldman-Rakic PS, Antic SD (2004) Burst generation in rat pyramidal neurones by regenerative potentials elicited in a restricted part of the basilar dendritic tree. J. Physiol. 558: 193–211.

Milojkovic BA, Wuskell JP, Loew LM, Antic SD (2005) Initiation of sodium spikelets in basal dendrites of neocortical pyramidal neurons. J. Membr. Biol. 208: 155–169.

Miltner WHR, Braun C, Arnold M, Witte H, Taub E (1999) Coherence of gamma-band EEG activity as a basis for associative learning. Nature 397: 434–436.

Mima T, Steger J, Schulman AE, Gerloff C, Hallett M (2000) Electroencephalographic measurements of motor cortex control of muscle activity in humans. Clin. Neurophysiol. 111: 326–337.

Mima T, Matsuoka T, Hallett M (2001a) Information flow from the sensorimotor cortex to muscle in humans. Clin. Neurophysiol. 112: 122–126.

Mima T, Oluwatimilehin T, Hiraoka T, Hallett M (2001b) Transient interhemispheric neuronal synchrony correlates with object recognition. J. Neurosci. 21: 3942–3948.

Mirnics K, Middleton FA, Marquez A, Lewis DA, Levitt P (2000) Molecular characterization of schizophrenia viewed by microarray analysis of gene expression in prefrontal cortex. Neuron 28: 53–67.

Miyashita Y (1988) Neuronal correlate of visual associative long-term memory in the primate temporal cortex. Nature 335: 817–820.

Miyashita Y, Chang HS (1988) Neuronal correlate of pictorial short-term memory in the primate temporal cortex. Nature 331: 68–70.

Miyasho T, Takagi H, Suzuki H, Watanabe S, Inoue M, Kudo Y, Miyakawa H (2001) Low-threshold potassium channels and a low-threshold calcium channel regulate Ca^{2+} spike firing in the dendrites of cerebellar Purkinje neurons: a modeling study. Brain Res. 891: 106–115.

Mody I, Pearce RA (2004) Diversity of inhibitory neurotransmission through $GABA_A$ receptors. Trends Neurosci. 27: 569–575.

Mohn AR, Gainetdinov RR, Caron MG, Koller BH (1999) Mice with reduced NMDA receptor expression display behaviors related to schizophrenia. Cell 98: 427–436.

Molinari HH (1988) Ultrastructural heterogeneity of spinal terminations in the cat inferior olive. Neuroscience 27: 425–435.

Mölle M, Marshall L, Gais S, Born J (2002) Grouping of spindle activity during slow oscillations in human non-rapid eye movement sleep. J. Neurosci. 22: 10941–10947.

Montana V, Malarkey EB, Verderio C, Matteoli M, Parpura V (2006) Vesicular transmitter release from astrocytes. Glia 54: 700–715.

Monto S, Vanhatalo S, Holmes MD, Palva JM (2007) Epileptogenic neocortical networks are revealed by abnormal temporal dynamics in seizure-free subdural EEG. Cereb. Cortex 17: 1386–1393.

Monyer H, Sprengel R, Schoepfer R, Herb A, Higuchi M, Lomeli H, Burnashev N, Sakmann B, Seeburg PH (1992) Heteromeric NMDA receptors: molecular and functional distinction of subtypes. Science 256: 1217–1221.

Monyer H, Burnashev N, Laurie DJ, Sakmann B, Seeburg PH (1994) Developmental and regional expression in the rat brain and functional properties of four NMDA receptors. Neuron 12: 529–540.

Moore H, Grace AA (2002) A role for electrotonic coupling in the striatum in the expression of dopamine receptor-mediated stereotypies. Neuropsychopharmacology 27: 980–992.

Moortgat KT, Keller CH, Bullock TH, Sejnowski TJ (1998) Submicrosecond pacemaker precision is behaviorally modulated: The gymnotiform electromotor pathway. Proc. Natl. Acad. Sci. USA 95: 4684–4689.

Moortgat KT, Bullock TH, Sejnowski TJ (2000a) Precision of the pacemaker nucleus in a weakly electric fish: network versus cellular influences. J. Neurophysiol. 83: 971–983.

Moortgat KT, Bullock TH, Sejnowski TJ (2000b) Gap junction effects on precision and frequency of a model pacemaker network. J. Neurophysiol. 83: 984–997.

Moran NF, Poole K, Bell G, Solomon J, Kendall S, McCarthy M, McCormick D, Nashef L, Sander J, Shorvon SD (2004) Epilepsy in the United Kingdom: seizure frequency and severity, anti-epileptic drug utilization and impact on life in 1652 people with epilepsy. Seizure 13: 425–433.

Morgado-Valle C, Feldman JL (2007) NMDA receptors in preBötzinger complex neurons can drive respiratory rhythm independent of AMPA receptors. J. Physiol. 582: 359–368.

Mori K, Nagao M, Yamashita H, Morinobu S, Yamawaki S (2004) Effect of switching to atypical antipsychotics on memory in patients with chronic schizophrenia. Prog. Neuropsychopharmacol. Biol. Psychiatry 28: 659–665.

Morita K, Kalra R, Aihara K, Robinson HP (2008) Recurrent synaptic input and the timing of gamma-frequency-modulated firing of pyramidal cells during neocortical "UP" states. J. Neurosci. 28: 1871–1881.

Morris BJ, Cochran SM, Pratt JA (2005) PCP: from pharmacology to modelling schizophrenia. Curr. Opin. Pharmacol. 5: 101–106.

Mosbacher J, Schoepfer R, Monyer H, Burnashev N, Seeburg PH, Ruppersberg JP (1994) A molecular determinant for submillisecond desensitization in glutamate receptors. Science 266: 1059–1062.

Moss BL, Fuller AD, Sahley CL, Burrell BD (2005) Serotonin modulates axo-axonal coupling between neurons critical for learning in the leech. J. Neurophysiol. 94: 2575–2589.

Mueser KT, Jeste DV (2008) Clinical Handbook of Schizophrenia. Guildford, New York.

Murakami S, Zhang T, Hirose A, Okada YC (2002) Physiological origins of evoked magnetic fields and extracellular field potentials produced by guinea-pig CA3 hippocampal slices. J. Physiol 544: 237–251.

Murphy AD, Hadley RD, Kater SB (1983) Axotomy-induced parallel increases in electrical and dye coupling between identified neurons of Helisoma. J. Neurosci. 3: 1422–1429.

Murthy VN, Fetz EE (1992) Coherent 25- to 35–Hz oscillations in the sensorimotor cortex of awake behaving monkeys. Proc. Natl. Acad. Sci. USA 89: 5670–5674.

Murthy VN, Fetz EE (1996a) Oscillatory activity in sensorimotor cortex of awake monkeys: synchronization of local field potentials and relation to behavior. J. Neurophysiol. 76: 3949–3967.

Murthy VN, Fetz EE (1996b) Synchronization of neurons during local field potential oscillations in sensorimotor cortex of awake monkeys. J. Neurophysiol. 76: 3968–3982.

Nagy GS (1978) Evaluation of carbenoxolone sodium in the treatment of duodenal ulcer. Gastroenterology 74: 7–10.

Nagy JI, Dudek FE, Rash JE (2004) Update on connexins and gap junctions in neurons and glia in the mammalian nervous system. Brain Res. Rev. 47: 191–215.

Nakamura Y, Katakura N, Nakajima M (1999) Generation of rhythmic ingestive activities of the trigeminal, facial, and hypoglossal motoneurons in the in vitro CNS preparations isolated from rats and mice. J. Med. Dent. Sci. 46: 63–73.

Nakase T, Naus CC (2004) Gap junctions and neurological disorders of the central nervous system. Biochim. Biophys. Acta 1662: 149–158.

Nakazawa K, Quirk MC, Chitwood RA, Watanabe M, Yeckel MF, Sun LD, Kato A, Carr CA, Johnston D, Wilson MA, Tonegawa S (2002) Requirement for hippocampal CA3 NMDA receptors in associative memory recall. Science 297: 211–218.

Nakazawa K, Sun LD, Quirk MC, Rondi-Reig L, Wilson MA, Tonegawa S (2003) Hippocampal CA3 NMDA receptors are crucial for memory acquisition of one-time experience. Neuron 38: 305–315.

Naquet R, Silva-Barrat C, Menini C (1995) Reflex epilepsy in the Papio-papio baboon, particularly photosensitive epilepsy. Ital. J. Neurol. Sci. 16: 119–125.

Neale JM, Oltmanns TF (1980) Schizophrenia. John Wiley & Sons, New York.

Nealis JGT, Duffy FH (1978) Paroxysmal beta actvity in the pediatric electro-encephalogram. Ann. Neurol. 4: 112–116.

Neckelmann D, Amzica F, Steriade M (1998) Spike-wave complexes and fast components of cortically generated seizures. III. Synchronizing mechanisms. J. Neurophysiol. 80: 1480–1494.

Nesvåg R, Lawyer G, Varnäs K, Fjell AM, Walhovd KB, Frigessi A, Jönsson EG, Agartz I (2008) Regional thinning of the cerebral cortex in schizophrenia: effects of diagnosis, age and antipsychotic medication. Schizophr. Res. 98: 16–28.

Neuper C, Pfurtscheller G (2001) Event-related dynamics of cortical rhythms: frequency-specific features and functional correlates. Int. J. Psychophysiol. 43: 41–58.

Neushul P (1993) Science, government, and the mass production of penicillin. J. Hist. Med. Allied Sci. 48: 371–395.

Nevian T, Larkum ME, Polsky A, Schiller J (2007) Properties of basal dendrites of layer 5 pyramidal neurons: a direct patch-clamp recording study. Nat. Neurosci. 10: 206–214.

Neville KR, Haberly LB (2003) Beta and gamma oscillations in the olfactory system of the urethane-anesthetized rat. J. Neurophysiol. 90: 3921–3930.

New PS, Wells CE (1965) Cerebral toxicity associated with massive intravenous penicillin therapy. Neurology 15: 1053–1058.

Newman MEJ, Strogatz SH, Watts DJ (2001) Random graphs with arbitrary degree distributions and their applications. Phys. Rev. E 64: 026118.

Ng SK, Hauser WA, Burst JC, Susser M (1988) Alcohol consumption and withdrawal in new-onset seizures. N. Engl. J. Med. 319: 666–673.

Ni X, Valente J, Azevedo MH, Pato MT, Pato CN, Kennedy JL (2007) Connexin50 gene on human chromosome 1q21 is associated with schizophrenia in matched case control and family-based studies. J. Med. Genet. 44: 532–536.

Niedermeyer E (2003) Electrophysiology of the frontal lobe. Clin. Electroencephalogr. 34: 5–12.

Niedermeyer E (2005) "Special Issue": Ultrafast Frequencies and Full-Band EEG Ultrafast EEG Activities and Their Significance. Clinical EEG and Neuroscience 36(4). With articles by Ozaki I, Hashimoto I, Mochizuki H, Ugawa Y, Okada Y et al, Lopes da Silva FH et al.

Niedermeyer E, Lopes da Silva F (1999) Electroencephalography. Basic Principles, Clinical Applications, and Related Fields, 4th ed. Lippincott Williams & Wilkins, Philadelphia.

Nikam SS, Awasthi AK (2008) Evolution of schizophrenia drugs: a focus on dopaminergic systems. Curr. Opin. Investig. Drugs 9: 37–46.

Nilsen KE, Kelso AR, Cock HR (2006) Antiepileptic effect of gap-junction blockers in a rat model of refractory focal cortical epilepsy. Epilepsia 47: 1169–1175.

Nimmrich V, Maier N, Schmitz D, Draguhn A (2005) Induced sharp wave-ripple complexes in the absence of synaptic inhibition in mouse hippocampal slices. J. Physiol. 563: 663–670.

Nini A, Feingold A, Slovin H, Bergman H (1995) Neurons in the globus pallidus do not show correlated activity in the normal monkey, but phase-locked oscillations appear in the MPTP model of parkinsonism. J. Neurophysiol. 74: 1800–1805.

Noebels JL (2003a) The biology of epilepsy genes. Annu. Rev. Neurosci. 26: 599–625.

Noebels JL (2003b) Exploring new gene discoveries in idiopathic generalized epilepsy. Epilepsia 44 Suppl. 2: 16–21.

Noebels JL, Prince DA (1977) Presynaptic origin of penicillin afterdischarges at mammalian nerve terminals. Brain Res. 138: 59–74.

Noebels JL, Prince DA (1978) Development of focal seizures in cerebral cortex: role of axon terminal bursting. J. Neurophysiol. 41: 1267–1281.

Noga JT, Bartley AJ, Jones DW, Torrey EF, Weinberger DR (1996) Cortical gyral anatomy and gross brain dimensions in monozygotic twins discordant for schizophrenia. Schizophr. Res. 22: 27–40.

Noguchi H, Moore JW (1913) A demonstration of Treponema pallidum in the brain in cases of general paralysis. J. Exp. Med. 17: 232–238.

Nolan MF, Logan SD, Spanswick D (1999) Electrophysiological properties of electrical synapses between rat sympathetic preganglionic neurones in vitro. J. Physiol. 519: 753–764.

Nowak L, Bregestovski P, Ascher P, Herbet A, Prochiantz A (1984) Magnesium gates glutamate-activated channels in mouse central neurons. Nature 307: 462–465.

Nowak LG, Munk MHJ, Nelson JI, James AC, Bullier J (1995) Structural basis of cortical synchronization. I. Three types of interhemispheric coupling. J. Neurophysiol. 74: 2379–2400.

Nowak LG, Azouz R, Sanchez-Vives MV, Gray CM, McCormick DA (2003) Electrophysiological classes of cat primary visual cortical neurons in vivo as revealed by quantitative analyses. J. Neurophysiol. 89: 1541–1566.

Nuñez A, Garcia-Austt E, Buño W (1990) In vivo electrophysiological analysis of Lucifer yellow-coupled hippocampal pyramids. Exp. Neurol. 108: 76–82.

Nuñez A, Amzica F, Steriade M (1992) Voltage-dependent fast (20–40 Hz) oscillations in long-axoned neocortical neurons. Neuroscience 51: 7–10.

Oades RD, Halliday GM (1987) Ventral tegmental (A10) system: neurobiology. 1. Anatomy and connectivity. Brain Res. 434: 117–165.

O'Connor SM, Berg RW, Kleinfeld D (2002) Coherent electrical activity between vibrissa sensory areas of cerebellum and neocortex is enhanced during free whisking. J. Neurophysiol. 87: 2137–2148.

O'Donnell P, Grace AA (1995) Different effects of subchronic clozapine and haloperidol on dye-coupling between neurons in the rat striatal complex. Neuroscience 66: 763–767.

Ohara S, Mima T, Baba K, Ikeda A, Kunieda T, Matsumoto R, Yamamoto J, Matsuhashi M, Nagamine T, Hirasawa K, Hori T, Mihara T, Hashmoto N, Salenius S, Shibasaki H (2001) Increased synchronization of cortical oscillatory activities between human supplementary motor and primary sensorimotor areas during voluntary movements. J. Neurosci. 21: 9377–9386.

Ohara PT, Granato A, Moallem TM, Wang BR, Tillet Y, Jasmin L (2003) Dopaminergic input to GABAergic neurons in the rostral agranular insular cortex of the rat. J. Neurocytol. 32: 131–141.

Ohno-Shosaku, Yamamoto C (1992) Identification of an ATP-sensitive K^+ channel in rat cultured cortical neurons. Pflügers Arch. 423: 260–266.

Ohnuma T, Augood SJ, Arai H, McKenna PJ, Emson PC (1999) Measurement of GABAergic parameters in the prefrontal cortex in schizophrenia: focus on GABA content, $GABA_A$ receptor alpha-1 subunit messenger RNA and human GABA transporter-1 (HGAT-1) messenger RNA expression. Neuroscience 93: 441–448.

Okada Y, Ikeda, Zhang T, Wang Y (2005) High-frequency signals (>400 Hz): a new window in electrophysiological analysis of the somatosensory system. Clin. EEG Neurosci. 36: 285–292.

Okada YC, Xu Chibing (1996) Single-epoch neuromagnetic signals during epileptiform activities in guinea pig longitudinal CA3 slices. Neurosci. Lett. 211: 155–158.

Okazaki MM, Evenson DA, Nadler JV (1995) Hippocampal mossy fiber sprouting and synapse formation after status epilepticus in rats: visualization after retrograde transport of biocytin. J. Comp. Neurol. 352: 515–534.

Onn S-P, Grace AA (1994) Dye coupling between rat striatal neurons recorded in vivo: compartmental organization and modulation by dopamine. J. Neurophysiol. 71: 1917–1934.

Onn S-P, Grace AA (1999) Alterations in electrophysiological activity and dye coupling of striatal spiny and aspiny neurons in dopamine-denervated rat striatum recorded in vivo. Synapse 33: 1–15.

Oren I, Mann EO, Paulsen O, Hájos N (2006) Synaptic currents in anatomically identified CA3 neurons during hippocampal gamma oscillations *in vitro*. J. Neurosci. 26: 9923–9934.

Orthmann-Murphy JL, Enriquez AD, Abrams CK, Scherer SS (2007) Loss-of-function GJA12/Connexin47 mutations cause Pelizaeus-Merzbacher-like disease. Mol. Cell. Neurosci. 34: 629–641.

Osawa M, Uemura S, Kimura H, Sato M (2001) Amygdala kindling develops without mossy fiber sprouting and hippocampal neuronal degeneration in rats. Psychiatry Clin. Neurosci. 55: 549-557.

Osten P, Stern-Bach Y (2006) Learning from stargazin: the mouse, the phenotype and the unexpected. Curr. Opin. Neurobiol. 16: 275–280.

Otani S, Blond O, Desce JM, Crépel F (1998) Dopamine facilitates long-term depression of glutamatergic transmission in rat prefrontal cortex. Neuroscience 85: 669–676.

Ottman R, Risch N, Hauser WA, Pedley TA, Lee JH, Barker-Cummings C, Lustenberger A, Nagle KJ, Lee KS, Scheuer ML, Neystat M, Wilhelmsen KC (1995) Localization of a gene for partial epilepsy to chromosome 10q. Nat. Genet. 10: 56–60.

Pais I, Hormuzdi SG, Monyer H, Traub RD, Wood IC, Buhl EH, Whittington MA, LeBeau FEN (2003) Sharp wave-like activity in the hippocampus *in vitro* in mice lacking the gap junction protein connexin 36. J. Neurophysiol. 89: 2046–2054.

Palanca BJA, DeAngelis GC (2005) Does neuronal synchrony underlie visual feature grouping? Neuron 46: 333–346.

Palay SL, Chan-Palay V (1974) Cerebellar Cortex: Cytology and Organization. Springer Verlag, New York.

Palhalmi J, Paulsen O, Freund TJ, Hajos N (2004) Distinct properties of carbachol- and DHPG-induced network oscillations in hippocampal slices. Neuropharmacology 47: 381–389.

Palmer LM, Stuart GJ (2006) Site of action potential initiation in layer 5 pyramidal neurons. J. Neurosci. 26: 1854–1863.

Palva JM, Lamsa K, Lauri SE, Rauvala H, Kaila K, Taira T (2000) Fast network oscillations in the newborn rat hippocampus *in vitro*. J. Neurosci. 20: 1170–1178.

Palva S, Palva JM, Shtyrov Y, Kujala T, Ilmoniemi RJ, Kaila K, Näätänen R (2002) Distinct gamma-band evoked responses to speech and non-speech sounds in humans. J. Neurosci. 22: RC211 (1–5).

Palva JM, Palva S, Kaila K (2005) Phase synchrony among neuronal oscillations in the human cortex. J. Neurosci. 25: 3962–3972.

Pan Z, Kao T, Horvath Z, Lemos J, Sul JY, Cranstoun SD, Bennett V, Scherer SS, Cooper EC (2006) A common ankyrin-G-based mechanism retains KCNQ and NaV channels at electrically active domains of the axon. J. Neurosci. 26: 2599–2613.

Panayiotopoulos CP (2001) Treatment of typical absence seizures and related epileptic syndromes. Paediatr. Drugs 3: 379–403.

Panchin Y, Kelmanson I, Matz M, Lukyanov K, Usman N, Lukyanov S (2000) A ubiquitous family of putative gap junction molecules. Curr. Biol. 10: R473–474.

Papatheodoropoulos C (2008) A possible role of ectopic action potentials in the *in vitro* hippocampal harp wave-ripple complexes. Neuroscience 157: 495–501.

Pappas GD, Asada Y, Bennett MVL (1971) Morphological correlates of increased coupling resistance at an electrotonic synapse. J. Cell Biol. 49: 173–188.

Parent JM, Janumpalli S, McNamara JO, Lowenstein DH (1998) Increased dentate granule cell neurogenesis following amygdala kindling in the adult rat. Neurosci. Lett. 247: 9–12.

Parkinson J (1817) An Essay on the Shaking Palsy. Sherwood, Neely and Jones, London. Reprinted in J. Neuropsychiatry Clin. Neurosci. (2002) 14: 223–236.

Parra P, Gulyás AI, Miles R (1998) How many subtypes of inhibitory cells in the hippocampus? Neuron 20: 983–993.

Paspalas CD, Goldman-Rakic PS (2005) Presynaptic D1 dopamine receptors in primate prefrontal cortex: target-specific expression in the glutamatergic synapse. J. Neurosci. 25: 1260–1267.

Passaglia CL, Dodge FA, Barlow RB (1998) Cell-based model of the *Limulus* eye. J. Neurophysiol. 80: 1800–1815.

Paternain AV, Rodríguez-Moreno A, Villarroel A, Lerma J (1998) Activation and desensitization properties of native and recombinant kainate receptors. Neuropharmacology 37: 1249–1259.

Paulsen O, Sejnowski TJ (2000) Natural patterns of activity and long-term synaptic plasticity. Curr. Opin. Neurobiol. 10: 172–179.

Payne BR, Peters A (2002) The concept of cat primary visual cortex. In: Payne BR, Peters A (Eds), The Cat Primary Visual Cortex. Academic Press, San Diego, CA, pp. 1–129.

Pearce RA (1993) Physiological evidence for two distinct $GABA_A$ responses in rat hippocampus. Neuron 10: 189–200.

Pearce RK, Hawkes CH, Daniel SE (1995) The anterior olfactory nucleus in Parkinson's disease. Mov. Disord. 10: 283–287.

Pedersen SB, Petersen KA (1998) Juvenile myoclonic epilepsy: clinical and EEG features. Acta Neurol. Scand. 97: 160–163.

Peinado A, Yuste R, Katz LC (1993) Extensive dye coupling between rat neocortical neurons during the period of circuit formation. Neuron 10: 103–114.

Pellerin JP, Lamarre Y (1997) Local field potential oscillations in primate cerebellar cortex during voluntary movement. J. Neurophysiol. 78: 3502–3507.

Penfield W, Roberts L (1966) Speech and Brain-Mechanisms. Atheneum, New York.

Penfield W, Welch K (1949) Instability of response to stimulation of the sensorimotor cortex of man. J. Physiol. 109: 358–365.

Pennacchio LA, Lehesjoki AE, Stone NE, Willou VL, Virtaneva K, Miao J, D'Amato E, Ramirez L, Faham M, Koskiniemi M, Warrington JA, Norio R, de la Chapelle A, Cox DR, Myers RM (1996) Mutations in the gene encoding cystatin B in progressive myoclonus epilepsy (EPM1) Science 271: 1731–1734.

Penney JB Jr, Young AB (1981) GABA as the pallidothalamic neurotransmitter: implications for basal ganglia function. Brain Res. 207: 195–199.

Penttonen M, Kamondi A, Acsády L, Buzsáki G (1998) Gamma frequency oscillation in the hippocampus: intracellular analysis *in vivo*. Eur. J. Neurosci. 10: 718–728.

Peracchia C (1991) Effects of the anesthetics heptanol, halothane and isoflurane on gap junction conductance in crayfish septate axons: a calcium- and hydrogen-independent phenomenon potentiated by caffeine and theophylline, and inhibited by 4–aminopyridine. J. Membr. Biol. 121: 67–78.

Pereda A, Triller A, Korn H, Faber DS (1992) Dopamine enhances both electrotonic coupling and chemical excitatory postsynaptic potentials at mixed synapses. Proc. Natl. Acad. Sci. USA 89: 12088–12092.

Pereda AE, Bell TD, Chang BH, Czernik AJ, Nairn AC, Soderling TR, Faber DS (1998) Ca^{2+}/calmodulin-dependent kinase II mediates simultaneous enhancement of gap-junctional conductance and glutamatergic transmission. Proc. Natl. Acad. Sci. USA 95: 13272–13277.

Perez Velazquez JL, Carlen PL (2000) Gap junctions, synchrony and seizures. Trends Neurosci. 23: 68–74.

Perez Velazquez JL (2003) Bicarbonate-dependent depolarizing potentials in pyramidal cells and interneurons during epileptiform activity. Eur. J. Neurosci. 18: 1337–1342.

Perez Velazquez JL, Han D, Carlen PL (1997) Neurotransmitter modulation of gap junctional communication in the rat hippocampus. Eur. J. Neurosci. 9: 2522–2531.

Perez-Velazquez JL, Valiante TA, Carlen PL (1994) Modulation of gap junctional mechanisms during calcium-free induced field burst activity: a possible role for electrotonic coupling in epileptogenesis. J. Neurosci. 14: 4308–4317.

Perouansky M, Yaari Y (1993) Kinetic properties of NMDA receptor-mediated synaptic currents in rat hippocampal pyramidal cells *versus* interneurones. J. Physiol. 465: 223–244.

Perreault P, Avoli M (1991) Physiology and pharmacology of epileptiform activity induced by 4–aminopyridine in rat hippocampal slices. J. Neurophysiol. 65: 771–785.

Perrins R, Roberts A (1995) Cholinergic and electrical synapses between synergistic spinal motoneurones in the *Xenopus laevis* embryo. J. Physiol. 485: 135–144.

Pesaran B, Pezaris JS, Sahani M, Mitra PP, Andersen RA (2002) Temporal structure in neuronal activity during working memory in macaque parietal cortex. Nat. Neurosci. 5: 805–811.

Pfurtscheller G, Neuper C, Kalcher J (1993) 40–Hz oscillations during motor behavior in man. Neurosci. Lett. 164: 179–182.

Phelan KD, Twery MJ, Gallagher JP (1993) Morphological and electrophysiological evidence for electrotonic coupling of rat dorsolateral septal nucleus neurons in vitro. Synapse 13: 39–49.

Phelan P (2005) Innexins: members of an evolutionarily conserved family of gap-junction proteins. Biochim. Biophys. Acta 1711: 225–245.

Phillips HA, Marini C, Scheffer IE, Sutherland GR, Mulley JC, Berkovic SF (2000) A de novo mutation in sporadic nocturnal frontal lobe epilepsy. Ann. Neurol. 48: 264–267.

Piccolino M, Neyton J, Gerschenfeld HM (1984) Decrease of gap junction permeability induced by dopamine and cyclic adenosine 3′:5′-monophosphate in horizontal cells of turtle retina. J. Neurosci. 4: 2477–2488.

Pichon Y (1995) Pharmacological induction of rhythmical activity and plateau action potentials in unmyelinated axons. J. Physiol. Paris 89: 171–180.

Pietrobon D (2002) Calcium channels and channelopathies in the central nervous system. Mol. Neurobiol. 25: 31–50.

Pinault D (1990) Antidromic firing occurs spontaneously on thalamic relay neurons: triggering of ectopic action potentials by somatic intrinsic burst discharges. Neuroscience 34: 281–292.

Pinault D (1995) Backpropagation of action potentials generated at ectopic axonal loci: hypothesis that axon terminals integrate local environmental signals. Brain Res. Rev. 21: 42–92.

Pinault D (2008) *N*-methyl D-aspartate receptor antagonists ketamine and MK-801 induce wake-related aberrant γ oscillations in the rat neocortex. Biol. Psychiatry 63: 730–735.

Pinault D, Pumain R (1989) Antidromic firing occurs spontaneously on thalamic relay neurons: triggering of somatic intrinsic burst discharges by ectopic action potentials. Neuroscience 31: 625–637.

Pinault D, Bourassa J, Deschênes M (1995) The axonal arborization of single thalamic reticular neurons in the somatosensory thalamus of the rat. Eur. J. Neurosci. 7: 31–40.

Pinault D, Smith Y, Deschênes M (1997) Dendrodendritic and axoaxonic synapses in the thalamic reticular nucleus of the adult rat. J. Neurosci. 17: 3215–3233.

Pinault D, Leresche N, Charpier S, Deniau JM, Marescaux C, Vergnes M, Crunelli V (1998) Intracellular recordings in the thalamic neurones during spontaneous spike and wave discharges in rats with absence epilepsy. J. Physiol. 509: 449–456.

Pinheiro P, Mulle C (2006) Kainate receptors. Cell Tissue Res. 326: 457–482.

Pinsky PF, Rinzel J (1994) Intrinsic and network rhythmogenesis in a reduced Traub model for CA3 neurons. J. Comput. Neurosci. 1: 39–60.

Piskulic D, Olver JS, Norman TR, Maruff P (2007) Behavioural studies of spatial working memory dysfunction in schizophrenia: a quantitative literature review. Psychiatry Res. 150: 111–121.

Pitts W, McCulloch WS (1947) How we know universals: the perception of auditory and visual forms. Bull. Math. Biophys. 9: 127–147.

Placantonakis DG, Bukovsky AA, Zeng XH, Kiem HP, Welsh JP (2004) Fundamental role of inferior olive connexin36 in muscle coherence during tremor. Proc. Natl. Acad. Sci. USA 101: 7164–7169.

Plenz D, Kitai ST (1999) A basal ganglia pacemaker formed by the subthalamic nucleus and external globus pallidus. Nature 400: 677–682.

Polack P-O, Guillemain I, Hu E, Deransart C, Depaulis A, Charpier S (2007) Deep layer somatosensory cortical neurons initiate spike-and-wave discharges in a genetic model of absence seizures. J. Neurosci. 27: 6590–6599.

Pollok B, Südmeyer M, Gross J, Schnitzler A (2005) The oscillatory network of simple repetitive bimanual movements. Brain Res. Cogn. Brain Res. 25: 300–311.

Poolos NP, Migliore M, Johnston D (2002) Pharmacological upregulation of h-channels reduces the excitability of pyramidal neuron dendrites. Nat. Neurosci. 5: 767–774.

Porter LL, Rizzo E, Hornung JP (1999) Dopamine affects parvalbumin expression during cortical development in vitro. J. Neurosci. 19: 8990–9003.

Prasad KM, Patel AR, Muddasani S, Sweeney J, Keshavan MS (2004) The entorhinal cortex in first-episode psychotic disorders: a structural magnetic resonance imaging study. Am. J. Psychiatry 161: 1612–1619.

Prechtl JC (1994) Visual motion induces synchronous oscillations in turtle visual cortex. Proc. Natl. Acad. Sci. USA 91: 12467–12471.

Prescott SA, Ratté S, De Koninck Y, Sejnowski TJ (2006) Nonlinear interaction between shunting and adaptation controls a switch between integration and coincidence detection in pyramidal neurons. J. Neurosci. 26: 9084–9097.

Price CJ, Cauli B, Kovacs ER, Kulik A, Lambolez B, Shigemoto R, Capogna M (2005) Neurogliaform neurons form a novel inhibitory network in the hippocampal CA1 area. J. Neurosci. 25: 6775–6786.

Prince DA (1968) The depolarization shift in "epileptic" neurons. Exp. Neurol. 21: 467–485.

Prince DA (1999) Epileptogenic neurons and circuits. Adv. Neurol. 79: 665–684.

Prince DA, Farrell D (1969) "Centrencephalic" spike-wave discharges following parenteral penicillin injection in the cat. Neurology 19: 309–310.

Prince DA, Wilder BJ (1967) Control mechanisms in cortical epileptogenic foci. "Surround" inhibition. Arch. Neurol. 16: 194–202.

Prole DL, Marrion NV (2004) Ionic permeation and conduction properties of neuronal KCNQ2/KCNQ3 potassium channels. Biophys. J. 86: 1454–1469.

Prole DL, Lima PA, Marrion NV (2003) Mechanisms underlying modulation of neuronal KCNQ2/KCNQ3 potassium channels by extracellular protons. J. Gen. Physiol. 122: 775–793.

Pumain R (1981) Electrophysiological abnormalities in chronic epileptogenic foci: an intracellular study. Brain Res. 219: 445–450.

Putnam TJ, Merritt HH (1937) Experimental determination of the anti-convulsant properties of some phenyl derivatives. Science 85: 525–526.

Quesney LF, Gloor P, Kratzenberg E, Zumstein H (1977) Pathophysiology of generalized penicillin epilepsy in the cat: the role of cortical and subcortical structures. I.

Systemic application of penicillin. Electroencephalogr. Clin. Neurophysiol. 42: 640–655.

Raastad M, Shepherd GMG (2003) Single-axon action potentials in the rat hippocampal cortex. J. Physiol. 548: 745–752.

Rácz A, Ponomarenko AA, Fuchs EC, Monyer H (2009) Augmented hippocampal ripple oscillations in mice with reduced fast excitation onto parvalbumin-positive cells. J. Neurosci. 29: 2563–2568.

Radnikow G, Misgeld U (1998) Dopamine D1 receptors facilitate GABA. A synaptic currents in the rat substantia nigra pars reticulata. J. Neurosci. 18: 2009-2016.

Rall W (1962) Theory of physiological properties of dendrites. Ann. NY Acad. Sci.. 96: 1071–1092.

Ranganath C, Minzenberg MJ, Ragland JD (2008) The cognitive neuroscience of memory function and dysfunction in schizophrenia. Biol. Psychiatry 64: 18–25.

Rapisarda C, Simonelli G, Monti S (1985) Cells of origin and topographic organization of corticospinal neurons in the guinea pig by the retrograde HRP method. Brain Res. 334: 85–96.

Rash JE, Dillman RK, Bilhartz BL, Duffy HS, Whalen LR, Yasumura T (1996) Mixed synapses discovered and mapped throughout mammalian spinal cord. Proc. Natl. Acad. Sci. USA 93: 4235–4239.

Rash JE, Staines WA, Yasumura T, Patel D, Furman CS, Stelmack GL, Nagy JI (2000) Immunogold evidence that neuronal gap junctions in adult rat brain and spinal cord contain connexin-36 but not connexin-32 or connexin-43. Proc. Natl. Acad. Sci. USA 97: 7573–7578.

Rash JE, Yasumura T, Dudek FE, Nagy JI (2001) Cell-specific expression of connexins and evidence of restricted gap junctional coupling between glial cells and between neurons. J. Neurosci. 21: 1983–2000.

Rasminsky M, Sears TA (1972) Internodal conduction in undissected demyelinated nerve fibres. J. Physiol. 227: 323–350.

Rathelot JA, Strick PL (2006) Muscle representation in the macaque motor cortex: an anatomical perspective. Proc. Natl. Acad. Sci. USA 103: 8257-8262.

Ray A, Zoidl G, Weickert S, Wahle P, Dermietzel R (2005) Site-specific and developmental expression of pannexin1 in the mouse nervous system. Eur. J. Neurosci. 21: 3277–3290.

Reilly JL, Harris MS, Keshavan MS, Sweeney JA (2006) Adverse effects of risperidone on spatial working memory in first-episode schizophrenia. Arch. Gen. Psychiatry. 63: 1189–1197.

Reilly JL, Harris MS, Khine TT, Keshavan MS, Sweeney JA (2007) Antipsychotic drugs exacerbate impairment on a working memory task in first-episode schizophrenia. Biol. Psychiatry 62: 818–821.

Reiner A, Jiao Y, Del Mar N, Laverghetta AV, Lei WL (2003) Differential morphology of pyramidal tract-type and intratelencephalically projecting-type corticostriatal neurons and their intrastriatal terminals in rats. J. Comp. Neurol. 457: 420–440.

Rekling JC, Shao XM, Feldman JL (2000) Electrical coupling and excitatory synaptic transmission between rhythmogenic respiratory neurons in the preBötzinger complex. J. Neurosci. 20: RC113: 1–5.

Ren J, Greer J (2003) Ontogeny of rhythmic motor patterns generated in the embryonic rat spinal cord. J. Neurophysiol. 89: 1187–1195.

Revel JP, Karnovsky MJ (1967) Hexagonal array of subunits in intercellular junctions of the mouse heart and liver. J. Cell Biol. 33: C7–C12.

Rho JM, Szot P, Tempel BL, Schwartzkroin PA (1999) Developmental seizure susceptibility of kv1.1 potassium channel knockout mice. Dev. Neurosci. 21: 320–327.

Ribary U, Ioannides AA, Singh KD, Hasson R, Bolton JPR, Lado F, Mogilner A, Llinás R (1991) Magnetic field tomography of coherent thalamocortical 40–Hz oscillations in humans. Proc. Natl. Acad. Sci. USA 88: 11037–11041.

Richards AN (1964) Production of penicillin in the United States (1941–1946). Nature 201: 441–445.

Rigas P, Castro-Alamancos MA (2007) Thalamocortical up states: differential effects of intrinsic and extrinsic cortical inputs on persistent activity. J. Neurosci. 27: 4261–4272.

Ripps H, Qian H, Zakevicius J (2002) Pharmacological enhancement of hemi-gap-junctional currents in Xenopus oocytes. J. Neurosci. Methods 121: 81–92.

Risch SC (1996) Pathophysiology of schizophrenia and the role of newer antipsychotics. Pharmacotherapy 16: 11–14.

Ristic A, Marinkovic J, Dragasevic N, Stanisavljevic D, Kostic V (2002) Long-term prognosis of vascular hemiballismus. Stroke 33: 2109–2111.

Rivlin-Etzion M, Marmor O, Heimer G, Raz A, Nini A, Bergman H (2006) Basal ganglia oscillations and pathophysiology of movement disorders. Curr. Opin. Neurobiol. 16: 629–637.

Rivlin-Etzion M, Marmor O, Saban G, Rosin B, Haber SN, Vaadia E, Prut Y, Bergman H (2008) Low-pass filter properties of basal ganglia-cortical-muscle loops in the normal and MPTP primate model of Parkinsonism. J. Neurosci. 28: 633–649.

Robbins TW (1996) Dissociating executive functions of the prefrontal cortex. Philos. Trans. R. Soc. Lond. B Biol. Sci. 351; 1463-1470.

Robertson JD (1963) The occurrence of a subunit pattern in the unit membranes of club endings in Mauthner cell synapses in goldfish brains. J. Cell Biol. 19: 201–221.

Rodgers KM, Benison AM, Barth DS (2006) Two-dimensional coincidence detection in the vibrissa/barrel field. J. Neurophysiol. 96: 1981–1990.

Rodriguez E, George N, Lachaux J-P, Martinerie J, Renault B, Varela FJ (1999) Perception's shadow: long-distance synchronization of human brain activity. Nature 397: 430–433.

Rodriguez R, Kallenbach U, Singer W, Munk MHJ (2004) Short- and long-term effects of cholinergic modulation on gamma oscillations and response synchronization in the visual cortex. J. Neurosci. 24: 10369–10378.

Rodriguez-Moreno A, Lerma J (1998) Kainate receptor modulation of GABA release involves a metabotropic function. Neuron 20: 1211–1218.

Rodriguez-Moreno A, Lerma J (2007) Kainate receptors with a metabotropic modus operandi. Trends Neurosci. 30: 630–637.

Roelfsema PR, König P, Engel AK, Sireteanu R, Singer W (1994) Reduced synchronization in the visual cortex of cats with strabismic amblyopia. Eur. J. Neurosci. 6: 1645–1655.

Roelfsema PR, Engel AK, König P, Singer W (1997) Visuomotor integration is associated with zero time-lag synchronization among cortical areas. Nature 385: 157–161.

Rols G, Tallon-Baudry C, Girard P, Bertrand O, Bullier J (2001) Cortical mapping of gamma oscillations in areas V1 and V4 macaque monkey. Vis. Neurosci. 18: 527–540.

Roopun A, Middleton SJ, Cunningham MO, LeBeau FEN, Bibbig A, Whittington MA, Traub RD (2006) A beta-2–frequency (20–30 Hz) oscillation in non-synaptic networks of somatosensory cortex. Proc. Natl. Acad. Sci. USA 103: 15646–15650.

Roopun AK, Cunningham MO, Racca C, Alter K, Traub RD, Whittington MA (2008a) Region-specific changes in gamma and beta-2 rhythms in NMDA receptor dysfunction models of schizophrenia. Schizophr. Bull. 34: 962–973.

Roopun AK, Kramer MA, Carracedo LM, Kaiser M, Davies CH, Traub RD, Kopell NJ, Whittington MA. (2008b) Period concatenation underlies interactions between gamma and beta rhythms in neocortex. Front. Cell. Neurosci. 2:1. Epub 2008 Apr 8.

Roopun AK, Traub RD, Baldeweg T, Cunningham MO, Whittaker RG, Trevelyan A, Duncan R, Russell AJC, Whittington MA (2009) Detecting seizure origin using basic, multiscale population dynamic measures: Preliminary findings. Epilepsy Behav. 14 Suppl. 1: 39–46.

Rörig B, Sutor B (1993) Serotonin regulates gap junction coupling in the developing rat somatosensory cortex. Eur. J. Neurosci. 8: 1685–1695.

Rosenblith, W.A. (1961) Sensory Communication. MIT Press, Cambridge, MA.

Ross CA, Margolis RL, Reading SAJ, Pletnikov M, Coyle JT (2006) Neurobiology of schizophrenia. Neuron 52: 139–153.

Ross FM, Gwyn P, Spanswick D, Davies SN (2000) Carbenoxolone depresses spontaneous epileptiform activity in the CA1 region of rat hippocampal slices. Neuroscience 100: 789–796.

Ross NR, Porter LL (2002) Effects of dopamine and estrogen upon cortical neurons that express parvalbumin in vitro. Dev. Brain Res. 137: 23–34.

Roth A, Häusser M (2001) Compartmental models of rat cerebellar Purkinje cells based on simultaneous somatic and dendritic patch-clamp recordings. J. Physiol. 535: 445–472.

Rothwell JC, Gandevia SC, Burke D (1990) Activation of fusimotor neurones by motor cortical stimulation in human subjects. J. Physiol. 431: 743–756.

Roy SA, Dear SP, Alloway KD (2001) Long-range cortical synchronization without concomitant oscillations in the somatosensory system of anesthetized cats. J. Neurosci. 21: 1795–1808.

Rozas JL, Paternain AV, Lerma J (2003) Noncanonical signaling by ionotropic kainate receptors. Neuron 39: 543–553.

Rumelhart DE, McClelland JL, PDP Research Group (1986) Parallel Distributed Processing. 2 Volumes. MIT Press, Cambridge, MA.

Sabatini BL, Oertner TG, Svoboda K (2002) The life cycle of Ca^{2+} ions in dendritic spines. Neuron 33: 439–452.

Saez JC, Spray DC, Nairn AC, Hertzberg E, Greengard P, Bennett MVL (1986) cAMP increases junctional conductance and stimulates phosphorylation of the 27–kDa principal gap junction polypeptide. Proc. Natl. Acad. Sci. USA 83: 2473–2477.

Saha S, Chant D, Welham J, McGrath J (2005) A systematic review of the prevalence of schizophrenia. PloS Med. 2(5): e141 doi:10.1371/journal.pmed.0020141.

Sakatani K, Chesler M, Hassan AZ (1991a) $GABA_A$ receptors modulate axonal conduction in dorsal columns of neonatal rat spinal cord. Brain Res. 542: 273–279.

Sakatani K, Hassan AZ, Ching W (1991b) Age-dependent extrasynaptic modulation of axonal conduction by exogenous and endogenous GABA in the rat optic nerve. Exp. Neurol. 114: 307–314.

Sakatani K, Black JA, Kocsis JD (1992) Transient presence and functional interaction of endogenous GABA and GABA_A receptors in developing rat optic nerve. Proc. R. Soc. Lond. B 247: 155–161.

Sakatani K, Chesler M, Hassan AZ, Lee M, Young W (1993) Non-synaptic modulation of dorsal column conduction by endogenous GABA in neonatal rat spinal cord. Brain Res. 622: 43–50.

Sakatani K, Hassan AZ, Chesler M (1994) Effects of GABA on axonal conduction and extracellular potassium activity in the neonatal rat optic nerve. Exper. Neurol. 127: 291–297.

Sakmann B, Neher E (1984) Patch clamp techniques for studying ionic channels in excitable membranes. Annu. Rev. Physiol. 46: 455–472.

Sampson JR, Harris PC (1994) The molecular genetics of tuberous sclerosis. Hum. Mol. Genet. 3 Spec. No.: 1477–1480.

Sanchez-Vives MV, McCormick DA (2000) Cellular and network mechanisms of rhythmic recurrent activity in neocortex. Nat. Neurosci. 3: 1027–1034.

Santoro B, Chen S, Lüthi A, Pavlidis P, Shumyatsky GP, Tibbs GR, Siegelbaum SA (2000) Molecular and functional heterogeneity of hyperpolarization-activated pacemaker channels in the mouse CNS. J. Neurosci. 20: 5264–5275.

Sarkar P (2000) A brief history of cellular automata. ACM Computing Surveys 32: 80–107.

Sato F, Parent M, Levesque M, Parent A (2000) Axonal branching pattern of neurons of the subthalamic nucleus in primates. J. Comp. Neurol. 424: 142–152.

Sato K, Momose-Sato Y (2008) Optical imaging analysis of neural circuit formation in the embryonic brain. Clin. Exp. Pharmacol. Physiol. 35: 706–713.

Sauer B (1993) Manipulation of transgenes by site-specific recombination: use of Cre recombinase. Methods Enzymol. 225: 890–900.

Sawa M, Kaji S, Usuki K (1965) Intracellular phenomena in electrically induced seizures. Electroencephalogr. Clin. Neurophysiol. 19: 248–255.

Sawa M., Maruyama N, Kaji S (1963) Intracellular potential during electrically induced seizures, Electroencephalogr. Clin. Neurophysiol. 15: 209–220.

Sawa M, Nakamura K, Naito H (1968) Intracellular phenomena and spread of epileptic seizure discharges, Electroenceph. Clin. Neurophysiol. 24: 146–154.

Schapira AH, Olanow CW (2004) Neuroprotection in Parkinson disease: mysteries, myths, and misconceptions. JAMA 291: 358–364.

Scheffer IE, Jones L, Pozzebon M, Howell RA, Saling MM, Berkovic SF (1995) Autosomal dominant rolandic epilepsy and speech dyspraxia: a new syndrome with anticipation. Ann. Neurol. 38: 633–642.

Schevon CA, Ng SK, Cappell J, Goodman RR, McKhann G Jr, Waziri A, Branner A, Sosunov A, Schroeder CE, Emerson RG (2008) Microphysiology of epileptiform activity in human neocortex. J. Clin. Neurophysiol. 25: 321–330.

Schiffmann SN, Cheron G, Lohof A, d'Alcantara P, Meyer M, Parmentier M, Schurmans S (1999) Impaired motor coordination and Purkinje cell excitability in mice lacking calretinin. Proc. Natl. Acad. Sci. USA 96: 5257–5262.

Schiller J, Schiller Y, Stuart G, Sakmann B (1997) Calcium action potentials restricted to distal apical dendrites of rat neocortical pyramidal neurons. J. Physiol. 505: 605–616.

Schiller J, Major G, Koester HJ, Schiller Y (2000) NMDA spikes in basal dendrites of cortical pyramidal neurons. Nature 404: 285–289.

Schiller Y (2002) Inter-ictal- and ictal-like epileptic discharges in the dendritic tree of neocortical pyramidal neurons. J. Neurophysiol. 88: 2954–2962.

Schiller Y, Cascino GD, Busacker NE, Sharbrough FW (1998) Characterization and comparison of local onset and remote propagated electrographic seizures recorded with intracranial electrodes. Epilepsia 39: 380–388.

Schmitz D, Schuchmann S, Fisahn A, Draguhn A, Buhl EH, Petrasch-Parwez RE, Dermietzel R, Heinemann U, Traub RD (2001) Axo-axonal coupling: a novel mechanism for ultrafast neuronal communication. Neuron 31: 831–840.

Schneider K (1958) Clinical Psychopathology, 5th ed., translated by M.W. Hamilton, Grune & Stratton, New York.

Schnitzler A, Timmermann L, Gross J (2006) Physiological and pathological oscillatory networks in the human motor system. J. Physiol. Paris 99: 3–7.

Scholz A, Reid G, Vogel W, Bostock H (1993) Ion channels in human axons. J. Neurophysiol. 70: 1274–1279.

Schuchmann S, Schmitz D, Rivera C, Vanhatalo S, Salmen B, Mackie K, Sipilä ST, Voipio J, Kaila K (2006) Experimental febrile seizures are precipitated by a hyperthermia-induced respiratory alkalosis. Nat. Med. 12: 817–823.

Schwartzkroin PA (1997) Origins of the epileptic state. Epilepsia 38: 853–858.

Schwartzkroin PA, Haglund MM (1986) Spontaneous rhythmic synchronous activity in epileptic human and normal monkey temporal lobe. Epilepsia 27: 523–533.

Schwartzkroin PA, Prince DA (1976) Microphysiology of human cerebral cortex studied in vitro. Brain Res. 115: 497–500.

Schwartzkroin PA, Prince DA (1977) Penicillin-induced epileptiform activity in the hippocampal in vitro preparation. Ann. Neurol. 1: 463–469.

Schwarz JR, Glassmeier G, Cooper EC, Kao TC, Nodera H, Tabuena D, Kaji R, Bostock H (2006) KCNQ channels mediate Iks, a slow K^+ current regulating excitability in the at node of Ranvier. J. Physiol. 573: 17–34.

Scott BW, Wang S, Burnham WM, De Boni U, Wojtowicz JM (1998) Kindling-induced neurogenesis in the dentate gyrus of the rat. Neurosci. Lett. 248: 73–76.

Seamans KB, Gloor P, Dobell ARC, Wyant JD (1968) Penicillin-induced seizures during cardiopulmonary bypass. A clinical and electroencephalographic study. N. Engl. J. Med. 278: 861–868.

Seamans JK, Yang CR (2004) The principal features and mechanisms of dopamine modulation in the prefrontal cortex. Prog. Neurobiol. 74: 1–57.

Sedgwick EM, Williams TD (1967) Responses of single units in the inferior olive to stimulation of the limb nerves, peripheral skin receptors, cerebellum, caudate nucleus and motor cortex. J. Physiol. 189: 261–279.

Seeburg PH (1993) The TINS/TiPS Lecture. The molecular biology of mammalian glutamate receptor channels. Trends Neurosci. 16: 359–365.

Seeman P (2006) Targeting the dopamine D2 receptor in schizophrenia. Expert Opin. Ther. Targets 10: 515–531.

Seeman P, Ko F, Tallerico T (2005a) Dopamine receptor contribution to the action of PCP, LSD and ketamine psychotomimetics. Mol. Psychiatry 10: 877–883.

Seeman P, Weinshenker D, Quirion R, Srivastava LK, Bhardwaj SK, Grandy DK, Premont RT, Sotnikova TD, Boksa P, El-Ghundi M, O'Dowd BF, George SR, Perreault ML, Männistö PT, Robinson S, Palmiter RD, Tallerico T (2005b) Dopamine supersensitivity correlates with $D2^{High}$ states, implying many paths to psychosis. Proc. Natl. Acad. Sci. USA 102: 3513–3518.

Seeman P, Schwarz J, Chen JF, Szechtman H, Perreault M, McKnight GS, Roder JC, Quirion R, Boksa P, Srivastava LK, Yanai K, Weinshenker D, Sumiyoshi T (2006) Psychosis pathways converge via D2high dopamine receptors. Synapse 60: 319–346.

Segal M, Barker JL (1984) Rat hippocampal neurons in culture: voltage-clamp analysis of inhibitory synaptic connections. J. Neurophysiol. 52: 469–487.

Seino S, Miki T (2003) Physiological and pathophysiological roles of ATP-sensitive K+ channels. Prog. Biophys. Molec. Biol. 81: 133–176.

Selverston AI (2005) A neural infrastructure for rhythmic motor patterns. Cell. Mol. Neurobiol. 25: 223–244.

Semyanov A, Kullmann DM (2001) Kainate receptor-dependent axonal depolarization and action potential initiation in interneurons. Nat. Neurosci. 4: 718–723.

Servais L, Cheron G (2005) Purkinje cell rhythmicity and synchronicity during modulation of fast cerebellar oscillation. Neuroscience 134: 1247–1259.

Servais L, Bearzatto B, Schwaller B, Dumont M, De Saedeleer C, Dan B, Barski JJ, Schiffmann SN, Cheron G (2005) Mono- and dual-frequency fast cerebellar oscillation in mice lacking parvalbumin and/or calbindin D-28k. Eur. J. Neurosci. 22: 861–870.

Shambes GM, Gibson JM, Welker W (1978) Fractured somatotopy in granule cell tactile areas of rat cerebellar hemispheres revealed by micromapping. Brain Behav. Evol. 15: 94–140.

Shannon CE, McCarthy J (Eds) (1956) Automata Studies, Annals of Mathematics Studies Number 34, Princeton University Press, Princeton.

Shao LR, Halvorsrud R, Borg-Graham L, Storm JF (1999) The role of BK-type Ca^{2+}-dependent K+ channels in spike broadening during repetitive firing in rat hippocampal pyramidal cells. J. Physiol. 521: 135–146.

Sharif Z (2008) Side effects as influencers of treatment outcome. J. Clin. Psychiatry 69 Suppl. 3: 38–43.

Sharifullina E, Ostroumov K, Nistri A (2005) Metabotropic glutamate receptor activity induces a novel oscillatory pattern in neonatal rat hypoglossal motoneurones. J. Physiol. 563: 139–159.

Sharrott A, Magill PJ, Harnack D, Kupsch A, Meissner W, Brown P (2005) Dopamine depletion increases the power and coherence of β-oscillations in the cerebral cortex and subthalamic nucleus of the awake rat. Eur. J. Neurosci. 21: 1413–1422.

Sherer MA (1988) Intravenous cocaine: psychiatric effcts, biological mechanisms. Biol. Psychiatry 24: 865–885.

Shergill SS, Cameron LA, Brammer MJ, Williams SC, Murray RM, McGuire PK (2001) Modality specific neural correlates of auditory and somatic hallucinations. J. Neurol. Neurosurg. Psychiatry 71: 688–690.

Sherrington C (1947) The Integrative Action of the Nervous System, 2nd ed. Yale University Press, New Haven.

Shi ZQ, Chang TM (1984) Amino acid disturbances in experimental hepatic coma in rats. Int. J. Artif. Organs 7: 197–202.

Shin HS, Lee J, Song I (2006) Genetic studies on the role of T-type Ca^{2+} channels in sleep and absence epilepsy. CNS Neurol. Disord. Drug Targets 5: 629–638.

Shinoda Y, Kakei S, Futami T, Wannier T (1993) Thalamocortical organization in the cerebello-thalamo-cortical system. Cereb. Cortex 3: 421–429.

Shiosaka S, Yamamoto T, Hertzberg EL, Nagy JI (1989) Gap junction protein in rat hippocampus: correlative light and electron microscope immunohistochemical localization. J. Comp. Neurol. 281: 282–297.

Shu Y, Yu Y, Yang J, McCormick DA (2007a) Selective control of cortical axonal spikes by a slowly inactivating K⁺ current. Proc. Natl. Acad. Sci. USA 104: 11453–11458.

Shu Y, Duque A, Yu Y, Haider B, McCormick DA (2007b) Properties of action-potential initiation in neocortical pyramidal cells: evidence from whole cell axon recordings. J. Neurophysiol. 97: 746–760.

Sidiropoulou K, Lu F-M, Fowler MA, Xiao R, Phillips C, Ozkan ED, Zhu MX, White FJ, Cooper DC (2009) Dopamine modulates an mGluR5–mediated depolarization underlying prefrontal persistent activity. Nat. Neurosci. 12: 190–199.

Silberberg G, Markram H (2007) Disynaptic inhibition between neocortical pyramidal cells mediated by Martinotti cells. Neuron 53: 735–746.

Sillanpää M, Jalava M, Kaleva O, Shinnar S (1998) Long-term prognoiss of seizures with onset in childhood. N. Engl. J. Med. 338: 1715–1722.

Silver RA, Momiyama A, Cull-Candy SG (1998) Locus of frequency-dependent depression identified with multiple-probability fluctuation analysis at rat climbing fibre-Purkinje cell synapses. J. Physiol. 510: 881–902.

Simon A, Oláh S, Molnár G, Szabadics J, Tamás G (2005) Gap-junctional coupling between neurogliaform cells and various interneuron types in the neocortex. J. Neurosci. 25: 6278–6285.

Singer W (1990) The formation of cooperative cell assemblies in the visual cortex. J. Exp. Biol. 153: 177–197.

Singer W (1999) Neuronal synchrony: a versatile code for the definition of relations? Neuron 24: 49–65.

Singer W, Gray CM (1995) Visual feature integration and the temporal correlation hypothesis. Annu. Rev. Neurosci. 18: 555–586.

Sirven JI (2002) Classifying seizures and epilepsy: a synopsis. Semin. Neurol. 22: 237–246.

Sloper JJ, Powell TPS (1978a) Dendro-dendritic and reciprocal synapses in the primate motor cortex. Proc. R. Soc. Lond. B 203: 23–38.

Sloper JJ, Powell TPS (1978b) Gap junctions between dendrites and somata of neurons in the primate sensori-motor cortex. Proc. R. Soc. Lond. B 203: 39–47.

Sloper JJ, Powell TPS (1979) A study of the axon initial segment and proximal axon of neurons in the primate motor and somatic sensory cortices. Philos. Trans. R. Soc. Ser. B 285: 173–197.

Sloviter RS (1999) Status epilepticus-induced neuronal injury and network reorganization. Epilepsia 40 Suppl. 1: S34–S39.

Smart SL, Lopantsev V, Zhang CL, Robbins CA, Wang H, Chiu SY, Schwartzkroin PA, Messing A, Tempel BL (1998) Deletion of the K(V)1.1 potassium channel causes epilepsy in mice. Neuron 20: 809–819.

Smiley JF, Williams SM, Szigeti K, Goldman-Rakic PS (1992) Light and electron microscopic characterization of dopamine-immunoreactive axons in human cerebral cortex. J. Comp. Neurol. 321: 325–335.

Soares-Weiser K, Fernandez HH (2007) Tardive dyskinesia. Semin. Neurol. 27: 159–169.

Sobolevsky AI, Prodromou ML, Yelshansky MV, Wollmuth LP (2007) Subunit-specific contribution of pore-forming domains to NMDA receptor channel structure and gating. J. Gen. Physiol. 129: 509–525.

Söhl G, Güldenagel M, Beck H, Teubner B, Traub O, Gutierrez R, Heinemann U, Willecke K (2000) Expression of connexin genes in hippocampus of kainate-treated and kindled rats under conditions of experimental epilepsy. Brain Res. 83: 44–51.

Söhl G, Maxeiner S, Willecke K (2005) Expression and functions of neuronal gap junctions. Nat. Rev. Neurosci. 6: 191–200.

Soleng AF, Chiu K, Raastad M (2003) Unmyelinated axons in the rat hippocampus hyperpolarize and activate an H current when spike frequency exceeds 1 Hz. J. Physiol. 552: 459–470.

Soltesz I, Deschênes M (1993) Low- and high-frequency membrane potential oscillations during theta activity in CA1 and CA3 pyramidal neurons of the rat hippocampus under ketamine-xylazine anesthesia. J. Neurophysiol. 70: 97–116.

Soltesz I, Lightowler S, Leresche N, Jassik-Gerschenfeld D, Pollard CE, Crunelli V (1991) Two inward currents and the transformation of low-frequency oscillations of rat and cat thalamocortical cells. J. Physiol. 441: 175–197.

Sommer B, Burnashev N, Verdoorn TA, Keinänen K, Sakmann B, Seeburg PH (1992) A glutamate receptor channel with high affinity for domoate and kainate. EMBO J. 11: 1651–1656.

Sommer W (1880) Erkrankung des Ammonshorns als aetiologisches Moment der Epilepsie. Arch. Psychiatr. Nervenkr. 10: 631–675.

Somogyi P, Klausberger T (2005) Defined types of cortical interneurone structure space and spike timing in the hippocampus. J. Physiol. 562: 9–26.

Somogyi P, Freund TF, Cowey A (1982) The axo-axonic interneuron in the cerebral cortex of the rat, cat and monkey. Neuroscience 7: 2577–2607.

Somogyi P, Tamás G, Lujan, R, Buhl EH (1998) Salient features of synaptic organization in the cerebral cortex. Brain Res. Rev. 26: 113–135.

Song C, Murray TA, Kimura R, Wakui M, Ellsworth K, Javedan SP, Marxer-Miller S, Lukas RJ, Wu J (2005) Role of alpha7–nicotinic acetylcholine receptors in tetanic stimulation-induced gamma oscillations in rat hippocampal slices. Neuropharmacology 48: 869–880.

Song WJ, Baba Y, Otsuka T, Murakami F (2000) Characterization of Ca^{2+} channels in rat subthalamic neurons. J. Neurophysiol. 84: 2630–2637.

Sotelo C (2003) Viewing the brain through the master hand of Ramón y Cajal. Nat. Rev. Neuroscience 4: 71–77,

Sotelo C, Llinás R (1972) Specialized membrane junctions between neurons in the vertebrate cerebellar cortex. J. Cell Biol. 53: 271–289.

Sotelo C, Taxi J (1970) Ultrastructural aspects of electrotonic junctions in the spinal cord of the frog. Brain Res. 17: 137–141.

Sotelo C, Llinás R, Baker R (1974) Structural study of inferior olivary nucleus of the cat: morphological correlates of electrotonic coupling. J. Neurophysiol. 37: 541–559.

Sotelo C, Rethelyi M, Szabo T (1975) Morphological correlates of electrotonic coupling in the magnocellular mesencephalic nucleus of the weakly electric fish *Gymnotus carapo*. J. Neurocytol. 4: 587–607.

Soteropoulos DS, Baker SN (2006) Cortico-cerebellar coherence during a precision grip task in the monkey. J. Neurophysiol. 95: 1194–1206.

Spence SJ, Saint-Cyr JA (1988) Comparative topography of projections from the mesodiencephalic junction to the inferior olive, vestibular nuclei, and upper cervical cord in the cat. J. Comp. Neurol. 268: 357–374.

Spencer KM, Nestor PG, Niznikiewicz MA, Salisbury DF, Shenton ME, McCarley RW (2003) Abnormal neural synchrony in schizophrenia. J. Neurosci. 23: 7407–7411.

Spencer KM, Nestor PG, Perlmutter R, Niznikiewicz MA, Klump MC, Frumin M, Shenton ME, McCarley RW (2004) Neural synchrony indexes disordered perception and cognition in schizophrenia. Proc. Natl. Acad. Sci. USA 101: 17288–17293.

Spencer WA, Kandel ER (1961) Electrophysiology of hippocampal neurons IV. Fast prepotentials. J. Neurophysiol. 24: 272–285.

Spray DC, Harris AL, Bennett MVL (1981) Gap junctional conductance is a simple and sensitive function of intracellular pH. Science 211: 712–715.

Spruston N, Schiller Y, Stuart G, Sakmann B (1995) Activity-dependent action potential invasion and calcium influx into hippocampal CA1 dendrites. Science 268: 297–300.

Srinivas M, Hopperstad MG, Spray DG (2001) Quinine blocks specific gap junction channel subtypes. Proc. Natl. Acad. Sci. USA 98: 10942–10947.

Srinivas M, Rozental R, Kojima T, Dermietzel R, Mehler M, Condorelli DF, Kessler JA, Spray DC (1999) Functional properties of channels formed by the neuronal gap junction protein connexin36. J. Neurosci. 19: 9848–9855.

Staba RJ, Wilson CL, Bragin A, Fried I, Engel J Jr (2002) Quantitative analysis of high-frequency oscillations (80–500 Hz) recorded in human epileptic hippocampus and entorhinal cortex. J. Neurophysiol. 88: 1743–1752.

Staba RJ, Bret-Green B, Paulsen M, Barth DS (2003) Effects of ventrobasal lesion and cortical cooling on fast oscillations (>200 Hz) in rat somatosensory cortex. J. Neurophysiol. 89: 2380–2388.

Staba RJ, Bergmann PC, Barth DS (2004a) Dissociation of slow waves and fast oscillations above 200 Hz during GABA application in rat somatosensory cortex. J. Physiol. 561: 205–214. [has halothane data]

Staba RJ, Wilson CL, Bragin A, Jhung D, Fried I, Engel J Jr (2004b) High-frequency oscillations recorded in human medial temporal lobe during sleep. Ann. Neurol. 56: 108–115.

Staba RJ, Ard TD, Benison AM, Barth DS (2005) Intracortical pathways mediate nonlinear fast oscillation (>200 Hz) interactions within rat barrel cortex. J. Neurophysiol. 93: 2934–2939.

Staley KJ, Longacher M, Bains JS, Yee A (1998) Presynaptic modulation of CA3 network activity. Nat. Neurosci. 1: 201–209.

Stanford IM (2003) Independent neuronal oscillators of the rat globus pallidus. J. Neurophysiol. 89: 1713–1717.

Stasheff SF, Wilson WA (1990) Increased ectopic action potential generation accompanies epileptogenesis in vitro. Neurosci. Lett. 111: 144–150.

Stasheff SF, Hines M, Wilson WA (1993a) Axon terminal hyperexcitability associated with epileptogenesis in vitro. I. Origin of ectopic spikes. J. Neurophysiol. 70: 960–975.

Stasheff SF, Mott DD, Wilson WA (1993b) Axon terminal hyperexcitability associated with epileptogenesis in vitro. II. Pharmacological regulation by NMDA and GABA$_A$ receptors. J. Neurophysiol. 70: 976–984.

Stefansson H, Rujescu D, Cichon S, Pietiläinen OPH, Ingason A, Steinberg S, Fossdal R, Sigurdsson E, Sigmundsson T, Buizer-Voskamp JE, Hansen T, Jakobsen KD, Muglia P, Francks C, Matthews PM, Gylfason A, Halldorsson BV, Gudbjartsson D,

Thorgeirsson TE, Sigurdsson A, Jonasdottir A, Jonasdottir A, Bjornsson A, Mattiasdottir S, Blondal T, Haraldsson M, Magnusdottir BB, Giegling I, Möller HJ, Hartmann A, Shianna KV, Ge D, Need AC, Crombie C, Fraser G, Walker N, Lonnqvist J, Suvisaari J, Tuulio-Henriksson A, Paunio T, Toulopoulou T, Bramon E, Di Forti M, Murray R, Ruggeri M, Vassos E, Tosato S, Walshe M, Li T, Vasilescu C, Mühleisen TW, Wang AG, Ullum H, Djurovic S, Melle I, Olesen J, Kiemeney LA, Franke B, Sabatti C, Freimer NB, Gulcher JR, Thorsteinsdottir U, Kong A, Andreassen OA, Ophoff RA, Georgi A, Rietschel M, Werge T, Petursson H, Goldstein DB, Nöthen MM, Peltonen L, Collier DA, St Clair D, Stefansson K. (2008) Large recurrent microdeletions associated with schizophrenia. Nature 455: 232–236.

Steinlein OK (2004) Genes and mutations in human idiopathic epilepsy. Brain Dev. 26: 213–218.

Steinlein OK, Noebels JL (2000) Ion channels and epilepsy in man and mouse. Curr. Opin. Genet. Dev. 10: 286–291.

Steinlein OK, Mulley JC, Propping P, Wallace RH, Phillips HA, Sutherland GR, Scheffer IE, Berkovic SF (1995) A misssense mutation in the neuronal nicotinic acetylcholine receptor α4 subunit is associated with autosomal dominant nocturnal frontal lobe epilepsy. Nat. Genet. 11: 201–203.

Stenkamp K, Palva JM, Uusisaari M, Schuchmann S, Schmitz D, Heinemann U, Kaila K (2001) Enhanced temporal stability of cholinergic hippocampal gamma oscillations following respiratory alkalosis in vitro. J. Neurophysiol. 85: 2063–2069.

Stephani U (2006) The natural history of myoclonic astatic epilepsy (Doose syndrome) and Lennox-Gastaut syndrome. Epilepsia 47 Suppl 2: 53–55.

Steriade M (2001) The Intact and Sliced Brain. MIT Press, Cambridge, MA.

Steriade M (2003) Neuronal Substrates of Sleep and Epilepsy. Cambridge University Press, Cambridge, U.K.

Steriade M (2005) Grouping of brain rhythms in corticothalamic systems. Neuroscience 137: 1087–1106.

Steriade M, Amzica F (1994) Dynamic coupling among neocortical neurons during evoked and spontaneous spike-wave seizure activity. J. Neurophysiol. 72: 2051–2069.

Steriade M, Amzica F (1996) Intracortical and corticothalamic coherency of fast spontaneous oscillations. Proc. Natl. Acad. Sci. USA 93: 2533–2538.

Steriade M, Amzica F (1998) Coalescence of sleep rhythms and their chronology in corticothalamic networks. Sleep Res. Online 1: 1–10.

Steriade M, Contreras D (1995) Relations between cortical and thalamic cellular events during transition from sleep patterns to paroxysmal activity. J. Neurosci. 15: 623–642.

Steriade M, Contreras D (1998) Spike-wave complexes and fast components of cortically generated seizures. I. Role of neocortex and thalamus. J. Neurophysiol. 80: 1439–1455.

Steriade M, Timofeev I (2003) Neuronal plasticity in thalamocortical networks during sleep and waking oscillations. Neuron 37: 563–576.

Steriade M, Domich L, Oakson G, Deschênes, M (1987) The deafferented reticular thalamic nucleus generates spindle rhythmicity. J. Neurophysiol. 57: 260–273.

Steriade M, Curró Dossi R, Paré D, Oakson G (1991) Fast oscillations (20–40 Hz) in thalamocortical systems and their potentiation by mesopontine cholinergic nuclei in the cat. Proc. Natl. Acad. Sci. USA 88: 4396–4400.

Steriade M, Nuñez A, Amzica F (1993a) A novel slow (<1 Hz) oscillation of neocortical neurons *in vivo*: depolarizing and hyperpolarizing components. J. Neurosci. 13: 3252–3265.

Steriade M, Nuñez A, Amzica F (1993b) Intracellular analysis of relations between the slow (<1 Hz) neocortical oscillation and other sleep rhythms of the electroencephalogram. J. Neurosci. 13: 3266–3283.

Steriade M, Amzica F, Nuñez A (1993c) Cholinergic and noradrenergic modulation of the slow (~0.3 Hz) oscillation in neocortical cells. J. Neurophysiol. 70: 1385–1400.

Steriade M, Contreras D, Curró Dossi R, Nuñez A (1993d) The slow (<1 Hz) oscillation in reticular thalamic and thalamocortical neurons: scenario of sleep rhythm generation in interacting thalamic and neocortical networks. J. Neurosci. 13: 3284–3299.

Steriade M, Curró Dossi R, Contreras D (1993e) Electrophysiological properties of intralaminar thalamocortical cells discharging rhythmic (~40 Hz) spike-bursts at ~1000 Hz during waking and rapid eye movement sleep. Neuroscience 56: 1–9.

Steriade M, Amzica F, Contreras D (1995) Synchronization of fast (30–40 Hz) spontaneous cortical rhythms during brain activation. J. Neurosci. 16: 392–417.

Steriade M, Contreras D, Amzica F, Timofeev I (1996) Synchronization of fast (30–40 Hz) spontaneous oscillations in intrathalamic and thalamocortical networks. *J. Neurosci.* 16: 2788–2808.

Steriade M, Amzica F, Neckelmann D, Timofeev I (1998a) Spike-wave complexes and fast components of cortically generated seizures. II. Extra- and intracellular pattens. J. Neurophysiol. 80: 1456–1479.

Steriade M, Timofeev I, Dürmüller N, Grenier F (1998b) Dynamic properties of corticothalamic neurons and local cortical interneurons generating fast rhythmic (30–40 Hz) spike bursts. J. Neurophysiol. 79: 483–490.

Steriade M, Timofeev I, Grenier F, Dürmüller N (1998) Role of thalamic and cortical neurons in augmenting responses and self-sustained activity: dual intracellular recordings *in vivo*. J. Neurosci. 18: 6425–6443.

Steriade M, Timofeev I, Grenier F (2001) Natural waking and sleep states: a view from inside neocortical neurons. J. Neurophysiol. 85: 1969–1985.

Stilling B (1864) Untersuchungen über den Bau des kleinen Gehirns des Menschen. Kassel.

Stopfer M, Laurent G (1999) Short-term memory in olfactory network dynamics. Nature 402: 664–668.

Stopfer M, Bhagavan S, Smith BH, Laurent G (1997) Impaired odour discrimination on desynchronization of odour-encoding neural assemblies. Nature 390: 70–74.

Strange PG (2008) Antipsychotic drug action: antagonism, inverse agonism or partial agonism. Trends Pharmacol. Sci. 29: 314–321.

Stuart G, Häusser M (1994) Initiation and spread of sodium action potentials in cerebellar Purkinje cells. Neuron 13: 703–712.

Stuart GJ, Sakmann B (1994) Active propagation of somatic action potentials into neocortical pyramidal cell dendrites. Nature 367: 69–72.

Stuart G, Schiller J, Sakmann B (1997) Action potential initiation and propagation in rat neocortical pyramidal neurons. J. Physiol. 505: 617–632.

Suffcyzynski P, Lopes da Silva F, Parra J, Velis D, Kalitzin S (2005) Epileptic transitions: model predictions and experimental validation. J. Clin. Neurophysiol. 22: 288–299.

Suh BC, Hille B (2002) Recovery from muscarinic modulation of M current channels requires phosphatidylinositol 4,5–bisphosphate synthesis. Neuron 3: 507–520.

Sukov W, Barth DS (2001) Cellular mechanisms of thalamically evoked gamma oscillations in auditory cortex. J. Neurophysiol. 85: 1235–1245.

Sullivan PF, Kendler KS, Neale MC (2003) Schizophrenia as a complex trait: evidence from a meta-analysis of twin studies. Arch. Gen. Psychiatry 60: 1187–1192.

Sun HS, Feng ZP, Barber PA, Buchan AM, French RJ (2007) Kir6.2–containing ATP-sensitive potassium channels protect cortical neurons from ischemic/anoxic injury in vitro and in vivo. Neuroscience 144: 1509–1515.

Sutton S, Braren M, Zubin J, John ER (1965) Evoked-potential correlates of stimulus uncertainty. Science 150: 1187–1188.

Sutula TP, Dudek FE (2007) Unmasking recurrent excitation generated by mossy fiber sprouting in the epileptic dentate gyrus: an emergent property of a complex system. Prog. Brain Res. 163: 541–563.

Suzuki T, Delgado-Escueta AV, Aguan K, Alonso ME, Shi J, Hara Y et al. (2004) Mutations in EFHC1 cause juvenile myoclonic epilepsy. Nat. Genet. 36: 842–849.

Svoboda K, Helmchen F, Denk W, Tank DW (1999) Spread of dendritic excitation in layer 2/3 pyramidal neurons in rat barrel cortex in vivo. Nat. Neurosci. 2: 65–73.

Swartz BE, Goldensohn ES (1998) Timeline of the history of EEG and associated fields. Electroenceph. Clin. Neurophysiol. 106: 173–176.

Szabadics J, Varga C, Molnár G, Oláh S, Barzó P, Tamás G (2006) Excitatory effect of GABAergic axo-axonic cells in cortical microcircuits. Science 311: 233–235.

Szabadics J, Tamás G, Soltesz I (2007) Different transmitter transients underlie presynaptic cell type specificity of $GABA_{A,slow}$ and $GABA_{A,fast}$. Proc. Natl. Acad. Sci. USA 104: 14831–14836. [neurogliaform cells are responsible for GABA-(A,slow) in neocortex]

Szabó CA, Leland MM, Knape K, Elliott JJ, Haines V, Williams JT (2005) Clinical and EEG phenotypes of epilepsy in the baboon (Papio hamadryas spp.). Epilepsy Res. 65: 71–80.

Szente M, Gajda Z, Ali KS, Hermesz E (2002) Involvement of electrical coupling in the in vivo ictal epileptiform activity induced by 4–aminopyridine in the neocortex. Neuroscience 115: 1067–1078.

Tallon-Baudry C, Bertrand O (1999) Oscillatory gamma activity in humans and its role in object representation. Trends Cogn. Sci. 3: 151–162.

Tallon-Baudry C, Bertrand O, Delpuech C, Pernier J (1997) Oscillatory γ-band (30–70 Hz) activity induced by a visual search task in humans. J. Neurosci. 17: 722–734.

Tallon-Baudry C, Bertrand O, Peronnet F, Pernier J (1998a) Induced γ-band activity during the delay of a visual short-term memory task in humans. J. Neurosci. 18: 4244–4254.

Tallon-Baudry C, Bertrand O, Peronnet F, Pernier J (1998b) Induced γ-band activity during the delay of a visual short-term memory task in humans. J. Neurosci. 18: 4244–4254.

Tallon-Baudry C, Bertrand O, Delpuech C, Pernier J (1999a) Stimulus specificity of phase-locked and non-phase-locked 40 Hz visual responses in human. J. Neurosci. 16: 4240–4249.

Tallon-Baudry C, Bertrand O, Pernier J (1999b) A ring-shaped distribution of dipoles as a source model of induced gamma-band activity. Clin. Neurophysiol. 110: 660–665.

Tallon-Baudry C, Kreiter A, Bertrand O (1999c) Sustained and transient oscillatory responses in the gamma and beta bands in a visual short-term memory task in humans. Vis. Neurosci. 16: 449–459.

Tallon-Baudry C, Bertrand O, Fischer C (2001) Oscillatory synchrony between human extrastriate areas during visual short-term memory maintenance. J. Neurosci. 21: RC177, 1–5.

Tallon-Baudry C, Bertrand O, Hénaff MA, Isnard J, Fischer C (2005) Attention modulates gamma-band oscillations differently in the human lateral occipital cortex and fusiform gyrus. Cereb. Cortex 15: 654–662.

Tamás G, Buhl EH, Somogyi P (1997) Fast IPSPs elicited via multiple synaptic release sites by different types of GABAergic neurone in the cat visual cortex. J. Physiol. 500: 715–738.

Tamás G, Somogyi P, Buhl EH (1998) Differentially interconnected networks of GABAergic interneurons in the visual cortex of the cat. J. Neurosci. 18: 4255–4270.

Tamás G, Buhl EH, Lörincz A, Somogyi P (2000) Proximally targeted GABAergic synapses and gap junctions precisely synchronize cortical interneurons. Nat. Neurosci. 3: 366–371.

Tamás G, Lörincz A, Simon A, Szabadics J (2003) Identified sources and targets of slow inhibition in the neocortex. Science 299:1902–1905.

Tan HY, Callicott JH, Weinberger DR (2007) Dysfunctional and compensatory prefrontal cortical systems, genes and the pathogenesis of schizophrenia. Cereb. Cortex. 17 Suppl. 1: i171–i181.

Tauck DL, Nadler JV (1985) Evidence of functional mossy fiber sprouting in hippocampal formation of kainic acid-treated rats. J. Neurosci. 5: 1016–1022.

Tellez-Zenteno JF, Pondal-Sordo M, Matijevic S, Wiebe S (2004) National and regional prevalence of self-reported epilepsy in Canada. Epilepsia 45: 1623–1629.

Teubner B, Odermatt B, Güldenagel M, Söhl G, Degen J, Bukauskas FF, Kronengold J, Verselis VK, Jung YT, Kozak CA, Schilling K, Willecke K (2001) Functional expression of the new gap junction gene connexin47 transcribed in mouse brain and spinal cord neurons. J. Neurosci. 21: 1117–1126.

Thach WT (1987) Cerebellar inputs to motor cortex. Ciba Found. Symp. 132: 201–220.

Thiele A, Stoner G (2003) Neuronal synchrony does not correlate with motion coherence in cortical area MT. Nature 421: 366–370.

Thierry A-M, Blanc G, Sobel A, Stinus L, Golwinski J (1973) Dopaminergic terminals in the rat cortex. Science 183: 499–501.

Thomas B, Beal MF (2007) Parkinson's disease. Hum. Mol. Genet. 16 Spec No. 2: R183–R194.

Thompson RJ, Zhou N, MacVicar BA (2006) Ischemia opens neuronal gap junction hemichannels. Science 312: 924–927.

Timmann D, Diener HC (2003) Coordination and ataxia. In: Fotez CG (Ed), Textbook of Clinical Neurology, 2nd ed. Saunders, Philadelphia, pp. 299–315.

Timofeev I, Steriade M (1996) Low-frequency rhythms in the thalamus of intact-cortex and decorticated cats. J.Neurophysiol. 76: 4152–4168.

Timofeev I, Steriade M (1997) Fast (mainly 30–100 Hz) oscillations in the cat cerebellothalamic pathway and their synchronization with cortical potentials. J. Physiol. 504: 153–168.

Timofeev I, Steriade M (2004) Neocortical seizures: initiation, development and cessation. Neuroscience 123: 299–336.

Timofeev I, Contreras D, Steriade (1996) Synaptic responsiveness of cortical and thalamic neurones during various phases of slow sleep oscillation in cat. J. Physiol. 494: 265–278.

Timofeev I, Grenier F, Steriade M (1998) Spike-wave complexes and fast components of cortically generated seizures. IV. Paroxysmal fast runs in cortical and thalamic neurons. J. Neurophysiol. 80: 1495–1513.

Timofeev I, Grenier F, Bazhenov M, Sejnowski TJ, Steriade M (2000) Origin of slow cortical oscillations in deafferented cortical slabs. Cereb. Cortex 10: 1185–1199.

Tölle R (2008) Eugen Bleuler (1857–1939) und die deutsche Psychiatrie. Der Nervenarzt 78: 90–98.

Topolnik L, Steraide M, Timofeev I (2003) Partial cortical deafferentation promotes development of paroxysmal activity. Cereb. Cortex 13: 883–893.

Tóth TI, Bessaïh T, Leresche N, Crunelli V (2007) The properties of reticular thalamic neuron $GABA_A$ IPSCs of absence epilepsy rats lead to enhanced network excitability. Eur. J. Neurosci. 26: 1832–1844.

Towers SK, Hestrin S (2008) D_1-like dopamine receptor activation modulates GABAergic inhibition but not electrical coupling between neocortical fast-spiking interneurons. J. Neurosci. 28: 2633–2641.

Towers SK, LeBeau FEN, Gloveli T, Traub RD, Whittington MA, Buhl EH (2002) Fast network oscillations in the rat dentate gyrus in vitro. J. Neurophysiol. 87: 1165–1168.

Towers SK, Gloveli T, Traub RD, Driver JE, Engel D, Fradley R, Rosahl TW, Maubach K, Buhl EH, Whittington MA (2004) $\alpha 5$ subunit-containing $GABA_A$ receptors affect the dynamic range of hippocampal gamma frequency oscillations in vitro. J. Physiol. 559: 721–728.

Toyoshima K, Sakai H (1982) Exact cortical extent of the origin of the corticospinal tract (CST) and the quantitative contribution to the CST in different cytoarchitectonic areas. A study with horseradish peroxidase in the monkey. J. Hirnforsch. 23: 257–269.

Trantham-Davidson H, Kröner S, Seamans JK (2008) Dopamine modulation of prefrontal cortex interneurons occurs independently of DARPP-32. Cereb. Cortex 18: 951–958.

Traub RD (1977) Motorneurons of different geometry and the size principle. Biol. Cybernetics 25: 163–176.

Traub RD, Bibbig A (2000) A model of high-frequency ripples in the hippocampus, based on synaptic coupling plus axon-axon gap junctions between pyramidal neurons. J. Neurosci. 20: 2086–2093.

Traub R, Llinás R (1977) The spatial distribution of ionic conductances in normal and axotomized motorneurons. Neuroscience 2, 829–849.

Traub RD, Miles R (1991) Neuronal Networks of the Hippocampus. Cambridge University Press, New York.

Traub RD, Miles R (1995) Pyramidal cell-to-inhibitory cell spike transduction explicable by active dendritic conductances in inhibitory cell. J. Comput. Neurosci. 2: 291–298.

Traub RD, Wong RKS (1982) Cellular mechanism of neuronal synchronization in epilepsy. Science 216, 745–747.

Traub RD, Duncan R, Russell AJC, Baldeweg T, Tu Y, Cunningham MO, Whittington MA (in press) Spatiotemporal patterns of electrocorticographic very fast oscillations (>80 Hz) consistent with a network model based on electrical coupling between principal neurons. Epilepsia.

Traub RD, Miles R, Wong RKS (1987) Models of synchronized hippocampal bursts in the presence of inhibition. 1. Single population events. J. Neurophysiol. 58: 739–751.

Traub RD, Middleton SJ, Knöpfel T, Whittington MA (2008) Model of very fast (>75 Hz) network oscillations generated by electrical coupling between the proximal axons of cerebellar Purkinje cells. Eur. J. Neurosci. 28: 1603-1616.

Traub RD, Wong RKS, Miles R, Michelson H (1991) A model of a CA3 hippocampal pyramidal neuron incorporating voltage-clamp data on intrinsic conductances. J. Neurophysiol. 66: 635–650.

Traub RD, Miles R, Buzsáki G (1992) Computer simulation of carbachol-driven rhythmic population oscillations in the CA3 region of the in vitro rat hippocampus. J. Physiol. 451: 653–672.

Traub RD, Miles R, Jefferys JGR (1993) Synaptic and intrinsic conductances shape picrotoxin-induced synchronized after-discharges in the guinea-pig hippocampal slice. J. Physiol. 461: 525–547.

Traub RD, Jefferys JGR, Whittington MA (1994a) Enhanced NMDA conductances can account for epileptiform activities induced by low Mg^{2+} in the rat hippocampal slice. J. Physiol. 478: 379–393.

Traub RD, Jefferys JGR, Miles R, Whittington MA, Tóth K (1994b) A branching dendritic model of a rodent CA3 pyramidal neurone. J. Physiol. 481: 79–95.

Traub RD, Colling SB, Jefferys JGR (1995) Cellular mechanisms of 4–aminopyridine-induced synchronized after-discharges in the rat hippocampal slice. J. Physiol. 489: 127–140.

Traub RD, Borck C, Colling SB, Jefferys JGR (1996a) On the structure of ictal events in vitro. Epilepsia 37: 879–891.

Traub RD, Whittington MA, Colling SB, Buzsáki G, Jefferys JGR (1996b) Analysis of gamma rhythms in the rat hippocampus in vitro and in vivo. J. Physiol. 493: 471–484. Traub RD, Whittington MA, Stanford IM, Jefferys JGR (1996c) A mechanism for generation of long-range synchronous fast oscillations in the cortex. Nature 383: 621–624.

Traub RD, Whittington MA, Jefferys JGR (1997) Gamma oscillation model predicts intensity coding by phase rather than frequency. Neural Comput 9: 1251–1264.

Traub RD, Jefferys JGR, Whittington MA (1999a) Fast Oscillations in Cortical Circuits. MIT Press, Cambridge, MA.

Traub RD, Schmitz D, Jefferys JGR, Draguhn A (1999b) High-frequency population oscillations are predicted to occur in hippocampal pyramidal neuronal networks interconnected by axoaxonal gap junctions. Neuroscience 92: 407–426.

Traub RD, Whittington MA, Buhl EH, Jefferys JGR, Faulkner HJ (1999c) On the mechanism of the $\gamma \rightarrow \beta$ frequency shift in neuronal oscillations induced in rat hippocampal slices by tetanic stimulation. J. Neurosci. 19: 1088–1105.

Traub RD, Bibbig A, Fisahn A, LeBeau FEN, Whittington MA, Buhl EH (2000) A model of gamma-frequency network oscillations induced in the rat CA3 region by carbachol in vitro. Eur. J. Neurosci. 12: 4093–4106.

Traub RD, Bibbig A, Piechotta A, Draguhn A, Schmitz D (2001a) Synaptic and nonsynaptic contributions to giant IPSPs and ectopic spikes induced by 4–aminopyridine in the hippocampus in vitro. J. Neurophysiol. 85: 1246–1256.

Traub RD, Kopell N, Bibbig A, Buhl EH, LeBeau FEN, Whittington MA (2001b) Gap junctions between interneuron dendrites can enhance long-range synchrony of gamma oscillations. J. Neurosci. 21: 9478–9486.

Traub RD., Whittington MA, Buhl EH, LeBeau FEN, Bibbig A, Boyd S, Cross H, Baldeweg T (2001c) A possible role for gap junctions in generation of very fast EEG oscillations preceding the onset of, and perhaps initiating, seizures. Epilepsia 42: 153–170.

Traub RD, Draguhn A, Whittington MA, Baldeweg T, Bibbig A, Buhl EH, Schmitz D (2002) Axonal gap junctions between principal neurons: A novel source of network oscillations, and perhaps epileptogenesis. Rev. Neurosci. 13: 1–30.

Traub RD, Buhl EH, Gloveli T, Whittington MA (2003a) Fast rhythmic bursting can be induced in layer 2/3 cortical neurons by enhancing persistent Na+ conductance or by blocking BK channels. J. Neurophysiol. 89: 909–921.

Traub RD, Cunningham MO, Gloveli T, LeBeau FEN, Bibbig A, Buhl EH, Whittington MA (2003b) GABA-enhanced collective behavior in neuronal axons under-lies persistent gamma-frequency oscillations. Proc. Natl. Acad. Sci. USA 100: 11047–11052.

Traub RD, Pais I, Bibbig A, LeBeau FEN, Buhl EH, Hormuzdi SG, Monyer H, Whittington MA (2003c) Contrasting roles of axonal (pyramidal cell) and dendritic (interneuron) electrical coupling in the generation of gamma oscillations in the hippocampus in vitro. Proc. Natl. Acad. Sci. USA 100: 1370–1374.

Traub RD, Bibbig A, LeBeau FEN, Buhl EH, Whittington MA (2004) Cellular mechanisms of neuronal population oscillations in the hippocampus in vitro. Annu. Rev. Neurosci. 27: 247–278.

Traub RD, Contreras D, Cunningham MO, Murray H, LeBeau FEN, Roopun A, Bibbig A, Wilent WB, Higley MJ, Whittington MA (2005a) Single-column thalamocortical network model exhibiting gamma oscillations, sleep spindles and epileptogenic bursts. J. Neurophysiol. 93: 2194–2232.

Traub RD, Contreras D, Whittington MA (2005b) Combined experimental/simulation studies of cellular and network mechanisms of epileptogenesis in vitro and in vivo. J. Clin. Neurophysiol. 22: 330–342.

Traub RD, Pais I, Bibbig A, LeBeau FEN, Buhl EH, Monyer H. Whittington MA (2005c) Transient depression of excitatory synapses on interneurons contributes to epileptiform bursts intermixed with gamma oscillations in the mouse hippocampal slice. J. Neurophysiol. 94: 1225–1235.

Traub RD, Contreras D, Whittington MA (2008a) Cellular and network mechanisms of oscillations preceding and perhaps initiating epileptic discharges. In: Soltesz I, Staley K. (Eds), Computational Neuroscience in Epilepsy. Academic Press, San Diego, pp 335–355.

Traub RD, Cunningham MO, Whittington MA (2008b) Fast oscillations in activated brain slices: an in vitro continuation of the pioneering in vivo studies of Mircea Steriade and colleagues. Thalamus Relat Systems 4: 79–86

Traub RD, Middleton SJ, Knöpfel T, Whittington MA (2008c) Model of very fast (>75 Hz) network oscillations generated by electrical coupling between the proximal axons of cerebellar Purkinje cells. Eur. J. Neurosci. 28: 1603–1616.

Trétiakoff C (1919) Contribution à l'étude de l'anatomie pathologique du locus niger de Soemmering avec quelques déductions relatives à la pathogénie des troubles du tonus musculaires et de la maladie de Parkinson. University of Paris.

Treviño M, Gutiérrez R (2005) The GABAergic projection of the dentate gyrus to hippocampal area CA3 of the rat: pre- and postsynaptic actions after seizures. J. Physiol. 567: 939–949.

Trexler EB, Bukauskas FF, Bennett MVL, Bargiello TA, Verselis VK (1999) Rapid and direct effects of pH on connexins revealed by the connexin46 hemichannel preparation. J. Gen. Physiol. 113: 721–742.

Trinka E (2005) Absence in adult seizure disorders. Acta Neurol. Scand. 112 Suppl. 182: 12–18.

Trump DL, Hochberg MC (1976) Bromide intoxication. Johns Hopkins Med. J. 138: 119–123.

Tsakiridou E, Bertollini L, de Curtis M, Avanzini G, Pape HC (1995) Selective increase in T-type calcium conductance of reticular thalamic neurons in a rat model of absence epilepsy. J. Neurosci. 15: 3110–3117.

Tsuang M (2000) Schizophrenia: genes and environment. Biol. Psychiatry 47: 210–220.

Tsujimoto T, Gemba H, Sasaki K (1993) Effect of cooling the dentate nucleus of the cerebellum on hand movement of the monkey. Brain Res. 629: 1–9.

Tukker JJ, Fuentealba P, Hartwich K, Somogyi P, Klausberger T (2007) Cell type-specific tuning of hippocampal interneuron firing during gamma oscillations in vivo. J. Neurosci. 27: 8184–8189.

Turabelidze G, Schootman M, Zhu BP, Malone JL, Horowitz S, Weidinger J, Williamson D, Simoes E (2008) Multiple sclerosis prevalence and possible lead exposure. J. Neurol. Sci. 269: 158–162.

Turin L, Warner A (1977) Carbon dioxide reversibly abolishes ionic communication between cells of early amphibian embryo. Nature 270: 56–69.

Uhlhaas PJ, Mishara AL (2007) Perceptual anomalies in schizophrenia: integrating phenomenology and cognitive neuroscience. Schizophr. Bull. 33: 142–156.

Uhlhaas PJ, Silverstein SM (2005) Perceptual organization in schizophrenia spectrum disorders: empirical research and theoretical implications. Psychol. Bull. 131: 618–632.

Uhlhaas PJ, Linden DE, Singer W, Haenschel C, Lindner M, Maurer K, Rodriguez E (2006) Dysfunctional long-range coordination of neural activity during Gestalt perception in schizophrenia. J. Neurosci. 26: 8168–8175.

Uhlhaas PJ, Haenschel C, Nikolić D, Singer W (2008) The role of oscillations and synchrony in cortical networks and their putative relevance for the pathophysiology of schizophrenia. Schizophr. Bull. 34: 927–943.

Ulrich D, Bettler B (2007) GABA$_B$ receptors: synaptic functions and mechanisms of diversity. Curr. Opin. Neurobiol. 17: 298–303.

Unger VM, Kumar NK, Gilula NB, Yeager M (1999) Three-dimensional structure of a recombinant gap junction membrane channel. Science 283: 1176–1180.

Urbano FJ, Leznik E, Llinás RR (2007) Modafinil enhances thalamocortical activity by increasing neuronal electrotonic coupling. Proc. Natl. Acad. Sci. USA 104: 12554–12559.

Urrestarazu E, Jirsch JD, Le Van P, Hall J (2006) High-frequency intracerebral EEG activity (100 – 500 Hz) following interictal spikes. Epilepsia 47: 1465–1476.

Usowicz MM, Gallo V, Cull-Candy SG (1989) Multiple conductance channels in type-2 cerebellar astrocytes activated by excitatory amino acids. Nature 339: 380–383.

Usui N, Kotagal P, Matsumoto R, Kellinghaus C, Lüders HO (2005) Focal semiologic and electroencephalographic features in patients with juvenile myoclonic epilepsy. Epilepsia 46: 1668–1676.

Uusisaari M, Knöpfel T (2008) GABAergic synaptic communication in the GABAergic and non-GABAergic cells in the deep cerebellar nuclei. Neuroscience 156: 537–549.

Uusisaari M, Obata K, Knöpfel T (2007) Morphological and electrophysiological properties of GABAergic and non-GABAergic cells in the deep cerebellar nuclei. J. Neurophysiol. 97: 901–911.

Vaadia E, Haalman I, Abeles M, Bergman H, Prut Y, Slovin H, Aertsen A (1995) Dynamics of neuronal interactions in monkey cortex in relation to behavioural events. Nature 373: 515–518.

Valiante TA, Perez Velazquez JL, Jahromi SS, Carlen PL (1995) Coupling potentials in CA1 neurons during calcium-free-induced field burst activity. J. Neurosci. 15: 6946–6956.

Van Amelsvoort T, Daly E, Henry J, Robertson D, Ng V, Owen M, Murphy KC, Murphy DG (2004) Brain anatomy in adults with velocardiofacial syndrome with and without schizophrenia: preliminary results of a structural magnetic resonance imaging study. Arch. Gen. Psychiatry 61: 1085–1096.

Vandecasteele M, Glowinski J, Deniau J-M, Venance L (2008) Chemical transmission between dopaminergic pairs. Proc. Natl. Acad. Sci. USA 105: 4904–4909.

van Gijn J (2007) From the archives. Brain 130: 4–7.

van Groen T, Miettinen P, Kadish I (2003) The entorhinal cortex of the mouse: organization of the projection to the hippocampal formation. Hippocampus 13: 133–149.

van Huffelen AC (1989) A tribute to Martinus Rulandus. A 16th-century description of benign focal epilepsy of childhood. Arch. Neurol. 46: 445–447.

van Vreeswijk C, Abbott LF, Ermentrout GB (1994) When inhibition not excitation synchronizes neural firing. J. Comput. Neurosci. 1: 313–321.

Venance L, Glowinski J, Giaume C (2004) Electrical and chemical transmision between striatal GABAergic output neurones in rat brain slices. J. Physiol. 559: 215–230.

Venance L, Rozov A, Blatow M, Burnashev N, Feldmeyer D, Monyer H (2000) Connexin expression in electrically coupled postnatal rat brain neurons. Proc. Natl. Acad. Sci. USA 97: 10260–10265.

Vernon D, Haenschel C, Dwivedi P, Gruzelier J (2005) Slow habituation of induced gamma and beta oscillations in association with unreality experiences in schizotypy. Int. J. Psychophysiol. 56: 15–24.

Victor M, Ropper AH (2001) Adams and Victor's Principles of Neurology. McGraw-Hill, New York.

Villars PS, Kanusky JT, Dougherty TB (2004) Stunning the neural nexus: mechanisms of general anesthesia. AANA J. 72: 197–205.

Vogt A, Hormuzdi SG, Monyer H (2005) Pannexin1 and Pannexin2 expression in the developing and mature rat brain. Brain Res. Mol. Brain Res. 141: 113–120.

Volk DW, Austin MC, Pierri JN, Sampson AR, Lewis DA (2000) Decreased glutamic acid decarboxylase67 messenger RNA expression in a subset of prefrontal cortical gamma-aminobutyric acid neurons in subjects with schizophrenia. Arch. Gen. Psychiatry 57: 237–245.

Volkmann J, Joliot M, Mogilner A, Ioannides AA, Lado F, Fazzini E, Ribary U, Llinás R (1996) Central motor loop oscillations in parkinsonian resting tremor revealed by magnetoencephalography. Neurology 46: 1359–1370.

von der Malsburg C, Buhmann J (1992) Sensory segmentation with coupled neural oscillators. Biol. Cybern. 67: 233–242.

von der Malsburg C, Schneider W (1986) A neural cocktail-party processor. Biol. Cybern. 54: 29–40.

von Krosigk M, Bal T, McCormick DA (1993) Cellular mechanisms of a synchronized oscillation in the thalamus. Science 261: 361–364.

Voogd J (2003) The human cerebellum. J. Chem. Neuroanat. 26: 243–252.

Voruganti LP, Baker LK, Awad AG (2008) New generation antipsychotic drugs and compliance behaviour. Curr. Opin. Psychiatry 21: 133–139.

Vreugdenhil M, Jefferys JGR, Celio MR, Schwaller B (2003) Parvalbumin-deficiency facilitates repetitive IPSCs and gamma oscillations in the hippocampus. J. Neurophysiol. 89: 1414–1422.

Wada JA, Sato M, Corcoran ME (1974) Persistent seizure susceptibility and recurrent spontaneous seizures in kindled cats. Epilepsia 15: 465–478.

Waddington JL, O'Tuathaigh C, O'Sullivan G, Tomiyama K, Koshikawa N, Croke DT (2005) Phenotypic studies on dopamine receptor subtype and associated signal transduction mutants: insights and challenges from 10 years at the psychopharmacology-molecular biology interface. Psychopharmacology 181: 611–638.

Walker AE, Johnson HC, Kollros JJ (1945) Penicillin convulsions. The convulsive effects of penicillin applied to the cerebral cortex of monkey and man. Surg. Gynecol. Obstet. 81: 692–701.

Walker MC, Ruiz A, Kullmann DM (2001) Monosynaptic GABAergic signaling from dentate to CA3 with a pharmacological and physiological profile typical of mossy fiber synapses. Neuron 29: 703–715.

Waller JC (2002) 'The illusion of an explanation': the concept of hereditary disease, 1770 – 1870. J. Hist. Med. Allied Sci. 57: 410–448.

Walsh T, McClellan JM, McCarthy SE, Addington AM, Pierce SB, Cooper GM, Nord AS, Kusenda M, Malhotra D, Bhandari A, Stray SM, Rippey CF, Roccanova P, Makarov V, Lakshmi B, Findling RL, Sikich L, Stronberg T, Merriman B, Gogtay N, Butler P, Eckstrand K, Noory L, Gochman P, Long R, Chen Z, Davis S, Baker C, Eichler EE, Meltzer PS, Nelson SF, Singleton AB, Lee MK, Rapoport JL, King MC, Sebat J (2008) Rare structural variants disrupt multiple genes in neurodevelopmental pathways in schizophrenia. Science 320: 539–543.

Wang CS, Yang SF, Xia Y, Johnson KM (2007) Postnatal phencyclidine administration selectively reduces adult cortical parvalbumin-containing interneurons. Neuropsychopharmacology, doi: 10.1038/sj.npp.1301647.

Wang HS, Pan Z, Shi W, Brown BS, Wymore RS, Cohen IS, Dixon JE, McKinnon D (1998) KCNQ2 and KCNQ3 potassium channel subunits: molecular correlates of the M-channel. Science 282: 1890–1893.

Wang J, O'Donnell P (2001) D(1) dopamine receptors potentiate NMDA-mediated excitability increase in layer V prefrontal cortical pyramidal neurons. Cereb. Cortex 11: 452–462.

Wang LY, Gan L, Forsythe ID, Kaczmarek LK (1998) Contribution of the Kv3.1 potassium channel to high-frequency firing in mouse auditory neurones. J. Physiol. 509: 183–194.

Wang X-J (1993) Ionic basis for intrinsic 40 Hz neuronal oscillations. NeuroReport 5: 221–224.

Wang X-J (1999a) Fast burst firing and short-term plasticity: a model of neocortical chattering neurons. Neuroscience 89: 347–362.

Wang X-J (1999b) Synaptic basis of cortical persistent activity: the importance of NMDA receptors to working memory. J. Neurosci. 19: 9587–9603.

Wang X-J (2001) Synaptic reverberation underlying mnemonic persistent activity. Trends Neurosci. 24: 455–463.

Wang X-J, Rinzel J (1993) Spindle rhythmicity in the reticularis thalami nucleus: synchronization among mutually inhibitory neurons. Neuroscience 53: 899–904.

Wang X-J, Golomb D, Rinzel J (1995) Emergent spindle oscillations and intermittent burst firing in a thalamic model: specific neuronal mechanisms. Proc. Natl. Acad. Sci. USA 92: 5577–5581.

Wang X-J, Tegner J, Constantinidis C, Goldman-Rakic PS (2004) Division of labor among distinct subtypes of inhibitory neurons in a cortical microcircuit of working memory. Proc. Natl. Acad. Sci. USA 101: 1368–1373.

Wang Y, Markram H, Goodman PH, Berge TK, Ma J, Goldman-Rakic PS (2006) Heterogeneity in the pyramidal network of the medial prefrontal cortex. Nat. Neurosci. 9: 534–542.

Wang Y, Toldedo-Rodriguez M, Gupta A, Wu C, Silberberg G, Luo J, Markram H (2004) Anatomical, physiological and molecular properties of Martinotti cells in the somatosensory cortex of the juvenile rat. J. Physiol. 561: 65–90.

Wanner SG, Koch RO, Koschak A, Trieb M, Garcia ML, Kaczorowski GJ, Knaus HG (1999) High-conductance calcium-activated channels in rat brain: pharmacology, distribution, and subunit composition. Biochemistry 38: 5392–5400.

Warner AE, Guthrie SC, Gilula NB (1984) Antibodies to gap-junctional protein selectively disrupt junctional communication in the early amphibian embryo. Nature 311: 127–131.

Wehr M, Laurent G (1996) Odour encoding by temporal sequences of firing in oscillating neural assemblies. Nature 384: 162–166.

Weickert CS, Ligons DL, Romanczyk T, Ungaro G, Hyde TM, Herman MM, Weiberger DR, Kleinman JE (2005a) Reductions in neurotrophin receptor mRNAs in the prefrontal cortex of patients with schizophrenia. Mol. Psychiatry 10: 637–650.

Weickert R, Ray A, Zoidl G, Dermietzel R (2005b) Expression of neural connexins and pannexin1 in the hippocampus and inferior olive: a quantitative approach. Brain Res. Mol. Brain Res. 133: 102–109.

Weinberger M, Mahant N, Hutchison WD, Lozano AM, Moro E, Hodaie M, Lang AE, Dostrovsky JO (2006) Beta oscillatory activity in the subthalamic nucleus and its relation to dopaminergic response in Parkinson's disease. J. Neurophysiol. 96: 3248–3256.

Welsh JP, Llinás R (1997) Some organizing principles for the control of movement based on olivocerebellar physiology. Prog. Brain Res. 114: 449–461.

Werkman TR, Glennon JC, Wadman WJ, McCreary AC (2006) Dopamine receptor pharmacology: interactions with serotonin receptors and significance for the aetiology and treatment of schizophrenia. CNS Neurol. Disord. Drug Targets 5: 3–23.

West WJ (1840–1841) On a peculiar form of infantile convulsions. Lancet I: 724–725.

Westfall JA, Elliott CF, Carlin RW (2002) Ultrastructural evidence for two-cell and three-cell neural pathways in the tentacle epidermis of the sea anemone Aiptasia pallida. J. Morphol. 251: 83–92.

White HS, Patel S, Meldrum BS (1992) Anticonvulsant profile of MDL 27,266: an orally active, broad-spectrum anticonvulsant agent. Epil. Res. 12: 217–226.

White HS, Smith MD, Wilcox KS (2007) Mechanisms of action of antiepileptic drugs. Int. Rev. Neurobiol. 81: 85–110.

White JA, Chow CC, Ritt J, Soto-Treviño C, Kopell N (1998) Synchronization and oscillatory dynamics in heterogeneous, mutually inhibited neurons. J. Comput. Neurosci. 5: 5–16.

White TW, Bruzzone R, Goodenough DA, Paul DL (1992) Mouse Cx50, a functional member of the connexin family of gap junction proteins, is the lens fiber protein MP70. Mol. Biol. Cell 3: 711–720.

Whittington MA, Traub RD (2003) Inhibitory interneurons and network oscillations in vitro. Trends Neurosci. 26: 676–682.

Whittington MA, Traub RD, Jefferys JGR (1995) Synchronized oscillations in interneuron networks driven by metabotropic glutamate receptor activation. Nature 373: 612–615.

Whittington MA, Jefferys JGR, Traub RD (1996) Effects of intravenous anaesthetic agents on fast inhibitory oscillations in the rat hippocampus in vitro. Br. J. Pharmacol. 118: 1977–1986.

Whittington MA, Stanford IM, Colling SB, Jefferys JGR, Traub RD (1997a) Spatiotemporal patterns of γ frequency oscillations tetanically induced in the rat hippocampal slice. J. Physiol. 502: 591–607.

Whittington MA, Traub RD, Faulkner HJ, Stanford IM, Jefferys JGR (1997b) Recurrent excitatory postsynaptic potentials induced by synchronized fast cortical oscillations. Proc. Natl. Acad. Sci. USA 94: 12198–12203.

Whittington MA, Traub RD, Kopell N, Ermentrout B, Buhl EH (2000) Inhibition-based rhythms: experimental and mathematical observations on network dynamics. Int. J. Psychophysiol. 38: 315–336.

Whittington MA, Doheny HC, Traub RD, LeBeau FEN, Buhl EH (2001) Differential expression of synaptic and non-synaptic mechanisms during stimulus-induced gamma oscillations in vitro. J. Neurosci. 21: 1727–1738.

WHO (World Health Organization) (1979) Schizophrenia: an International Follow-Up Study. John Wiley & Sons, Geneva.

Wichmann T, Bergman H, DeLong MR (1994) The primate subthalamic nucleus. I. Functional properties in intact animals. J. Neurophysiol. 72: 494–506.

Wiener N (1961) Cybernetics. MIT Press & John Wiley & Sons, Cambridge and New York.

Wigmore MA, Lacey MG (2000) A Kv3–like persistent, outwardly rectifying, Cs$^+$-permeable, K$^+$ current in rat subthalamic nucleus neurones. J. Physiol. 527: 493–506.

Wilder RM (1921) The effects of ketonemia on the course of epilepsy. Mayo Clin. Proc. 2: 307–308.

Wilding TJ, Huettner JE (1997) Activation and desensitization of hippocampal kainate receptors. J. Neurosci. 17: 2713–2721.

Williams CA (2005) Neurological aspects of the Angelman syndrome. Brain Dev. 27: 88–94.

Williams D, Tijssen M, Van Bruggen G, Bosch A, Insola A, Di Layyaro V, Mayyone P, Oliviero A, Quartarone A, Speelman H, Brown P (2002) Dopamine-dependent changes in the functional connectivity between basal ganglia and cerebral cortex in humans. Brain 125: 1558–1569.

Williams SR, Stuart GJ (1999) Mechanisms and consequences of action potential burst firing in rat neocortical pyramidal neurons. J. Physiol. 521: 467–482.

Williams SR, Stuart GJ (2000a) Action potential backpropagation and somato-dendritic distribution of ion channels in thalamocortical neurons. J. Neurosci. 20: 1307–1317.

Williams SR, Stuart GJ (2000b) Site independence of EPSP time course is mediated by dendritic I_h in neocortical pyramidal neurons. J. Neurophysiol. 83: 3177–3182.

Williams SR, Stuart GJ (2000c) Backpropagation of physiological spike trains in neocortical pyramidal neurons: implications for temporal coding dendrites. J. Neurosci. 20: 8238–8246.

Williams SR, Turner JP, Anderson CM, Crunelli V (1996) Electrophysiological and morphological properties of interneurones in the rat dorsal lateral geniculate nucleus *in vitro*. J. Physiol. 490: 129–147.

Williams SR, Tóth TI, Turner JP, Hughes SW, Crunelli V (1997a) The "window" component of the low threshold Ca^{2+} current produces input signal amplification and bistability in cat and rat thalamocortical neurones. J. Physiol. 505: 689–705.

Williams SR, Turner JP, Hughes SW, Crunelli V (1997b) On the nature of anomalous rectification in thalamocortical neurones of the cat ventrobasal thalamus *in vitro*. J. Physiol. 505: 7727–747.

Williamson LC, Fitzgerald SC, Neale EA (1992) Differential effects of tetanus toxin on inhibitory and excitatory neurotransmitter release from mammalian spinal cord cells in culture. J. Neurochem. 59: 2148–2157.

Willmore LJ, Sypert GW, Munson JV, Hurd RW (1978) Chronic focal epileptiform discharges induced by injection of iron into rat and cat cortex. Science 200: 1501–1503.

Wilson CL, Puntis M, Lacey MG (2004) Overwhelmingly asynchronous firing of rat subthalamic nucleus neurones in brain slices provides little evidence for intrinsic interconnectivity. Neuroscience 123: 187–200.

Wilson M, Bower JM (1992) Cortical oscillations and temporal interactions in a computer simulation of piriform cortex. J. Neurophysiol. 67: 981–995.

Wingeier B, Tcheng T, Koop MM, Hill BC, Heit G, Bronte-Stewart HM (2006) Intra-operative STN DBS attenuates the prominent beta rhythm in the STN in Parkinson's disease. Exp. Neurol. 197: 244–251.

Wirrell EC (2003) Natural history of absence epilepsy in children. Can. J. Neurol. Sci. 30: 184–188.

Wisden W, Seeburg P (1993) A complex mosaic of high-affinity kainate receptors in rat brain. J. Neurosci. 13: 3582–3598.

Witham CL, Wang M, Baker SN (2007) Cells in somatosensory areas show synchrony with beta oscillations in monkey motor cortex. Eur. J. Neurosci. 26: 2677–2686.

Wolters EC, Braak H (2006) Parkinson's disease: premotor clinicopathological correlations. J. Neural Transm. Suppl. 70: 309–319.

Wong RKS, Prince DA (1981) Afterpotential generation in hippocampal pyramidal cells. J. Neurophysiol. 45: 86–97.

Wong RKS, Traub RD (1983) Synchronized burst discharge in the disinhibited hippocampal slice. I. Initiation in the CA2–CA3 region. J. Neurophysiol. 49: 442–458.

Wong RKS, Prince DA, Basbaum AI (1979) Intradendritic recordings from hippocampal neurons. Proc. Natl. Acad. Sci. USA 76: 986–990.

Woo TU, Walsh JP, Benes FM (2004) Density of glutamic acid decarboxylase 67 messenger RNA-containing neurons that express the N-methyl-D-aspartate receptor subunit NR2A in the anterior cingulate cortex in schizophrenia and bipolar disorder. Arch. Gen. Psychiatry 61: 649–657.

Woo TU, Kim AM, Viscidi E (2008) Disease-specific alterations in glutamatergic neurotransmission on inhibitory interneurons in the prefrontal cortex in schizophrenia. Brain Res. 1218: 267–277.

Woo TU, Whitehead RE, Melchitzky DS, Lewis DA (1998) A subclass of prefrontal gamma-aminobutyric acid axon terminals are selectively altered in schizophrenia. Proc. Natl. Acad. Sci. USA 95: 5341–5346.

Worrell GA, So EL, Kazemi J, O'Brien TJ, Mosewich RK, Cascino GD, Meyer FB, Marsh WR (2002) Focal ictal β discharge on scalp EEG predicts excellent outcome of frontal lobe epilepsy surgery. Epilepsia 43: 277–282.

Worrell GA, Parish L, Cranstoun SD, Jonas R, Baltuch G, Litt B (2004) High-frequency oscillations and seizure generation in neocortical epilepsy. Brain 127: 1496–1506.

Wouterlood FG, Barbas-Henry HA, Lohman AH (1988) Mixed-junction axon terminals on identified principal abducens motoneurons in the monitor lizard. Brain Res. 463: 198–203.

Wulff P, Ponomarenko AA, Bartos M, Korotkova TM, Fuchs EC, Bähner F, Both M, Tort ABL, Kopell NJ, Wisden W, Monyer H (2009) Hippocampal theta rhythm and its coupling with gamma oscillations require fast inhibition onto parvalbumin-positive interneurons. Proc. Natl. Acad. Sci. USA. 106: 3561–3566.

Xiang Z, Huguenard JR, Prince DA (2002) Synaptic inhibition of pyramidal cells evoked by different interneuronal subtypes in layer V of rat visual cortex. J. Neurophysiol. 88: 740–750.

Yamada MK, Nakanishi K, Ohba S, Nakamura T, Ikegaya Y, Nishiyama N, Matsuki N (2002) Brain-derived neurotrophic factor promotes the maturation of GABAergic mechanisms in cultured hippocampal neurons. J. Neurosci. 22: 7580–7585.

Yamamoto C, McIlwain H (1966) Electrical activites in thin sections from the mammalian brain maintaned in chemically-defined media in vitro. J. Neurochem. 13: 1333–1343.

Yamamoto T, Shiosaka S, Whittaker ME, Hertzberg EL, Nagy JI (1989) Gap junction protein in rat hippocampus: light microscope immunohistochemical localization. J. Comp. Neurol. 281: 269–281.

Yamawaki N, Stanford IM, Hall SD, Woodhall GL (2008) Pharmacologically induced and stimulus evoked rhythmic neuronal oscillatory activity in the primary motor cortex in vitro. Neuroscience 151: 386–395.

Yang CR, Seamans JK (1996) Dopamine D1 receptor actions lin layers V-VI rat prefrontal cortex neurons in vitro: modulation of dendritic-somatic signal integration. J. Neurosci. 16: 1922–1935.

Yang X-D, Korn H, Faber DS (1990) Long-term potentiation of electrotonic coupling at mixed synapses. Nature 348: 542–545.

Yaqub BA (1993) Electroclinical seizures in Lennox-Gastaut syndrome. Epilepsia 34: 120–127.

Yeager M, Harris AL (2007) Gap junction channel structure in the early 21st century: facts and fantasies. Curr. Opin. Cell Biol. 19: 521–528.

Ylinen A, Bragin A, Nádasdy Z, Jandó G, Szabó I, Sik A, Buzsáki G (1995a) Sharp wave-associated high frequency oscillation (200 Hz) in the intact hippocampus: network and intracellular mechanisms. J. Neurosci. 15: 30–46.

Ylinen A, Soltész I, Bragin A, Penttonen M, Sik A, Buzsáki G (1995b) Intracellular correlates of hippocampal theta rhythm in identified pyramidal cells, granule cells and basket cells. Hippocampus 5: 78–90.

Yoon JH, Minzenberg MJ, Ursu S, Walters R, Wendelken C, Ragland JD, Carter CS (2008) Association of dorsolateral prefrontal cortex dysfunction with disrupted coordinated brain activity in schizophrenia: relationship with impaired cognition, behavioral disorganization, and global function. Am. J. Psychiatry 165: 1006–1014.

Yue C, Yaari Y (2004) KCNQ/M channels control spike afterdepolarization and burst generation in hippocampal neurons. J. Neurosci. 24: 4614–4624.

Yue C, Yaari Y (2006) Axo-somatic and apical dendritic Kv7/M channels differentially regulate the intrinsic excitability of adult rat CA1 pyramidal cells. J. Neurophysiol. 95: 3480–3495.

Zhu JJ, Uhlrich DJ (1998) Cellular mechanisms underlying two muscarinic receptor-mediated depolarizing responses in relay cells of the rat lateral geniculate nucleus. Neuroscience 87: 767–781.

Zhu ZT, Shen KZ, Johnson SW (2002) Pharmacological identification of inward current evoked by dopamine in rat subthalamic neurons in vitro. Neuropharmacology 42: 772–781.

Zhu ZT, Munhall A, Shen KZ, Johnson SW (2004) Calcium-dependent subthreshold oscillations determine bursting activity induced by N-methyl-D-aspartate in rat subthalamic neurons in vitro. Eur. J. Neurosci. 19: 1296–1304.

Zhu L, Blethyn KL, Cope DW, Tsomaia V, Crunelli V, Hughes SW (2006) Nucleus- and species-specific properties of the slow (<1 Hz) sleep oscillation in thalamocortical neurons. Neuroscience 141: 621–636.

Zirh TA, Lenz FA, Reich SG, Dougherty PM (1998) Patterns of bursting occurring in thalamic cells during parkinsonian tremor. Neuroscience 83: 107–121.

Zimmerman RS, Sirven JI (2003) An overview of surgery for chronic seizures. Mayo Clin. Proc. 78: 109–117.

Zsiros V, Maccaferri G (2008) Noradrenergic modulation of electrical coupling in GABAergic networks of the hippocampus. J. Neurosci. 28: 1804–1815.

Zuberi SM, Eunson LH, Spauschus A, De Silva R, Tolmie J, Wood NW, McWilliam RC, Stephenson JB, Kullmann DM, Hanna MG (1999) A novel mutation in the human voltage-gated potassium channel gene (Kv1.1) associates with episodic ataxia type 1 and sometimes with partial epilepsy. Brain 122: 817–825.

Index